CIVIL WAR
Flags
of
ILLINOIS

BY

JOHN D. SCHMALE MD

ISBN: 978-1-962568-28-9

Dedication

Nothing but flags, but simple flags,
Torn and tattered and hanging in rags,
And we pass by them with a careless tread
Nor think of the hosts of the mighty dead
Who marched beneath them in the days gone by
With a burning cheek and a kindling eye.
And have bathed each fold with their life's young tide
And dying blessed them and blessing died.

Nothing but flags? Yet me thinks at night
They tell each other their tales of fright,
That dim specters rise and their thin arms twine
'Round each standard torn as it stands in line,
As the word is given they charge they form
And the dim hall rings with the battle storm,
As once again through smoke and strife
Those colors lead to a Nation's life.

Nothing but flags? Yet they're bathed with tears,
They tell of triumphs, of hopes and fears,
Of mothers' prayers, of a boy away,
Of a serpent crushed! Of the coming day.
Silent, they speak and the tear will start
As we stand beneath them with throbbing heart
And think of those who are near forgot,
Their flags come home, why come they not?

Nothing but flags? Yet we hold our breath
As we gaze in awe on those types of death.
Nothing but flags? Yet the thought will come,
The heart must pray though the lips are dumb;
They are sacred, pure, and we see no stain
On their dear loved folds come home again,
Baptized in the blood of our purest and best
Tattered and torn they are now at rest.[1]

This book is dedicated to the brave soldiers of Illinois who carried these flags and saved the Union. And it is dedicated to my dear wife Joyce who put up with my spending so many days, weeks, months and years putting this material together and going to see 'Just one more flag' and to my son Michael Schmale who with his expertise spent many hours turning my "notes" into a book on flags.

---

[1] Anonymous author on viewing the retired Illinois flags. Quoted in Hubert, Charles F., *History of the Fiftieth Regiment Illinois Volunteer Infantry in the War of the Union.* Western Veteran Publishing Company, Kansas City, Mo. (1894) p 439

About the Author

The author is a retired practicing physician who spent the last thirty years researching the history of the flags carried by the Illinois Infantry, Cavalry and Artillery Regiments during the Civil War. His material has been gleaned from newspapers, regimental histories, soldiers' diaries, various museum files and personal communications. The book contains the history of each regiment, the home county of the men enlisting in each regiment and the history and photos of any flags known to have been carried by the regiment.

Acknowledgements

This book would not have been possible without the contributions of Charley Mooney, the curator of the Illinois State Military Museum. He introduced me to the romance of civil war flags. I was first introduced to the history of the Civil War when I was in high school. My parents gifted me a set of books by Bruce Catton. The rest is history.

I spent the next thirty years researching and writing this book. Over the years, I have sought and received help from innumerable contributors, including the county historical societies and museums across Illinois. I am deeply grateful to them for sharing information and photos of the flags they hold in their inventories.

I would specifically like to acknowledge the contributions of Howard Madaus, Greg Biggs, Steve Hill, Robert Behrens, Richard Sauers, Del Thomasson, Jonathan Deiss, David Gugerty, Jean Kay, Chris Bremner, Jim Rafferty, Mike Venso, Josh Phillips, Carla Smith, Amy Meyer, John Fraley, Maureen Naughtin, and Mark Whitlock. Thank you all for your contributions to the book.

Introduction

– JOHN SCHMALE MD ILLINOIS FLAGS BOOK

The history of Civil War military unit colors is linear, and they did not appear in a vacuum; hundreds of years brought changes, use in battle and more to what the flags of the American Civil War would be and stand for. Thus, flags underwent a learning curve. Concepts would change making flags easier to use and easier to make in large numbers. Additionally, flags in battle would become the heart and soul of a unit symbolizing the unit itself being consecrated with men who would be killed or wounded as the colors were flying in the heat of combat. But the goals never changed; create and use a flag for a military unit in combat that stood for the unit, added to its morale and allowed its commanders to note its place on a field of battle.

The earliest known idea for military flags was first put into print over 2000 years ago when Sun Tzu wrote *The Art of War*. The noted military theorist created this concept to help identify friendly troops and for commanders to be able to control them in the heat of battle as well as mark positions on the field. This was the first case of using flags as an item of command and control and it would not be the last.

The Legions of Rome marched into battle behind a legionary staff. These were decorated with items important to that legion and toped often with animals, zodiac signs and other devices. With the reforms of Gaius Marius, the legion staff was topped with an Eagle. Below this on its mount were the letters S.P.Q.R (Senatus Populus Que Romanus) which translated from Latin means, *For the Senate and the Roman People.* This solidified exactly who that legion was fighting for. Legion sub-units followed a staff bearing a piece of cloth hanging from it via a crossbar called a Vexillum. This further enabled a legion and sub-unit commander to be able to identify and control his troops in battle. Eagles topping flag staffs would return in Napoleon's Grande Armee and with Hitler's legions.

The fall of the Western Roman Empire destroyed the concept of the nation-state helping to usher in the Dark Ages. Gradually, small kingdoms and dukedoms sprang up each with its own military to protect its citizens. By the Middle Ages, with most armies now being fully armored or very close to that, a need to identify troops came out of Medieval Heraldry. Each of these small political entities created a heraldic symbol which was used on its flags. Accordingly, this symbol would be placed on a cloth covering that went over the armor, was painted on the shields and was carried as a battle flag. The concept was the same as with Sun Tzu; command and control and well as a unit identifier. As nation states were again created many heraldic symbols became incorporated into them or new ones were created. For example, an old heraldic symbol for England was the Lion and later it was the Cross of St. George.

The battle flags of the British Army influenced American battle flags. The basis of the British Army of the 1700s into the 1900s was created by Oliver Cromwell and continued later by Sir John Churchill, the Duke of Marlborough. In the early 1700s after defeating the French of the Sun King, Louis XIV, Churchill put his stamp onto his Redcoats. Before Churchill's rise, British regiments were raised and paid for by their colonels under a royal warrant. Initially, the regiment flew a colonel's color but there were also other colors for the unit; a lieutenant colonel's color, major's color and captain's colors, one for each of the ten (or more) companies. This was soon found unwieldy and by 1707 this was changed to three flags; colonel's, lieutenant colonel's and major's. By 1747, British regiments flew only two flags; the King's color, until 1801 a blue field with the red cross of St. George on top of the white cross of St. Andrew and the unit designation in the center. After 1801 the red Irish Cross of St. Patrick was added. The Regimental color was a particular solid color (red, blue, yellow, buff, etc.) with the unit designation in the center and the British Union cross in the upper left corner. The color of this flag dictated the color of the uniform facings the regiment wore which aided in unit identification, blue flag, blue facings etc. This system of regimental colors might have been inspired by those used by the Swedish Army of Gustavus Adolphus who used only flags of a solid color with uniform facings to match.

Churchill's foe, the French Army of the Sun King bore colors bearing the Sun on them. Later, the Fleur De Lis appeared on regimental flags of the Bourbon Ancien Regime, which had a cross in the center and four different colors of the field, one in each corner bisected by the cross. With the coming of the French Revolution, the blue, white and red flags began to appear on the colors of the Demi-Brigades of the republican revolutionary army. With the rise of Napoleon, the modern French Tri-color was born and his Grande Armee would use three different types (1804, 1812 and 1815) before his reign ended at Waterloo.

Within this time frame, the American Continental Army had formed and fought in our Revolutionary War. Its first drill master, Prussian Baron Friedrich Wilhelm Von Steuben, wrote the *Blue Book* which described the training of troops as well as declaring only two flags per regiment. One was "the standard of the United States," while the second was a regimental color which would, "vary according to the facings of the regiment." These flags were to be five feet square. The Prussian Army of Frederick the Great bore five flags per regiment so it is curious why Von Steuben, despite this background, chose to copy the British in terms of flags. However, very few flags seem to have been made along these lines. Part of the problem is that the "standard of the United States" had still not been decided upon by war's end even though the Stars and Stripes had been created in 1777. The national Coat of Arms was created in 1782 but it would not be until later that it would appear on a U.S. Army flag.

After the new nation was formed, the first national regiments were raised starting in 1784 with the 1st American Regiment. Its flag was blue, with the national Coat of Arms in the center surmounted by thirteen stars. In the upper left former was the unit designation. In 1808, with the Army's expansion, two flags would represent the unit. The National colors were to be blue with the national Coat of Arms with the unit designation placed upon it and measuring two yards by two and a half yards (hoist to fly). The regimental colors were to be white or buff bearing only a blue ribbon with the unit designation upon it. This flag was to be five by six feet (hoist to fly). These were the colors used in the War of 1812.

General Winfield Scott placed an important imprint on the U.S. Army. Seeking professionalism and intense training, he wrote several drill manuals and regulations books during his command career. One manual, in 1834, changed the National standard for the artillery regiments to the Stars and Stripes. The regimental colors remained the same. In 1841, the infantry regiments, via new regulations, carried the Stars and Stripes as their National colors and the older National standard of blue with the Coat of Arms became their regimental colors. All flags were to be 72 inches on the hoist by 76 inches on the fly. Scott's books also made specifications for the color guard and color company. These were the flags under which the Army would fight in the Mexican War – the first time the Stars and Stripes were used as battle flags on land. The same flags were codified in the 1861 Army regulations but with embroidered unit designations.

The mounted arm of the Army carried a blue standard which was a smaller version of the infantry's regimental colors. Starting with the two regiments of dragoons and the Regiment of Mounted Rifles, each mounted company carried a small swallow-tailed guidon being red over white with the company letter placed upon it. However, in January 1862, General Orders No. 4 allowed the Stars and Stripes to be used replacing the red over white guidons for the various companies but retained the blue regimental standard. So many of the Stars and Stripes guidons would be produced that they were also issued as infantry regiment flank markers as well as to artillery units. These flags saw use well after the Civil War as the Army fought with the mounted nations of Plains Indians. This long line of flag history set the tone for the flags of the Union Army as used in the Civil War.

Fortunately for today's vexillologists, author and historian John Schmale has dedicated many years of his life to the study of the flags of Illinois as used in the Civil War. The Prairie State ranked Number Four in terms of putting men into blue uniforms (after New York, Pennsylvania and Ohio respectively) and Illinois troops fought in all three of the major theaters of war. The core of one famous Union army, the Army of the Tennessee, was built around a large cadre of Illinois regiments and artillery batteries. Accordingly, a lot of flags were made for the units raised within the state. Some, in the first blush of patriotism realized after the firing on Fort Sumter in April 1861, received home made flags from the local communities or by paid subscription to a professional flag maker. Either way, that flag was the direct connection to the hometown or county a unit was raised in. It was also a direct connection to a soldier's home. With companies being raised on a local basis, the ranks were often fathers and sons, brothers, school mates, other family members and all were there to defend their nation and support each other when the shooting started. Often elaborate presentation ceremonies were the social events of the day where the flag was present followed by a picnic where hundreds of people attended. Local newspapers were there to cover the event and details printed within a day or two.

Companies went off to training camps and to form regiments around the state which mixed companies from one part of Illinois with others from a different part. Some regiments recruited entirely from a region. The huge number of companies that were raised in 1861 (ten per regiment, infantry or cavalry) needed flags although after early 1862 most of these flags had fallen by the wayside. Each regiment needed two sets of flags per regiment initially but by war's end a combat veteran unit had often gone through anywhere from three to five sets of colors. Also included were a regiment's camp colors, garrison colors and more. All of this meant that author Schmale's task was daunting indeed. The search was on to track untold hundreds of flags.

Casting his research net far and wide, Schmale found a plethora of first-person accounts, post-war accounts, official reports and more regarding the flags of Illinois units. Most of the flags depicted in this book still exist, the bulk in the collections of the Illinois National Guard in the state capital of Springfield. Others can be found in smaller museums in the state and a few more in private hands. Sadly, like so many other Civil War flags of both sides, many are missing today, in particular early war company level colors. What the author was able to find is depicted in this book.

For whatever reason, the study of Union flags of the Civil War is lacking behind that of Confederate colors in terms of published books and even journal articles. Thus, this work on Illinois flags will join the ranks of modern era books produced for the Civil War flags of Pennsylvania, Connecticut and New Jersey. There are some books for Indiana, New York and Michigan and others but these are over 100 years old and were often done to commemorate the turning in of the flags to the respective state and they being placed in display in the state capital building. The late flag historian Howard Madaus, the dean of Civil War flags scholars, wrote a book on the flags of the famous Iron Brigade along with scholarly articles on camp and garrison colors, headquarters flags and others. Devereaux Cannon, also sadly deceased, wrote a fine book on Union flags that serves as a general overview of the many types of flags the Union Army used. This book has been updated by another author. The excellent Time-Life book *Arms and Equipment of the Union* features a detailed chapter on Union colors written by Madaus.

Other than these works, the shelves are slim for books on Union flags which adds to the importance of John Schmale's work. The *Civil War Flags of Illinois* book will find a cross-section of interested readers from flag buffs to reenactors, those interested in Illinois' role in the Civil War and more. Museums holding flag collections will find it a welcome addition to their libraries. Not many books find an audience of so many interests.

Sit back, relax and dig into each chapter of the valiant units raised in Illinois and the flags they followed. Smell the smoke of battle, hear the battle cries of men charging or howls of pain from men being wounded and killed. You can visualize the color bearer waving his flag to rally a broken regiment or carrying the colors forward to victory. The barked orders of commanding officers will come through loudly as the pages are turned. Modern Americans know little of the importance of the regimental color guard and the reader will come away with a greater appreciation of the heroism of these men who would be decimated in battle and yet more and more men came forward to carry the flags and protect them. Besides the flags themselves, Schmale reports on the flag makers, from individuals to the large depot systems that ramped up in 1862 who would sub-contract flags to professionals in the cities where the depots were located.

It is all here and more. Civil War buffs and flag historians are indeed very grateful.

Keep them flying.

Greg Biggs

Military Historian

December 2023

God save our country's flag!
Long may it wave.
In triumph o'er the free
The noble, brave—
The sign of liberty
Our fathers gave.
God save our country's flag!
No traitor hand
Shall hurl it from its place;
No rebel band
Shall trail it in disgrace
O'er this fair land.
God save our country's flag!
With colors bright
Still, may it float afar
In heaven's pure light!
By every stripe and star,
God speed the right![2]

[2] *Chicago Daily Tribune,* June 24, 1861. Original ode composed to be sung at the presentation of colors to the Chicago University Cadets from the young ladies of the Dearborn Female Seminary.

Union Flags

Although legislation establishing the national flag was enacted on June 14, 1777, the national flag was not commonly displayed. Today flags are displayed on schools, post offices, businesses, homes, and city and government buildings. This was not the case in the 19th century. Only four branches of the US government were required to display the flag: the Army, the Navy, the Revenue Service, and the Indian Department. Here in Illinois, routine display of a national flag would have been uncommon except perhaps on the 4th of July or at a political rally.

The Quartermaster's department considered a "stand of colors" to consist not only of a military unit fag, but also the separate flag staff with its brass fittings, the set of cords and tassels with which the flag the flag could be secured when furled, often a water proof case to slip over the furled flag in wet weather, and occasionally a flag belt or sling which permitted the color bearer to comfortably handle the unfurled flag with one hand.[3]

At the time of the Civil War, the Quartermaster of the Union Army furnished the following flags/colors:[4] (the term colors and flags will be used interchangeably)

Garrison colors-the National flag

Designating Flags[5]-Headquarters, Division, Brigade, Chief Quartermaster flags

Infantry colors–National and Regimental flag, guidons, flank markers, camp colors[6]

Cavalry flags-Standard flag, guidons

Artillery colors-National and Regimental flags, Battery flags

Miscellaneous

Field Hospital Flag, Ambulance guidon,

Signal Corps flags-seven red, white and black flags with a contrasting center square

Engineer Battalion flags

Recruiting flags

There is no evidence that the state of Illinois furnished flags to its Civil War Volunteer Regiments. Rather, flags were supplied to the Illinois Volunteer Regiments through the Quartermaster Corps of the Union Army; therefore, the national and regimental colors of most of Illinois' units are of

---

[3] Madaus, Howard Michael. "The Acessessories for a "Stand of Colors" the Madaus Collection of Staffs, Cords & Tassels, and Flag Slings". U.S Army Heritage & Education Center

[4] Cannon, Devereaux D .*The Flags of the Union an Illustrated History,* Gretna, Pelican Publishing Company,1994.

[5] *Military Commanders and Designating Flags of the United States Army 1861-1865.* Philadelphia, Burk & McFetridge, 1888. Reprinted as *Civil War Battle Flags of the Union Army and Order of Battle,* .Compiled by General C. McKeever, Quartermaster General of the U.S. Army, Forward by Steven J. Wright, New York, Knickerbocker Press, 1997

[6] Madaus, Howard Michael. "Camp Colors, General Guide Flags and Flank Markers in the United States Army, 1861-1865: Part I-IV," *Military Collector & Historian,* 48, No.2 (Summer 1996), 48, No.3 (Fall 1996), 48, No. 4 (Winter 1996), 49, No.1 (Spring 1997).

typical Federal manufacture. In contrast, some states, such as Pennsylvania,[7] supplied their volunteer regiments with flags bearing the 'coat of arms of the Commonwealth' in the canton of the national flag. While the colors of each Illinois Regiment were similar, most Illinois Volunteer Companies left home with a banner supplied by the 'ladies' of the community.[8] Naturally, these homemade flags were of various sizes and shapes and carried a myriad of mottoes. The majority of Illinois Regiments marched off to war with national, regimental, and company flags flying.

Early in 1862, the Federal government became responsible for supplying State Volunteer troops with all necessary equipment, including flags. Prior to that time, each state was required to recruit and supply its state's troops. Thus, early in the war, many states purchased or manufactured the flags for their own troops, and this resulted in a variety of patterns and sizes of flags carried by their men. At the time of the Civil War, by Union army regulations, each infantry regiment was furnished a stand of colors consisting of a national flag and a regimental flag. The flag of each cavalry regiment was called the 'standard' and was similar to the infantry regimental flag, only much smaller in size. The cavalry regiment was not required to display a national flag. The artillery regiments had both an artillery regimental flag and a national flag; however, since the artillery batteries rarely fought together as a regiment but rather as individual batteries, each battery usually had a battery flag that was a smaller version of the artillery regimental flag.

Prior to the Civil War, all the army's flags were supplied from the Philadelphia depot at the Schuylkill Arsenal. The army purchased cloth for the flags but transferred the cloth to a Philadelphia contractor, Samuel Brewer, to sew and paint the flags. In addition to Brewer, the Philadelphia depot also let contracts to Horstmann Brothers, Evans & Hassall, William Scheible, and Albert Ertle.

However, with the influx of thousands of new troops and hundreds of new regiments at the outset of the Civil War, the Philadelphia depot was overwhelmed, and additional flag depots were established in New York and Cincinnati. Although all the contractors for these depots produced flags in keeping with the Federal regulations, the regulations did not define such items as the arrangement of the stars and shape of the canton on the national flag or the shape and position of the eagle on the regimental flag. Thus, by examining these features, it is usually possible to tell the city of manufacture of the various regimental colors.

At the time of firing on Fort Sumter, with the recent addition of Oregon to the Union, there were 33 stars on the flag; on July 4, 1861, the flag had 34 stars with the addition of the State of Kansas; in June 1863 a 35th star was added to the flag as a result of the formation of the state of West Virginia; late in the War, a 36th star was added to honor the admission of Nevada to the Union. All flags had seven red and six white stripes.

---

[7] Sauers, Richard A. *Advance the Colors! Pennsylvania Civil War Battle Flags, Vol. I & Vol. II,* Lebanon, Pa. Sowers Printing Col., 1987.

[8] "How to Make An American Flag-Notwithstanding the number in use, few persons know how to make an American flag properly. The rule is as follows: The flag should be one half longer than wide, viz., if six feet wide, nine feet long—if eight feet wide, twelve feet long. The stripes should be alternately red and white, seven red and six white, top and bottom both being red. The field should be blue, and extend over seven stripes, commencing at the top, four red and three white, and should be a perfect square. The stars signifying the number of States, should be white, arranged in the form of a five-pointed star." *Chicago Weekly Democrat,* May 25, 1861

Illinois Civil War Flags

Come Tigers, let us raise our flag,
And stand beneath its stars,
With hands upraised and eyes fast fixed
Upon its crimson bars;
And let us swear by all that's right,
By all that's good and true,
That in the din and heat of fight
We'll keep its folds in view.
Then raise a rousing shout, my boys'
For the Girls of Old Champaign,
Whose fingers worked our Stars and Stripes,
For the bloody battle plain

The above stanzas reflect the feeling of the men from Illinois who marched off carrying flags, many of them made by their sweethearts and wives. However, at the regimental level, flags carried by the Illinois Volunteers were made under Federal contract and supplied through the Union army Quartermaster. Early in 1862, the Federal government became responsible for supplying state volunteer troops with all necessary equipment, including flags. Prior to that, each state recruited and supplied its own troops. Thus, early in the war, many states purchased or manufactured the flags for their state's troops. This resulted in a variety of patterns and sizes of flags carried by their men

At the time of the Civil War, by Union army regulations, each infantry regiment was to carry a stand of colors consisting of a national flag and a regimental flag. Each cavalry regiment was to carry a 'standard'. The 'standard' was similar to the infantry regimental flag of the infantry, only smaller in size since the cavalry color had to be carried on horseback. A cavalry regiment was not required to display a national flag. The artillery regiments had both an artillery regimental flag and a national flag. However, since the artillery batteries rarely fought together as a regiment but rather as individual batteries each battery usually displayed only a battery flag which was a smaller version of the artillery regimental flag.

There is no record as to how many flags each of the Illinois regiments received. Each regiment was entitled to colors as prescribed in the army regulations. However, because of weather and battle damage, flags often had to be replaced and many regiments may have had six or more flags in the course of the regiment's existence. At the end of the war, all flags carried by the Union volunteers were to be turned in to the Governor of the respective states pursuant to a War Department order of June 13, 1865. In Illinois, most of the regiments turned in a regimental or national flag or both. Together with guidons and captured Confederate flags, Illinois has one of the largest collections of Civil War flags (392) in the United States.

ILLINOIS CIVIL WAR FLAGS

For the most part, the flags in the Illinois collection are made of silk and painted with oil-based paints. Over the years, the flags have deteriorated due to natural processes and poor storage facilities. There have been two previous attempts at the preservation of these colors. In 1878 the flags were removed from their staffs, stitched to pieces of white tarlatan cut to the size of the flag, pressed, and placed back on their respective staffs. In 1924, the tarlatan was removed, and the flags were cleaned, brushed, pressed, and stitched between two fine silk nettings and affixed to their original staffs with strips of leather.

Up until 2004, the flags were on display in the Hall of Flags in the Howlett Building in Springfield. However, they have now been removed from the display cases and placed 'at rest' in stainless steel trays in a temperature and humidity-controlled environment at the Illinois State Military Museum at Camp Lincoln in Springfield, where they remain under the custody and protection of the Adjutant General of the State of Illinois.

In addition to the flags at Camp Lincoln, there are other Illinois Civil War flags in courthouses, historical museums, libraries, and private collections. The majority of these 'other' flags were not issued to the Illinois regiments through the Quartermaster Department but were presented to companies and regiments by local communities and community organizations.

A search through local newspapers of the Civil War-era documents the presentation of many flags whose existence is no longer known. Many of these probably are in public or private collections or remain with a descendent of the soldier who brought them home, but many are no longer in existence. An unknown number of Illinois Civil War flags were brought back to and deposited in Chicago. All these flags were lost in the Great Chicago Fire of 1871. An effort has been made to find and document all existing Illinois Civil War flags and include them in this book.

Many Confederate flags captured by Illinois soldiers are still held by the State of Illinois. Included in this book are photos and information about each of these captured flags.

Artillery Colors

"Each regiment of artillery shall have two silken colors. The first or the national color, of stars and stripes as described for the garrison flag. The number and name of the regiment are to be embroidered with gold on the center stripe. The second or the regimental color, to be yellow, of the same dimensions as the first, bearing in the center two cannon crossing, with the letters U. S. above and the number of the regiment below: fringe, yellow. Each color to be six feet six inches fly and six feet deep on the pike. The pike, including the spear and ferrule, is nine feet ten inches in length. Cords and tassels, red and yellow silk intermixed."

In the artillery, the regiments were divided into batteries rather than companies. An artillery battery consisted of four to six guns with guns served by 110 horses, 70 -100 men and commanded by a captain. Batteries were divided into sections of two guns apiece, each section normally under the command of a lieutenant. In addition to the two guns each section also had a limber and caisson for each gun. The limber was a two-wheeled cart that was attached to the trail piece of the gun and a six-horse hitch for travel. Artillery regiments rarely fought together as regiments, but rather as individual batteries and were often attached to infantry brigades, divisions, or cavalry brigades. Thus, artillery batteries usually had their own colors. This contrasts with the infantry or cavalry regiments in which there were regimental colors but no company colors. Therefore, in the Illinois CW flag collection, there are flags representing Illinois batteries but none for Artillery Regiments. The artillery also used guidons like those of the Cavalry.

Very few Illinois Civil War artillery flags survive; therefore, the section of Artillery flags is limited.

1st Regiment Illinois Light Artillery-Battery A[9]

"Receive this Flag and Banner, then, brave men, as slight tokens of the gratitude, the admiration, and the unwavering confidence of your Chicago friends. On the glorious blue of the banner, we have inscribed the words, 'Fort Donelson,' because there you covered yourselves with glory, and, by driving back by deeds of valor, an enemy flushed with the certainty of victory, you won in that memorable conflict the applause and the approbation of a grateful Nation."[10]

The Chicago Light Artillery had its beginnings as the *Chicago Hussars and Light Artillery* in Chicago in November 1847 during the U. S--Mexican War. On May 5, 1854, the Battery was reorganized as the *Chicago Light Artillery* with James Smith as Captain and Ezra Taylor as 1st Lieutenant. Following the shots on Fort Sumter in April 1861, Lincoln called for 75,000 troops for 90 days. Within days the *Chicago Light Artillery* was ready, and at 11 AM on Sept 21, 1861, the Battery left for Cairo, Illinois, with four six-pounder pieces of artillery under the command of Captain James Smith and 1st Lieutenant Charles Willard. One gun, commanded by Lieutenant Willard, was placed to defend the bridge at Big Muddy, while the remainder of the Battery, was assigned the duty, at Cairo, of bringing all passing boats on the Mississippi and Ohio Rivers to shore for inspection. The Battery did not muster in for three months of service. However, on Jul 16, 1861, while stationed at Cairo, 87 members of the *Chicago Light Artillery,* Company A were mustered in for three years' service as First Regiment Illinois Light Artillery, Battery A. Within the next two weeks 75 additional recruits were added. On Sept 6, General Grant ordered the Battery and two regiments of infantry up the Ohio River to Paducah, Kentucky, to fortify and guard the town. During November, two new howitzers were added, bringing Battery A to a full six-gun Battery.[11] In early February 1862, the Battery was transported by steamer up the Tennessee River and landed below Fort Heiman. Union gunboats captured both Fort Heiman and Fort Henry (on the opposite shore) on Feb 6, and the rebel forces escaped to nearby Fort Donelson on the Cumberland River.[12] On Feb 14, the Battery, together with Company A of the Thirty Second Illinois Infantry, marched cross-country and joined the Union forces in the siege of Fort Donelson. The following morning the men of the Battery had their first experience under fire and, while holding the right of the Union line, succeeded in repelling a rebel charge. After the fall of Fort Donelson, the Battery marched back to Fort Heiman, where they remained until March. On Mar 6, the Battery departed Fort Heiman and moved up the Tennessee River to Crump's Landing, Tennessee, in an attempt the destroy the Mobile & Ohio Railroad supplying Corinth, Mississippi. Mar 27, the Battery moved to Pittsburg Landing, Tennessee, and on April 6 and 7, it participated in the Battle of Shiloh.[13] The Battery returned to camp, and on Apr 15, they received a stand of colors from friends in Chicago. At the end of April, Artillery Battery A moved toward Corinth, Mississippi and from April 29-May 30, were at the siege of Corinth. From Corinth, the Battery moved to Memphis, passing through Purdy, Somerville, and Bolivar en route, and

---

[9] The Battery was also known as *Morgan's Battery, Smith's Battery, Wood's Battery, Willard's Battery*, Chicago Light Artillery-Company A, and *Wilcox's Battery.*

[10] On April 15,1862 the Battery received a stand of colors from "friends at home." The above quotation is from a portion of the letter accompanying the colors. Kimbell, Charles B., *History of Battery "A" First Illinois Light Artillery Volunteers,* Cushing Printing Company, Chicago (1899) page 48-9.

[11] While at Paducah the Battery along with the men of the Eighth and Ninth Illinois Infantry printed a newspaper entitled the *Union Picket Guard* on the presses of the defunct *Paducah Herald.*. Ibid, Kimbell, page 37.

[12] February 11,1862 2nd Lt. Peter P. Wood assumed command of the Battery; Capt. Smith having resigned due to illness.

[13] On April 6 the Battery fought in the "Hornets Nest" and lost 30 men killed and wounded and 48 horses killed or disabled but did not lose any guns.

went into camp at Union Station. They remained on guard duty at Memphis through the summer of 1862. They were on an expedition to Coldwater and Hernando, Mississippi, in September and were part of Grant's "Tallahatchie March" in November and December. On Dec 20, the Battery boarded the steamer *City of Memphis* and moved down the Mississippi River and up the Yazoo River, disembarking and fighting at Chickasaw Bayou and Chickasaw Bluff. Battery A next moved up the Mississippi River to the White River, through the "cut off" to the Arkansas River, and engaged in battle at Arkansas Post—Fort Hindman—January 10 and 11, 1863. The Battery spent the winter at Young's Point, Louisiana, where in addition to their own guns, they also manned two 30 lb Parrott guns. In March, a detachment from the Battery was sent with Union infantry to Deer Creek for relief of Porter's trapped gunboats. Following demonstrations against Haines' and Drumgould's Bluffs north of Vicksburg on April 29-May 2, the Battery moved via Grand Gulf, Mississippi, to join Grant's army in the rear of Vicksburg. On May 14, Battery A was engaged in fighting at Jackson, Mississippi, and on May 16 at Champion's Hill. In the subsequent siege of Vicksburg, the Battery manned three 30 lb Parrott guns in addition to their own guns. Vicksburg surrendered on Jul 4, and the following morning, Battery A was ordered to Jackson to besiege the city until Jul 17, when rebel General Johnston evacuated the city. The Battery camped at the Big Black River Bridge on the Vicksburg and Jackson Railroad until Sept 26, then moved by boat to Memphis, by train to Corinth, Mississippi, and finally marched to Chattanooga through northern Alabama; on arrival, Battery A fought at Tunnel Hill, Georgia November 23-24 and Mission Ridge Nov 25. The Battery joined in the pursuit of Bragg as far as Dalton, Georgia, returned to Chattanooga, and then went into winter quarters at Larkinsville, Alabama.[14] First Illinois Light Artillery Battery A joined Sherman's Atlanta campaign in May 1864. The Battery was in action at Resaca, Dallas, New Hope Church, and Allatoona Hills in May, Kenesaw Mountain in June, Nickajack Creek, Chattahoochee River, and Atlanta in July[15], and the battles at Jonesboro and Lovejoy Station in August and September. In the Battle of Atlanta, on Jun 22, the Battery was surrounded. Many of the men were captured, and the rebel forces took four of the Battery's guns. Captain Smith was one of the prisoners taken; therefore, Lieutenant Echte of a Missouri battery was placed in temporary command until September, when Captain Edward Wilcox, formerly of Battery B, was placed in command. After the pursuit of Hood into North Georgia and Alabama, Battery A was ordered to Nashville, Tennessee, in October but was held in reserve at the battle of Nashville. The Battery was then ordered to Chattanooga, where they remained until June 1865 and then ordered home to Chicago for muster out on July 3,1865.[16]

The only color in the collection of the State of Illinois from Battery A of the First Illinois Regiment of Light Artillery is a guidon. However, colors were presented to the Battery on at least two occasions.

The first flag was presented to the Battery while it was stationed in Cairo, Illinois, in the Fall of 1861. "While stationed at Cairo, the Battery was presented with a beautiful silk flag by Miss Katie Sturgis of Chicago. This flag was carried by the Battery through all its subsequent campaigns. It was destroyed by the great fire of Oct 10, 1871, in the rooms of the Chicago Historical Society, where it had been placed for safe keeping.[17]"

---

[14] En route to Larkinsville, the Battery lost its mascot, a "thorough-bred" mongrel dog named Tony. Ibid, Kimbell, page 80-1.

[15] July 12, 1864 Captain Peter Wood together with all of the men who had served three years were mustered out, their term of service having expired. The members of First Illinois Light Artillery, Batteries A and B whose terms had not expired were consolidated, under the designation "Battery A—New," under command of Capt. Samuel Smith who had been a First Lieutenant in Battery F of the First Illinois Light Artillery.

[16] Ibid, Kimbell. The Battery history by Kimbell is an excellent source or information concerning this battery.

[17] Ibid, Kimbell, page 23.

The second set of colors was presented to the Battery at Pittsburg Landing, Tennessee, on Apr 15, 1862.

The Chicago Daily Tribune reported, "THE COLORS FOR COMPANY A, LIGHT ARTILLERY—The flags will be sent to the South by two members of the Company today. Subscribers to the fund are requested to call and see them before they go, at No. 116 Clark Street.[18]" Kimbell, in the Battery's history, notes that on Apr 15, 1862, the Battery received a Flag and a Banner and that the blue of the banner was inscribed "Fort Donelson.[19]" The colors are not further described, nor is there any further mention of the colors in Kimbell's book. It is possible that this set of colors was also lost in the Chicago fire.

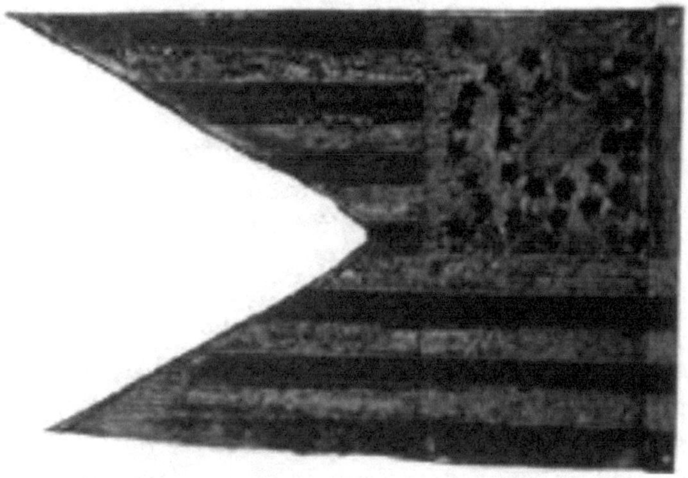

This guidon of Battery A is not inscribed.

[18] *Chicago Daily Tribune*, April 8, 1862.
[19] Ibid, Kimbell, page 48-9.

1st Regiment Illinois Light Artillery, Battery B[20]

"Long years have passed since last we met,
Bearing the colors true,
With canteen black and old knapsack
And our uniforms of blue;
Which told the noble story
That artillery boys were we,
And belonged to Captain Taylor's
Chicago Battery B.
Three long years we fought together
In the great and bloody strife,
And to save our country's honor
We each laid down our life.
To defend our starry emblem,
The banner of the free,
None quicker rallied to her aid
Then the boys of Battery B.[21]"

The men of Battery B of the First Regiment Illinois Light Artillery volunteered from Chicago, Cook County, and St. Louis, Missouri. The Battery was mustered in at Chicago for three months of service on May 2, 1861, and mustered in for three years' service on Jul 16, 1861, under the command of Captain Ezra Taylor. [22]On Jun 1, the Battery departed for Cairo, Illinois, and was on duty at Cairo and Bird's Point, Missouri[23], during June and July. In August, a section of Battery B was detailed to Paducah, Kentucky, to join in defense of the city against threats from rebel forces. Following the return of the detail to camp near Cape Girardeau, Missouri, on Oct 18, the Battery, along with infantry, started toward Fredericktown, Missouri, in pursuit of Rebel forces commanded by General Jeff Thompson. The Union forces attacked and routed the rebel forces on Oct 21 and returned to Cape Girardeau on Oct 25. Nov 6, Battery B boarded the steamer *Chancellor* and moved downriver to Lucas Bend, Missouri, three miles above Columbus, Kentucky. The following morning Union forces, including Battery B, engaged Rebel forces at Belmont, Missouri, directly across the river from Columbus. When the rebels sent reserve forces across the river from Columbus, the Union troops were forced to withdraw, and Battery B embarked with their guns along with two captured guns of Watson's New Orleans Battery. The Battery went into winter quarters at Bird's Point until Jan 14, 1862, when ordered down the river on the *City of Memphis* to Camp Jefferson, Kentucky, marched into Kentucky as far as Mayfield Creek, and returned to Bird's Point. On Feb 2, aboard the steamer *Aleck Scott*, Battery B, together with other troops, were ordered up the Ohio River and

[20] Also known as *Chicago Light Artillery, Battery B, Barrentt's Battery, Taylor's Battery, White's Battery, Bridges' Battery, Ramsey's Battery.*

[21] Portion of a poem entitled "A Tribute to The Boys of Chicago Battery B." by Mrs. Mary J. Peckham. *Reunions of Taylor's Battery, 18th Anniversary of the Battle of Fort Donelson, February 14, 1880, 25th Anniversary of the Battle of Belmont, November 6, 1886,* Craig Press, Chicago (1890), pages 3-4.

[22] Ezra Taylor was born in New York in 1819 and came to Chicago in 1839. He was in the "packing" business until the outbreak of the civil war.

[23] Bird's Point lay directly across the Mississippi River from Cairo just below the confluence of the Ohio and Mississippi rivers.

Tennessee Rivers and disembarked four miles above Fort Henry. While marching toward the fort, the forces inside the fort surrendered to Union gunboats and retreated cross-country to Fort Donelson on the Cumberland River. After four days of heavy fighting at Fort Donelson, February 12-16, the fort surrendered, and Battery B was given the honor of saluting the Union flag as it was raised over the fort. The Battery was ordered to Savannah, Tennessee, and on Mar 7, while aboard the steamer *Silver Moon* a committee from the Board of Trade of Chicago presented the Battery with a flag, on which was inscribed "Presented to Taylor's Chicago Battery, by the Chicago Board of Trade, for gallant deeds performed on the fields of Fredericktown, Belmont, and Donelson.[24]" From Savannah, the Battery moved up river to Pittsburg Landing, Tennessee, where on Apr 6, the Union forces were attacked in the Battle of Shiloh. At the end of April, the Union troops marched south to attack the rebel forces at Corinth, Mississippi. A one-month siege resulted in the evacuation of the rebel forces, and Battery B was ordered to Memphis, Tennessee for duty there until November. November 25-December 14, the Battery was ordered into northern Mississippi as part of Grant's "Tallahatchie March" but made no contact with the enemy and returned to Memphis. Dec 20, Battery B, on board the steamer *City of Memphis,* moved down the Mississippi River and up the Yazoo River to engage the enemy at Chickasaw Bayou and Chickasaw Bluff in late December. January 1863, the Battery moved from the Mississippi River into the White River, through the cut off to the Arkansas River to Fort Hindman (Arkansas Post), where on January 10 and 11, the Battery participated in the actions resulting in the surrender of the fort. January through March, the Battery remained in camp at Young's Point, Louisiana, participated in an expedition to Deer Creek for relief of Porter's trapped gunboats in mid-March, and then demonstrations against Haines' and Drumgould's Bluffs in early May. Battery B crossed the Mississippi River at Grand Gulf, Mississippi, to join the Union Army positioned in the rear of Vicksburg, fighting at Jackson, Mississippi, on May 14 at Champion's Hill on May 16. Battery B also fought in the assault on the Vicksburg lines on May 19 and 22 and was present when Vicksburg surrendered on Jul 4, 1863. The Battery remained in camp at Big Black River until late September and then moved to Chattanooga, Tennessee arriving there in late November. In the Ringgold Campaign, Battery B fought at Tunnel Hill on November 23-24 and at Mission Ridge the following day. The Battery was ordered to the relief of Burnside at Knoxville, Tennessee, but the rebel forces lifted the siege before the Union relief forces arrived. Battery B was sent to winter quarters at Larkinsville, Alabama and in May 1864, joined in Sherman's Atlanta campaign. In that campaign, the Battery fought at Resaca on May 14-15, Dallas on May 25-June 5, Kenesaw Mountain on Jun 27, Nickajack Creek during July 2-5, and the Chattahoochee River from July 6-12. On Jul 12, Battery B was ordered to the rear for muster out on Jul 23, 1864. Those men whose term of service had not expired and those men re-enlisting as veterans were transferred to Battery A of the First Regiment Illinois Light Artillery.[25] [26]

There are no colors in the collection of the State of Illinois identified as having been carried by Battery B. However, on the frontispiece of the reunion book for Taylor's Battery[27], there is a photo of the national flag carried by the Battery. The canton of the flag contains a painting of a cannon on a carriage with a scroll above the cannon reading "Chicago Light Artillery, Comp. B." Thirty-five stars surround the cannon. At the time the photo was taken, the flag (presumably silk) was already in poor condition. The remnants of this flag are now in the Special Collections Department of the Chicago Public Library. The

---

[24] Dudley, Henry Walbridge, *Autobiography of Henry Walbridge Dudley,* George Banta Publishing Co. , Menasha, Wisconsin (1914), page 45. *Chicago Daily Tribune, February 28, 1862, March 6, 1862, March 20, 1862.*

[25] Ibid, Walbridge

[26] Dyer, Frederick H., *A Compendium of the War of the Rebellion,* Thomas Yoseloff, New York (1959), Vol III, page 1035-6.

[27] Ibid, *Reunions of Taylor's Battery*

library describes the remnants of the flag as follows: Canton of a flag from Battery B, 1st Illinois Light Artillery. Originally, the flag also had seven red and six alternating white stripes, but these are not present and probably disintegrated. Canton is blue silk with gold decorations painted on it. Decorations consist of a cannon mounted on a carriage in the center of the canton with a banner above the cannon which reads, "CHICAGO LIGHT ARTILLERY COMP. B." This central image is surrounded by a circle of 30 five-pointed stars outlined in dark brown and a similar star in each corner, totaling 34. GAR records indicate that this regimental flag was presented to Battery B by Miss Kate Sturgiss in May 1861 and was with the Battery to the close of the war in 1865." The presentation of this flag to Taylor's Battery is detailed in May 15, 1861, issue of the Chicago Evening Journal.

In 1862 the Chicago Board of Trade purchased a "stand of colors for Taylor's Battery at the cost of $120[28]" The colors were sent down to Fort Henry, where they were received by Captain Taylor. The newspaper does not further identify the makeup of the colors.

No other colors carried by this battery are known to survive.

[28] *Chicago Daily Tribune,*March 6,1862.

1st Regiment Illinois Light Artillery, Battery C[29]

Battery C, 1st Regiment Illinois Light Artillery was organized in Ottawa, Illinois, and mustered in for three years' service on Oct 31, 1861. Most of the men volunteered from LaSalle County and Pulaski County, Illinois. Captain Charles Houghtaling of Ottawa was in command of the Battery. He had served in the Mexican War under Zachery Taylor. The Battery provided the following military service. Ordered to Cairo, Illinois. Duty in the Military District of Cairo until February 1862. Operations against New Madrid, Missouri, and Island No. 10, Mississippi River, February 28-April 8. Moved to Pittsburg Landing, Tennessee, April Moved to Iuka, Mississippi, Jul 21, thence to Courtland, Alabama, March to Nashville, Tennessee. Siege of Nashville September 12-November 7. Battle of Stones River December 30–31, 1862 and January 1–3, 1863 Battle of Chickamauga, September 19–20. Siege of Chattanooga, Tennessee, Sept 24- Missionary Ridge November 24–25. Atlanta Campaign May 1 to Sept 8. Operations. Assault on Kennesaw Jun 27. Ruff's Station and Vining Station Jul 4. Chattahoochee River July 6–17. Peachtree Creek July 19–20. Siege of Atlanta July 22-August 25. Battle of Jonesboro August 31-September 1. Lovejoy's Station September 2–6. Pursuit of Hood into Alabama October 1–26. March to the Sea November 15-December 10. Siege of Savannah December 10–21. Campaign of the Carolinas January to April 1865. Battle of Bentonville, North Carolina, March 19–21. Occupation of Goldsboro Mar 24. Occupation of Raleigh Apr 14. Bennett's House Apr 26. Surrender of Johnston and his army. March to Washington, DC, via Richmond, Virginia, April 29-May 20. Grand Review of the Armies, May 24. Moved to Louisville, Kentucky, in June.[30]

There are no known existing flags from this Battery in any collection, and no references have been found reporting the presentation of a flag to this Battery.

[29] This Battery was also known as *Houghtaling's Battery, Channel's Battery, Ottowa Light Artillery* and *Prescott's Battery.*
[30] This information is from a text now in the public domain: Dyer, Frederick H. (1908). *A Compendium of the War of the Rebellion.* Des Moines, IA: Dyer Publishing Co.

1st Regiment Illinois Light Artillery, Battery D[31]

Captain Edward McAllister, when 24 years old, came to Plainfield Township in Will County, Illinois, and purchased 160 acres of land for farming. In 1856 he assisted in the organization of a militia force subsequently known as the *Plainfield Artillery*. At the outbreak of the Civil War, he was one of the first to enlist and was elected captain of the artillery. He first served as captain of a three-month artillery battery, but when the three months had expired, he was authorized to raise an artillery Battery in Cairo, Illinois. The men for this Battery volunteered from several counties in Illinois as well as from Michigan, Indiana, New York, Kentucky, Ohio, Pennsylvania, and Texas. In 1862 the Battery was ordered south and fought in the battles at Fort Henry and Fort Donelson. They were next transported to Pittsburg Landing and subsequently were in the fighting at Shiloh and the siege of Corinth. Captain McAllister resigned on May 5, 1862, and Captain Henry Rodgers became commander of the Battery. During the summer of 1862, the Battery was part of Grant's Central Mississippi Campaign, and in 1863 the Battery was ordered to the Vicksburg area. Captain Rodgers was killed in action May 29, 1863, during the siege. During 1864 and 1865, the Battery under the command of Captain Edgar Cooper was in action in Tennessee, Georgia, and Alabama. The Battery was mustered out of service in Chicago on July 28, 1865.

The are no known surviving colors from this Battery.

[31] Also known as *McAllister's Battery* and *Cooper's Battery*

1st Regiment Illinois Light Artillery, Battery E[32]

Battery E was organized at Camp Doulas in Chicago in the fall of 1861 and mustered in under the command of Captain Allen Waterhouse on Dec 19, 1861. The men who volunteered to serve in the Battery were all from Cook County. In February 1862, the men were ordered to Cairo, Illinois, where they were supplied with military equipment, including guns and horses. With less than a month of training, in March, the Battery was transported to Pittsburg Landing and subsequently fought in the Battle of Shiloh. Although the Battery lost three of its four guns during the battle, the lost guns were replaced by guns captured from the Confederates during the battle. As part of operations in the Western Theatre, Battery E skirmished at Corinth, White Station, and finally, the Battle at Jackson and the siege of Vicksburg. In February 1864, the Battery was assigned to the Meridian expedition. In June 1864, Battery E was mauled in the Battle of Brice's Crossroads, salvaging only 2 of its six guns. During the remainder of the War, Battery E fought in the campaign at Nashville and Memphis, and Chattanooga. The men were mustered out at Louisville, Kentucky, on Dec 24, 1864.

No surviving flags from this Battery are known to exist.

[32] Also known as *Waterhouse Battery*

1st Regiment Illinois Light Artillery, Battery F[33]

Battery F was recruited in Dixon, Lee County, then moved to Camp Butler in Springfield for organization and training. The men volunteered from Brown, Ogle, Morgan, Lee, and Whiteside Counties. On Feb 25, 1862, the Battery was mustered in for a three-year enlistment under the command of Captain John T. Cheney.[34] On Mar 15, the Battery moved to Benton Barracks in St. Louis and, in April, was at the siege of Corinth. In the summer of 1862, the Battery was part of Grant's Central Mississippi Campaign. In 1863 the Battery participated in action at the siege of Vicksburg and the siege of Jackson, Mississippi, and in 1864 participated in the campaigns related to the Battle of Atlanta.

The Battery was discontinued on Feb 22, 1865, and the men transferred to other batteries.

There are no known surviving flags of this Battery. However, John Cheney's war diary is in the possession of Gordon Armstrong; he shared the following entries from Cheney's diary with me in an email dated Jul 14, 2003.

"Mar 12, 1862, It is reported that the ladies of Springfield are going to present our company with a banner. I do not know that it is true, but if it is, it will be a very pretty compliment for our company, among entire strangers as we are.

On Feb 2, 1863, he mentions that he wrote to his brother in Dixon, IL, to purchase a flag and send it to him. John Cheney makes several references to hoping it will arrive in time to be presented to the Battery on the 1st anniversary of their service. It arrives late.

Mar 17, 1863, at 3 PM, the Battery mounted was drawn up in line, and the flag presented by Capt. Cheney with a few remarks, which were eloquently responded to by S. E. Parker on Behalf of the Battery. Capt. Cogswell being present, was called upon and very beautifully and eloquently paid a just tribute to the American flag. After a short drill, all came back to camp apparently well pleased with the exercises.

On Mar 22, 1863. Our flag came to us on the 16th instant and was presented to the Battery on the 17th. It is a very pretty one, and the boys are much pleased with it. Walter Little is the color bearer.

Aug 11, 1863, I have today applied for the privilege of inscribing on our flag "Vicksburg" and Jackson." I hope we may be allowed to do so and believe we will."

There is no further history of this flag and no other flags are known to have been presented to this battery.

---

[33] Also known as *Cheney's Battery* and *Burton's Battery*

[34] John T. Cheney was born in New Hampshire in 1830. At the outbreak of the civil war he was a grain dealer living in Dixon, Illinois.

1st Regiment Illinois Light Artillery, Battery G[35]

Battery G of the 1st Regiment of Illinois Light Artillery was organized in Cairo, Illinois in 1861 and mustered February 28, 1862. The Battery was under the command of Captain Arthur O'Leary of Sangamon County, Springfield, Illinois. The men of the battery primarily volunteered from Cairo in Alexander County and from St. Louis, Missouri but men also volunteered from counties scattered throughout Illinois. Following muster, the battery was first in operations against Island No. 10 in the Mississippi followed by garrison duty at New Madrid, Missouri. In April 1862 they took part in the siege and occupation of Corinth, Mississippi and remained on duty at Corinth until January 1864. In early 1864 the battery was ordered to Memphis where they remained on guard duty until mustered out on July 24, 1865. The only known surviving color of this battery is a guidon held by the Illinois State Military Museum in Springfield, Illinois.

[35] This battery was also known as *O'Leary's Battery* and later *Rombauer's Battery*

1st Regiment Illinois Light Artillery, Battery H[36]

Battery H was recruited and organized by the efforts of Axel Silversparre, a former Swedish artillery officer who while living at Cape Girardeau, Missouri became acquainted with a local jeweler, Francis DeGress. DeGress sold his jewelry store and moved to Chicago with Silversparre, agreeing to use the money from the sale of his business to recruit a battery. In Chicago they met John A. Anderson and asked him to help them recruit an artillery battery from the Swedish population in northern Illinois. In January and February of 1862 men were recruited from Cook County, Winnebago County, DuPage County, Kane County, DeKalb County, Bureau County, Henry County, Knox County and Rock Island County. The battery was mustered in on February 20, 1862 under the command of Captain Axel Silversparre. In March they were ordered to St. Louis where they were equipped as a four 20-pound Parrott gun battery and shipped down river to Pittsburg Landing to fight in the Battle of Shiloh. In June they fought in the siege and occupation of Corinth followed by guard duty at Memphis until November of 1862. It was at Memphis that DeGress took command of the battery. The battery served the remainder of the war in General Sherman's command and participated in battles at Vicksburg, Mission Ridge, Atlanta and in the March to the Sea. After marching in the Grand Review in Washington DC, the battery mustered out June 14,1865 at Louisville, Kentucky. No record was found re presentation of a flag to this battery. However, presumably the battery would have carried colors per military regulations. The colors would have returned with the battery to Chicago and if so, were lost in the Great Chicago fire of 1871. However, a guidon carried by this battery survives at the Illinois State Military Museum in Springfield, Illinois.

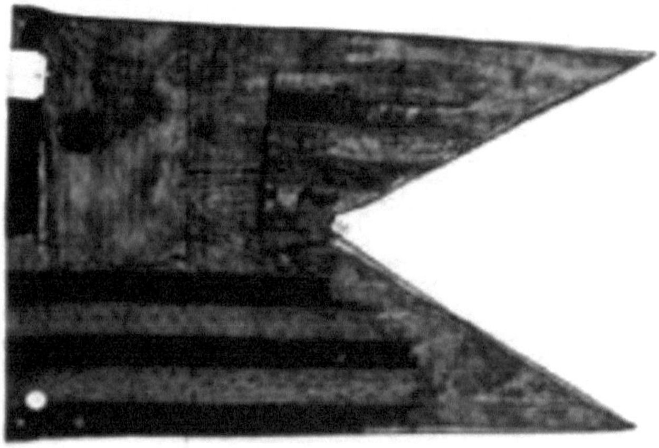

[36] This battery was also known as *DeGress' Battery, Silversparre's Battery, Hart's Battery*

1st Regiment Illinois Light Artillery, Battery I[37]

In 1859 Edward Bouton moved to Chicago from Bath, New York where he had had a thriving business is shipping grain, wool and produce. In Chicago he developed a lucrative business as a grain commission merchant with ownership of shipping vessels. At the outbreak of the Civil War, he sold some of his holdings to raise money to recruit and equip an artillery battery. Many of the men volunteered from out of state—Ohio, Iowa, New York, Michigan, Maryland, Wisconsin, and Pennsylvania. The remainder of the men volunteered from Cook County. The battery was mustered in at Camp Douglas in Chicago on February 10,1862 under the command of Captain Bouton. On March 1 the battery was transported by rail to Benton Barracks in St. Louis for equipping and training; the cost of the equipment and guns was largely borne by Captain Bouton. In early April the battery was moved to Pittsburg Landing and fought in battle of Shiloh on April 6 and 7. In May Battery I was engaged at the siege and occupation of Corinth, then afterward was ordered to guard duty at Memphis. The battery was ordered to Vicksburg in June 1863 and participated in the siege and fall of Vicksburg in July. During the remainder of the war the battery fought under general Sherman at Chattanooga, Ringgold and Nashville. After the battery returned from furlough they were engaged in Alabama and Eastport Mississippi and then ordered to Chicago for muster out on July 26,1865. There is no record of flags having been presented to this battery but there is a guidon from this Battery at the Illinois State Military Museum in Springfield, Illinois.

[37] Also known at *Bouton's Battery*, *Cudney's Battery, Neeley's Battery*

1st Regiment Illinois Light Artillery, Battery K[38]

Battery K was raised in "Little Egypt" in Shawneetown, Illinois and mustered into service January 9,1862 under the command of Angrean Franklin, a 45-year-old farmer from Vienna Illinois. (Captain Franklin resigned March 31, 1862, and was replaced by 59-year old Rev Jason Smith) The men of the battery volunteered from Johnson County, Massac County and Pope County. The battery spent most of 1862 and 1863 skirmishing with guerillas and standing guard duty at Louisville. From 17 April to 2 May 1863, a detachment from the battery with six 2-pounder Woodruff guns took part in Grierson's Raid from La Grange Tennessee to Baton Rouge, Louisiana. In 1863 and 1864 the battery was in East Tennessee and Knoxville. A portion of the battery was mustered out at Springfield, Illinois in June 1865 and the remainder of the battery mustered out in Chicago the following month. The only known surviving color is a guidon held by the Illinois State Military Museum in Springfield, Illinois

[38] Also known as *Franklin's Battery, Smith's Battery* and *Colvin's Battery*

1st Regiment Illinois Light Artillery, Battery L[39]

The battery was mustered February 22, 1862 at Chicago, under the command of Captain John Rourke, The battery guarded prisoners at Camp Douglas until June 1862 and then spent the remainder of the war on guard duty in West Virginia without seeing any major actions. The soldiers were mustered out July 10, 1865. The only known surviving color from this battery is a guidon at the Illinois State Military Museum in Springfield, Illinois.

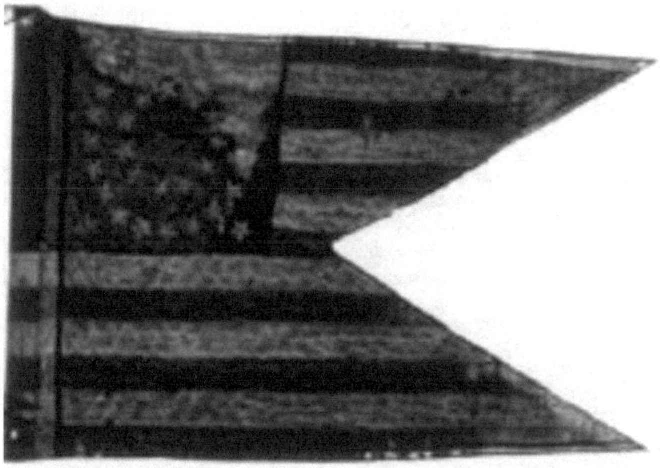

[39] Also known as *Mulligan's Battery, Oconto Irish Guards,* and *Rourke's Battery*

1st Regiment Illinois Light Artillery, Battery M[40]

Battery M of the 1st Regiment Illinois Light Artillery was organized at Camp Douglas and mustered August 12,1862 under the command of Captain John Miller of Ottawa. The men of the battery volunteered from Cook County and La Salle County. In late 1862 and 1863 Battery M skirmished and fought in Kentucky and Tennessee including major engagements at Chickamauga, Rossville Gap, Chattanooga, and Mission Ridge. During 1864 the battery saw action in north Georgia, Resaca, Kenesaw Mountain at the battle of Peach Tree Creek and Atlanta. In early 1865 they served garrison duty in Tennessee and in early July moved to Chicago where they mustered out July 24, 1865. The only known surviving colors of this battery are two guidons at the Illinois State Military Museum in Springfield, Illinois.

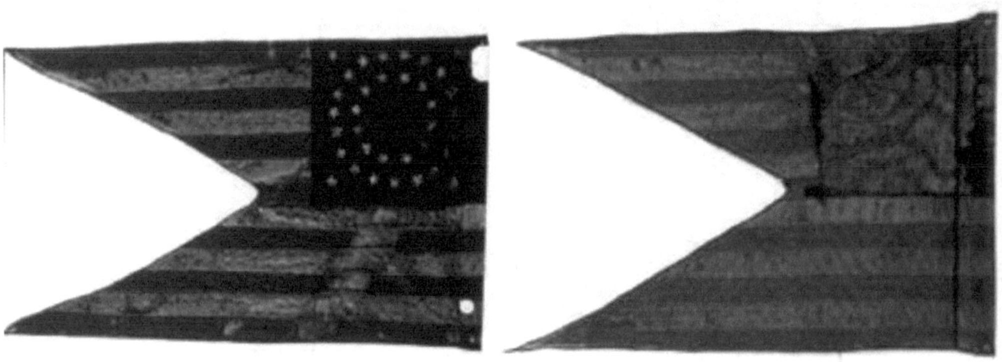

[40] Also known as *Miller's Battery, Spencer's Battery,* and *Cogswell's Battery*

2nd Regiment Illinois Light Artillery, Battery A[41]

Battery A of the 2nd Regiment of Illinois Light Artillery was organized at Peoria, Illinois, and was mustered in May 23, 1861 for state service under the command of Captain Peter Davidson of Peoria, and subsequently mustered into federal service for a three years-service on August 17, 1861 at Jefferson Barracks in St. Louis. Most of the men in battery A volunteered from Peoria County. In 1861 the battery fought in Fremont's Missouri campaign and in 1862 campaigned primarily in Arkansas--fought at Pea Ridge, Sugar Creek and Helena. Later in the year they were engaged at Thompson's Hill, Port Gibson and Champion's Hill. In 1863 the battery was in operations in Louisiana including New Orleans and remained at New Orleans until late 1864. Following sieges at Fort Gaines and Fort Morgan the battery was mustered out on July 28, 1865. The only known surviving color carried by this battery is a 31-star guidon held by the Illinois State Military Museum in Springfield, Illinois.

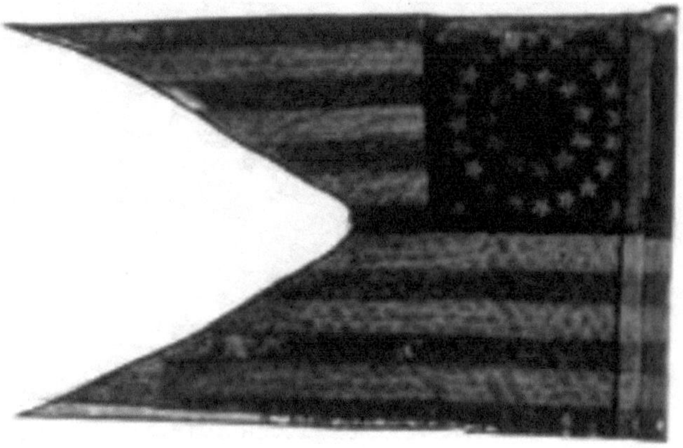

[41] This battery was alternately known as *Davidson's Battery, Borris' Battery*, and *Peoria Light Artillery*.

2nd Regiment Illinois Light Artillery, Battery B[42]

Battery B, 2nd Illinois Artillery Regiment was organized at Springfield, Illinois and mustered into Federal service on 20 June 1861 under the command of Relly Madison of Marshall Illinois. The men of the battery volunteered from Clark County and Coles County. The battery was assigned to the Department of Missouri and in April 1862 was engaged at the Battle of Shiloh. (Battery B named four of its guns Old Abe, Dick Yates, Secretary Hatch, and Jesse DuBois). [43] Later in 1862 battery B was engaged at the Siege and Battle of Corinth and remained on garrison duty at Corinth through January 1864. Later that year they fought at Brice's Crossroads and then were at post in Memphis until mustered out July 15,1865. The Illinois State Military Museum has a guidon carried by this battery. No other flags carried by this battery are known to survive.

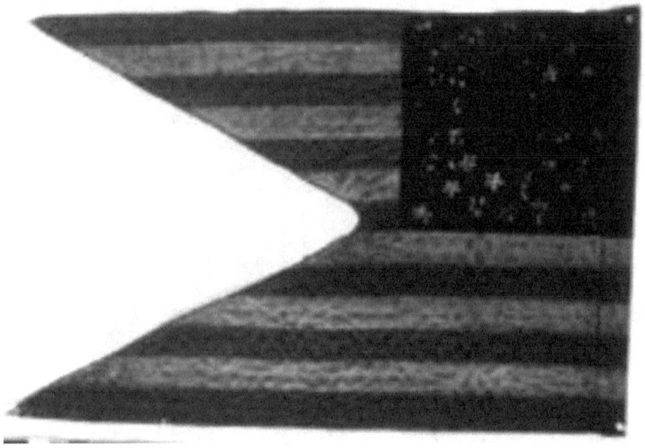

[42] Also known as *Chapman's Peoria Independent and Light Artillery* and *Madison's Battery*

[43] *Wikepedia:* Battery "B" 2nd Illinois Light Artillery

2nd Regiment Illinois Light Artillery, Battery C[44]

Battery C was organized at Cairo Illinois and mustered in August 19, 1861 under command of Captain Caleb Hopkins of Springfield. The majority of men volunteered from Sangamon County and Alexander County. The battery spent the war on guard duty at Fort Donelson and Clarkesville, Tennessee. The only know surviving color carried by this battery is a guidon held by the Illinois State Military Museum.

[44] Also known as *Flood's Battery* and *Springfield Artillery*

2nd Regiment Illinois Light Artillery, Battery D[45]

Battery D was organized at Cairo, Illinois and mustered for three years' service December 17,1861 under the command of Captain Jasper Dresser. The battery was on duty at Cairo until February 1862 when they were ordered to engagements at Fort Henry, Fort Donelson, battle of Shiloh and siege of Corinth. They were then on duty at Jackson, Mississippi until the end of the year. In 1863 and until January 1864 they were at post duty at Grand Junction and then on guard duty at Decatur, Alabama until muster out November 21, 1864. No flags of this battery are known to survive.

---

[45][45] This battery was also known as *Dresser's Battery, Cooper's Battery*

2nd Regiment Illinois Light Artillery, Battery E[46]

Battery E of the 2nd Regiment Illinois Light Infantry was organized at St. Louis, as *Schwartz's Missouri Battery* and mustered in for a three year enlistment on August 20, 1861. It was under the command of Captain Adolph Schwartz of New York. Most of the men forming this battery volunteered from St. Louis, Missouri and from Belleville in St. Clair County, Illinois The battery was first on duty in northern Missouri and then ordered to Cairo, Illinois. In 1862 battery E participated in action at Fort Henry, Fort Donelson, Battle of Shiloh and the siege and battle of Corinth. In 1863 the battery was at Moscow Tennessee guarding the *Memphis & Charleston RR.* In 1864 they were in the Western Louisiana campaign in Baton Rouge and then New Orleans where they mustered out on September 29, 1864. There are guidons at the Illinois State Military Museum carried by this battery. No other colors carried by this battery are known to survive.

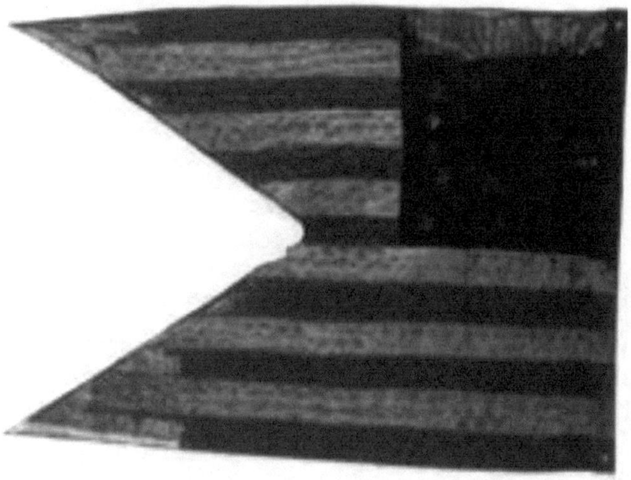

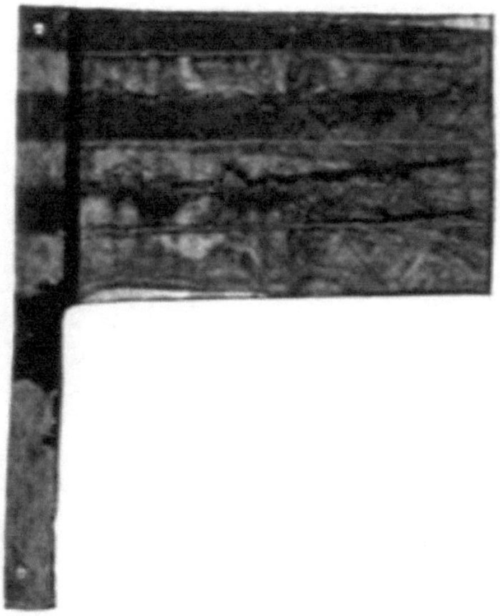

46 Also known as *Schwartz's Missouri Battery, Cram's Battery, Fitch's Battery, Waterhouse's Battery*

2nd Regiment Illinois Light Artillery, Battery F[47]

Battery F was organized at Cape Girardeau Missouri in late 1861 and mustered on December 11, 1861, under the command of Captain John W. Powell. Most men in the battery volunteered from Missouri. The battery was on garrison duty at Forts A and B at Cape Girardeau until March 1862 when ordered to Pittsburg Landing to fight in the Battle of Shiloh in early April of that year. In later April and May the battery fought in the siege of Corinth and then remained there on garrison duty through November. During January through August 1863 the battery, under Grant's command was in action in the Vicksburg campaign, at Jackson Mississippi and then moved to Natchez Mississippi where they remained until February 1864. Later in 1864 Battery F was active in the Atlanta campaign including battles at Kennesaw Mountain, Atlanta, Jonesborough, and Nashville. In 1865 the battery served in Louisiana and in July was ordered to Springfield Illinois and mustered out July 27, 1865. There is a guidon carried by this battery in the Illinois State Military Museum. A photo is not available.

---

[47] Battery F was also known as *Powell's Battery, Richardson's Battery*

2nd Regiment Illinois Light Artillery, Battery G[48]

Battery G of the 2nd Regiment Illinois Artillery was organized at Camp Butler and mustered on December 31,1861 under the command of Captain Charles Stolbrand. The men of the battery volunteered from Cook County and DeKalb County and were initially identified as *DeKalb Horse Artillery.* [49] In early 1862 the battery was stationed at Fort Holt Kentucky until March and then moved to Jackson Mississippi and joined Grant's Central Mississippi campaign. Battery G continued under Grant's command in 1863 and was engaged in the battle of Vicksburg and after the fall of Vicksburg remained on duty there until November of 1863. In 1864 the battery skirmished around Columbus Kentucky until June then in the pursuit of Price in Missouri. In December 1864 at the battle of Nashville, Corporal Samuel J. Churchill, a cannoneer, single-handedly continued firing his cannon despite the remainder of his crew having become panic-stricken deserting their post[50]. For his heroism, Churchill received the Medal of Honor. In 1865 battery G campaigned in Louisiana and took part in the siege of Spanish Fort and Fort Blakely. The battery mustered out September 4. 1865. There is a guidon carried by Battery G in the Illinois State Military Museum in Springfield, Illinois. No other colors carried by this battery are known to survivc.

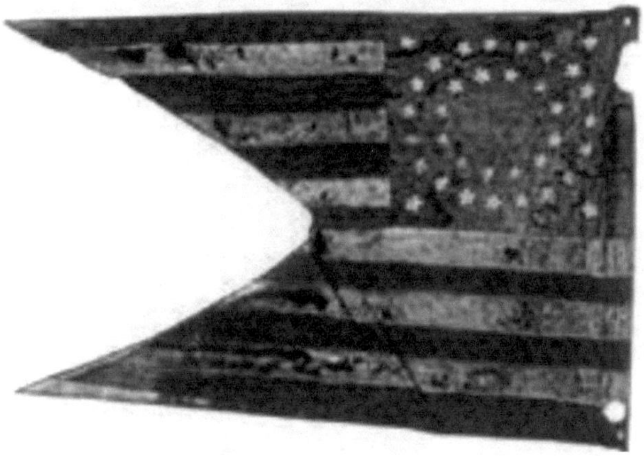

---

48 Also known as *DeKalb Horse Artillery, Stolbrand's Battery, Sparrestrom's Battery, Lowell's Battery*

49 Barnickel, Linda A., We Listed as Patriots: The Civil War Records of Battery G, 2nd Illinois Light Artillery, Bowie Maryland, Heritage Books Inc., 1998, p118-119

50 Deeds of Valor, from the Records in the Archives of the United States Government. Vol. 1, The Perrien-Keydel Company, Detroit, Mich. 1907, p 466.

2nd Regiment Illinois Light Artillery, Battery H[51]

Battery H was organized at Camp Butler and mustered December 31,1861 under the command of Captain Andrew Stenbeck of Galesburg. Most of the men in the battery enlisted from Adams County. In February 1862 the battery was transported to Cairo and stationed at Fort Holt and then at Columbus Kentucky until August. The battery was divided into two sections with one section sent to Fort Pillow and the other to Fort Donelson. The battery subsequently skirmished in Tennessee and Kentucky. In 1863 the battery was ordered to the relief of Fort Donelson and then moved to Clarksville, Tennessee. In January 1864, sixty-five of the men reenlisted as veterans and were sent home on furlough. During the summer of 1864 the battery was mounted and equipped as cavalry. The Battery was mustered out at Springfield, Illinois on March 29, 1865. The Illinois

The Illinois State Military Museum in Springfield has a guidon from Batter H that has the battery designation on the field of the guidon. The guidon is inscribed: "BATTERY H"; "ILL L ART" and there are crossed cannons on the field. No other colors carried by this battery are known to survive.

[51] Also known as *Stenbeck's Battery* and *Whittmore's Battery*

2nd Regiment Illinois Light Artillery, Battery I[52]

Hurrah! Hurrah! Our wish shall ever be,
That right prevail—our land be ever free--
Our flag, "Old Glory," still upheld, that all may in it see
The banner of the right and the Union[53]

Battery I was recruited from Will County and mustered on December 31, 1861 at Camp Butler under command of Captain Charles W. Keith. The battery was transported to Cairo and remained there until February 1862. During the next 4 years-the Battery fought at the Battle of Island NO. 10, Siege of Corinth, Battle of Perryville, Tullahoma Campaign, Battle of Chickamauga, Siege of Chattanooga, Battle of Lookout Mountain, Battle of Missionary Ridge, Battle of Resaca, Battle of Kennesaw Mountain, Siege of Atlanta, Battle of Jonesboro, and in Sherman's March to the Sea. Following the surrender of Confederate General Johnson, the Battery was sent to Washington DC and marched in the Grand Review. They then were ordered to Springfield, Illinois for muster out June 14, 1865. The only known surviving color carried by this battery is a guidon at the State of Illinois Military Museum in Springfield, Illinois.

[52] Also known as *Barnet's Battery, Keith's Battery, Rich's Battery*

[53] Chorus of a song entitled *A Song of Battery I,* Dedicated to Battery I, 2nd Illinois Light Artillery; written by a member of the battery, J.R.Beford of Verona, Illinois. Brown, Thaddeus C.S,, Murphy, Samuel., Putney, William G; Behind the Guns, The History of Battery I 2nd Regiment, Illinois Light Artillery, Carbondale and Edwardsville, Illinois; Southern Illinois University Press, 1965, p165

2nd Regiment Illinois Light Artillery, Battery K[54]

Battery K of the 2nd Regiment Illinois Light Artillery was organized at Camp Butler and mustered December 31, 1861. The Battery was under the command of Captain Benjamin Franklin Rodgers from Jacksonville, Illinois[55] Most of the men in the Battery were from Cook County. In late 1862 the Battery joined Grant's Central Mississippi Campaign, designed to push into Mississippi in preparation for capturing Vicksburg. Therefore in 1863 Battery K was at the siege and capture of Vicksburg. Later that year, in the Jackson Campaign, Battery K fought at the siege and capture of Jackson Mississippi. Battery K, in 1864, moved to Natchez, Mississippi and was on garrison duty there until October. Under the command of Brigadier General Benjamin Grierson, on December 28, 1864, battery K aided in the defeat of the Confederate forces at Egypt Station, Mississippi. The battery closed out the war at Memphis and was then ordered to Chicago for muster out on July 11, 1865. A guidon from Battery K is in the civil war flag collection at the State of Illinois Military Museum in Springfield, Illinois. No other colors carried by this battery are known to survive.

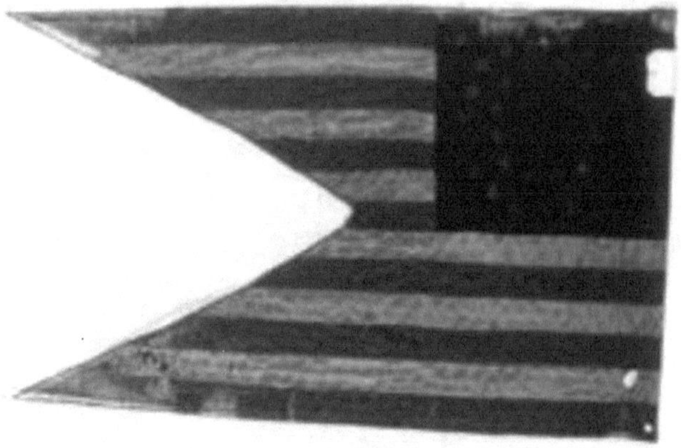

---

[54] Also known as *Barber's Battery, Roger's Battery*

[55] At time of enlistment Rodgers' occupation was listed as shoemaker. Before then his occupation had been listed as tailor. Rodgers became a licensed physician in 1878 and practiced until his death at Patoka in Marion county, Illinois.

2nd Regiment Illinois Light Artillery, Battery L[56]

Battery L of 2nd Regiment Illinois Artillery was organized at Camp Douglas in Chicago and mustered on February 28, 1862, under the command of Captain William H. Bolton. Most of the men in the regiment volunteered from Cook County. Shortly after muster the battery was ordered to Pittsburg Landing and on April 6-7 fought at Shiloh and then advanced to the siege and capture of Corinth; following the fall of Corinth, Battery L was ordered on to Holly Springs, Mississippi; then to Germantown, Tennessee. On October 5, at the battle of Hatchie's Bridge, Battery L claimed the capture of a Confederate flag. "I commanded all the artillery at the battle of Hatchie. I saw this flag with my field glasses unfurled and borne aloft amidst three columns of infantry, advancing to take Spears' battery. I being at the time within 300 yards of the enemy, I read the inscription 'Clark County Farmers' and immediately after giving my orders I saw that flag stricken down by a shot fired by Lieut. E. A. James of Bolton's battery, who at the same time poured such an incessant charge of grape and cannister on the advancing foe that they were driven from the field leaving their colors behind them."[57] The flag was subsequently presented to the city of Chicago. "Coroner James then presented the flag to the city in a neat and appropriate speech.... The flag is six feet long and four feet wide made of red merino with a blue silk field. There four stars upon it, one in each corner of the red. The inscription across the centre reads 'Clark County Farmers' which is the name of the company from which it was captured."[58] The whereabouts of this flag today is unknown. Subsequently Battery L fought in the actions at Port Gibson, Raymond, Jackson, Champion Hill, Siege of Vicksburg and in the Yazoo expedition. The battery was mustered out of service in Chicago on August 9, 1865. The only known surviving color carried by this battery is a guidon held at the Illinois State Military Museum.

[56] This battery was also known as *Bolton's Battery, Hulaniski's Battery,* and *Nichols' Battery.*
[57] *Chicago Tribune,* Dec. 23, 1862.
[58] *Chicago Daily Tribune,* Nov. 21, 1862

2nd Regiment Illinois Light Artillery, Battery M[59]

Battery M was recruited in Chicago and most of the volunteers came from Cook County. On June 6, 1862, the battery was mustered in for three years' of service under the command of Irish born Captain John C. Phillips of Chicago. The battery was ordered to Martinsburg, Virginia in June and then on to the siege of Harper's Ferry in September. On September 15 Battery M surrendered to Confederate forces and returned to Camp Douglas. After parole the battery served the remainder of the war in the Western Theatre. Following the siege of Knoxville, they were ordered to Camp Nelson, Kentucky. They remained there until April 1864 when the Battery mustered out and men transferred to other batteries. There are two colors carried by the battery at the Illinois State Military Museum—a guidon and probable flank marker. No other colors carried by the battery are known to survive.

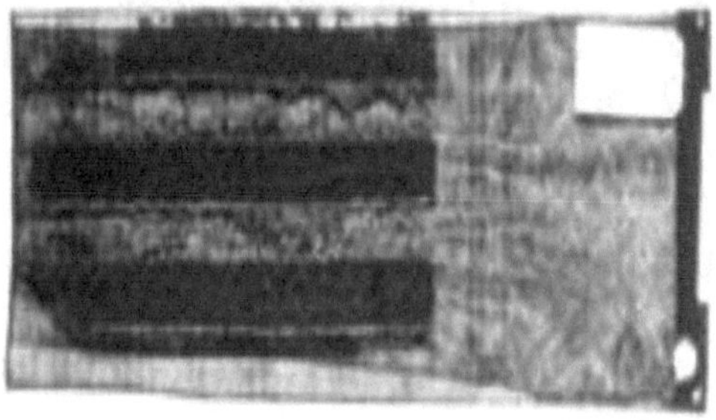

[59] Also known as *Phillips Battery*

Bridges' Independent Battery Light Artillery [60]

Bridges' Battery began as Company G of the Nineteenth Regiment Illinois Infantry. In 1863, when the Regiment was left as part of the garrison at Nashville, Tennessee, General Negley, commanding the post, ordered Captain Lyman Bridges, who was commanding Company G of the Nineteenth, to organize a Battery using captured guns from the Ordnance Department at Nashville. Following organization, the Battery was placed near the city hospital between Franklin and George Washington Pikes as part of Nashville's defenses. However, when the siege was lifted, Bridges' Company (Battery) was detached and sent to Gallatin, Tennessee, with 1,100 prisoners and orders to construct a fort. On completion of the fort, the guns of Bridges' Company were placed in the fort, and on Dec 20, Bridges was ordered to turn over the guns to another battery that had lost its guns at Hartsville, Alabama. Bridges' Company was then armed with muskets and rejoined the Nineteenth Illinois Infantry on the battlefield at Murfreesboro (Stone River) on the evening of Jan 2, 1863. On Jan 14, 1863, by order of the War Department, Company G was permanently detached as an independent battery of light artillery—Bridges' Battery—and furnished with new guns at Nashville. The Battery was assigned to the Pioneer Brigade, Army of the Cumberland, and on Feb 20, was ordered to Murfreesboro, Tennessee. In June, Bridges' Battery, as part of Middle Tennessee or Tullahoma Campaign, marched to Manchester and the Elk River. At Bridges' request for a more active campaign, the Battery was ordered to duty with the 14th Army Corps and in July crossed the Cumberland, Sand, and Lookout Mountains to fight at Davis' Cross Roads near Dug Gap, Georgia, on Sept 11 and in the Battle of Chickamauga[61] September 19-21. The Battery was part of the defense of Chattanooga, positioned at Fort Negley near the Rossville Pike until Oct 12 when ordered to the defense of Fort Wood in northeast Chattanooga. In late November, as part of the Ringgold Campaign, the Battery saw action at Orchard Knob and Missionary Ridge. On Nov 28, the Battery was ordered to Knoxville for relief of Union General Burnside and, from Dec 8, 1863, until February 1864, campaigned in East Tennessee. In April 1864, the Battery marched to Knoxville and then Cleveland, Tennessee, where the Battery rejoined the 14th Army Corps. From May to September, Bridges' Battery was part of Sherman's Atlanta Campaign and was engaged at Tunnel Hill, Buzzard's Roost Gap, Resaca, Adairsville, Kingston, Cassville, Dallas, New Hope Church, and Allatoona Hill in May, Marietta, Kenesaw Mountain, and Lost Mountain in June[62], Ruff's Station, Chattahoochee River, Peachtree Creek and the siege of Atlanta in July and August and Jonesboro and Lovejoy Station in September. Following the pursuit of Hood into North Georgia and Alabama from Sept 29—Nov 3, the Battery engaged in the Nashville Campaign in November and December, fighting at Columbia and Duck River November 24-27, the Battle of Franklin on Nov 30, and in the Battle of Nashville December 15-16. On Dec 21, 1864, Bridges' Battery was transferred to the 1st Regiment Illinois Light Artillery and designated as "Battery B—New" in the records of the Adjutant General of the State of Illinois. The Battery was attached to garrison artillery at Nashville, where it remained until mustered out on Jul 6, 1865[63].

There are three colors of Bridges' Battery in the flag holdings of the State of Illinois—one national flag, one battery flag, and one guidon. The origin of the national flag and battery flag is described in the

---

[60] After December 21, 1864 the Battery was attached to the First Regiment Illinois Light Artillery Battery B—New and was also known as *White's Battery*.

[61] The battery lost 26 men killed or wounded and four captured along with loss of 46 horses.

[62] June 5, 1864 the Battery marched to Morris Hill Church near Ackworth, George where all of the non-veterans were mustered out of service and the command of the battery devolved upon 1st Lieut. Lyman White.

[63] Eddy, T. M., *The Patriotism of Illinois,* Clarke & Co., Publishers, Chicago (1865), Volume I, page 420-24

Chicago Evening Journal. "COLORS FOR BRIDGES' BATTERY—At the office of W. W. Boyington, the well-known architect, we saw this morning a set of colors that have been gotten up by a number of the personal friends of Captain Lyman Bridges, formerly of the 19th Illinois Infantry, but now of the Battery bearing his name. Two elegant silk flags, one, the regulation flag of yellow silk six feet by six feet six inches, in one solid piece, and the other, the standard, the same size, containing a constellation of golden stars upon an azure field, and the white and crimson stripes. The regulation of the battery flag is the most beautiful one that ever went out of Chicago. It is mounted upon a staff, the spearhead of which is of solid silver. Two bronze cannons crossed in the center are surrounded by the inscription—'Bridges' Battery Illinois Light Artillery, Presented by Friends in Chicago.' Thirty-four stars, in gold, describing the arc of a circle, are suspended over the inscription. Both flags are fringed with rich yellow silk and, taken altogether, surpass anything of the kind that we have any remembrance of. They were manufactured and painted by Mr. Chas. L. Dubois of the Custom House Place.[64]"

A copy of a letter thanking the donors for the colors appeared in the Nashville Daily Union. The letter was dated Jul 20, 1863, and sent to "Wm. W. Boyington, John V. Farwell, John W. Roberts, Chas. N. Holden, Milo J. Warren, and others," presumably the men responsible for the purchase of the colors.[65]

The national flag is inscribed as follows, BRIDGES' BATTERY; CHATAOOGA (sp?), JUNE 1862; STONE RIVER; CHICKAMAUGA; MISSION RIDGE; TUNNEL HILL; RESACA; CASSVILLE; MOUNT HOPE CHURCH; PINE MOUNTAIN; KENESAW MOUNTAIN; CHATTAHOOCHEE RIVER; PEACH TREE CREEK; ATLANTA; JONESBORO; FRANKLIN; NASHVILLE

[64] *Chicago Evening Journal,* July 15, 1863

[65] *Nashville Daily Union ,* July 31, 1863.

The Battery color is inscribed: BRIDGES' BATTERY; ILLINOIS LIGHT ARTILLERY; PRESENTED BY FRIENDS IN CHICAGO.

This Artillery guidon measures approximately 24" x 36" and has 35 stars in the canton. It is inscribed- -STONE RIVER, CHICKAMAUGA, BRIDGES' BATTERY, B 1st ILLS. LA, MISSION RIDGE, ATLANTA, FRANKLIN, NASHVILLE. *The 35 stars date this guidon as having been manufactured in mid-1863 or later.* No other flags carried by Bridges' Battery are known to survive.

Busteed's Independent Battery Light Artillery

The history of this battery is best told by Chicago newspapers. "Captain Richard Busteed, who for some three or four weeks has been at work in this city forming a company of light artillery, having his headquarters at the Wigwam, has just completed his organization, and received orders to march for the Potomac. This battery now consists of one hundred and fifty-six men…"[66] Seven weeks later, this item was posted in the Chicago Times: "This company of 150 men was formed in Chicago during this summer and early in September was ordered to Washington…They received their clothing, six pieces of artillery, horses, and other equipment and settled down to the work of making a reputation for themselves and the State they came from. They were from Illinois and were proud of it. At least, this was the way the men of the company felt. But it seems the officers or a portion of them are great loafers, and in order to convince this world that they had it in them to become distinguished, two lieutenants and one non-commissioned office have managed to get arrested for stealing horses. Capt. Busteed has resigned, leaving the Illinois boys under his command at the mercy of strangers in a strange land. The company is completely broken up."[67] There is no further history of this battery.

[66] *Chicago Tribune,* Oct 1, 1861 This information was sent to the *Chicago Tribune* by C.C. Campbell, Major and Chief of Artillery on Major General Hurlbut's staff.
[67] *Chicago Times,* Nov 23, 1861.

Campbell's Independent Battery Light Artillery

Campbell's Battery was a three-month of service battery from Ottawa, Illinois, under the command of Captain Charles Campbell, that became part of Battery D of the 2nd Regiment of Illinois Light Artillery. The is no record of the enlisted men of this battery and no flags carried by this battery are known to survive.

Chapman's Independent Battery Light Artillery [68]

Chapman's Battery was organized with the 14th Illinois Infantry and mustered in under the command of Captain Fletcher H. Chapman on January 1, 1862. It was subsequently consolidated with the 2nd Illinois Light Artillery Battery B. There are no know surviving flags of this battery.

[68] Also known as *Chapman's Peoria Independent Battery* Light Artillery

Chicago Board of Trade Independent Battery Light Artillery [69]

The Chicago Board of Trade was established in Chicago in 1848 as a voluntary organization to facilitate grain sales. By the onset of the Civil War, the State of Illinois had granted a charter to the Board of Trade to standardize the quality of grain and regulate grain sale futures. At a meeting of the Board on July 16, 1862, the members of the Board "pledge ourselves to use our influence and money to recruit a battery to be known as the Board of Trade Battery."[70] The men were rapidly recruited from Cook County, and the battery was mustered on August 1, 1862. The battery was under the command of James H. Stokes, a 45-year-old insurance agent. Soon after muster, the battery was involved in the pursuit of Bragg in Kentucky, then were ordered to occupy Bowling Green through mid-December. In late December, the battery was engaged in the Battle at Stone's River north of Murfreesboro, Tennessee. The battery remained at Murfreesboro through June of 1863. At that time, the battery was changed from Mounted Field Artillery to Flying Horse Artillery and was assigned to the 2nd Cavalry Division. In the latter months of 1863, the battery rode in the Tullahoma campaign and fought at the battle at Chickamauga, Georgia, in September. In 1864 – as part of the Atlanta Campaign-they fought at Resaca, Tanner's bridge, Dallas, Big Shanty, Kenesaw Mountain, Marietta, Chattahoochee River, and Atlanta. Later in 1864, the battery was in the Nashville campaign and fought at the Battle of Nashville. During 1865 they saw action in Alabama and Georgia, then moved to Nashville, Tennessee, and finally to Chicago to muster out on June 30, 1865. The history of the flags carried by the Board of Trade Battery is summarized in the history of the Battery[71]: "By special orders, our flags (these flags being the first carried by an artillery company, and then by special orders) had placed on them 'Stone River', 'Elk River', 'Chickamauga', 'Farmington', 'Dallas', 'Decatur', 'Atlanta', 'Lovejoy', 'Nashville', 'Selma'. The flags (two) were placed in the vault of the Chamber of Commerce in Chicago and, in the great fire of 1871, were destroyed. Nevertheless, fortunately, two guidons carried by the battery survive. The first is at the National Park Service-Stones River National Battlefield. According to their records, the guidon is loosely woven red bunting that is 21-5/8" in length, 19 ½" in width and width of letters 2" and height of letters 5". The letters are painted on with white enamel and shadowed with black paint.

There is also a guidon carried by the Board of Trade Battery in the State of Illinois Military Museum. It comes with the following story.

[69] Also known as *Stokes' Independent Battery of Light Artillery*

[70] Historical Sketch of the Chicago Board of Trade Battery, Horse Artillery, Illinois Volunteers. Chicago, 1902 page 17-18.

[71] Ibid. Historical Sketch of the Chicago Board of Trade etc. page 33

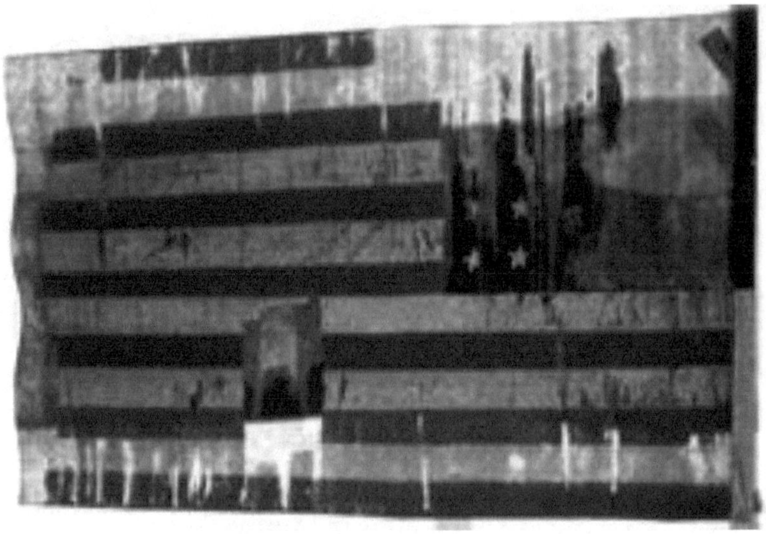

" I have today the honor and pleasure of placing in Grand Army Memorial Hall two relics dear to all, and so far as I know, the only relics of our Battery now in existence. One, the Bugle used by our late Comrade William Berdel...The other relic is one of the Guidons that was furnished in Chicago before you went to the front. It was with you in all your marches and battles, receiving its first baptism of fire and bullets in the battle of Stone River.... When our late Comrade Frederick G. Deane was mustered out of service, he secured this guidon and placed it with his mother, who could not be induced to part with it and cared for it until her death. After that, it was cared for by her son D. H. Deane who has now presented it to your association."72 No other flags carried by the Chicago Board of Trade Battery are known to survive.

[72] Ibid. Historical Sketch of the Chicago Board of Trade. Page 81-2

Chicago Mercantile Independent Battery Light Artillery

The Mercantile Exchange was formed in the mid-1840s in Chicago by businessmen who were involved in the billing and handling of large shipments of grain on Lake Michigan, through the Illinois & Michigan canal, and on the Galena & Chicago Union RR. At a meeting of the Mercantile Association in July 1862, they decided to not only recruit an artillery battery but also pay for the battery's equipment and any bounties. The battery was tentatively named the *Doggett Guard Battery* in honor of William E. Doggett, who was president of the Mercantile Association. The battery was recruited with men from Cook County and mustered in for three years' service on August 29, 1862, under the command of Captain Charles G. Cooley. The battery moved to Memphis in November. In 1863 they were in the assault and capture of Fort Hindman and then moved to Young's Point, Louisiana, on January 17 and remained on duty there until March.[73] On May 1, the battery fought at the Battle of Champion Hill; on May 16. Big Black River Bridge May 17. The battery participated in the Siege of Vicksburg, Mississippi, May 18-July 4. For actions at Vicksburg on May 17, 1861, five members of the battery were awarded the Medal of Honor.[74] In 1864 the battery saw action in Texas and Louisiana. In 1865 they were ordered to Baton Rouge, Louisiana, and then on to Chicago, where they were mustered out on July 10, 1865.

The Chicago Tribune reported that the battery was presented with a fine set of colors, but those colors are not further described. The only known surviving color carried by this battery is held in the State of Illinois Military Museum. No other flags carried by this battery are known to survive.

This national flag is inscribed "Chicago Mercantile Battery"

[73] Due to criticism re his leadership, Captain Cooley resigned on February 24,1863 and was replaced by Captain Patrick H. White.

[74] Cpl James Dunne, Pvt Charles Kloth, Pvt George Kretsinger, Pvt. Patrick McGuire, Capt. Patrick White received the MOH for "Carried with others, by hand, a cannon up to and fired it through an embrasure of the enemy's works."; Civil War Medal of Honor Winners from Illinois-published by Civil War Commission of Illinois, July 1962.

Colvin's Independent Battery Light Artillery

Colvin's Battery was organized in October 1863 from men detached from the 107th Illinois Infantry Regiment, 33rd Kentucky Infantry Regiment, and 22nd Indiana Battery; the battery was under the command of Captain John Colvin. The men served in the Knoxville campaign and in eastern Tennessee. In March 1865, the unit was re-designated Battery K of the 1st Illinois Light Artillery Regiment. The Illinois Adjutant General's records do not contain a roster of this battery, and there are no known flags associated with this battery.

Henshaw's Independent Battery Light Artillery [75]

Henshaw's Battery was recruited in Ottawa, Illinois, in 1862 and mustered in for three years' service on December 3, 1862, under the command of Captain Edward H. Henshaw of Ottawa. Most of the men in the battery volunteered from LaSalle County. The battery served much of its early service on guard duty in Kentucky. However, in 1863 they were engaged in actions at the Battle of Buffington Island, Battle of Salineville, in the Knoxville campaign, Battle of Campbell's Station, and Siege of Knoxville in 1863. It spent the remainder of the war on garrison duty in east Tennessee. The battery mustered out in July 1865. No flags have been found that were carried by Henshaw's Battery.

[75] Also known as *Ottawa Light Artillery*

Renwick's Elgin Independent Battery Light Artillery [76]

Renwick's Battery was organized in Elgin, Illinois and the men of the battery volunteered from Kane County. Captain George W. Renwick commanded the battery when it was mustered in on November 15, 1862. The battery was initially assigned to guard duty in Kentucky, but in August 1863, the battery joined in Burnside's campaign in east Tennessee, including the Knoxville campaign. In February 1865, the battery fought at Wyse Fork and was then mustered out on July 18, 1865. No flags carried by Renwick's Elgin Battery are known to exist.

[76] Also known as *Elgin Battery Light Artillery*

Springfield Independent Battery Light Artillery [77]

On August 21, 1862, the Springfield Independent Light Artillery was mustered in for three years of service under the command of Captain Thomas F. Vaughn[78]. The men of the battery volunteered from Sangamon County and St. Clair County. Following muster, the battery was ordered to Bolivar, Tennessee, where they were on garrison duty until June 1863. The battery was divided into three sections with orders to guard the *Memphis and Charleston Railroad*, with one section each at Moscow, Collierville, and Germantown through August 1863. Following the capture of Little Rock, Arkansas, Vaughn's Battery served garrison duty there until March 1864. Later in 1864, the battery was in action at the Battle of Prairie D'Ane and the Battle of Jenkins' Ferry. The battery was ordered to Springfield, Illinois, in June and mustered out 25, 1865. No flags carried by battery are known to be in existence today.

---

[77] Also known as *Vaughn's Independent Illinois Artillery Battery*

[78] "Vaughn previously commanded Battery B, 1st Rhode Island Light Artillery from 25 August 1861 until 1 December 1861 when he resigned his commission." Rhodes, John H. (1894). "The History of Battery B, First Regiment Rhode Island Light Artillery". Providence, R.I.: Snow & Farnham Printers.

Cogswell's Independent Battery Light Artillery

An artillery battery was recruited in Ottawa, Illinois, and mustered in for three years' service on November 11, 1861, under the command of Captain William Cogswell. At the time of muster, the battery was Company A, Artillery, 53rd Illinois Infantry. In February 1862, the battery was ordered to Chicago, and in March, the battery was detached from the 53rd Illinois Infantry and became known as Cogswell's Battery Light Infantry. The battery was ordered to the Siege of Corinth in May and then removed to and remained at Memphis until June 1863. In June and July, the battery was active at the siege and capture of Vicksburg. During 1964 the battery performed garrison duty until December, when they fought at Nashville. Eighteen sixty-five saw the battery fighting at Spanish Fort, Fort Blakely, and Mobile, Alabama. The battery was mustered out on August 14, 1865.

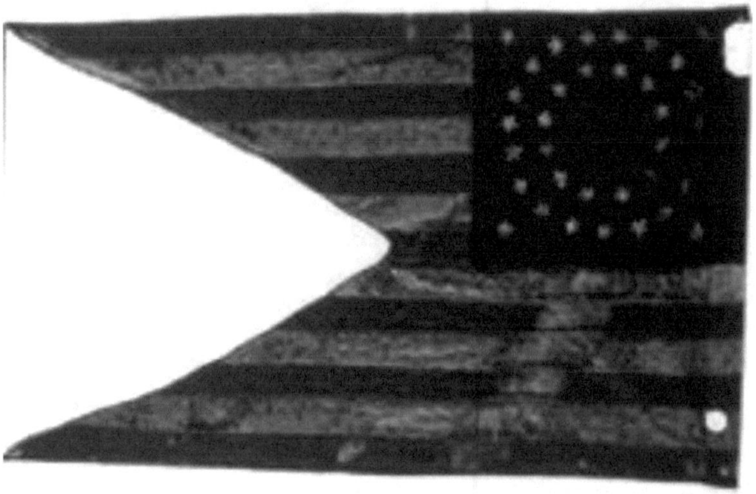

The only known surviving color carried by this battery is a guidon at the State of Illinois Military Museum.

Cavalry Colors

The Cavalry color or flag was called a "standard." By regulation, the cavalry used only a single flag which was a smaller version of the Regimental flag used by the Infantry. The cavalry did not officially carry a national flag. According to the Army regulations

"Each regiment shall have a silken standard, and each company a silken guidon. The standard to bear the arms of the United States, embroidered in silk, on a blue ground, with the number and name of the regiment, in a scroll underneath the eagle. The flag of the standard is to be two feet five inches wide, two feet three inches on the lance, and to be edged with yellow silk fringe. The flag of the guidon is swallow-tailed, three feet five inches from the lance to the end of the swallow-tail, fifteen inches to the fork of the swallow-tail, and two feet three inches on the lance. To be half red and half white, dividing at the fork, the red above. On the red, the letters U. S. in white; and on the white, the letter of the company in the red. The lance of the standards and guidons to be nine feet long, including spear and ferrule." The cords and tassels of the Cavalry standard were made of silk, red and white intermixed.

On Jan 18, 1862, General Order #4:

"Under instructions from the Secretary of War, dated Jan 7, 1862, guidons and camp colors for the Army will be made like the United States flag, with stars and stripes." In the Illinois, Civil War flag collection are several national flags that were carried by cavalry regiments reflecting the fact that many cavalry regiments carried a national flag in addition to the "standard" or regimental flag.

1st Illinois Regiment Cavalry

The First Illinois Cavalry Regiment was organized at Alton, Illinois, and mustered in for three years' service on Jul 3, 1861, under the command of Colonel Thomas Marshall[79]. The men volunteered from the following counties: Company A (*Loomis Dragoons*)-McLean County; Company B (*Gallatin County Cavalry*)-Gallatin County; Company C-Coles County; Company D-Knox and Peoria Counties; Company E-Montgomery County; Company F-Sangamon County and Company G-Warren County. Company H from Marion County (*Centralia Cavalry or Nobleman's Cavalry*); Company I from White County (*White County Cavalry*), and Company K from Winnebago County (*Light Dragoons* or *Steele's Guards* or *Huntley's Cavalry Company*) did not muster in with the other companies nor did they serve with them in the field. From Alton, Illinois, the seven companies (companies A thru G) of the Regiment moved to St. Louis, Missouri, where they received their arms and uniforms. The Regiment then moved to Jefferson City, Missouri, and on to Hannibal, and finally to Lexington, Missouri. The Battle of Lexington began on Sept 11, 1861; a much larger Confederate force attacked the First Illinois Cavalry, Twenty-third Illinois Infantry, and the Thirteenth Missouri Infantry. After several days of hard fighting, the Union troops were forced to surrender on Sept 20[80]. The officers were paroled, and the noncommissioned officers and privates were given the option to either take an oath not to take up arms until exchanged or to become prisoners of war. They elected to take the oath, and the non-commissioned officers and privates returned to St. Louis to await the return of their officers. The men of the First Illinois Cavalry were paid off at the end of September and returned home to await exchange. However, by November 1861, the Government had failed to arrange an exchange. Therefore, the non-commissioned officers and privates were ordered to Springfield, where they were discharged. In December 1861, the commissioned officers were exchanged and were ordered to reorganize the First Illinois Cavalry Regiment. The non-commissioned officers and privates were encouraged to return to their previous companies, and if they did, they would receive pay as if they had remained continuously in service. The Regiment was reorganized at Benton Barracks in St. Louis in January 1862, using members of the original Regiment as well as new recruits. For the next four months, the Regiment guarded railroads and stations in southeast Missouri. Unfortunately, at the time of re-organization, there were vacancies of officers in every company due to promotions or assignments to other regiments. There was dissatisfaction on the part of the men on the filling of many of these vacancies. Due to the growing discord in the Regiment, on order from the war department, the Regiment was mustered out at Benton Barracks on Jul 14, 1862. Companies I, H and K had not campaigned with the other companies of the Regiment. Company I was mustered in as cavalry on Jun 12, 1861, and ordered to Bird's Point, Missouri, on July 12. Company H was ordered to Bird's Point in August and joined Company I. Both companies fought at the battle of Belmont and the capture of New Madrid. Company I, in addition, assisted in the capture of Island No. 10. Both companies were on duty at New Madrid until June 1862, were at Memphis, Tennessee, until July, and were mustered out Jul 15, 1862. Company K was organized at Alton and mustered on Dec 9, 1861; it never served with the First Illinois Cavalry; instead, was

[79] 57 year old Thomas Marshall was an attorney from Charleston, Illinois.. He was a member of the Whig Party, was a delegate to the Illinois constitutional convention in 1847, helped to establish the Illinois Repulican Party and attended the 1860 Republican National Convention as a delegate. Marshall was elected to the Illinois Senate in 1858 to 1862 and in 1861 served as president pro tem of the Senate. (From the Papers of Abraham Lincoln digital Library)

[80] George H. Palmer, Bugler of Company G from Monmouth, Illinois, received the Medal of Honor for actions at Lexington, Missouri on September 20, 1861, for: "Volunteered to fight in the trenches and also led a charge which resulted in the recapture of a Union hospital, together with Confederate sharpshooters then occupying the same."

transferred to the Fifteenth Illinois Cavalry Regiment and became Company M.[81]

There are no flags in the collection of the Adjutant General of the State of Illinois identified as flags carried by the First Illinois Cavalry. It is possible that no colors were issued to the Regiment; it is also possible that the colors of the Regiment were captured with the capture of the Regiment at the battle of Lexington on Sept 20, 1861.

At least Company A of the First Illinois Cavalry was presented a flag, and the flag did return home with the Company after the battle of Lexington. A Bloomington newspaper reported[82] on Sept 27, 1861: "MCLEAN COUNTY. The *Chicago Tribune* of the 25th says: "On the train this afternoon were portions of Captain Graham's company, belonging to Gen. Lane's brigade, and a cavalry company from Bloomington. The latter brought their flag with them, which private Carico[83] saved by winding it about his body after the surrender. It bears the marks of severe usage, having two cannon ball holes and various bullet perforations, and is fearfully torn and stained. It was presented to them by the ladies of Bloomington."

In another article from the *Illinois Statesman*, Oct 18, 1861, "On Saturday last Captain McNulty and his company of *Loomis Dragoons* returned home. They were received at the Western Depot by the Zouave Cadets and a large number of citizens, accompanied by the Bloomington Brass Band and were escorted to the Court House Square, where they were welcomed by hearty cheers. They brought with them the beautiful flag presented to the company before they left for the war. It was sadly mutilated by the rough usage it received at the battle of Lexington, but we take great pleasure in being able to state that it was returned without any taint of dishonor upon its bright silken folds." This flag is not in the Illinois Military Museum in Springfield, nor was it deposited with the McLean County Historical Museum. It may either be in a private collection or has been lost. No other flags carried by this regiment are known to exist.

---

[81] *Report of the Adjutant General of the State of Illinois,* H. W. Rokker, State Printer and Binder, Springfield, 1886, page484-5.

[82] *Illinois Statesman, September 27, 1861*

[83] Henry C. Carico of Bloomington, Illinois

2nd Illinois Regiment Cavalry

"Nothing good can be said of them. Take officers and men, and they are the worst looking military organization I have ever seen. The only clean or respectable place I saw in camp was that occupied by their animals[84] ..."

The Second Illinois Cavalry Regiment was organized at Camp Butler and mustered in on Aug 12, 1861, under the command of Colonel Silas Noble[85]. (Company M did not join the Regiment until after its muster in on Dec 30, 1861.) The men volunteered from the following counties: Company A (also known as *Logan's Escort*, *Ord's Escort,* or *Stevenson's Escort*) from Ogle County; Company B (also known as *Quinby's Escort*) from Logan County; Company C from Mason County; Company D from Jersey and Madison Counties; Company E (also known as *Quinby's Escort*)) from St. Clair County; Company F from Douglas, Macon and Piatt Counties; Company G from Hancock County; Company H from McDonough County; Company I from Champaign County; Company K from Pike County; Company L (Ad*ams County Cavalry*) from Adams and Hancock Counties and Company M (*McKean's Escort*)from Mason County. The Regiment was split into battalions—First Battalion--Companies A, B, C, D, I, and L moved to Cairo, Illinois, on Oct 3, while—Second Battalion –Companies E, F, G, H, and K were ordered to Paducah, Kentucky on Nov 1, 1861. Company C moved on from Cairo to Caledonia, Kentucky, for winter encampment, and in January 1862, Company M joined the Second Battalion stationed at Paducah. Also, in early 1862 Companies, A and B were transferred to the Army of Tennessee and fought at Fort Donelson in February and at Shiloh in April. In March 1862, the Second Battalion (Companies E, F, G, H, K, and M) commanded by Lieutenant Colonel Harvey Hogg advanced on Columbus, Kentucky but arrived after the rebels had evacuated the town. In addition, Companies G, H, K, and I were in action against New Madrid, Island No. 10, and Fort Pillow from February through April, while Companies D and L remained in Cairo. In May 1862, Companies A, B, C, and D of the First Battalion were ordered south and participated in the siege of Corinth, Mississippi, and after the fall of Corinth on May 30, guarded the *Mobile & Ohio Railroad* in western Tennessee. On Aug 15, 1862, Lieutenants Calvin Terry and John Goodheart and Private Jacob Stone of Company C were killed in battle on the Obion River near Union City, Tennessee. At Bolivar, Tennessee, on Aug 30, four companies commanded by Lieutenant Colonel Hogg met a larger Confederate; force. Hogg, along with lieutenants Neil Shannon and Levi Leib of Company F, were killed. During the latter half of 1862, the companies of the Regiment were on patrol, scouting, and guard duties in Tennessee and Kentucky. In December 1862, Companies C, F, G, H, I, and K were sent to Holly Springs, Mississippi, and on Dec 20, Confederate forces overran the depot; several of the officers and men were killed, wounded, or taken prisoner, 150 horses were lost along with all of the regiment's records and equipment. In 1863 various companies saw action in fighting near Vicksburg, Mississippi, and in eastern Louisiana. In August, seven companies moved to Carrollton, Louisiana, just above New Orleans, and skirmished with enemy forces in that area for the remainder of the year. In January 1864, one hundred and fifty men of the six companies stationed in New Orleans re-enlisted as veterans, and in February, they went

---

[84] Wiley, Bell Irvin, *The Life of Billy Yank*, Doubleday, Garden City, New York, 1971, page 222. The report of Colonel James Hardie of the Inspector-General's Department to his superiors on the condition of the Second Illinois Cavalry—September 1864. Hardie, an Easterner, was shocked at the casual appearance of the Western soldiers.

[85] 43 year old Noble came to Illinois in 1841 and became one of six attornies in Dixon. From 1846-1848 he represented Lee, Henry, Rock Island and Whiteside counties in the Illinois Senate as a Whig. He became a prominent farmer and landowner before becoming Colonel of the 2nd Regiment Illinois Cavalry. (Papers of Abraham Lincoln Digital Library.)

home on furlough. The non-veterans during that time took fought in the ill-fated Red River campaign. They skirmished regularly with Texas Cavalry, and on Apr 8, 1864, the Second Cavalry was in the battle at Sabine Crossroads (also known as the battle of Mansfield, Louisiana). There the Union forces marched into a rebel trap. The Regiment lost 42 men killed and wounded, ten were taken prisoner, and the Regiment lost all its equipment. In the retreat to the Mississippi River, the Second Regiment skirmished near Cloutierville, Alexandria, Bayou De Glaze, and at the Atchafalaya River, finally arriving back in New Orleans on Jun 10. On Jul 6, 1864, the veterans and new recruits rejoined the Second Illinois Cavalry at Baton Rouge. In August, the men who had completed their three-year term but had not re-enlisted were mustered out, and five companies of the Regiment that had been in Tennessee rejoined the Regiment, bringing the Regiment back to full company strength. For the remainder of 1864, the Second Cavalry, using Baton Rouge as a base, scouted, and skirmished in Louisiana and Mississippi. In March 1865, the Regiment was transferred to Pensacola, Florida, and then up the Blackwater River to Crigler's Mills, Florida, and using this as a base, raided into Alabama[86]. After the fall of Mobile, Alabama, Union cavalry units were sent toward Tuscaloosa and then ordered to ride cross-country to Vicksburg, Mississippi. From Vicksburg, the Regiment was transported up the Red River to Shreveport, Louisiana, and due to the reduction of the number of men, the twelve companies were consolidated into six. On Jul 10, 1865, the Consolidated Second Illinois Cavalry Regiment was ordered to ride from Shreveport to San Antonio, Texas. The Regiment mustered out at San Antonio on Nov 22, 1865.[87]

The records of the Second Illinois Cavalry relate two occasions at which the "Equipment of the regiment was lost"—the first at Holly Springs and the second at Sabine Crossroads. Although the loss of the Regiments colors is not mentioned, this may have occurred. Currently, the only flag in the holdings of the State of Illinois Military as a flag of the Second Illinois Cavalry Regiment is a guidon.

The inscription on this guidon reads: PRESENTED TO Co H, 2nd ILL CAV VETERAN VOLUNTEERS BY CITIZENS OF McDONOUGH COUNTY

[86] "On the 1st Colonel Spurling's command was sent ahead of the column to ascertain the best route to Holyoke to communicate with headquarters in regard to our movements &c. About four miles and a half from Blakely and one mile from where the road forks toward Holyoke he found the road barricaded, and a strong picket or outpost, composed of cavalry and infantry, which he charged, capturing the battle-flag of the Forty sixth Mississippi Infantry and 74 prisoners." O.R. Series 1, Volume XLIX/1 page 281.

[87] Behrens, Robert H., *From Salt Fork to Chickamauga, Champaign County Soldiers in the Civil War.* The Urbana Free Library, Urbana, Illinois, 1988, pages 233-53.

2ND ILLINOIS REGIMENT CAVALRY

There is a handmade cavalry standard in the holdings of the Abraham Lincoln Presidential Library and Museum in Springfield[88]

The flag is paint on silk. It is homemade and the flag might have been cut to fit in a frame. It measures 14 in x 22 ½ in and the label attached to the flag reads:

"This is a handmade flag for the 2nd IL Cavalry. Flag is a rectangle of blue silk panels, onto which handpainted fabric or paper elements have been affixed. Such elements comprise a left-facing spread eagle, gripping a pink banner in its beak that reads "UNUM." Eagle has a crest on its chest and carried arrows and laurel (holly) in its talons.

Paper tags attached to the flag indicate the following: "Presented by the family of Maj. Frank Moore." "Battleflag of the Second Illinois Cavalry." "Framed battleflag of 2nd Illinois Cavalry. Donors-- Family of Frank Moore."

A portion of Company I of this regiment volunteered from Champaign County. The following relates the story of a cavalry standard presented to company I. Might this be the same flag?

A cavalry Standard was presented to Co I by the ladies of the city of Champaign. In an article entitled "Meeting of Cavalry Company at Camp Butler," The following appears: "A beautiful standard of colors in silk was formally presented to the company in the name and behalf of the ladies of Champaign city, (by) Lieut. H. Bartling with patriotic remarks. Lieut. Cacy was then called out, who responded in a few words expressing his purpose in following this flag, to write his record with his sword, and not with his pen"[89]. In the same issue of the newspaper is another story about the flag presentation: "The Champaign Cavalry Co., through some mistake or misunderstanding, was named letter I, at which our boys feeling themselves agrieved (sp?), and justly too, as they had put on their best behalf; and did not consider themselves second to any, and our noble Col. a gentleman and soldier in very sense, willing to do us justice, rode out on parade, and in a feeling speech, said that, as we were all superior in drill and good behavior, any position he could give after A should be ours; not that the Co. stood behind any in his regiment but simply as an act of justice, we were entitled to the best in his gift; but circumstances beyond his control prevented his giving letter A, and in as much as the ladies of Champaign had inscribed upon our Standard the letter "B" we consented to take that, and make our Co the first in the regiment, in all respects except the letter." Based on this report, it

---

[88] The photo and information re this flag are courtesy of the Abraham Lincoln Presidental Library and Museum in Springfield, Illinois

[89] *Central Illinois Gazette,* September 4, 1861

appears that the standard was inscribed with "B," but actually, the company designation was "I." The whereabouts of this standard is unknown today.

At least one National flag was presented to Company A of the Regiment. "A silk flag was presented by the ladies of the town, but by whom the presentation was made or who was responsible for the gift is not now known. The flag was accepted by David B. Dewey, who, on behalf of the company, made an earnest and appropriate speech.[90]" No other flags carried by this regiment are known to survive.

---

[90] Fletcher, Samuel H., *The History of Company A, Second Illinois Cavalry,* Chicago, 1912, page 7

3rd Illinois Regiment Cavalry

The Third Illinois Cavalry Regiment was organized at Camp Butler, near Springfield, Illinois, and mustered in for three years' service on Aug 27, 1861, under the command of Colonel Eugene A. Carr[91]. The Regiment was organized from the following counties—Company A from Sangamon County (*Morgan's Escort*), Company B from Tazewell County, Company C from Cass County, Company D from Bond County (*Steele's Escort*), Company E from Saline County, Company F from Adams County, Company G from Brown County, Company H from Fulton and Schuyler Counties, Company I from Mc Lean County (*Scott's Dragoons*), Company K from Livingston County, Company L from Macoupin and Montgomery Counties (*McClernand's Escort*) and Company M from Christian and Shelby Counties. Following instruction and drill at Camp Butler, the Regiment was ordered to St. Louis on Sept 25; on Oct 1, the Regiment moved to Jefferson City, Missouri, and then on to Warsaw, Missouri. On arrival at Warsaw on Oct 11, the Third Illinois Cavalry captured a large store of enemy supplies. While in camp, the Regiment assisted in building a bridge across the Osage River; following completion of the bridge, the men advanced toward Springfield, Missouri, as part of forces commanded by Major General John Fremont. Following the engagement at Springfield, the First and Second Battalion of the Third Cavalry (Companies A-H) retreated toward Rolla in November, while the Third Battalion (Companies I-M) remained at Springfield under the command of Major James Ruggles, protecting the rear of the retreating army. The entire Regiment was reunited at Rolla in late November and remained there on duty until January 1862, when it moved in advance of General Samuel R. Curtis' advance on Springfield, Missouri. The Union force arrived at Marshfield on Feb 11 and went into camp eight miles from Springfield on Feb 13. Major Ruggles received permission to advance four miles further with the Third Battalion and met and defeated enemy forces in the first battle of the campaign. The following day, General Sterling Price, commander of the Missouri State Militia, abandoned Springfield, fleeing south into Arkansas. March 6-8, the Third Illinois Cavalry joined the Union forces in the battle against the Confederate forces at Elkhorn Tavern (Pea Ridge), Arkansas. In the ensuing battle, the Third Cavalry lost ten men killed and forty wounded. Among the wounded was Colonel Carr[92]. Command of the Regiment devolved to Lt. Col. Lafayette McCrillis. A week later, the Third Cavalry raided Fayetteville, Arkansas, and drove the enemy from that area. The Third Cavalry accompanied the army forces north back into southern Missouri in March but in April returned to Arkansas near Batesville to control enemy activity. On Apr 19, near Talbot's Ferry on the White River, the Third Cavalry joined in the destruction of the enemy's saltpeter works. In early June, the Regiment fell back to Fairview, Arkansas, and on Jun 7, Captain David Sparks, Company L, and 66 of his men were surrounded but fought their way out, losing four wounded and four taken prisoner. Later in June, the Regiment was again ordered to Batesville for general scouting, and then with the remainder of the army was ordered on to Helena, Arkansas; the Union forces followed the course of the White River and experienced ten days of fighting, starving, and lack of water en route. During the stay at Helena, the Regiment lost many men due to disease, demoralization, and desertion. It was during this encampment that the Third Cavalry participated in raids to the St. Francis River, to Grenada, Mississippi, and to Arkansas Post. On Dec 23, 1862, the Regiment (except for Companies E and G that remained at Helena) boarded transports and joined the forces moving against Vicksburg under

---

[91] Carr was an officer in the Regular Army and a West Point graduate in the class of 1850.

[92] Colonel Carr was one of the wounded and was subsequently awarded the Medal of Honor—"Directed the deployment of his command and held his ground, under a brisk fire of shot and shell in which he was several times wounded at Pea Ridge, Arkansas, March 7, 1862." *Civil War Medal of Honor Winners from Illinois,* Published by the Civil War Centennial Commission of Illinois, July 1962, page 6.

the command of Major General William Sherman. In the Vicksburg Campaign, the Regiment saw action at Chickasaw Bayou, Haines' Bluff, Arkansas Post, Milliken's Bend, and Richmond. In February 1863, the Companies B, C, F, H, and I, under command of Colonel McCrillis, were ordered to Memphis, Tennessee, and until December 1863, guarded the line of the *Memphis & Charleston Railroad* and fought at the battles of Tupelo, Okolona, Guntown and Salem. The other Companies remained on duty with the 13th Army Corps until August 1863. During those months, Companies A, E, G, K, L, and M or detachments of these Companies were engaged at Milliken's Bend, Bruinsburg, Grand Gulf, Port Gibson, Champion's Hill, Big Black River Bridge, the Siege of Vicksburg, Edwards' Station, Jackson, Bolton's Depot, and Brookhaven; these Companies were then assigned to Brig. General Albert Lee's Cavalry Division took part in the Western Louisiana campaign. From February through December 1863, Company L was assigned as escort to Major General McClernand. The Regiment was reunited at Memphis in December 1863 and began recruitment of new troops. In February 1864, the Regiment joined Smith's Expedition from Colliersville, Tennessee, to Holly Springs, Pontotoc, Houlka Swamp Okolona, and New Albany, Mississippi. In June, the Regiment was part of Smith's Expedition to Tupelo, Mississippi. In July 1864, a portion of the Regiment re-enlisted as Veterans, and the non-Veterans were placed in charge of the garrison at Germantown, Tennessee. Those who chose not to re-enlist and whose term of service had expired left for their homes in Illinois in August and were mustered out of service. On Aug 24, 1864, the Veterans were consolidated into six companies under the command of Colonel Robert H. Carnahan. The Consolidated Third Cavalry Regiment left Memphis in late September and saw subsequent action in the Battle of Franklin in November and the Battle of Nashville in December. The Regiment went into camp at Gravelly Springs, Alabama, in January 1865 and in May was ordered to St. Paul, Minnesota. On Jul 4, 1865, the Regiment embarked on an Indian expedition across Minnesota and the Dakota Territory and then returned via Devil's Lake, Fort Berthold, and Fort Snelling, arriving at St. Paul on Oct 1. After fifty months of service, the Regiment was mustered out at Springfield, Illinois, on Oct 18, 1865.[93]

The military flags collection at the State of Illinois Military Museum holds one standard and one guidon carried by the Third Illinois Cavalry Regiment.

The standard contains 34 stars above the eagle and no battle honors.

[93] *Report of the Adjutant General of the State of Illinois*, H.W.Rokker, State Printer and Binder, Springfield, Ill, (1886), Volume VII, pages 574-8.

The guidon bears no inscriptions.

The Cavalry Regiment also carried a flag presented to Company F by the "ladies of Carrolton."

"OUR BOYS AT PEA RIDGE—The correspondent of the St. Louis Democrat writes from Pea Ridge on the 1st inst., with reference to acts of bravery of some of the boys in the cavalry company raised by Capt. (now Lieut. Col.) Lewis. The flag he gallantly defended was made and presented by the ladies of Carrollton: Corporal Bennett, of company F, flag bearer, had his horse shot through with twelve bullets. The horse falling caused a dislocation of the corporal's ankle. In this condition, he was surrounded by three companies of rebel cavalry, who ordered him to surrender the flag. Orderly Sergeant (William?) Ray, seeing the unfortunate plight of his comrade and determined to rescue the flag charged, alone, on the rebel cavalry, opening a way for Bennett to make his escape[94]." This flag is not described, and the whereabouts of this flag today are unknown. No other flags carried by the 3rd Illinois Regiment Cavalry are known to survive.

[94] *Carrollton Gazette,* April 12, 1862.

4th Illinois Regiment Cavalry

As he stood near the Provost Marshal's tent, barefooted, bare-headed ragged, and dirty, he saw the flag floating in front of the General's tent and remarked: "You can't imagine the feeling that came over me when I saw that flag as I came into our lines. I couldn't laugh, and tears fell in spite of me[95]."

In the late summer of 1861, Judge T. Lyle Dickey[96] organized the Fourth Illinois Cavalry Regiment at Camp Hunter in Ottawa, LaSalle County, Illinois. The men of the Regiment volunteered from the following counties, Company A (*Grant's Escort)* from Cook and LaSalle Counties; Company B (*Carmichael's Cavalry)* from Cook and LaSalle, Counties; Company C (*Dollins' Cavalry)* from Kendall County; Company D(*Quinby's Escort)* from Will County; Company E from Marshall, Ogle and Putnam Counties; Company F from Vermilion County; Company G from McLean and Woodford Counties; Company H from Logan and Tazewell Counties; Company I from LaSalle County; Company K from Kankakee County; Company L from De Witt and Mc Lean Counties and Company M from Henry and Rock Island Counties. The Regiment was mustered in for three years' service on Sept 21, 1861, and on Nov 4, departed Camp Hunter for Cairo, Illinois. Soon after arrival in Cairo on Dec 6, Company C was detached to guard the bridge at Big Muddy, north of Cairo, and Company A was detached as an escort to General U. S. Grant[97] [98]. On Feb 2, 1862, the remaining companies boarded the Steamer *Keystone*, steamed up the Tennessee River, were deboarded at Parson's Ferry, and marched to camp 3 miles below Fort Henry. On the morning of Feb 6, the Fourth led the advance of General McClernand's Division, drove in the rebel pickets, and took possession of the fort. A Battalion led by Colonel Dickey went on reconnaissance to Fort Donelson on February 11-13, and in the assault on the fort, the Battalion was in support of Taylor's Battery. Following the surrender of Fort Donelson, the regiment moved to a camp at Randolph Forges near Dover, Tennessee. During March, the Regiment moved to the Tennessee River and on to Pittsburg Landing. On the morning of Apr 6, rebel forces commanded by Generals Johnson and Beauregard attacked the Regiment and other Union forces in the Battle of Shiloh. The Regiment was driven back, and the camp pillaged. The following day the Union forces regained the lost ground, and the Cavalry pursued the retreating rebel forces. Companies E, G, H, L, and M were detached on an expedition to Bear Creek, Alabama, April 12-13. As the Union Army advanced toward Corinth, Companies E, F, G, H, I, K, L, and M, commanded by General Sherman, occupied the right of the advancing army; following the occupation of Corinth, these eight companies rode to Memphis while the remaining four companies (A, B, C, D) operated near Jackson, Tennessee until November. During this time, the detachments saw action at Tallahatchie Bridge, Mississippi, Hatchie Bottom, and in Grant's Central Mississippi Campaign and scouted along the lines of the *Mobile & Ohio* and the *Mississippi Central Railroad.* In December, as part of Grant's Campaign, the Fourth Illinois Cavalry and the Seventh Kansas Cavalry attacked the rebel forces of Van Dorn and Price at Holly Springs and then followed them toward Coffeeville; on Dec 5; Lt. Col. William McCullough of Fourth

---

[95] *Kendall County Record,* October 13, 1864. Excerpt of a letter from a soldier of the 20th Illinois Infantry who relates the return of a soldier from the 4th Illinois Cavalry after his release from Andersonville Prison.

[96] 50 year old Theophilus Lyle Dickey was admitted to the Illinois bar in 1835 and moved to Ottawa, Illinois in 1839. He raised a company of volunteers and served as captain in the Mexican-American War. Dickey was elected to the Illinois Ninth Judicial Circuit in 1848; and he was a supporter of Stephen A. Douglas. In 1861 he was authorized to raise a company of cavalry. (Wikipedia)

[97] Company C continued as Grant's Escort until August 1863.

[98] On January 31 and February 1, 1862 twenty-five men from the Regiment were discharged to enter the naval service.

Illinois was killed[99]. The Regiment remained in Coffeeville on daily scouting until Dec 20 when ordered back to Holly Springs in pursuit of Van Dorn. By the time the Union troops reached Holly Springs, Van Dorn had retreated, and the Fourth Illinois Cavalry was ordered to move north along the Wolf River to clear the area of Confederate Cavalry. Companies were detached from the Regiment to collect horses, mules, saddles, and bridles from the nearby plantations. After a stay at Moscow, Tennessee, the Regiment moved to Colliersville, Tennessee, twenty-five miles east of Memphis[100]. In late April, the Regiment rode from Colliersville in a circuit to LaGrange, Tennessee, and then south into Mississippi to Ripley, New Albany, and Chesterville and back through Ripley and LaGrange to Colliersville, arriving there on May 6[101]. Until ordered to Vicksburg in August, the Regiment scouted and foraged in western Tennessee and northern Mississippi and saw action at Senatobia, Quinn's Mill, Coldwater, Holly springs, Clinton, Germantown, and Grenada. On Sept 15, the Regiment rode to Memphis and embarked on the steamships *Illinois* and *Thomas Tutt* for the trip downriver to Vicksburg, Mississippi; the men went into camp on the Big Black River east of Vicksburg. During the winter of 1863-64, detachments of the Regiment took part in expeditions and actions at Bayou Macon, Yazoo City, Morris Ford, Port Gibson, Canton, Bogue Chitto Creek, Satartia, Tallulah Court House, and Independence. In December 1863, the Regiment was ordered to Natchez, Mississippi, and remained on duty there with scouting in Southern Mississippi and Eastern Louisiana until December 1864. During that time, the detachments of the Regiment saw action in the Meridian Campaign, at Chunky Station, Washington, Concordia, Gillespie's Plantation, Cross Bayou, Bullitt's Bayou, and an expedition to Woodville, Louisiana. In October 1864, those members of the Regiment who had not re-enlisted were ordered to Springfield, Illinois, to be mustered out of service, and the remainder of the Regiment was reorganized into five companies[102] with Major Anthony Search in command. Until June 1865, the Fourth Consolidated Cavalry Battalion was brigaded with Third United States Colored Cavalry[103] and with them were ordered to Memphis in December for Grierson's Raid on the *Mobile & Ohio Railroad.* In late January, the Brigade was on expedition to Marion, Arkansas, in February into Northern Mississippi, and in April to Brownsville, Mississippi. On Jun 14, 1865, the remaining veterans and recruits of the Fourth Cavalry Brigade were transferred to and consolidated with the Twelfth Illinois Cavalry[104].

The military flags collection in the State of Illinois Military Museum contains a standard carried by the Fourth Illinois Cavalry Regiment.

---

99 The Regiment lost 64 men killed, wounded or taken prisoner at Coffeeville.

100 At this time the Regiment was down to 176 men and 100 non-commissioned officers fit for duty. Company A was still with General Grant as his bodyguard. Avery, Phineas O. *History of the Fourth Illinois Cavalry Regiment,* The Enterprise: A Print Shop, Humboldt, Nebraska (1903), page 118.

101 The object of this expedition was to draw the attention of the rebel cavalry in that area from the raid through central Mississippi by the Seventh Illinois Cavalry under Colonel Grierson that had started at this same time. *Ibid,* Avery, page 126-7.

102 Special Order No. 89, VIII and Special Order No. 161, I. At the time of the Consolidation 340 men remained of the 1,100 that had left Cairo in February 1862.

103 The following officers and men of the Fourth Illinois Cavalry organized and provided officers to the 1st Mississippi Colored Cavalry which became the 3rd US Colored Cavalry: William T. Beadles as Surgeon, Edward Main as Major, Charles M. Parks as 1st. Lieut., Embury D. Osband as Colonel, Jessie Brainard as Capt., Frank Cook as Capt., Jasper Jennings as 1st Lieut., Marshal Moon as 1st Lieut., John Shotzell as 1st Lieut., Harrison White as Captain, William Weber as Capt., Andrew Haynes as Capt., James S. Matthews as 1st Lieut., E. Chase McKibbon as 1st Lieut., Charles H. Chapin as Major, Philip Wolfersberger as 1st Lieut., William Hyland as Capt., Frank Calais as 1st Lieut., Andrew Emery as Capt., Leland Hall as 2nd Lieut., Joseph R. Randal as 1st Lieut., Seward H. Pettengill as 1st Lieut., Jeremiah Cook as Major.

104 *Ibid,* Avery. Avery's history of the Fourth Illinois Cavalry is an excellent detailed and accurate history of the Regiment.

The Standard measures approximately 30" x 20". Based on the rays above the stars and the shape of the eagle, this flag was manufactured by Gilbert and Hubbard of Chicago.

According to the diary of Thomas Mitchell of Company G, "Oct 28--The Regiment went to Ottawa today and was presented with a flag by the citizens. They also presented a saber to Col. Dickey. Speeches were made by Judge Caton and R.C. Cook. The flag is very nice and is said to have cost $75.00.[105]." The design of the flag presented is not detailed in the diary. Since this flag presentation was early in the war, the standard shown above may have been purchased from Gilbert and Hubbard of Chicago and is the flag described above as being presented to the Regiment.

The presentation of a sword and flag is also reported in the *Bloomington Daily Pantagraph,* "Just before departure of the 4th Cavalry Regiment from Ottawa for Camp Butler, a splendid sword was presented to Col. T. L. Dickey, and a rich and beautiful flag to the regiment[106]." No other flags carried by this regiment are known to survive.

---

[105] *Civil War Diary of Thomas K. Mitchell"* transcribed by his grandson Harry J. Mitchell, privately published. The entry is for October 28, 1861.

[106] *Bloomington Daily Pantagraph*, Thursday Evening November 7, 1861.

5th Illinois Regiment Cavalry

The Fifth Illinois Cavalry Regiment was organized in the fall of 1861 at Camp Butler near Springfield, Illinois, and mustered in for three years' service in November 1861 under the command of Colonel Hall Wilson.[107] The men volunteered from the following counties, Company A from Cumberland County; Company B from Coles, Moultrie, and Piatt Counties; Company C from McLean County; Company D from Wayne County; Company E from Coles County; Company F from Crawford County; Company G from Pike and Shelby Counties; Company H from St. Clair and Washington Counties; Company I from Cumberland County; Company K from Randolph County,[108] Company L from Effingham and Jasper Counties and Company M from Wayne County. The regiment trained at Camp Butler until February 20, 1862, when it departed for Benton Barracks in St. Louis. On March 3, the Fifth Cavalry moved to Pilot Knob in southern Missouri. The men saw their first action in a skirmish with the enemy near Doniphan, Missouri, near the Arkansas border on April 4. For the next six weeks, the regiment remained near the area of the Arkansas-Missouri border, skirmishing, and scouting. A detachment containing Companies D, F, and L was sent to Smithville in northwest Missouri on June 17, and on June 27, the Fifth Cavalry rode east toward the Mississippi River. The regiment arrived at Jacksonport, Arkansas, on the White River on June 29 and at Helena, Arkansas, on the Mississippi River on July 13. While en route, the regiment fought at Hill's Plantation on the Cache River on July 7 where it defeated an enemy force three times its size. The regiment remained headquartered at Helena until May 29, 1863. While at Helena,[109] the regiment participated in an expedition to Clarendon, Arkansas, in August and an expedition up the St. Francis River to Jeffersonville and Mariana, Arkansas, on September 26. In late November, the Fifth Cavalry scouted to Arkansas Post and to Grenada, Mississippi. In May 1863, Grant surrounded Vicksburg, and he requested cavalry to guard the Union Army's rear. The Fifth Illinois Cavalry reported to Grant at Chickasaw Bayou on May 31 and moved up the Yazoo River to Snyder's Bluff. During the first week in June, the regiment was part of a cavalry action against Satartia/Mechanicsburg.[110] On July 6, the Fifth Cavalry marched with Sherman's Army toward Jackson, Mississippi. In the siege of Jackson, the regiment saw action near Canton, Holton's Depot, Grant's Ferry on the Pearl River, and at Briar Creek. In August, the Fifth Cavalry, along with the Third and Fourth Iowa Cavalry, was ordered on a raid to Grenada, Mississippi, to destroy the *Mississippi Central Railroad.* Having accomplished destruction of the railroad, engines, cars, and bridges, the regiment rode toward Memphis. En route, the regiment met and defeated enemy cavalry at Coldwater, Mississippi, and arrived at Memphis on August 22, 1863. In late August, the regiment was ordered back to Vicksburg and went into camp at Black River. The regiment

---

[107] The original Colonel was John J. Updegraff, but he was court-martialed for drunkenness and dismissed from service before the Regiment was mustered. English born, 30-year old Hall Wilson from Fairfield, Illinois first joined the 27th Illinois Infantry Regiment as Major but transferred to and assumed command of the 5th Regiment of Illinois Cavalry with the rank of Colonel.

[108] Many of the men came from Liberty in Randolph County. The name of the town today is Rockwood. Slaves who came through the town on the Underground Railroad called the town Liberty.

[109] While in Helena 74 members of the Regiment died of sickness or wounds. Kohl, Rhonda, "On Grant's Front Line, The Fifth Illinois Cavalry in Mississippi," *Illinois Historical Journal*, Vol. 90 (Spring, 1998).

[110] The Cavalry forces were commanded by Lt. Col. Benjamin Wiley of the Fifth supported by Infantry forces that moved up the Yazoo River. The expedition is detailed in Kohl's paper in the *Illinois Historical Journal.* A 39-man detachment of the Fifth Illinois, attempting to destroy Confederate rail and communication lines, was captured near Ellisville, Mississippi on June 25.

was headquartered at Black River until May 1, 1864, when the regiment moved to Vicksburg.[111] In late September 1863, a regimental detachment participated in a raid from camp at Big Black River to Yazoo City and in late October moved with the XVII Corps in a raid on Jackson, Mississippi. As part of that action, the regiment took part in a cavalry charge that routed Confederate General Wirt Adams. During February 1864, as part of Sherman's Meridian, Mississippi campaign, the Fifth Cavalry saw action at Champion's Hill, Jackson, Brandon, Morton, and Meridian. In early May, a detachment went on scout from Vicksburg to Yazoo City, and on May 27, 1864, Colonel John McConnell, formerly a Major in the Third Illinois Cavalry, took command of the Fifth Cavalry. Due to the regiment's lack of horses, Companies A, B, C and D were completely mounted and equipped while the other eight companies were dismounted. On July 1, 1864, the mounted Battalion, along with members of the Eleventh Illinois Cavalry, Second Wisconsin Cavalry, and the Third US Colored Cavalry, moved to Jackson, Mississippi. On the return from Jackson, on July 5, the expedition was attacked, but the enemy repulsed. The mounted detachment of the Fifth Illinois Cavalry participated in an expedition in late September down river to Port Gibson, drove the enemy from that place, and moved on to Natchez, Mississippi, and finally to Tonica Bend, Louisiana. The men fought at Woodville, Mississippi, drove the enemy off, and returned to Vicksburg on October 11. The entire regiment participated in operations in Issaqueena and Washington Counties, north of Vicksburg, in late October and then participated in expeditions from Vicksburg to Gaines' Landing, Arkansas, and Bayou Macon, Louisiana, in early November. In late November, the regiment was part of an expedition east of Vicksburg that successfully destroyed the *Mississippi Central Railroad,* thus denying the transport of supplies to Confederate General Hood. In January 1865, the regiment moved to Memphis, Tennessee, and then rode through Southern Arkansas and Northern Louisiana during January and early February. The Fifth Cavalry was part of another expedition in Northern Mississippi in early March, moved to Alexandria, Louisiana, on July 1, and then marched to Hempstead, Texas, arriving in late August. The regiment remained on duty at Hempstead until October when ordered to Springfield, Illinois, for muster out on October 27, 1865.[112]

The military flag collection of the State of Illinois Military Museum contains two flags—a national and regimental--carried by the Fifth Regiment Illinois Cavalry. Interestingly, both are "full size" flags usually carried by infantry regiments. Presumably, these two flags were issued to the regiment when six of the ten companies were dismounted in May 1864. There are no cavalry standards or guidons from this regiment in the Illinois collection.

[111] On January 1, 1864, many men of the Regiment re-enlisted as veterans; on March 10 the veterans were sent home on furlough to return May 10.

[112] *Report of the Adjutant General of the State of Illinois*, H. W. Rokker, State Printer and Binder (1886), Volume VII, page 691-2.

The national flag shows a star pattern typical of those flags manufactured for the Philadelphia Depot by either Evans & Hassall or by Horstmann Brothers & Company. There are no battle honors or unit designation of the flag. It appears to be a 34-star flag.

The regimental flag is typical of flags manufactured by Horstmann Brothers & Company for the Philadelphia Depot. No other flags carried by this regiment are known to survive. There is a presentation sword presented to Major Hall Wilson by the "SC Guard" in a private collection. Since he was presented the sword when he was a "Major" it was presumably presented when he was in the 27th Regiment Illinois Infantry.

No other flags carried by the 5th Illinois Cavalry are known to survive.

6th Illinois Regiment Cavalry

Through their cities and over their streams
The flag of the Union once more gleams.
There's a curse on the air, but in under breath,
As the troopers go on their work of death.
Like lightning flashes, each loyal blade
To light the path of Grierson's raid.[113]

The Sixth Illinois Cavalry was organized at Camp Butler in November 1861.[114] The men of the regiment, under the command of Colonel Thomas H. Cavanaugh[115], volunteered from the following counties in Southern Illinois—Company A from Massac and Pope Counties; Company B from Johnson County and Pope County; Company C from Cass, Morgan and Scott Counties; Company D from Hamilton County; Company E from Edwards County and Richland County; Company F from Gallatin County and Saline County; Company G from Pope County; Company H from Hamilton County; Company I from Perry County; Company K from Hamilton and Saline Counties; Company L from Gallatin County; Company K from Alexander County and Union County. On November 25, 1861, the regiment was ordered to Camp Yates, Shawneetown, Illinois. Following training, in February, the Sixth Cavalry first moved to Paducah, Kentucky and then on to Columbus, Kentucky, where the regiment was divided into five companies (A, C, D, E and F) going to Trenton, Tennessee, 5 companies remaining at Columbus (G, H, I K and L) and two companies guarding Paducah and Bird's Point (B and M).[116] In June, the companies Stationed at Columbus were ordered to Coldwater Station, Mississippi and fought at White Oak Bayou. The regiment was reunited at Memphis in the fall of 1862 and, in November, led the advance of Sherman's expedition towards Grenada, Mississippi. In December, the regiment was part of the Union force that pursued Confederate General Earl Van Dorn's forces after their raid on Holly Springs, Mississippi. The Sixth Cavalry went into winter quarters at La Grange, Tennessee, in January 1863 and operated against local partisans during the next three months. On March 9, the regiment attacked Colonel Robert Richardson's First Tennessee Partisan Rangers near Covington, Tennessee and captured in addition to prisoners, stores, ammunition, and company records. On March 29, Richardson attacked a detachment of the Sixth Cavalry at night; although outnumbered, the Sixth repulsed the attack. On April 17, 1863, the Sixth and Seventh Illinois Cavalry and the Second Iowa Cavalry departed La Grange for a raid through Mississippi. The Second Iowa fought Confederate Colonel Barteau's Second Tennessee Cavalry south of Pontotoc, Mississippi, allowing the Sixth and Seventh Illinois Cavalry to move south on Grierson's Raid.[117] The Illinois Regiments arrived safely at Baton Rouge, Louisiana, on May 2, 1863. In the actions against Port Hudson, Louisiana, between May 24 and June 9, the regiment saw action at Clinton and

---

[113] Woodward, Lt-Col., L. L., "Grierson's Raid, April 17th to May 2d, 1863" *Journal of the United States Cavalry Association*, Vol. XIV, No. 2, April 1904.

[114] Although organized at Camp Butler in November, the Regiment was not mustered in for three years' service until January 7, 1862, at Camp Yates. Grierson, Benjamin H. *Record of Service Rendered the Government 1862.* page 36 (related in a letter from Grierson to his wife) This private printing is in the Illinois State Historical Library, Springfield, Illinois

[115] 52 year old Thomas Cavanaugh was a physician who had served in the Black Hawk War and Mexican-American Wars.

[116] On March 28, 1862, Colonel Thomas H. Cavanaugh resigned, and Benjamin H. Grierson was promoted from Major to Colonel to become commander of the Regiment. Grierson was a music teacher and band leader from Jacksonville, Illinois.

[117] Brown, D. Alexander *Grierson's Raid*, University of Illinois Press, Urbana, Illinois, 1954. This book gives a detailed account of Grierson's famous raid.

Jackson Cross Roads while a detachment of the Sixth skirmished near Holly Springs, Mississippi. Following the fall of Port Hudson, the regiment moved by steamer up the Mississippi River to Memphis, arriving in late July 1863. The Sixth Cavalry encamped at Germantown, Tennessee, east of Memphis, until November 1863. During that time, detachments of the regiment scouted and skirmished in the nearby area of west Tennessee and Mississippi. On December 4, the Sixth Cavalry engaged the enemy at Moscow, Tennessee, routed the enemy and returned to Germantown until February 17, 1864. In early February, the regiment was part of General William Smith's expedition against the enemy at West Point, Mississippi and engaged in three days' battle with Forrest's command. In March, the three-year term of the regiment expired, and many of the regiment re-enlisted as veterans and were ordered back to Illinois on furlough. The returning men rendezvoused at Mattoon, Illinois and then moved back to Memphis, where they remained until July 1864. A detachment of the regiment was ordered to Colliersville, Tennessee, in August to guard the *Memphis & Charleston RR.* Later in August, the remainder of the regiment joined in Smith's expedition to Oxford, Mississippi and engaged Forrest's command at Hurricane Creek. The regiment encamped at White Station, Tennessee, until October 1, 1864, when it joined Hatch's march through west Tennessee to Clifton on the Tennessee River. There the regiment joined Washburne's Infantry (45th Illinois Infantry) and served as the advance guard to Lawrenceburg and then on to Savannah, Tennessee and then back to Clifton; with the base at Clifton, the regiment scouted between Clifton and Pulaski, Tennessee. In the Nashville campaign of November and December, the Sixth marched down Shoal River to Florence, Alabama and skirmished daily with Confederate General Hood's forces. The regiment then fell back to Lawrenceburg on Hood's advance. In late November, the Sixth Cavalry was ordered to Shelbyville, Tennessee, crossed the Duck River, moved down river and crossed again at Pike Ford and returned to Columbia. En route, the regiment encountered Forest's entire command and escaped by swimming the Duck River and rejoined the command at Franklin, Tennessee, in time to fight in the Battle of Franklin on November 30. Following the battle, the regiment marched to Nashville. The Sixth Cavalry was ordered on an expedition to Glasgow, Kentucky, on December 5-13 and then returned to Nashville and fought in the Battle of Nashville on December 15-16. Following the repulse of Hood at Nashville, the Sixth joined in the pursuit of Hood's forces to Florence, Alabama, on the Tennessee River, then abandoned the pursuit and marched to Gravelly Springs, Alabama. In February 1865, the Sixth Illinois Cavalry moved to Eastport, Mississippi and remained there until ordered to Nashville and then Decatur and Montgomery, Alabama. On November 1, 1865, the regiment was ordered to Selma, Alabama and was mustered out on November 5, 1865.[118]

The holdings of the Adjutant General of the State of Illinois do not contain any Quartermaster

[118] *Report of the Adjutant General of the State of Illinois,* H. W. Rokker, State Printer and Binder, Springfield, ILL, (1886) Volume VIII, pages 50-2.

6TH ILLINOIS REGIMENT CAVALRY

Department flags carried by the Sixth Regiment Illinois Cavalry. The only flag in the collection ascribed to the regiment is a homemade 34-star national flag. The flag measures 3' x 4', does not bear any inscriptions, and the origin of the flag is unknown.

No other flags carried by the 6th Regiment Illinois Cavalry are known to survive.

7th Illinois Regiment Cavalry

Mount to horse, mount to horse
Forward battalion!
Gallop the gallant force;
Down with rebellion!
Over hill, creek and plain,
Clatter the fearless.
Dash away, splash away,
Led by the peerless!
Hark to the rifle crack!
Wild conflagration
Marks where the brave bear
The flag of the nation.
Down on the rebel hordes,
Gaping in wonder.
Dash they with gleaming swords
And war cries of thunder.[119]

The Seventh Illinois Cavalry was organized at Camp Butler and mustered in for three years' service on October 13, 1861, under the command of Colonel William Pitt Kellogg.[120] The men of the regiment volunteered from the following counties; Company A from Edgar County; Company B from Carroll, Ogle and Stephenson Counties; Company C from Lee County; Company D from Knox County; Company E from White County; Company F from Fayette County; Company G from Gallatin and White Counties; Company H from Shelby County; Company I from Macon County; Company K (*Canton Cavalry)*from Fulton County; Company L from Fulton, Henry and McDonough Counties and Company M from Randolph County. At the end of October, Companies A, C, G, and I were ordered to Bird's Point, Missouri; the remainder of the regiment remained at Camp Butler until ordered to Bird's Point in late December. In January 1862, Companies B, C, L, and I remained at Bird's Point while the remainder of the regiment moved to Cape Girardeau, Missouri. In March 1862, those Companies at Cape Girardeau joined the Union forces commanded by Brig. General John Pope at Commerce, Missouri and led the advance against New Madrid, Missouri. The regiment captured a battery of steel breech-loading artillery from Confederate forces commanded by Jeff Thompson. The regiment was reunited at New Madrid and operated south of New Madrid on the banks of the Mississippi River until the fall of Island No. 10 on April 8. The Seventh Cavalry then boarded steamers for the trip up the Mississippi River to the Ohio River, then up the Tennessee River and landed at Hamburg Landing, Tennessee. Along with the forces commanded by General Pope, the regiment moved toward Corinth, Mississippi. Between April 29 and

---

119 From a poem entitled "Grierson's Raid" by Gordon and published in *The De Witt County Public,* August 20, 1863.

120 Kellogg was a lawyer from Canon, Illinois who had been a presidential elector on the Republican ticket of 1860. In 1861 he was appointed by Lincoln as chief justice of the supreme court of the Nebraska Territory but returned to Illinois to command the Seventh Illinois Cavalry. Post CW he served in the US Senate from Louisiana and as Governor of Louisiana. He died in Washington DC in 1918 and was interred at Arlington National Cemetery.

May 30, the regiment took part in the siege of Corinth[121] and saw action at Sharp's Mills, Iuka, and Tuscumbia Creek. The Confederate forces abandoned Corinth on May 29 before the arrival of the Union troops; the Seventh Cavalry was then ordered to guard the *Memphis & Charleston RR* between Tuscumbia and Decatur, Alabama. While guarding the railroad, the regiment also participated in the battle of Iuka in September, the battle of Corinth on October 3-4 and Grant's Central Mississippi campaign; a portion of the regiment[122] skirmished with Richardson's enemy forces near Somerville, Tennessee, in late November. In December, the Seventh was assigned to Colonel T. Lyle Dickey's command at Holly Springs, Mississippi. Dickey's forces pursued Price to Coffeeville, Mississippi, where Price made a stand and repulsed Dickey's Cavalry with severe losses. In late December, a detachment from the Seventh Illinois Cavalry and the Second Iowa Cavalry marched from Oxford to Pontotoc to Tupelo, Mississippi destroying stores and portions of the *Mobile & Ohio Railroad.* In early 1863, at La Grange, Tennessee, the Sixth and Seventh Illinois Cavalry and the Second Iowa Cavalry were brigaded together under the command of Colonel Benjamin Grierson and foraged western Tennessee for horses and supplies. On April 17, the brigade left La Grange on "Grierson's Raid", arriving at Baton Rouge, Louisiana, on May 2.[123] The regiment next moved north to invest in Port Hudson, Louisiana. Following the fall of Vicksburg, Grierson's brigade was ordered to Memphis to guard the *Memphis & Charleston RR.* The regiment had numerous skirmishes with enemy forces in the Fall of 1863, including significant actions at Byhalia, Mississippi and Collierville, Tennessee. On December 26, the Seventh Cavalry, commanded by Colonel Edward Prince, met Forrest's command near New Castle, Tennessee. However, the Seventh did not receive expected support and was forced to retire in the face of superior forces. In February 1864, the regiment marched with General William Sooy Smith's expedition from Collierville, Tennessee, to West Point and Okolona, Mississippi. While at Germantown, Tennessee, in March, the regiment's three-year term of service expired, and 289 officers and men re-enlisted and were granted an April-June furlough. While the veterans were on furlough, 120 non-veterans served with General Samuel Sturgis and suffered defeat by Forrest at Brice's Crossroads (Guntown, Mississippi) on June 10. When Forrest attacked Memphis on August 21, seven companies of the Seventh Cavalry were present and were instrumental in repelling Forrest's attack. In late September, the regiment left Memphis, moved eastward through Bolivar, and crossed the Tennessee River at Clifton, Tennessee, in pursuit of Forrest. The Seventh remained in that general area until ordered further east to Pulaski, Tennessee, on October 26 and then on to Shoal Creek, Alabama, in November. During Hood's retreat from the Atlanta Campaign, the Seventh saw action against Hood's forces at Lawrenceburg, Tennessee, on November 21 and was nearly surrounded by Forrest's army at Campbellsville three days later. The Seventh covered the rear and flanks of the Union infantry as it moved north through Columbia to Franklin and skirmished with the enemy at Duck River, Hurt's Crossroads and Mount Carmel. At the Battle of Franklin, November 30, the Seventh charged an enemy infantry division and drove it across the Harpeth River. In the Nashville campaign, the regiment fought at Brentwood Hills and in the Battle of Nashville. Following Hood's rout, the cavalry followed him to the Tennessee River. The Seventh Cavalry was at Gravelly Springs, Alabama, in mid-January 1865, but the officers and men fit for duty had declined to 199 compared to 450 in mid-December 1864. The regiment was dismounted and sent to Eastport, Alabama,

---

[121] Major Zenas Applington was killed while leading a charge against the enemy at Farmington, Mississippi on May 15,1862

[122] Portions of Companies A, B, D, E, F, G, H, I and K skirmished with enemy forces near Somerville, Tennessee on November 26 and 27, and captured "two very handsome colors," *O.R..—Series I—Volume SVII/1 [S#24]* One of the colors was captured by Company K and is described as a "silk" flag taken from "Jackson's Mississippi Cavalry" near Cannon's Mill, Tennessee. *Canton Weekly Register,* December 22, 1862.

[123] Brown, D. Alexander, *Grierson's Raid,* University of Illinois Press, Urbana, Illinois, 1954. This book gives a detailed account of Grierson's famous raid.

to receive new recruits. With the strength of the Seventh at 1,600 and after the surrender of the rebel armies, the Seventh Cavalry moved to Okolona, Mississippi and then Decatur, Alabama, where it was mounted. The regiment returned to Camp Butler and was mustered out of service on November 17, 1865.[124]

The flags carried by the Seventh Illinois Volunteer Cavalry Regiment that are in the collection of the State of Illinois Military Museum are ---a homemade national flag, a cavalry standard, the remnants of a regimental flag and twelve guidons. According to the records of the Adjutant General of the State of Illinois, in addition, a second national color was also deposited with the state but is no longer in the collection. That flag is reported as being inscribed with the following battle honors, Corinth, Nashville, Iuka, Grierson Raid, Franklin, Coffeeville, Island 10, Collierville, New Madrid, Port Hudson, Campbellsville, and Summerville.[125]

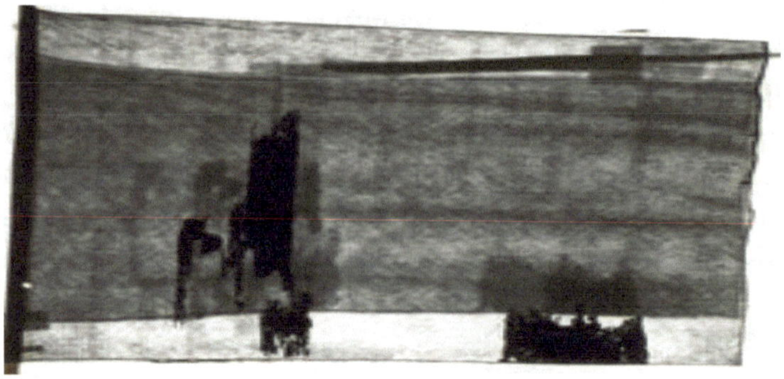

Although only shreds remain of this regimental color, it appears to have been a full-size 6-foot flag and not the smaller cavalry standard.

This color is the standard for the 7th Illinois Cavalry. It measures approximately 28" x 20". The battles inscribed on this flag are--N. MADRID; COLLIERVILLE; ISLAND NO. 10; COFFEEVILLE; FRANKLIN; GRIERSON'S RAID; IUKA; NASHVILLE; CORINTH; PORT HUDSON; CAMPBELLSVILLE; SOMERVILLE.

---

124 Eddy, T. M., *The Patriotism of Illinois,* Clarke & Co. Publishers, Chicago (1866), Volume II, pp 622-7.

125 *Report of the Adjutant General of the State of Illinois,* H.W.Rokker, State Printer and Binder, Springfield, IL, (1886) Volume I, page 163

The cavalry regiments were not issued an "official" national color. However, the Seventh Illinois Cavalry carried a "homemade" national color. It measures 45" X 81". This may be the flag referred to by Henry Eby in his history of the regiment. "About 11 o'clock, all who had enlisted were formed in two ranks in front of the Illinois Central freight house, facing toward it, when a Miss Davis, who stood upon a raised platform at the northwest corner of the building, delivered an appropriate address, presenting us with an elegant flag on behalf of the citizens of Mendota."126

Photos of the guidons carried by this regiment are not shown.

No other flags carried by this regiment are known to survive.

126 Eby, Henry H., *Observations of an Illinois Boy in Battle, Camp and Prisons—1861 to 1865,* Published by the author, Mendota, IL, (1910), page 16.

8th Illinois Regiment Cavalry

"By this time, a little after 7 o'clock, the enemy's advance—composed of Archer's Tennessee brigade, followed by Davis' and Brockenborough's brigades—appeared in sight on the hill west of Marsh Creek a mile or so away. Upon sighting our picket post, they deployed skirmishers on either side of the road with the precision of veterans marching steadily down the pike to the stone bridge...asking Sergeant Levi Shaffer of Naperville, who was in charge of the picket line on the pike, for his carbine, I took aim at an officer on a white or light gray horse and fired—The first shot at the Battle of Gettysburg."[127]

The Eighth Illinois Cavalry was organized in St. Charles[128], Kane County, Illinois, by Colonel John Farnsworth[129] and mustered in for three years' service on September 18, 1861. The men of the regiment volunteered from the following counties, Company A (*St. Charles Cavalry)* from Kane County; Company B (Syca*more Cavalry*) from De Kalb County; Company C (*Whiteside County Cavalry*) from Whiteside County; Company D (*Du Page County Cavalry*) from Cook and Du Page Counties; Company E (*Wheaton Dragoons or Milton Cavalry)* from Du Page County; Company F (*Evanston Dragoons)* from Cook and Will Counties; Company G (*Chicago Cavalry*) from Cook, Mc Henry and Peoria Counties; Company H (*Mc Henry County Cavalry)* from Mc Henry County; Company I (*Kane County Cavalry)* primarily from Kane County with a small number from Mc Henry County; Company K from De Kalb and Will Counties; Company L from De Kalb and Winnebago Counties and Company M (*Winnebago, Rockford Cavalry)* from Ogle and Winnebago Counties. On October 14, the regiment marched to the depot in Geneva and men and horses were loaded for the trip to Washington DC. At Washington, the regiment rode down Pennsylvania Avenue to pass in review before Abraham Lincoln[130] at the White House and then went into camp at Meridian Hill. The first action for the Eighth Illinois Cavalry occurred on March 10, 1862, when the regiment joined the general advance on Manassas as part of General Sumner's Division. The Eighth remained at Warrenton, Virginia, until April 12, but that time was instrumental in driving the enemy across the Rappahannock River on four occasions. On May 4, the Eighth moved to Williamsburg, Virginia and fought in the regiment's first major battle on May 5. The regiment then advanced toward Richmond and met the enemy again at Mechanicsville, Virginia, on May 24-26. While guarding the south bank of the Chickahominy River, the Illinois Eighth Cavalry again engaged the enemy May 31-June 1 in the Battle of Fair Oaks.[131] Later in June, as the Union Army lay siege to Richmond, the army of Northern Virginia, commanded by Robert E. Lee, attacked the Union forces, and drove them back. In that action, the Eighth Illinois Cavalry fought rear guard actions at Mechanicsville, Hundley's Corners, Gaines' Mill, White Oak Swamp, Savage Station, and Malvern Hill.[132] The regiment remained on picket duty at Harrison's Landing

---

[127] From the diary of Marcellus Jones as read by his wife at West Chicago, Illinois on May 27, 1897

[128] Camp Kane—now Langum Park in St. Charles, Illinois

[129] Farnsworth was a 41 year old, Quebec Canada born, attorney from St. Charles, Illinois. He had served two terms in the U.S. Congress from the state of Illinois

[130] "On this occasion, it is said, the President called us 'Farnsworth's Big Abolition Regiment.'" Hard, Abner MD, *History of the Eighth Cavalry Regiment Illinois Volunteers, During the Great Rebellion,* Aurora, IL (1868), page 43.

[131] Sergt. Horace J. Capron, Company G, was awarded the Medal of Honor for "Gallantry in Action" at Chickahominy and Ashland, Virginia, June 1862,

[132] During the retreat from the Chickahominy to the James River, ex-governor of Illinois, John Wood, exchanged his civilian clothes for "slouch hat, sabre, belt and spurs" and rode with the Eighth Cavalry and "gave courage and hope to hearts almost

on the James River until August 16 and, during that time, fought in the second battle of Malvern Hill. On August 20, the regiment embarked at Yorktown landed at Alexandria, Virginia and crossed into Maryland; in a skirmish near Barnesville, on September 9, Captain Elon Farnsworth's squadron (Companies K and D) captured the flag of the Twelfth Virginia Cavalry (Ashby Cavalry).[133] Later in September, the regiment fought at Middletown, Sugar Loaf Mountain, Frederick, South Mountain, and Boonsboro. At the Battle of Antietam, on the morning of September 17, the regiment was ordered to the front to support a battery near the center of the Union line and that evening marched half a mile back to rest. The following morning, when the Union forces advanced, they found that the Confederates had escaped back across the Potomac River. On October 1, a cavalry brigade consisting of the Eighth Illinois, a portion of the Eighth Pennsylvania and one battery of Regular Artillery was ordered on a reconnaissance towards Martinsburg, Virginia[134] , in search of Lee's retreating forces. The brigade met an overwhelming enemy force at Martinsburg, and though surrounded, the brigade managed to fight their way back across the Potomac River. During the remainder of October, the regiment saw action in Virginia at Purcellville, Upperville and Snickersville and in November at Philmont, Upperville, Barber's Cross Roads, Sperryville and Markham Station. In mid-November, Union General Ambrose Burnside received Lincoln's approval to shift the Army of the Potomac east to Fredericksburg, Virginia, on the Rappahannock River and from there advance on the Confederate capital at Richmond. This advance culminated in the Battle of Fredericksburg—December 12-15—a sound defeat for the Union Army; although the Eighth Cavalry was under fire during the Battle, the regiment sustained no losses. The remainder of 1862 was passed in King George County, performing picket duty, scouting, and foraging. In January 1863, the regiment continued guarding flanks of the Union Army, first at Port Conroy on the north bank of the Rappahannock River and later further north at Dumphries, Virginia. During February, the regiment operated in Westmoreland and Richmond Counties, and in March, April and May were engaged at Sulphur Springs, Warrenton, Rapidan Station, Northern Neck, Borstly Ford and Upperville, Virginia. In June, the Confederate Army began moving north into Pennsylvania, and on June 17th, the Eighth Illinois Cavalry, as part of the 1st Brigade, 1st Division Cavalry Corps, followed, skirmishing with the Confederate Cavalry at Aldie, Goose Creek and Upperville, Virginia. On June 27, the regiment crossed the Potomac and reached Middletown, Maryland, on June 28. The regiment rode east toward Gettysburg and reached the area of the Herr Tavern west of Willoughby Run on June 30. At 7:30 am, on July 1, Lt. Marcellus Jones, Company E, commanding a vedette post on the Chambersburg Pike, fired the first shot of the Battle at Gettysburg. The following morning the men of the Eighth Illinois were ordered to guard the army's supply trains that lay to the south of Gettysburg. The regiment heard the battle at Gettysburg for the next two days but did not participate further in active fighting.[135] During the remainder of 1863, the Eighth Illinois was in constant action and fought in many actions, including those at Boonsboro, Funktown, Falling Water, Chester Gap, Kelly's Ford, Brandy Station, Culpepper and Pony Mountain, Liberty Mills, Raccoon Ford and Morton's Ford, Manassas, Madison Court House, Sulphur Springs and Jennings' Farm. In January 1864, three-fourths of the regiment re-enlisted as veterans and, on January 11, left Culpepper, Virginia, on furlough. During the thirty days of furlough, the men spent time with family and recruited men to replenish the regiment's ranks. On return

---

ready to despair." Ibid, Hard, page 141.

[133] O.R., Series I, Volume SIS/1 [S#27], page 208. The Eighth Illinois Cavalry captured a total of 7 enemy colors during the War. Eddy, T.M, D.D., *The Patriotism of Illinois,* Clarke & Co., Chicago (1866) Vol. II, p 549.

[134] This area is now in West Virginia

[135] Elon Farnsworth, the nephew of General John Farnsworth, and Captain of Co. K had been promoted to the rank of Brig. General on June 28, 1863. He was ordered to lead an ill-advised attack on the rear of Confederate forces occupying Seminary Ridge on July 3 and was killed leading the charge. L. Doyl Twend, *Chicago Tribune Magazine,* October 11, 1964, page 18-27.

from furlough in February, the veterans patrolled Washington until ordered to Alexandria. There the regiment was divided--one company at Alexandria, two at Aquila Creek, five at Belle Plain and four in Washington. July found the headquarters of the regiment and Company F in Washington, and the remaining companies scattered at Alexandria, Muddy Creek, Maryland and at Camp Stoneman. For the next year, the Eighth Illinois Cavalry had numerous encounters with Mosby's guerrillas in the Loudon County area of Virginia. After the surrender of Lee on April 10, 1865, arrangements were made for a victory celebration at the 1st Brigade headquarters (the brigade included the Eighth Illinois Cavalry) to be held on April 15 and among those who had accepted an invitation was the actor John Wilkes Booth.[136] Ironically, following Lincoln's assassination, the Eighth Illinois Cavalry was one of the regiments ordered to search for Booth. A detachment of the Eighth Illinois took part in the Lincoln funeral ceremonies in Washington, and Lieutenant-Colonel Clendenin of the Eighth was a member of the military commission that tried the conspirators. The regiment was ordered to St. Louis in June and mustered out on July 17, 1865.[137]

There are two national flags carried by the Eighth Illinois Cavalry Regiment in the collection of the State of Illinois Military Museum. Both flags appear to have been "full size", i.e., 6' x 6' 6" when originally presented to the regiment.

The first flag shown here is typical of flags manufactured by Evans & Hassall and by Horstmann Brothers & Company for the Philadelphia Depot. It appears that the flag originally contained 34 stars. There are no inscriptions on the flag.

---

[136] Ibid, Hard, page 320.

[137] The most accurate and detailed history of the Regiment can be found in the Regimental history by Abner Hard MD.

Little remains of the second national flag. However, it appears that the flag originally had five horizontal rows of stars. This star pattern is typical of flags manufactured in New York.

The *Chicago Times* reported the presentation of flags to Companies D and E of the regiment at the DuPage County Agricultural Fair held in Wheaton (Du Page County) in September 1861. First, a "stand of colors" was presented to the *Milton Cavalry* (Company E), "At one o'clock Lieut.-Governor Hoffman made his appearance upon the platform and read an abolition harangue, intended as a presentation speech to the Milton Cavalry, at the request of the lady donors of a splendid stand of colors to that command. This address was received with the damning of faint applause. It was a real, undiluted political speech, abounding in inflammatory and uselessly harsh language, and totally unfit for the purpose intended." The report does not further describe the "stand of colors." There is no further history of the flag(s) presented to Company E. The story goes on to describe the presentation of a flag to Company D, "Immediately in the wake of Lieutenant-Governor Hoffman came Rev. D. Chapman, of Bloomingdale, who presented a fine silken banner to the *DuPage County Cavalry*, mainly enlisted from that place. This present was from the hands of the ladies of Bloomingdale. Upon receiving the banner, Captain Gearhart (sp?) made a few modest but appropriate remarks."[138]

The national flag presented to Company D is now in the DuPage County Historical Museum in Wheaton, Illinois. According to the records at the Museum, the citizens of Bloomingdale (DuPage County) presented the flag to Co D in 1861. The flag was made by Polly Mills, mother of L. W. Mills (no record of a volunteer named Mills in the company records) and carried by the company to Camp McAllister near Plainfield, Illinois. The flag measures 72" x 96" and is made of silk with 2 1/2" gold fringe. The stars are painted in gold. The flag is now in tatters, and no photos are available. A flag was also presented to the regiment by the women of St. Charles (Kane County). "As another token of their regard for the brave men who were about to risk their lives in defence(sp) of that country they loved so well, they prepared a splendid silk flag which they presented to the regiment on this never-to-be-forgotten day. After a few appropriate remarks, they proceeded to place it upon the flagstaff it was designed to grace. All eyes were at this moment gazing intently upon the scene, each individual seemingly anxious to catch the first gleam of brightness reflected from the beautiful banner when it was unfurled in the sunshine. Suddenly the wind caught it, but instead of floating gracefully in the air, as expected, it was rent almost in twain."[139] Further history of this flag is unknown.

No other flags carried by the 8th Regiment of Illinois Cavalry are known to survive.

---

[138] *Chicago Times*, September 28, 1861

[139] Ibid, Hard, page 35.

9th Illinois Cavalry—"*First Western Cavalry*" Or "*Brackett's Cavalry Regiment*"

"Smooth back the locks from his high, noble brow,
Take the last look as he lieth there now!
Noble, sad god-like, true-hearted and brave
He died as a martyr his country to save."[140]

Colonel Albert G. Brackett organized the Ninth Illinois Cavalry Regiment[141]at Camp Douglas in Chicago; the regiment was mustered in for three years' service on November 31, 1861. The men of the regiment volunteered from the following counties, Company A from Rock Island County; Company B and C from Henry County; Company D from Cook, Macoupin, and Will Counties, along with men from Michigan and Ohio; Company E from Logansport, Indiana, Company F from Cook County, Company G from Lake, La Porte, Marshall and Porter Counties, Indiana; Company H from Cook and Henry Counties; Company I from Boone County; Company K from Bureau, Knox and Stark Counties, Company L from Cook, Henry, Knox and Warren Counties and Company M from Iroquois County. On February 15, 1862, the regiment was ordered via *St. Louis & Alton RR* to Benton Barracks in St. Louis and then on to Pilot Knob, Missouri. In March, the Ninth moved south to Reeves Station and in April to Pocahontas, Arkansas, skirmishing with the enemy en route. The regiment spent May and June in Jacksonport, Arkansas skirmishing with the enemy at Cache River Bridge and on the Augusta Road. On June 12, a detachment of the regiment (Companies C, D, K and M) escorted a train of thirty-six wagons to Waddell to secure bacon and corn. Confederates attacked the detachment, but the rebels were driven off. Late in June, the Ninth was ordered to ride south and, by July 14, arrived at Helena, Arkansas and, while en route, fought at Stewart's Plantation. The regiment remained on duty at Helena until April 1863.[142] During that time, detachments of the regiment skirmished at Moro, Clarendon, La Grange and Marianna, Arkansas and were in expeditions into Mississippi. In April 1863, the regiment moved to Memphis and then on to Germantown, Tennessee. The Ninth engaged the rebels at Coldwater, Mississippi, on July 28 and in Grenada on August 18, then moving to La Grange, Tennessee. During October, they drove the enemy from positions at Salem and Wyatt, Mississippi. On November 8, the Ninth Illinois Cavalry rode to Corinth, Mississippi. Two weeks later, the men were ordered back north to Collierville, Tennessee, from where they made sorties to engage the enemy at Saulsbury and Moscow, Tennessee.

January 1864 found the regiment still headquartered at Colliersville as part of the Army of Tennessee. Under orders from General Sherman, on February 11, the Ninth, as part of the Union cavalry commanded by General William Sooy Smith, crossed the Tallahatchie River at New Albany, and moved to Okolona, Mississippi. En route, they engaged enemy forces at West Point, Okolona, and Mount Ivy. A portion of the Union cavalry was sent further south to threaten Columbus, Mississippi, while the main body destroyed corn, cotton and other stores needed by the Confederate forces. The regiment returned to camp

---

[140] From the obituary of Dr. Charles Brackett, Regimental Surgeon who died at Helena, Arkansas February, 1863 of typhoid fever. Wheaton, James W., Compiler, *Surgeon on Horseback,* Guild Press of Indiana, Carmel, 1998, page 231.

[141] Forty-two-year-old Colonel Brackett had fought in the Mexican-American war and had been a Captain in the Second US Cavalry. In August 1861 he was authorized by the Secretary of War to raise a cavalry regiment in the west.

[142] Dr. Charles Brackett, Regimental Surgeon, and brother of Colonel Albert Brackett, died at camp February 20, 1863.

at Germantown, Tennessee, on February 24. The three-year term of the men in the regiment expired in March, and the majority elected to re-enlist. Therefore, on March 16, the regiment was mustered in as a Veteran Regiment, and the following day those who re-enlisted proceeded home on a thirty-day furlough. The veterans and recruits reassembled at Camp Fry in Chicago and, on the evening of April 26, departed on the *Illinois Central Railroad* for Cairo, then boarded the *Belle of St. Louis* for the trip down the Mississippi River to Memphis. In June, a detail from the Ninth joined an expedition commanded by General Sturgis against General Forest. The Union forces were defeated at Brice's crossroads on June 10, and the detachment from the Ninth served as the rear guard of the demoralized Union troops. The Ninth fared better in July. In an expedition under General Andrew Smith, the Union cavalry scattered enemy forces at Pontotoc, Tupelo and Old Town Creek, Mississippi. A second expedition under General Smith took place in August, and the Ninth was again part of the cavalry division. During August, the regiment marched to Oxford, Mississippi and skirmished at the Tallahatchie River and Hurricane Creek. In late September, as part of General Hatch's cavalry division, the Ninth crossed the Wolf River at Germantown and spent October and November in operations against Hood in West Tennessee and Alabama. This included fighting at Shoal Creek, Alabama, on November 6, at Lawrenceburg, Tennessee, on November 19 and at Campbellsville, Tennessee, on November 24. On November 30, 1864, the cavalry command, of which the Ninth formed a part, went into camp three miles east of Franklin, Tennessee. At 3 P.M. Hood's forces attacked the Union forces in what was to become a disaster for the Confederate forces. Over the next six hours, the battle became Hood's greatest disaster.[143] During the night, the Union forces withdrew to Nashville, and in the ensuing Battle of Nashville on December 15 and 16, the Ninth "played a distinguished role."[144] The Ninth participated in the chase of the retreating army of General Hood and engaged them at Brentwood, Rutherford's Creek, Sugar Creek, Big Creek, and Ross' Farm. January 1865, the regiment received orders to march to Eastport, Mississippi, on the Tennessee River. During March, when plans were made for a large cavalry force to go on a grand raid into the heart of the Confederacy, the Ninth was not selected and therefore was obliged to turn over their horses and weapons to the troops selected for the raid. Following the surrender of Lee in April, the Ninth was ordered to Iuka, Mississippi and then to several towns in Alabama. The regiment was mustered out of service at Selma, Alabama, on October 31, 1865, and returned to Springfield, Illinois, for final pay and discharge.[145]

There are two flags of the Ninth Illinois Cavalry in the care of the State of Illinois Military Museum. There is a small shield attached to one of the staffs on which the following battles are inscribed: Jacksonport, Waddell Farm, Stewart's Plantation, Coldwater, Salem, Wyatt, Moscow, West Point, Okolona, Mount Ivy, Tishomingo Creek, Tupelo, Hurricane Creek, Old Town Creek, Franklin, Nashville.

"Before the Ninth left Chicago, Miss Florence D. Scammon, daughter of Hon. J. Young Scammon,[146] and the sister of lieutenant Scammon, of the Ninth, presented the regiment with a beautiful

---

[143] The Battle of Franklin cost the Confederate army 6,200 casualties including the deaths of six generals—Cleburne, Adams, Shaw, Strahl, Granbury and Gist and the capture of Major-General Gordon.

[144] "Participated in a well-directed charge upon a redoubt occupied by the enemy, which it carried and at the same time captured four pieces of artillery and took one hundred and fifty prisoners, sharing in the final rout of the enemy." *AGR*, Vol VIII, page 211.

[145] This summary is a compilation of the material in *Report of the Adjutant General of the State of Illinois*, H.W. Rokker, State Printer and Binder, Springfield, Ill, Volume VIII,209-12 and Davenport, Edward Adolphus, *History of the Ninth regiment Illinois cavalry volunteers,* Published under the auspices of the Historical committee of the Regiment, Chicago, Donahue & Henneberry, printers, 1888.

[146] Scammon was a prominent Chicago attorney, a friend of Lincoln, instrumental in founding the Chicago Historical Society and a supporter of the Underground Railroad.

standard." [147] The inscribed standard below may be this standard. It measures 36 x 20" and is made of silk.

This cavalry standard is inscribed: Salem, Miss., Oct 8, 1863; Tupelo, Miss., July 15, 1864; Hurricane Creek, Miss., Aug 14, 1864; Campbellville, Tenn., Nov. 14, 1864; Franklin, Nov. 17, 1864; Nashville, Dec. 15 & 16, 1864.

Also held in the Illinois collection is a guidon from the Ninth Cavalry Regiment. No other flags carried by the 9th Illinois Cavalry and known to survive.

[147] *History of the Ninth Regiment,* page 22.

10th Illinois Cavalry—*"Frontier Regiment"*

"An animated discussion grew out of the subject of uniforming the company. The suggestions of Mr. Finch met with most favor. He proposed that the company should adopt for head-dress the Turkish turban, which is cochlear form, with a red feather, tipped with white and that the cap should be blue...the coat should be light blue with brass buttons and silver epaulet (sp) on the left shoulder; the pantaloons gray, with boots to the knees, with heavy silver spurs."[148]

The Tenth Illinois Cavalry was organized at Camp Butler in September 1861 and was mustered into service for three years on November 25, 1861, under the command of Colonel James A. Barrett.[149] The men of the regiment served from the following counties:[150] Company A--Menard and Sangamon Counties (with a few from Montgomery and Piatt Counties); Company B—Sangamon County; Company C—Iroquois County and Champaign County; Company D—Cook County, Company E—Macoupin, Madison and Menard Counties; Company F—Madison County, Company G—Logan and Sangamon Counties; Company H—Coles and Sangamon Counties; Company I—Champaign County (*Champaign Tigers)*, Company K—Brown County; Company L—Sangamon, Effingham and Pike Counties and Company M--Coles and Moultrie Counties. In late January 1862, the regiment moved from Camp Butler to Quincy, Illinois; in March, the Tenth moved to Benton Barracks in St. Louis. In April, the three battalions moved by train to Rolla, Missouri and then rode to Springfield, Missouri. On June 15, the First Battalion of the Tenth (Companies A, D, G and K—commanded by Major Elvis Shaw) was detached and joined the forces commanded by Major General Curtis at Jacksonport, Arkansas. The battalion saw action at Cotton Plant, Arkansas, on July 7 and then rode to Old Town Landing, Arkansas. In October, the First Battalion was ordered to Helena and subsequently was with expeditions to the mouth of the White River in November and to Oakland, Mississippi, in December. On January 11, 1863, the battalion was engaged at Arkansas Post and then participated in the Yazoo Pass expedition to Fort Pemberton[151] in February, March, and April. On June 6 and 7, 1863, the men were engaged at Richmond, Louisiana, Milliken's Bend and Young's Point. Finally, the First Battalion moved back to Helena, Arkansas, in July and rejoined the remainder of the Tenth Cavalry in August near Wittsburg,[152] Arkansas.

In July 1862, the Second (Companies B, E, H and L) and Third Battalions (Companies C, F, I and M) marched to Keytesville, Missouri, and then returned to Springfield, Missouri. Between August and November 1862, the headquarters of the regiment remained at Marshfield, Missouri, while detachments were stationed at Sand Springs, Hartsville, and Clark's Mills, Missouri and patrolled southwest Missouri and Arkansas. On December 7, the battalions fought at Prairie Grove, Arkansas. The following day the Third Battalion moved to Fayetteville, Arkansas and remained there until the regiment was reunited in Missouri in 1863. Meanwhile, the Second Battalion moved south across the Boston Mountains to Van Buren, Arkansas arriving there in late December 1862. In January 1863, the Second

---

148 *Central Illinois Gazette*, November 8, 1861. Fortunately, Company I did not adopt this dress code. [ed]

149 James Barrett was from Sangamon County, Illinois. Chaplain of the Regiment was Rev. Francis Springer. His role in the War is detailed in *The Preacher's Tale* (The University of Arkansas Press; Fayetteville, 2001) edited by William Furry.

150 Immediately after organization a Regimental Band was formed.

151 Fort Pemberton was constructed of cotton bales covered with earth and was located near Richmond, Ms. on the Tallahatchie River.

152 At the time Wittsburg was a port on the St. Francis River.

Battalion joined General Totten at Huntsville, Arkansas and moved with the division to Flat Creek and then Rolla, Missouri. In March, the Second and Third Battalions joined at Rolla, Missouri, where they were finally issued carbines. In April, the two battalions were on an expedition to the relief of Cape Girardeau, Missouri. In July, the Battalions as part of the 1st Cavalry Division marched south to capture Little Rock, Arkansas.

On August 1, 1863, the three Battalions joined near Wittsburg, Arkansas. On September 10, the Tenth Illinois Cavalry, as part of the First Cavalry Division, fought at Bayou Fourche and the fall of Little Rock and then pursued Price's retreating rebel troops. The regiment remained headquartered at Little Rock until March 1864. The three-year enlistments expired in February 1864, and those men who re-enlisted were granted a thirty days of home furlough. They were welcomed home with a rousing speech by Governor Yates.[153] The regiment returned to service at Nashville, Tennessee, on March 17, 1864, but soon was ordered to Little Rock, Arkansas, for the defense of the city. The regiment arrived without weapons or horses but was supplied muskets and a six-gun battery. In July 1864, the regiment was ordered to Huntsville, Arkansas, in September to Austin, Arkansas and in November to Brownsville, Arkansas. On January 27, 1865, the veterans and recruits of the Tenth Illinois Cavalry were consolidated into nine companies and together with three companies from the Fifteenth Illinois Cavalry were then reorganized into the Tenth (Reorganized) Regiment Illinois Cavalry Volunteers. The regiment left Brownsville, Arkansas, on March 16 and was at Greenville, Louisiana, when word reached the regiment that President Lincoln had been assassinated. The Tenth was immediately ordered to New Orleans to act as provost guard and to prevent retaliation against the southern population. In June 1865, the Tenth Cavalry was sent by boat to Shreveport. From there until muster out, the regiment served in the area surrounding San Antonio, Texas. On November 22, 1865, the Tenth Illinois Cavalry was mustered out of service; the final discharge occurred at Camp Butler on January 6, 1866.[154]

There are two flags of the Tenth Illinois Cavalry in the holdings of the State of Illinois Military Museum.

This national color was made for Cavalry use since it measures approximately 36" x 24". It is inscribed with battles and was manufactured by Gilbert and Hubbard Co. of Chicago. This flag has portions missing, but the inscriptions appear to read: Prairie Grove; Van Buren; Milliken's Bend;

[153] Eddy, T. M, *The Patriotism of Illinois,* Clarke & Co., Chicago, 1866, Volume I, pages 394-6.

[154] *Report of the Adjutant General of the State of Illinois,* H.W.Rokker, State Printer and Binder, Springfield, IL. 1886, Vol. VIII, pages 254-8.

Richmond, La.; Vicksburg; Bayou Metoe; Little Rock; Cotton Plant; Saline; Cape Girard; Mobile; Chalk Bluffs; Telville. *According to the report of the Adjutant General of the State of Illinois,155 "January 12, 1864, the regiment received, by the hands of Lt. Robert J. Bellamy, Company D, from Mrs. E. S. Turner, of the Soldiers' Christian Association, Chicago, a beautiful silk national flag, which upon the disbandment of the regiment was deposited with the Adjutant General of the State." (I presume that this is the flag shown above. Ed.)*

Above is the cavalry standard of the 10th Illinois Cavalry. It measures approximately 24" x 20". The lower scroll contains the regimental ID.

A guidon was presented to Co I of this regiment. "Last Tuesday evening, a large and respectable audience convened at Bailey's Hall for the purpose of witnessing the ceremonies of presenting a beautiful guidon to Capt. Butterfield and his company. Rev. J. C. Stoughton was called to the chair; then, an address was delivered by the Rev. Mr. Thomas. His remarks were filled (with) patriotic, moral, and manly advice. At the close of his speech, he presented the banner to Capt. Butterfield, who responded on behalf of himself and company... The guidon was a very nice one and was made by the ladies themselves."[156] There is no further information re the fate of this guidon. No other flags carried by the 10th Illinois Cavalry are known to survive.

[155] *AGR,* Vol. VIII, page 256

[156] *Central Illinois Gazette*, October 16, 1861

11th Illinois Cavalry—"*McKinstry Guards*"[157]

Presumably referring to the "Union traitors" back home, a member of the regiment, Samuel A. Poland wrote, "The very idea of the thing is loathing to a patriot who has left home with all that home's call dear and is boiling for the mighty rallying under the folds of the glorious old flag of our country and has sworn never to sheath our sabers until that old flag shall wave over every state in the union."[158]

The 11th Illinois Cavalry Regiment was organized in Peoria, Illinois, during the summer of 1861 and mustered in for three years' service on December 20, 1861, with Colonel Robert G. Ingersoll in command.[159] The men in the regiment volunteered from the following counties: Company A—Peoria, Cook and Woodford County; Company B—Fulton County; Company C—mainly from Knox but also men from Peoria, Tazewell and Warren County; Company D—Peoria County; Company E—Peoria County; Company F—Peoria and Tazewell counties; Company G[160]--Fulton County; Company H—Henderson and Marshall counties; Company I—McDonough and Warren counties; Company K—Warren and Burlington Iowa; Company L—Mason and Peoria counties and Company M—Peoria County. The Eleventh remained at Camp Lyon in Peoria until February 1862 when ordered to Benton Barracks in St. Louis. On March 25, the regiment was ordered to Pittsburg Landing and on April 6 and 7 fought in the Battle of Shiloh. Following the battle, the Eleventh participated in the advance and siege of Corinth, Mississippi; when the enemy retreated, the regiment pursued the rebels as far as Booneville, Mississippi, then returned to camp at Corinth. Over the next six months, the Eleventh scouted in Tennessee and northern Mississippi.[161] During that time, the men were involved in actions at Bolivar, Tennessee, at Davis Bridge on the Hatchie River and a three-day battle at Corinth and Iuka in October. During the winter of 1862-63, the regiment was stationed at Jackson, Tennessee. They met Forrest's rebel forces at Lexington, Tennessee, on December 18 and in that action, 100 men of the regiment were killed or taken prisoner, and Colonel Ingersoll and several other officers were captured.[162] The remainder of the regiment participated in actions against Forrest at Jackson on December 19, at Humboldt on December 20 and at Parker's Cross Roads on December 31. By this time, the regiment was reduced to less than 300 effective men and 200 horses. January 1863, the Eleventh Illinois Cavalry moved from Jackson to Grand Junction, Tennessee—stationed along the *Memphis & Charleston Railroad.* During the next six months, detachments of the regiment made numerous raids in Tennessee and Mississippi. In an engagement with General James Chalmers cavalry at Hudsonville, Mississippi, on June 21, the regiment suffered a significant loss in men killed, wounded, and taken prisoner. On September 17, 1863, the Eleventh was ordered to Memphis, where men and horses boarded boats to move downriver to

---

[157] The Regiment was called the *McKinstry Guards* in honor of General Justus McKinstry, the Provost Marshal of St. Louis. *Historical Encyclopedia of Illinois and History of Peoria County,* Munsell Publishing Company, Chicago and Peoria, 1902, Volume II, Chapter 26. Unfortunately, in October 1862, McKinstry was convicted by court-martial of graft, corruption, and fraud in the quartermaster's office at St, Louis, Missouri. Following that debacle, the 11th Illinois Cavalry was no longer referred to as the McKinstry Guards

[158] *Surviving Dixie*, compiled by Kent Harris from the letters of Lt. SA Poland; Café Press, Louisville, Ky,2004

[159] Robert Green Ingersoll was a Peoria attorney and one of the recruiters and organizers of the Regiment. After the War he became a noted orator and in 1867 became the first Attorney General of Illinois.

[160] *Blair's Escort, Crocker's Escort, McArthur's Escort* from Fulton County were combined to form Company G

[161] Company G was detached from the Regiment in the summer of 1861 and served as Headquarters Guard for the 17th Corps. The company never rejoined the Regiment.

[162] O.R. Series I, Volume XVII/1 [S #24] page 593.

Vicksburg, Mississippi. During October, the regiment joined McPherson's expedition to Canton, Mississippi and skirmished with enemy forces along the Big Black River. The regiment moved south to Natchez, Mississippi, in December and, in February 1864, joined in the Meridian campaign commanded by General Sherman.[163] During that campaign, the Eleventh fought at Champion Hill on February 4, at Jackson on February 7, at Morton on February 8 and at Meridian on February 9-14. In March, the regiment was ordered to Vicksburg, and since most of the regiment had re-enlisted as veterans in December 1863, the veterans were then sent home on furlough. The veterans returned from furlough in April, and with a base at Vicksburg, the regiment and regimental detachments scouted and skirmished at Benton, Clinton, Jackson, Yazoo City, Rodney and Fayette, Mississippi, between April and October 1864. The Eleventh departed Vicksburg by boat and arrived at Natchez in early October. From Natchez, a detachment was ordered to Baton Rouge, Louisiana and then on to Woodville, Louisiana. At Woodville, the Union forces captured a Battery of 6 guns and 100 prisoners. Returning to Vicksburg, the Eleventh participated in expeditions to Gaines Landing, Arkansas, in early November and to Yazoo City in late November. The regiment moved to Memphis, Tennessee[164] , in December and during December 1864 and January 1865, was part of Grierson's Raid to destroy the *Mobile & Ohio RR* in Mississippi. In this raid, the Eleventh fought at Franklin Creek, Egypt Station and Franklin, Mississippi. In an expedition to Marion, Arkansas, in late January, the regiment fought at Marion on January 21 and 21. Until muster out on September 30, 1865, the Eleventh Illinois Cavalry guarded the *Memphis & Charleston RR* between Grand Junction and Memphis, Tennessee. The regiment returned to Springfield, Illinois and was discharged on October 20, 1865.[165]

The State of Illinois Military Museum holds two national flags and one regimental flag or standard of the Eleventh Illinois Cavalry Regiment.

This national color measures approximately 48"on the fly x 30"on the hoist Based on the star design, the color was manufactured in Philadelphia. There are no inscriptions on the color.

---

[163] During the campaign the Regiment captured one piece of artillery that was returned to Springfield, Illinois when the Regiment came home on Veteran furlough.

[164] At Memphis the non-Veterans and most of the officers were mustered out.

[165] Compilation of information derived from Dyer, Frederick H. *A Compendium of the War of the Rebellion,* Sagamore Press, New York, (1959) Volume III, 1028-9; Litvin, Marvin (ed) *Sergeant Allen and Private Renick, A Memoir of the Eleventh Illinois Cavalry Written by Henry A. Allen.,* Wagoner Printing Company, Galesburg (1971); *Report of the Adjutant General of the State of Illinois,* H. W. Rokker, State Printer and Binder, Springfield, (1886) Volume 8, page 330-1

This national color measures 6' x 4', which would have made it difficult to carry on horseback.

This regimental standard measures 6' x 4'. The standard is much larger than a regulation cavalry standard. It may have been used as a recruiting flag. There are no other known surviving flags that were carried by the 11th Illinois Cavalry.

12th Illinois Cavalry Regiment

"Now luck to the horseman who fastest can ride
Drawn sabre & pistols, our carbines aside,
Right onward & onward & onward we go
For he wins the palm who shall first smite the foe"[166]

With Colonel Arno Voss[167] of Chicago as commander, men from the following counties were recruited to form the Twelfth Illinois Cavalry Regiment: Company A—Winnebago and Ogle counties; Company B—Cook County[168]; Company C—DuPage County; Company D[169]--Cook County; Company E[170]--Kankakee County; Company F[171]—Greene County; Company G—Hancock County; Company H[172]--Cook County; Company I[173]--Cook County; Company K—Cook County; Company L—Warren County and Company M—Cook, Hancock, and Winnebago counties. Companies A, B, C, D, E, and G were organized at Camp Butler in Springfield and were mustered for three years' service on February 24, 1862. Company F, originally attached to the 32nd Illinois Infantry and was mustered in with that Regiment on December 31, 1861, but was assigned to the 12th Illinois Cavalry Regiment in February 1862. Companies H and I were organized as *McClellan's Dragoons* in Chicago, October 1861 and assigned to the 12th Illinois Cavalry in February 1862; however, they served detached until later in 1862. Companies K, L and M were organized at Springfield but not until December 1863. The seven Companies of the Regiment at Camp Butler in Springfield in February 1862 were anxious to be outfitted as cavalry; however, orders were issued from the Secretary of War to equip the men for guard duty—to guard the rebel prisoners held at Camp Butler. But finally, in June the Regiment was mounted and ordered to Martinsburg, Virginia. The Regiment's first action came on September 3 and 4 near Bunker Hill, Virginia[174] when on a scouting mission, men of the Twelfth met a superior number of rebel cavalry; nevertheless, the scouting party charged and scattered the rebel troopers several miles capturing many of the enemy and horses. On September 7 the Confederate cavalry attempted to capture Lt. Colonel Davis and his command. However, Davis led a detachment of 40 men from the Twelfth routing 800 rebel cavalry with a ferocious saber charge.[175] On September 13, the Regiment was ordered to the defense of Harper's Ferry. The town was soon surrounded by Confederate forces but during the night of

---

[166] From a poem entitled "The Cavalry Charge" written by Abraham Joseph Warner who was the Chaplain of the Regiment. Enderton, Col. Herbert B., *The Private Journal of Abraham Joseph Warner*, private printing by Col. Herbert B. Enderton, San Diego, Calif. (1973), page 150.

[167] Colonel Arno Voss was born in Prussia in 1821. He immigrated to the United States and in 1850 opened a law practice in Chicago. He was active in the Chicago German Off Battalion and the newly formed Washington Independent Regiment.

[168] The company also had men from Missouri and Wisconsin.

[169] 28 men from this company deserted during their first year of service.

[170] 27 men from this company deserted during their first year of service

[171] *Gilmore's (Ephraim M.) Cavalry* Company had been attached to the 32nd Illinois Infantry but became Company F of the 12th Illinois Cavalry.

[172] This company was originally attached to the 52nd Illinois Infantry and was composed of *Gilbert's (*Franklin T.) *Cavalry Company, McClellan Dragoon's, Co.* and *Barker's* (Charles W) *Dragoons.*

[173] Comprised of members of *Barker's Dragoons* and *McClellan Dragoons, Co. B.*

[174] Bunker Hill is now in West Virginia.

[175] Eddy, T.M., D.D., *The Patriotism of Illinois,* Chicago; Clarke & Co. (1866) Vol. II, p412-13.

September 14 *Coles Cavalry*[176], the 12th Illinois Cavalry, the 8th New York Cavalry and the 7th Squadron Rhode Island Cavalry escaped under cover of darkness. The 12th Illinois was ordered to Antietam, Maryland to support McClellan but arrived after the battle had ended. Later in September, while at Sharpsburg, *McClellan's Dragoons* (Companies H and I) rejoined the Regiment. In late September and October 1862, the Regiment was sent on picket to Williamsport Maryland and Dam No. 4 and then to Russell's Ferry near Green Spring Furnace on the upper Potomac River. The Twelfth acted as escort of Sigel's Army from Warrenton to Fredericksburg but did not see battle at either Fredericksburg or Marye's Heights. By December 20 the Twelfth Illinois Cavalry was camped at Dumfries, Virginia, a large Union supply depot.[177] The Twelfth remained in the Dumfries area until they embarked on the Chancellorsville Campaign in late April of 1863. From April 29 until May 4 the Regiment, part of a brigade led by Union General George Stoneman, rode south toward Richmond. The Twelfth played a notable role in the "Raid" cutting telegraph wires and destroying a bridge for the *Richmond, Fredericksburg & Potomac RR*, destroying an engine and wagon train and capturing 80 mules at Ashland. On May 5, the troopers struck the *Virginia Central RR at* Hanover Station, capturing thirty Confederates and destroying railroad and telegraph lines. The enemy attempted to intercept the Twelfth at Tunstall Station on the *Richmond and York River RR* east of Richmond, but the ensuing battle ended in a stalemate. Colonel Davis then led the Regiment to Gloucester Point across the York River from Yorktown; while enroute they captured rebels, destroyed cavalry equipment, and captured a wagon train loaded with provisions. Following the raid, while one hundred troopers remained at Gloucester Point, one Battalion was sent to Fortress Monroe and another to Alexandria. The four companies (A, E, F and H) at Alexandria became part of the Cavalry Corps of the Army of the Potomac, First Brigade, commanded by Colonel William Gamble. These companies were heavily involved in the fighting at Upperville, Virginia on June 21 and moved with the First Brigade north through Maryland to arrive on the field at Gettysburg, Pennsylvania in the late afternoon of June 30. The Eighth and Twelfth Illinois Cavalry were a part of the Union forces that felt the brunt of the Confederate attack on Gettysburg on July 1. The troopers at first defended McPherson's Ridge but were forced to retreat to Seminary Ridge. On July 2, the Brigade was ordered to Westminster, Maryland to guard an ammunition depot. Following the Battle of Gettysburg, the Regiment or detachments of the Regiment remained with the Army of the Potomac and fought at Williamsport, Beaver Creek, and Funkstown in Maryland, then at Falling Waters, Kelly's Ford and Brandy Station in Virginia. In September during the advance from the Rappahannock to the Rapidan River the Regiment fought at Culpeper Court House and at Raccoon Ford. During October they skirmished with Stuart's forces at Raccoon Ford and Morton's Ford on the Rapidan River and at the Second Battle of Brandy Station. Many of the men reenlisted in November 1863, turned in their horses and weapons and were sent to Chicago to recruit and reorganize. February 9, 1864, the Twelfth Illinois Cavalry left for St. Louis with a full regiment of 1,256 officers and men and arrived in New Orleans on April 15. By the end of April, they were traveling up the Red River to join General Banks and during April and May the Regiment carried out counter guerilla operations in the bayou area moving south from Alexandria down to Morganza, Louisiana. The expedition arrived back in New Orleans on June 1. June through October 1864 the Cavalry was based at Napoleonville, west of New Orleans, for picket duty and scouting.[178] The Regiment was ordered to Baton Rouge in November and in late November was involved in Davidson's failed expedition against the *Mobile and Ohio RR* in eastern

---

[176] 1st Maryland Cavalry, Cos. A, B, C, D.

[177] Stuart's Cavalry captured thirty troopers from the 12th, mainly form Company G, during an attack on Dumphries on December 27.

[178] During that time Companies I and L were sent on expedition to Natchez, Mississippi.

Mississippi.[179] Upon return to Baton Rouge the Twelfth moved up the Mississippi River to Vicksburg, Mississippi and then on to Memphis, Tennessee and finally to Gaines' Landing, Arkansas in January 1865. During late January and early February, a detachment of the Twelfth accompanied Colonel Embry Ostend's expedition through southern Arkansas and northern Louisiana; in early March[180] the Twelfth was part of Colonel John P. C. Shanks' expedition from Memphis into northern Mississippi in search of Nathan Bedford Forrest. With the surrender of the Confederate forces in April and May, the men assumed that they would be returning home within days, however they were ordered to Alexandria, Louisiana and then on to Hempstead, Texas in August. The Regiment was mustered out on May 29, 1866, at Houston, Texas.[181]

The flag holdings by the State of Illinois Military Museum include a cavalry standard and guidon identified as having been carried by the Twelfth Illinois Cavalry Regiment

The cavalry standard is in poor condition, and it appears that the men of the Regiment may have removed portions of the flag—perhaps as souvenirs.

[179] During this time the Regiment lost over 100 men to expiration of enlistments, sickness, and desertion. Blackwell, Samuel M. *In The First Line of Battle, The 12th Illinois Cavalry in the Civil War,* Northern Illinois University Press, DeKalb (2002), p 149.

[180] Pursuant to Special Order No. 92, on March 2, 1865, the Twelfth Illinois Cavalry was consolidated into an eight-company organization. On June 14, 1865, the remaining veterans and recruits of the Fourth Cavalry Brigade were transferred to and consolidated with the Twelfth Illinois Cavalry

[181] The best source of information on the Twelfth Illinois Cavalry is the book by Samuel Blackwell referenced above. Not only is the book concisely written—there is excellent documentation from a wide number of resources.

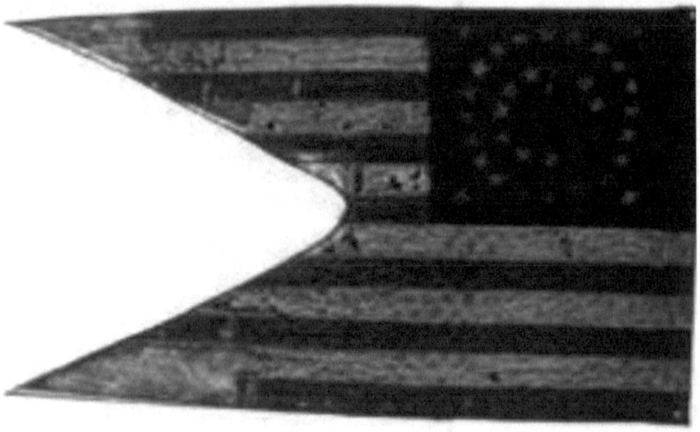

This Cavalry guidon of the 12th Illinois Cavalry measures approximately 24" x 20". There are no inscriptions on the flag.

The fourth reunion of the Regiment was held at the Chicago World's Fair in September 1893. According to the program from the reunion— "All comrades present desired to know the whereabouts of our last Regimental Colors. That color on exhibition in the state collection is but a guidon, and it is reported that the flag is not at the state capitol. The Battle Flag of the Twelfth bore a horse's head in gold, "rampant," in a blue field, and a scroll with the mottoe(sp) "I Like Your Style!"(the greeting the governor of the state met us with on our return from Virginia) The flag is all surrounded by heavy gold fringe. Who of the comrades or of the state officers of Illinois can give information as to its whereabouts?" The whereabouts of this flag today is unknown.

No other flags carried by the 12th Illinois Cavalry are known to survive.

13th Illinois Cavalry Regiment

"Grave deep his memory on your hearts
Keep ye his country free
Live for the flag for which he died—
This is his legacy[182]

Eight companies of the Thirteenth Illinois Cavalry were organized at Camp Douglas in Chicago and mustered in under command of Colonel Joseph W. Bell[183] for three years' service on December 31,1861. The First Battalion (known as *Hartmann's Dragoons*) consisted of Company A from Cook and Union counties; Company B and D (*Bell's Cavalry*) from Cook County and Company C from Cook and Will counties. The Second Battalion was comprised of Company E from Warren and Lawrence counties; Company F from Cook and Will counties; Company G from Douglas and Morgan counties and Company H from Douglas and Kane counties. The Regiment was ordered to Benton Barracks, Missouri in February 1862 where it was equipped and mounted. The Regiment then moved to southeast Missouri scouting and skirmishing with guerillas. In June the Thirteenth joined General Curtis' army at Jacksonport, Arkansas and skirmished with guerilla forces at Pitman's Ferry, Cotton Plant, Arkansas and Union City Camp Pillow, Bloomfield, Van Buren and Allen, Missouri. In 1863 the Thirteenth Cavalry saw further action at Eleven Point River, Jackson, White River, Union City and Bushy Creek, Missouri. During this time, on May 8, the original eight companies were consolidated into one battalion of three companies. In July 1863 the Regiment moved, as part of General Davidson's Cavalry Division, into Arkansas and fought in major actions at Brownsville, Bayou Metairie, and Austin, Arkansas. In September, the Thirteenth fought at Little Rock, Arkansas and was the first Regiment to enter the city on September 10, 1863; it then took part in the pursuit of Price's army. In February 1864, at Pine Bluff, Arkansas, nine new companies were added to the consolidated three companies of the First Battalion giving the Regiment a total of twelve companies. In the spring of 1864, as part of General Steele's army expedition to Camden into southern Arkansas, the Regiment fought at Arkadelphia, Ocolona, and Prairie du Ann on April 2, 3, and 4 and then Camden, Marksville, and Jenkins' Ferry on April 29 and 30. In May it returned to Little Rock and was active in fighting enemy forces commanded by Shelby and Marmaduke and defeated them in battles at Clarendon and Pine Bluff. The remainder of the 1864 was spent in scouting and picketing in Arkansas and Missouri with a base of operations at Pine Bluff. On January 24, 1865, the cavalry division of the Seventh Army Corps was discontinued, and the Thirteenth was assigned to post duty at Pine Bluff. Following the surrender of Lee and Johnston the Regiment took possession of Monticello, Camden and Washington, Arkansas. It was ordered to Springfield, Illinois where it was mustered out on August 31, 1865.[184],[185]

The current inventory of the State of Illinois Military Museum contains two cavalry standards, one national flag and one guidon belonging to the Thirteenth Illinois Cavalry.

---

[182] Theodore Secrist was a student at Monmouth College, Monmouth, Illinois and enlisted in *Bell's Cavalry* in 1862. He died at Helena, Arkansas on October 3, 1862 and was buried in the South Henderson Cemetery in Henderson County, Illinois. This is a portion of a poem read at his funeral. Monmouth College Oracle, Vol. XV, No 34, May 30, 1911, page 28.

[183] Born in Tennessee, Colonel Bell was a 47 year old Chicago attorney.

[184] Eddy, T.M., *The Patriotism of Illinois,* Clarke & Co., Publishers, Chicago, 1866, Volume II, pages 627-33

[185] *Report of the Adjutant General of the State of Illinois,* H.W.Rokker, State Printer and Binder, Springfield, Ill, 1886k, Volume VIII, pages 422-3

This color is the Standard of the 13th Illinois Cavalry. It measures approximately 30" x 22". There are no battle inscriptions.

This color is another Standard of the 13th Illinois Cavalry. It measures approximately 30" x 22". This flag is typical of the flags carried by the U.S. Cavalry at the outset of the Civil War and served as a model for the cavalry standards manufactured during the War.[186]

---

[186] *Echoes of Glory, Arms and Equipment of The Union,* Time-Life Books, Alexandria, Virginia, 1991, page 271.

13TH ILLINOIS CAVALRY REGIMENT

This National color is tagged "Recruiting Flag" and measures approximately 60" x 30". It appears to be "home made."

This Cavalry guidon measures approximately 30" x 36". Each company in the regiment, for its identification, would have carried a guidon. This guidon is of the "Stars and Stripes" design and is based on the regulations of January 1862.

No other flags carried by the 13th Illinois Cavalry are known to survive.

14th Illinois Cavalry Regiment

"Happy were the boys of the 14th Illinois Cavalry when Saturday, the 28th day of March 1863, came, bringing orders to march to join our noble comrades in the front, to aid in crushing treason, and in bearing aloft our glorious flag, the emblem of the proudest and most glorious nationality that ever blessed the earth."[187]

In the fall of 1862, a nucleus of three cavalry regiments gathered in Peoria, Illinois. All three were far from having an adequate number of volunteers to form a complete regiment. They were finally consolidated as the Fourteenth Illinois Cavalry Regiment and the first and second battalions were mustered on January 7, 1863, while the third battalion was mustered February 6, 1863. Colonel Horace Capron[188] of Peoria commanded the Regiment. The men volunteered from the following counties: Company A- Knox County; Company B-Peoria County; Company C- Woodford and Putnam[189] counties; Company D-Cook, Logan and Sangamon counties; Company E-Gallatin and Johnson counties; Company F-Fayette County; Company G-Johnson, Pope and Union counties; Company H-White County; Company I-Stephenson County and Gallatin County; Company K-Cook and Peoria counties; Company L-De Kalb and Rock Island counties; Company M-Cook and Peoria counties. The Regiment departed Peoria on March 28 and arrived in Glasgow, Kentucky on April 17, 1863. The Regiment was brigaded and immediately ordered south to attack the enemy forces at Celina, Tennessee on the Cumberland River. Following capture of the town and destruction of rebel stores, the Regiment returned to headquarters in Glasgow and remained there on scouting duties. On June 22 the Regiment was ordered to Tompkinsville, Kentucky and from there pursued Confederate Brigadier General John Hunt Morgan into Indiana and Ohio. After a month of constant pursuit, the Fourteenth Illinois aided in the capture of Morgan at Salineville, Ohio on July 26. By July 30, the Regiment had passed through Cincinnati and was enroute by train to Lexington, Kentucky. During August the men rode east as part of Burnside's campaign in east Tennessee and on September 1 occupied Knoxville, Tennessee capturing the rear guard of the retreating Confederate forces. The Fourteenth next rode in the attack of the enemy forces occupying Cumberland Gap. Surprisingly the rebel forces surrendered on September 9 and the Union forces captured prisoners, guns, ammunition, and supplies. The remainder of September the Union cavalry campaigned in eastern Tennessee and the Fourteenth was engaged at Rheatown, Kingsport, Bristol, Zollicoffer,[190] Jonesborough and Blountville. The Fourteenth Illinois Cavalry was a part of the Union Cavalry that fought in the Knoxville campaign of November and December. During that campaign the Fourteenth scouted and skirmished with rebel forces northeast of Knoxville at Maynardsville, at Walker's Ford on the Clinch River, at Bean's Station and at Blains's Crossroads. In late December the Regiment was ordered to Dandridge, Tennessee to look for Longstreet's cavalry. Longstreet's forces had retreated from Dandridge but engaged the Union cavalry northwest of Dandridge at New Market, Tennessee. Longstreet's forces were driven off and they lost more than 200 men. January 1864 the Fourteenth Illinois Cavalry was in camp in Dandridge and in mid-January Colonel

---

[187] Sanford, Washington L. *History of the Fourteenth Illinois Cavalry and the Brigades to which it belonged* Chicago, R. R. Donnelley & Sons Company, 1898, page 14.

[188] At the time, 53-year-old Colonel Capron was a farmer living near Peoria. He had been one of the inventors of barbed wire. Starr, Merritt, "General Horace Capron, 1804-1885." Jour. State Historical Society, XVIII (1925) 259-349.

[189] This company also had volunteers from Will, Lee, Macoupin, Richland, Henry, Marshall, McLean Perry, Rock Island, Clinton, Peoria, Bureau, La Salle, Stephenson, Hancock and Cook Counties.

[190] Now known as Bluff City.

Capron along with a recruiting sergeant from each company left for Illinois to raise recruits for the Fourteenth. During January, from the base at Dandridge the Regiment fought at Kimbrough's Crossroads, Dandridge, Fair Garden and Fain's Crossroads. In early 1864 word reached Union command that the "Thomas Legion"[191] was near the forks of the Little Tennessee and Tuckasegee Rivers in North Carolina. General Samuel Sturgis selected the Fourteenth Illinois Cavalry to ride south into North Carolina to attack these enemy troops. After a perilous ride over the mountains, the Regiment attacked and defeated "Thomas Legion" on February 2 [192] in a battle at Deep Creek on the Tuckasegee River. By February 16 the Regiment was back in Marysville, Tennessee marching toward Knoxville. The next four months[193] the Fourteenth remained in east Tennessee occupying a line of posts at fords and crossings of rivers and passes over the mountains from Loudon on the Tennessee River to Charleston on the Hiwassee River—a line of posts 40 miles long. On May 27, Confederate Colonel Thomas and a band of Cherokee Indians attacked the Fourteenth's post at Citico (or Cittico) Tennessee and were repulsed. On June 13 the Fourteenth Illinois Cavalry Regiment was ordered to join General George Stoneman's command for the Atlanta campaign.[194] General Sherman conceived a plan for a cavalry raid around and south of Atlanta to destroy three to five miles of the *Macon & Western RR* along with telegraph poles and wires, bridges, enemy stores and ammunition and then free Union prisoners held at Macon and Andersonville. One of the cavalry divisions involved in this plan was commanded by Major General George Stoneman and included the Fourteenth Illinois Cavalry. Stoneman's division left on July 27. On the afternoon of July 30, they reached Macon only to discover that the town was well defended and that the only route into the town was via a railroad bridge and pontoon bridge over the Ocmulgee River. Stoneman ordered an attack, but the Union Kentucky Brigade commanded by Colonel Silas Adams fell back soon after they were fired upon, and Stoneman began retreating back north. Rebel forces at Sunshine Church trapped the division on August 3 and most of the men including Stoneman were either captured or killed. Colonel Capron, his youngest son Osmond, and Lieutenant Calkins of the Eighth Michigan Cavalry managed to escape and made it back to the Union lines near Atlanta.[195] The remnants of the Fourteenth straggled in over the next month. The first inspection, after the raid, was held on August 17th and the Regiment had only 14 men mounted. On September 15, the remnants of the Regiment were ordered to Nicholasville, near Lexington, Kentucky for equipment and remount. In mid-November the Fourteenth was in mid Tennessee scouting and skirmishing with Hood's forces. On November 18 Capron's men occupied Waynesboro; then on November 23 fought Forrest's forces at Henryville and Mount Pleasant, Tennessee. The Fourteenth was ordered to guard the fords on the Duck River below Columbia on November 26 and on November 28 they opposed Forrest's forces as they attempted to cross the Duck River at the Lewisburg Pike crossing. On the evening of the 28th Capron's' command, including the Fourteenth was surrounded by rebel forces but fought its way out and headed toward Franklin, Tennessee. At the Battle of Franklin on November 30, the Fourteenth Illinois Cavalry fought on the Union left flank. Following the Confederate withdrawal, the Fourteenth rode to Nashville. On December 3 the Regiment was dismounted and in the fighting at Nashville on December 15 the Regiment supported Battery I of the 4th US Artillery. The remainder of December the Fourteenth joined in

---

[191] Confederate Colonel William Thomas, although white, was the chief of the Eastern Band of the Cherokees and before the War a US Senator from North Carolina. In April 1862 Thomas and two companies of Cherokees were mustered into the Confederate Army. Thomas' command expanded to finally become the "Thomas Legion" with 2,000 men—both Cherokees and whites.

[192] Colonel Thomas escaped capture. Lieutenant Horace Capron of Company A, the son of Colonel Capron was killed.

[193] In April the Regiment formed its own regimental band at Madisonville.

[194] On June 17, 298 new recruits were added to the Regiment

[195] Capron, Horace, "General Capron's Narrative of Stoneman's raid south of Atlanta," Prepared by Lieutenant-Colonel Gilbert C. Kniffin. 1899, MOLLUS-*DC, no*. 32.

the pursuit of Hood's forces to the Tennessee River. The Regiment was in camp near Nashville until April when ordered to Pulaski, Tennessee. In July the Fourteenth Illinois Cavalry was ordered to Nashville for muster out on July 31, 1865.[196]

There are two national flags identified as having been carried by the Fourteenth Illinois Cavalry in the Illinois State Military Museum flag collection. There is one record of the Regiment receiving a flag prior to leaving Peoria in 1863. "In the last of January, we received our horses and drew Springfield sabers. February 23rd, we received from the state a beautiful rich silk banner in a stirring and appropriate speech."[197] The flag is not further described but based on precedent would most likely have been a national flag. One of the two flags below may have been that flag.

This National color measures approximately 54" x 48". Based on the star pattern, this flag was probably manufactured by Wm. Scheible of Philadelphia. There are no inscriptions on the flag.

This National color is inscribed: 14th Regt. Ill Cav; Bean Station, Knoxville & Atlanta, *and measures approximately 60" x 30". The light blue square canton containing six horizontal rows of stars is consistent with flags manufactured in Louisville by Hugh Wilkins.*

No cavalry standards or other presentation flags carried by this regiment are known to survive.

---

[196] Ibid. Sanford. The Sanford Regimental history is one of the better Illinois regimental histories. The history is well documented and includes many personal observations of the Regiment's history.

[197] *Ibid,* Sanford, page 12.

15th Illinois Cavalry Regiment

Nail to the mast our dear old flag,
Long has it waved on high;
And many an eye has danced to see
That banner in the sky;
Beneath it rung the battle shout
And bust the cannon's roar;
Her folds once red with heroes' blood,
Where knelt the vanquished foe.[198]

The Fifteenth Illinois Cavalry Regiment has a unique history. When the regiment was formed in the spring of 1863 the companies that made up the new regiment had been operating independently for many months. *Colonel Warren Stewart*[199] *Independent Cavalry* served as the nucleus for the Regiment. Prior to the attack upon Fort Sumter, Stewart was living at Clear Creek Landing, Alexander County in southern Illinois across the Mississippi River from Cape Girardeau, Missouri. Hearing news of the fall of Fort Sumter, Stewart called a meeting in Alexander County and recruited one hundred and thirty men to serve in a local militia. Stewart was named Captain of the Company and Governor Yates immediately supplied arms. On August 3, 1861, General Fremont urged Stewart to enter the cavalry service and aid in defending the Cape Girardeau area from Confederate infantry and cavalry forces commanded by General Jeff Thompson. On August 10, 1861, the company was mustered as *Stewart's Independent Cavalry Company*. In September the Company was ordered by General US Grant to Norfolk[200], Missouri to scout the Confederate forces at nearby Belmont, Missouri. Under the overall command of Colonel Oglesby, *Stewart's Cavalry* scouted and skirmished around Sikeston, Fredericktown, and Charleston Missouri until ordered to Cairo, Illinois. There Colonel Warren Stewart was placed in command of a battalion known as *Stewart's Independent Battalion Cavalry*. In addition to *Stewart's Cavalry Company,* this battalion included *Hutchen's Cavalry* (from Mason County and originally mustered with the 27th Illinois Infantry), *Carmichael's Cavalry* Company (from Alexander and White Counties and originally mustered with the 29th Illinois Infantry), *Clinton County Cavalry* (originally mustered with the 30th Illinois Infantry) and *Dollins' Cavalry* (from Williamson County and originally mustered with the 31st Illinois Infantry. Under command of the now Major Stewart, the Battalion joined in the fighting at Columbus, Kentucky, at Fort Henry and in early February at Fort Donelson. Stewart's forces also scouted the position of enemy forces at Shiloh in April. During the Battle of Shiloh, Stewart was wounded and carried from the field. In June *Stewart's Cavalry* routed the Confederate cavalry at Jackson, Mississippi. He was ordered to Springfield in September 1862 with the authority to increase his Battalion to regimental strength. The Fifteenth Illinois Cavalry Regiment was organized and mustered into service on December 25, 1862 by the consolidation of the following companies into the Regiment: Company A-*Stewart's Independent Cavalry Company*, Company B—*Egleton Carmichael's Cavalry Company*,

[198] Written by Noah V. McDowell, private Company L. He was later Secretary of the Fifteenth Illinois Cavalry Association, and his poem is printed in the *First Re-Union Fifteenth Illinois Cavalry Volunteers,* held at St. Louis, Mo., September 27 and 28, 1887. Worthington, Minn., Advance Book and Job Printing House, 1888, page 10.

[199] Colonel Warren Stewart was born in New York in 1813. After living in St. Louis and Texas he settled in Clear Creek Landing in Alexander County, Illinois. He was killed near Vicksburg, Mississippe on January 23, 1863

[200] This is now known as Norfolk Landing on the Mississippi River.

Company C—*James Dollins' Cavalry Company*, Company D—*Clinton County Cavalry*, Company E—*William Hutchen's Cavalry Company*, Company F—recruits from Franklin County,[201] Company G—*Franklin Gilbert's Cavalry* from Kane County,[202]Company H—*Christian Dodson's Independent Cavalry Company* and *William Wilder's Independent Cavalry*,[203] Company I—*Albert Jenks Cavalry* Company,[204] Company K—*Henry Smith's* Cavalry Company,[205] Company L—*William Ford's Independent Cavalry*,[206] Company M—*Oscar Huntley's Independent Company A*, Cavalry from Winnebago County. [207]

In January 1863, under the command of Colonel George Bacon,[208] the Fifteenth Illinois Cavalry (comprised of companies A, B, C, D, E, F, G and I) was ordered to southwest Tennessee in the area of Clifton, Monterey (current day Michie) and Hamburg Landing and northern Mississippi for scouting, picket duty and skirmishing. On April 9, Company E was detached and sent to the area of Hickman Kentucky and saw action at Obion Plank Road Crossing, Tennessee in May. Company H, commanded by Capt. W. C. Wilder, was detached, and served as the escort Company to Maj. Gen. William Sherman, commander of the XV Army Corps. During June and July Company H was attached to Maj. Gen. F. Steele's XV Army Corps and fought in the Greenville, Mississippi Expedition and during the Vicksburg Siege. Company K served detached at the Headquarters, Right Wing of the XIV Army Corps in January 1863 and then in the Headquarters XX Army Cops, Army of the Cumberland until October 1863. During that time Company K was in action at Rover, and near Murfreesboro, Tennessee and in the campaign over the Cumberland Mountains and the Tennessee River in the Chickamauga Campaign and in the Battle of Chickamauga. From April 15 until May 8 companies A, B, C, D, F, G and I rode in the expedition from Corinth, Mississippi to Courtland, Alabama then on to Tupelo, Mississippi seeing action at Dickson's Station, Great Bear Creek and Cherokee Station, Tuscumbia, Florence and Tupelo. May 26-May 31, 1863, Companies A, B, C, D and G rode in the expedition from Corinth, Mississippi to Florence, Alabama and then north to Hamburg Landing, Tennessee. Most of the Regiment remained on duty along the Mississippi River between Columbus, Kentucky and Memphis, Tennessee from July 1863 until January 1864.

---

201 These men had been recruits to *Stewart's Independent Battalion* and were in companies A and E of the Battalion.

202 This Cavalry Company had been mustered in with the 52nd Illinois Infantry as attached Cavalry Company A. Afterward the Company was Company H of the 12th Illinois Cavalry Regiment.

203 *Dodson's Independent Cavalry Company* became *Wilder's Independent Cavalry* on the resignation of Christian Dodson. This Company was also known as the *Kane County Cavalry*.

204 Mustered in with the 36th Illinois Infantry as Cavalry Company A. The men of this company volunteered from Kane County. When Jenks was promoted to Lieut. Col of the 36th Illinois Infantry the Company was under command of George Willis and was referred to as *Willis's Cavalry*.

205 This company was mustered with the 36th Illinois Infantry as attached Cavalry Company B and known as *Henry Smith's Cavalry Company*. When Smith was cashiered, Samuel Sherrer became Captain of the Company, and the company was known as *Sherrer's Cavalry Company*. The men of this company volunteered from Kane County.

206 *William Ford's Cavalry* was recruited from LaSalle County and mustered in with the 53rd Illinois Infantry as Cavalry Company A. Prior to joining the 15th Illinois Cavalry the Independent Cavalry Company served as *General Halleck's Escort* and then *General Grant's Escort*.

207 This Company was assigned as Company M of the 15th Illinois Cavalry on December 25, 1862. However, the Company had been mustered out as Company K of the 1st Illinois Cavalry on December 27, 1862 before the new orders of assignment reached them. Therefore, this Company was never mustered with the 15th Illinois Cavalry Regiment. *First Re-Union Fifteenth Illinois Cavalry Volunteers,* held at St. Louis, Mo., September 27 and 28, 1887. Worthington, Minn., Advance Book and Job Printing House, 1888, page 10.

208 Colonel Warren Stewart was serving as an aide to General John McClernand in the Yazoo River Expedition. Stewart was killed on January 23, 1863, while reconnoitering with a small cavalry force.

However, companies F and I were detached and were engaged in outpost duties on Big Black River, in the siege of Jackson, Mississippi and then ordered to western Louisiana and campaigned in the area west of Baton Rouge at Opelousas and Barre's Landing on Bayou Courtableau (Company F) and at New Iberia, Louisiana (Company I). In January 1864, Company I was ordered as escort to Brigadier General Albert L. Lee, chief of cavalry, Department of the Gulf.

Companies A, B, C, D, E, G and K were ordered to Helena Arkansas in January 1864 for garrison duties and served there until January 1865. During that duty, detachments of the Regiment saw action during the White River expedition, the St. Francis River expedition, at Wallace's Ferry, at Lamb's Plantation and at Kendall's Grist Mill. Non-veterans were mustered out on August 25, 1864, and the Veterans and Recruits consolidated into a Battalion. In keeping with Special Order No. 24, in January 1865, at Little Rock, Arkansas, the Tenth and Fifteenth Illinois Cavalry Regiments were consolidated and re-organized as the Tenth Regiment Illinois Cavalry Volunteers.

Meanwhile in 1863, 1864 and 1865 Company K fought in the Atlanta Campaign. The company fought in the battles of Lookout Mountain, Mission Ridge and Ringgold Gap in 1863 and then at Resaca, Pumpkin Vine Creek, Dallas, New Hope Church, Marietta, Kenesaw Mountain, Chattahoochee River and Siege of Atlanta from May through August, 1864. Following the fall of Atlanta, Company K accompanied Sherman's forces in the March to the Sea and the Siege of Savannah in November and December 1864 and then in the Campaign of the Carolinas from January to April 1865. Following Confederate Johnston's surrender, the company was ordered to Washington D.C., via Richmond, Virginia and rode in the Grand Review on May 23, 1865.

There are no flags identified as belonging to the Fifteenth Illinois Cavalry Regiment in the State of Illinois flags collection. This is probably because each company of the Regiment began as an independent cavalry company. No flags carried by any of the companies in the Regiment are known to survive.

16th Illinois Cavalry

Thinking no less of them,
Loving our country the more,
We sent them forth to fight for the flag,
Their fathers before them bore.
Though the great tear-drops started,
This was our parting trust:
"God bless you boys! We'll welcome you home
When Rebels are in the dust."
Brave boys are they!
Gone at their country's call;
And yet, and yet, we cannot forget,
That many brave boys must fall.[209]

The Sixteenth Illinois Cavalry had the unique distinction of losing more men to death in Confederate prisons than any other Illinois Regiment. Of 1,462 men who served in the Sixteenth Illinois Cavalry, 157 men died in Confederate prisons.[210]

Christian Thielemann's Independent Cavalry[211] Company was organized in Chicago on July 2, 1861. The Company Served in the area from Paducah to Columbus, Kentucky, until December 1861 when at Smithland, Kentucky, *Mathew Marx's Independent Cavalry* Company (organized in Chicago in November 1861) was consolidated with Thielemann's Company to form *Thielemann's Independent Cavalry Battalion*, companies A and B under command of Colonel Christian Thielemann. The Battalion was on duty at Paducah and Smithland until March 1862 and was then ordered to Tennessee and fought at the Battle of Shiloh, Tennessee, on April 6-7. Following the Siege of Corinth, Mississippi, the Battalion marched to Memphis, Tennessee, and remained on duty there until November. (In January 1863, the Battalion was assigned to the Sixteenth Illinois Cavalry as companies A and B but continued to serve detached until joining the remainder of the Regiment at Cumberland Gap, Tennessee, in January 1864.) In the subsequent Vicksburg campaign of 1863, the Battalion was in action at Chickasaw Bayou, Chickasaw Bluff, Mississippi, at Arkansas Post (Fort Hindman) in Arkansas, and at Young's Point, Louisiana. The Battalion joined the Union forces at the rear of Vicksburg in May 1863 and fought at Jackson, Mississippi, Champion's' Hill, and the Siege of both Vicksburg[212] and Jackson. In September, the Battalion was ordered to Chattanooga, Tennessee, and the remainder of 1863 was in operations along the *Memphis & Charleston RR* in Alabama and then in the Chattanooga-Ringgold Campaign—fighting

---

[209] Excerpt from a song written at Andersonville Prison by Able "Nosey" Payne of Danville, Illinois—a fellow prisoner with John McElroy—both of Company L of the 16th Illinois Cavalry Regiment. McElroy, John, *Andersonville A Story of Rebel Military Prisons,* D. R. Locke, Toledo, 1879, pp 218-19.

[210] Fox, William F., *Regimental Losses in the American Civil War 1861-1865,* Albany Publishing Company, Albany, NY, 1889, p 524.

[211] 53 year old Thielemann was German born. In additiont to serving as Captain, Company M, 1st Illinois Cavalry and Major, Thielemann's Independent Cavalry, he was Colonel of the 16th Illinois Cavalry and Commander of Camp Butler. He worked as a theater owner and manager after the war.

[212] During the siege of Vicksburg, the Battalion performed escort duty for General W. T. Sherman and then accompanied him to Jackson and then to headquarters on the Big Black River.

at Chattanooga. Tennessee November 23-25, 1863. The Battalion was ordered to the relief of Knoxville, Tennessee, November 28-December 8 and then ordered to Cumberland Gap to join the Regiment.

Schaumbeck's [213]*Independent Cavalry* Company was organized in Chicago in July 1861. The Company was ordered to New Creek, West Virginia, and saw its first fighting on the Gauley River at Carnifex Ferry, West Virginia, on September 10. The Company, as part of McCook's Brigade, served in the District of the Kanawha, West Virginia, until August 1862. During that period, the Company was at New River, Gauley, and on the *Virginia & Tennessee RR,* Wolf Creek, Flat Top Mountain, and on the Blue Stone River, West Virginia. (Although the Company was assigned to Thielemann's Cavalry Battalion as Company C in December 1861 and subsequently to the Sixteenth Illinois Cavalry as Company C, in December 1862, Schambeck's Company served detached from both for their entire term of service.) Later in September, the Company fought in the Battle of Antietam and then returned to duty in the Kanawha Valley, West Virginia, until April 1863. April 20-May 14 Schambeck's Company was a part of the Union cavalry that fought Confederate General John Imboden' cavalry forces that ravaged the Kanawha Valley. Later in 1863, Schambeck's Company accompanied the Union raids into Western West Virginia and the *Virginia & Tennessee RR* led by General William Averill. The Company marched through Elk Mountain Pass to Beverly, West Virginia, in December and remained at Beverly until April 1864, at Harper's Ferry and the *Baltimore & Ohio RR* until mustered out on July 16, 1864.[214]

The Sixteenth Illinois Cavalry was organized at Camp Butler, near Springfield, Illinois, between January and April 1863. Companies A and B, recruited from Cook County, had served detached since July 1861 as *Thielemann's Independent Cavalry Battalion,* and did not physically join the Regiment until January 1864. Company C, also from Cook County, was *Schambeck's Independent Cavalry* Company and served detached, never physically associated with the remainder of the Regiment. Company D was organized in March 1863 with recruits from Cook County and Sangamon County in Illinois as well as men from St. Louis, Missouri, from Ohio, New York, and Louisiana. Company E, with recruits mainly from St. Louis, Missouri, was originally organized for the Seventeenth Illinois Cavalry but was instead assigned as Company E of the Sixteenth. Most men in Company F came East St. Louis in St. Clair County, Illinois. Company G was organized in May 1863, with majority of men from St. Clair County and a few from Cook County. Kane, McHenry, and Sangamon County supplied the men for Company H.

Seven Companies of the Regiment (D, G, H, I, K, L M) remained at Camp Butler, drilling and serving as guards for Confederate prisoners.

Companies E and F remained at Camp in Quincy, Illinois, under the command of Major Schambeck.

In October, the Regiment (Companies D, G, H, I, K, L M) was ordered to Cincinnati, Ohio, and then on to Cumberland Gap, Kentucky, where Companies A and B rejoined the Regiment. On December 1, 1863, a battalion of the Sixteenth attacked rebel forces near Jonesville, Virginia driving them off and capturing men and horses. One month later, on January 3, 1864, Confederate General Samuel Jones' cavalry attacked a battalion of the Sixteenth (Companies I, K, L, M) under command of Major Charles Beers near Jonesville, Virginia. Over 200 men of the Sixteenth were captured.[215]

---

[213] This name is spelled as either Schaumbeck or Schambeck

[214] Dyer, Frederick H., *A Compendium of the War of the Rebellion,* Thomas Yoseloff, New York, 1959, volume III, p 1034.

[215] O.R. Series I—Volume XXXII/1 [S#57] p57. The battle and subsequent interment of the men at Andersonville is detailed by in the book by John Mc Elroy of company L cited above. James D. Fox of Company H also details his capture and imprisonment. Fox, James D. *A True History of the Reign of Terror in Southern Illinois, A Part of the Campaign in Western Virginia and Fourteen Months of Prison Life,* J.D. Fox, Publisher, Aurora, Ill., 1885.

16TH ILLINOIS CAVALRY REGIMENT

Although there are no colors known to survive this cavalry regiment, the regiment did receive at least one guidon. "Yesterday afternoon, at 3 o'clock, Thielemann's cavalry company broke camp to march to St. Louis. Captain Webb of the U.S. Army inspected the company, called the roll, and signed in a few men who had enlisted in the meantime. Mr. A. Hesing presented the men a nice cavalry guidon with a fiery and patriotic speech, whereupon Captain Thielemann answered in short, matching words. Thereupon the company was set in motion, and when it rode through the lined companies of Knobelsdorf's regiment, there were hurrahs without end."[216]

Frederick Schambeck's Independent Cavalry Company, afterward *Hoffman's Dragoon's* Company B became Company C of the 16th Illinois Cavalry. This company was presented with a stand of colors in 1861. "Capt. Schambecks Dragoons were in readiness yesterday morning at 9 o'clock at the store of H.A. Kohn & Bro., Lake Street, and there received their uniforms. The company were then marched to the Washington street front of the Court House, where Lieut. Gov. Hoffman presented them with a fine stand of regimental colors of silk trimmed with gold fringe. On one side was "The Union forever—National Union." On the other, "Illinois Mounted Co. B Hoffman Dragoons." The Lieut. Governor delivered a neat address to the soldiers, praising them for their discipline and soldier-like appearance. The colors were received on behalf of the company by Arno Voss, Esq, a former member of the *Washington Independent Company* who responded in a very brief speech."[217] The whereabouts of these flags today is unknown. Many of the flags supplied to Chicago regiments were brought back to Chicago after the Civil War and were lost in the Chicago fire of 1871. No other colors carried by the 16th Illinois Cavalry Regiment are known to survive.

---

[216] *Tagliche Illinois Staat-Zeitungm* 22 August, 1861. (Provided by John Thielemann)

[217] *Chicago Tribune*; June 7, 1861

17th Illinois Cavalry

"One day, we met a large company of Indians crossing the "trail" in front of us; the officers stopped them when we had a chance to see and examine their equipments(sp), such as bows, scalping knives, arrows, tomahawks, lassoes &e. I remember Lieut. Alfred B. Amet had an arrow in his hand and tried to make the Indian understand that he would like to take it home with him, but no, they must have it back, and putting spurs to their ponies, they left in a hurry."[218]

The 17th Illinois Cavalry Regiment was organized in St. Charles, Illinois. Eight companies were mustered on January 28, 1864, and 4 companies were mustered February 12, 1864. Since this regiment was organized and mustered later in the War, several of its officers had served in other regiments. Colonel John Beveridge[219] had originally mustered as Major of the 8th Illinois Cavalry in September 1861. Other officers of the 17th included Lt. Col Dennis Hynes, who enlisted in the 8th Illinois Cavalry as a private and rose to the rank of captain; Major Hiram Hillard was formerly an officer in the 1st Wisconsin Cavalry; Major Lucius Matlock had been Chaplain in the 8th Illinois Cavalry, [220]but joined the 17th Cavalry as a Major. The men from the following counties formed the regiment; Company A from Cook county; Company B from Cook, McHenry and Kane counties; Company C from DeKalb county; Company D from DeWitt, Livingston and Schuyler counties; Company E from Champaign, Cole and Pulaski counties; Company F from Jo Daviess county; Company G from McHenry county; Company H from DuPage county; Company I from Lake county; Company K from Cook and Winnebago counties; Company L from Cook and Rock Island counties; and Company M from Ogle County. In May 1864, the regiment was ordered to St. Louis and then to Jefferson Barracks where they were supplied with horses and equipment. The regiment was divided into three battalions: 1st Battalion, companies A, B, C and D; 2nd Battalion, companies E, F, G and H; 3rd Battalion, companies I, K, L and M and in May was ordered to Alton, Illinois to relieve the 13th Illinois Cavalry in guarding at the Alton Military Prison. On August the 1st Battalion was ordered to St. Louis and then to the District of North Missouri where they provided escort and provost guard duty at Weston and St. Joseph Missouri until June 1865. Even though the rebel army under Price invaded Missouri in September 1864, the 1st Battalion was not involved in any major battles. In June 1864 the 2nd Battalion was ordered to Glasgow Missouri; this site was amid the strongholds of rebel sympathizers and guerilla bands. The Battalion was responsible for the protection of the telegraph lines within 60 miles of Glasgow and worked day and night to assure protection of these lines. In September 1864, due to reports of rebel Colonel Thornton's presence, with 1,500 men in the area, the Battalion was ordered north and west of Glasgow in search of the enemy.

---

[218] Carpenter, Edwin A., History of the17th Illinois Cavalry Volunteers, 1886. Apparently originally published weekly in a local newspaper. After the CW Carpenter became a physician and practiced in Stephenson County, Illinois.

[219] Colonel Beveridge was a native of New York. He was admitted to the Illinois Bar in 1851.At the outset of the Civil War became Captain Company F, 8th Illinois Cavalry. Two years later he returned to Illinois and recruited the 17th Illinois Cavalry and became Colonel of the Regiment. After the war he was elected sheriff of Cook County, Republican delegate to the US House of Representatives and in 1873 the Governor of Illinois.

[220] Lucius Matlock was born to a Quaker family in Baltimore in 1816. In 1831 he became a traveling Methodist minister but his credentials were revoked in 1837 due to his Abolitionist beliefs. He continued to be an outspoken Abolitionist, left the Methodist Church for the Wesleyan Methodist Church and in 1860 he became President of Illinois Institute (now Wheaton College). He joined the 8th Illinois Cavalry as Chaplain but one year later returned to Wheaton to recruit soldiers for the 17th Illinois Cavalry. He then joined the 17th Illinois Cavalry as a major and held that rank throughout the war.

Despite searching an area of over 300 miles, no evidence of the enemy was found, and the 2nd Battalion rejoined the 17th Illinois Cavalry Regiment at Jefferson City, Missouri. The 3rd Battalion remained at Alton until September 1864 when it was then ordered to Rolla, Missouri to join the 2nd Battalion. Early in October, the 17th Illinois Cavalry (2nd and 3rd Battalions together with Companies C and D of the 1st Battalion) under command of Colonel Beveridge aided in the defense of Jefferson City. In late October the Seventeenth was engaged with the enemy at Hickman's Mills and two days later was in the Division that captured Major General Marmaduke, Brigadier General Cabel, ten rebel cannon and more than a thousand prisoners with their arms. By mid-November 1864 the Regiment had returned to Rolla. In January 1865 the Regiment was in Pilot Knob and then ordered to Cape Girardeau, Missouri. From Cape Girardeau the Regiment was ordered to Kansas where it occupied detached posts along the plains. The Regiment mustered out Companies "C," "E," "I" and "M," November 23, and rest of Regiment December 15 to 22, 1865.[221]

Following the end of the Civil War a national and a standard were returned to the State of Illinois Adjutant General; at this time only, the national flag exists.

This national flag has 34 stars arranged in a pattern consistent with manufacture in Philadelphia. It measures 70 in. on the hoist by 77 in. on the fly.

No other flags carried by this regiment are known to survive.

[221] *Report of the Adjutant General of the State of Illinois,* H.W.Rokker, State Printer and Binder, Springfield, Ill, 1886k, Volume VIII, pages 562-93

Infantry Colors

Each Civil War Infantry regiment carried two flags; by military definition, these were called the regimental colors--a national color and a regimental color. According to the 1861 Army Regulations:

"Each regiment of Infantry shall have two silken colors. The first, or the national color, of stars and stripes, as described for the garrison flag; the number and name of the regiment to be embroidered with silver on the center stripe. The second, or regimental color, is to be blue, with the arms of the United States embroidered in silk on the center. The name of the regiment in a scroll underneath the eagle. The size of each color to be six feet six inches fly and six feet deep on the pike. The length of the pike, including the spear and ferrule, to be nine feet ten inches. The fringe yellow; cords and tassels, blue and white silk intermixed."[222]

National flag with Regimental designation on the center stripe and battle honors.

Regimental flag with Regimental designation on scroll beneath the eagle.

---

[222] "Colors of Infantry Regiments" *U.S.Army Regulations,* Article L. Flags, Colors, Standards, Guidons, p.475;1438.

After February 1862, also inscribed upon the colors of the regiment were the names of the battles in which the regiment "have borne a meritorious part." These "Battle Honors" could be inscribed on the flag itself or on a ribbon attached to the staff of the flag.

In addition to the colors described above, infantry regiments also carried camp, guide, or flank markers (also called guidons) --the terms are used interchangeably. These flags were usually eighteen inches square and mounted on an eight-foot staff. Early in the War, they were white with the number of the regiment in the center. Later in the War, they were similar in design to the national flag. These colors were used in camp to mark the color line and were used as markers during the dress parades[223]. In the Illinois flag collection, there are some Infantry regimental guidons with a swallowtail pattern, like that used by the cavalry.

Guidon of the 7th Illinois Infantry Regiment (18" x 18"

[223] Madaus, Howard Michael. "Camp Colors, General Guide Flags and Flank Markers in the United States Army, 1861-1865: Part I-IV," *Military Collector & Historian,* 48, No.2 (Summer 1996), 48, No.3 (Fall 1996), 48, No. 4 (Winter 1996), 49, No.1 (Spring 1997).

Guidon of the 96th Illinois Infantry Regiment--swallowtail pattern (24" x 30")

7th Illinois Infantry

"Bring out the old flags; their rents, and their scars,
Are as dear to our hearts as their stripes and their stars,
Wave your old flags, men, point them towards home,
Proudly in victory and honor, we come."[224]

"Let its stained and tattered mass,
Tell the story of the terror and the glory
Of the battle of the Allatoona Pass."[225]

Illinois sent six infantry regiments to the Mexican War. Thus, when the Civil War began, the first infantry regiment raised was numbered "7." The 7th was originally organized as a "three-month" regiment but was reorganized and mustered into the Union Army for three years' service on July 25, 1861, under the command of Colonel John Pope Cook[226]. The men of this Regiment served from nine Illinois counties: Company A from Kane County--*Elgin Continentals*; Company B from Coles County; Company C – *Aurora Guards* from Kane, DuPage, and Kendall Counties; the towns of Litchfield[227], Lincoln, Atlanta, and vicinity sent men to Companies D -*Litchfield Guards*-and Company E -*Lincoln Guards;* Company F -*Bunker Hill Company* primarily came from Macoupin County; Companies G and I were filled with men from Sangamon and Fayette Counties; Company H from Logan County and Company K – *Carlinville* or *Macoupin Invincibles* came from Macoupin County. The Seventh was the only regiment in the Union Army that purchased its own weapons and armed itself with the 16-shot Henry rifle. Early in 1862, the Regiment was sent to Fort Henry, then on to Fort Donelson, where it took part in the final charge on the Fort. The Seventh fought April 6 and 7 at Shiloh, then April 29-May 30 at the Siege of Corinth. Afterwards they pursued the enemy south to Booneville, Mississippi. In June they were on guard at Corinth and remained there until October 3 and 4, when they fought in the battle of Corinth. In 1863 the men skirmished in Alabama at Great Bear Creek and at Cherokee; for the remainder of the year, the Regiment scouted, patrolled, and skirmished in Tennessee. January 1864 the Seventh re-enlisted as veterans and reinforced with 200 recruits, became a mounted regiment patrolling the Tennessee River. They fought at Florence and Sweetwater and, in June, were dismounted and moved to Chattanooga, Tennessee. In July, the Seventh was ordered to Tifton, Georgia, then on to Rome, Georgia where they remained until October. The Regiment saw action at Etowah River on September 15; in a savage battle at Allatoona Pass on October 4 and 5, the Seventh held off a rebel Division of six thousand men. In November, the unit was remounted for Sherman's March to the Sea. Over the next five months, the Regiment fought at Savannah, Georgia, and in South Carolina at Hickory Hill, Salkehatchie Swamps, South and North Edisto Rivers, Columbia, Cheraw, and Florence. March 19-21, the Seventh was

---

224 Ambrose, Daniel Leib. *History of the Seventh Regiment Illinois Volunteer Infantry, from its First Muster into the U.S. Service, April 25, 1861, to its Final Muster Out, July 9, 1865.* Springfield, Ill., Illinois Journal Company, 1868. p312, attributed to Lt. Solomon F. Flint.

225 AGR, *History of the Seventh Infantry,* Vol. 1, p380

226 35 year old Colonel Cook was a member of the Illinois bar, elected mayor of Springfield in 1855, Sheriff of Sangamon County in 1856 and was the captain and commander of the *Springfield Grays.* He was the grandson of Ninian Edwards,

227 Note: this town of Litchfield was/is in Montgomery County. Today there is a second town of Litchfield in Kankakee County.

engaged at Bentonville, North Carolina; they marched to Goldsboro, then on to Raleigh on April 14. Following Johnston's surrender they marched to Washington DC via Richmond, Virginia to participate in the Grand Review on May 24. The Seventh Illinois Infantry Regiment mustered out on July 9, 1865, at Louisville, Kentucky.[228]

There were four National flags returned to the Governor by the 7th Illinois Infantry. One of the National flags of the Regiment is inscribed with battle honors. Although several of the battle honors on this flag can no longer be read because of deterioration of the silk and the paint, the original battle honors read as follows: TOWN CREEK, FORT DONELSON, SWALLOW BLUFF, PITTSBURG LANDING, BUTLERS CREEK, SIEGE OF CORINTH, WHIPPY SWAMP, NEUSE BRIDGE, CORINTH, ALLATOONA PASS, COLUMBIA, RALEIGH, SAVANNAH, BENTONVILLE. *Based on the shape of the canton and the arrangement of the stars, this flag was probably manufactured in New York by Paton & Co, Brandon Bros, or Albert Ertle.*

This national flag was photographed with two guidons lying on top of the flag. The National flag now measures 43 ½" x 72" but was probably a full-size National flag of the Regiment.

[228] AGR Vol 1, 378-81

The center portion of this 74 x 68" silk regimental flag appears to have been cut out.

Ambrose, in his book, relates the presentation of the first set of the Regiment's colors: "On the eleventh (July 11, 1862), our regiment received a new stand of colors. The colors we carried through the battles of Fort Donelson and Shiloh are now so mutilated that we are unable to carry them any longer. They will be sent to Springfield to be laid away in their glory, and while they thus rest from battle and storm, back with Illinois' great loyal people, may they ever remember as they gaze upon its hallowed ribbons, the noble ones who went down while its rents and scars were being made; whose lamps of life flickered out while wrathful storms were sweeping along the shores of the Cumberland and Tennessee, and over the fields of Mississippi; remember that while it was swung in its glory the noble hearts of Captains Mendell and Ward, Lieutenants Myers and Estabrook, Sergeants Wheeler, and Mitchell, Corporals William Boring, Seth Hamilton and Nixon, Privates Charles Newton, John Fifer, Andres McKinnon, John Teft, Richard Lamherdt, Isaac Britton, John H. Duff, John Gibland, Ole Porter, Peter Miller, John H. Hopper and others ceased their loyal throbbings forever. They loved that old banner, made so hallowed on fields of blood. It was the pride of their hearts; for it they lived, for it they died. Those shot-riven folds will speak to the loyal people in a silent language, telling them a thrilling story--a story, the letters of which have been written in blood. We send them back to the good people from whence they came, hoping that the story they tell will find an entrance into their loyal hearts and cause it to start a tear to the memory of those who went down beneath its folds."[229]

---

[229] Ambrose, p82-3

Two guidons or flank markers were retired by the Seventh. Each measure approximately 18 x 18".

Two other flags were presented to this Regiment, and perhaps the two flags were identified only as national "company" flags in the Illinois collection.

Another presentation was reported by the *Alton Telegraph* under the headline "Flag Presentation."[230] "On Tuesday, a large delegation arrived in the City from Carlinville, Macoupin County, for the purpose of presenting a flag to the *Macoupin Invincibles,* one of the companies composing Col. Cook's Regiment stationed here. The presentation took place in the City Hall and was witnessed by a large number of our citizens. The flag is a very beautiful one made of silk and splendidly ornamented. It bears the name of the company upon it in gold letters. It was made and presented by the ladies of Carlinville.... The occasion was one of unalloyed pleasure and called for many expressions of high resolve to uphold and maintain the "Stars and Stripes" at all hazards." The *Macoupin Invincibles* became Co. K of the 7th Illinois Infantry.

A flag was presented to the men of the *Aurora Guards* who became Company Co. C. of the "three months" 7th Illinois. The flag presentation is reported in two Aurora newspapers[231]. Capt. Lawyer accepted the flag from Mrs. Hawkins on behalf of the "*Guard.*" The flag itself is not described but was probably a national flag. There is no further history of this flag.

The following are probably company flags of the Seventh Illinois Infantry.

230 *The Alton Telegraph,* May 10,1861.

231 *The Beacon News (*Aurora), May 23, 1861 and the *Aurora Beacon,* May 30, 1861.

The first flag measures 69 x 78" and bears no inscription.

The second measures 52 x 76" and also bears no inscriptions.

No other flags carried by this Regiment are known to survive.

8th Illinois Infantry – "*Dick Oglesby's Regiment*"

"Flag of the good, the brave, the free,
Wide floating now or land and sea;
Who, who, would dare thy colors mar,
Who, who would pluck from thee a star?

Fling out thy red, thy white, thy blue,
We have fair flag thy colors true:
O'er all our land thou yet shall wave,
When traitors fill a traitor's grave.

Gem of the breeze! Thy stars shall gleam
When I have passed life's turbid stream.
Yet all I'd ask to mark my grave,
Is that thy folds may o'er me wave."[232]

The 8th Illinois Infantry Regiment mustered in as a "three-month" regiment on April 25, 1861. The men of this regiment had volunteered from Macon County (Cos. A and B -- *Decatur Light Infantry);* Coles County (Co. C – *Coles County Volunteers*); Richland County (Co. D --*Richland Volunteers*); Fulton and Peoria Counties (Co. E – *Peoria Blues*); Tazewell County(Co. F—*Pekin Guards*); Alexander and Pike Counties (Co. G –*Pike County Union Guards*); Fayette County(Co. H *Fayette Guards*); Lawrence County (Co. I - *Lawrence County Volunteers*); and McLean County(Co. K-*Bloomington Infantry*). Richard J. Oglesby[233] who was later to become Governor and then Senator from Illinois commanded the Regiment. On July 25, 1861, the regiment was reorganized at Cairo, Illinois, and mustered as a three-year regiment. In February 1862, the Eighth fought at Fort Henry and at Fort Donelson, Tennessee, and in March moved to Pittsburg Landing. They fought at the Battle of Shiloh on April 6 and 7 and in the siege of Corinth, Mississippi, from April 29 to May 30. In June, the Regiment was ordered to Jackson, Mississippi, to guard the rail stations at Toone's and Medon, Mississippi, and remained there until October when ordered to LaGrange, Tennessee. In January 1863, the men marched to Memphis, Tennessee, were moved to Lake Providence, Louisiana, in February, and remained there until April. In the subsequent Vicksburg campaign, the Eighth fought at Thompson's Hill and Port Gibson on May 1, Raymond on May 12, Champion's Hill on May 16, and in the siege until the surrender of Vicksburg on July 4. The Regiment remained on garrison duty at Vicksburg until July 1864. During that time on March 24, 1864, the regiment re-enlisted as a veteran regiment. The remainder of 1864, the Eighth skirmished with the enemy in Mississippi and Tennessee and, in January 1865, arrived at New Orleans, Louisiana. February 1865 the Eighth was transported to Dauphine Island in Mobile Bay and, in March, campaigned against Spanish Fort. On April 9, 1865, the regiment took part in the charge on Fort Blakely, Alabama, and was the first regiment to plant its colors on the earthworks, but in the assault,

---

232 *To The American Flag,* Canton Weekly Register, March 9, 1863

233 Kentucky born,Colonel Oglesby, was orphaned and raised by his uncle in Decatur, Illinois. He served in the 4th Illinois Volunteer Infantry Regiment in the Mexican American War; he studied at Louisville Law School in 1849 and then left for California in the Gold Rush. Oglesby was a member of the Republican Party and is said to be the one who applied the nickname "Railsplitter" to Abraham Lincoln. After the war he served as Governor of Illinois and as a Senator from Illinois.

color-bearer Sergt. Isaac H. Simonson was killed. Two men of the Eighth Illinois Infantry received the Medal of Honor for their actions on that day. Lloyd Wheaton, Lieutenant Colonel of the Regiment—"Led the right wing of his regiment and, springing through an embrasure, was the first to enter the enemy's works, against a strong fire of artillery and infantry[234][235]." and German-born Captain Henry Miller, Company B, of Decatur, received the Medal for "Capture of Flag." On April 12, the Regiment crossed Mobile Bay and was the first to enter the city of Mobile. In June 1865, the Eighth was ordered to Marshall, Texas and was there until the fall of that year when it was ordered to Alexandria, Louisiana. The men remained there until May 4, 1866, when the regiment was mustered out and sent to Springfield.[236]

The 8th Illinois Infantry returned both a National and Regimental color to the State, but they are now only shreds of silk.

The National flag measures 67" x 67". The women of Decatur in Macon County sewed the first flag carried by the Regiment. "As soon as the 8th Infantry was organized, work was stated on making a regimental flag. More than 100 women gathered in the parlor of the Revere House to work on the project. Only three sewing machines were available, and these were saved for the long seams. The women were so eager to work on the flag that it was ruled that no one should be allowed to finish a star. Some stars had each point done by a different loving hand. The Decatur companies were assembled at Central Park, where the flag was presented by Mrs. Cal Allen while Miss Hattie White held the staff. Col. Oglesby made an eloquent response."[237] The above national flag may or may not be the flag sewn by the women of Decatur.

[234] *Civil War Medal of Honor Winners,* published by the Civil War Centennial Commission of Illinois, Springfield, July 1962

[235] Beyer, WF and Keydel, OF, eds; *Deeds of Valor How America's Civil War Heroes Won the congressional Medal of Honor, reprinted by Longmeadow Press, (1993). P533-37.*

[236] AGR vol.1, p422-424.

[237] Banton, OT, Editor, *History of Macon County 1976,* printed by Macon County Historical Society, Decatur, Il. (1976), page 180.

This Regimental flag measures 70" on the fly x 56".on the hoist. Very little is left of the flag due to deterioration of the silk.

A flag was also presented to Captain Dennison of Company E. "During this day (Saturday) Captain Dennison's company were mustered in front of the residence of Senator Bestor on Main Street, opposite the Court House, where a splendid flag, the handiwork of Mrs. George C. Bestor and Mrs. William S. Gregg, was presented them. Prayer was first offered by Rev. J. M. Waite; the presentation speech was made by Mr. Bestor, the Star Spangled Banner was sung as a solo by Miss Harris; the audience joined in the chorus; the closing prayer was made by Rev. S.J.G. Worthington of the First Methodist Church after which three cheers were given for the Union, three for the ladies, three for the stars and stripes and three for the *Peoria Volunteers*."[238] There is no further record or description of this flag. No other flags carried by this Regiment are known to survive.

[238] *Historical Encyclopedia of Illinois and History of Peoria County,* Volume II, Chapter XXV, Munsell Publishing Company, Chicago and Peoria, 1902. This presentation occurred on April 20, 1861.

9th Illinois Infantry

"Then conquer we must, when our cause is so just,
And this be our motto—In God is our trust;
And the Star Spangled Banner in triumph shall wave,
O'er the land of the free and the home of the brave."[239]

The Ninth Illinois Volunteer Infantry was one of six of Illinois' "three-month" regiments organized at the first call of President Lincoln. This "three-month" regiment was mustered on April 26, 1861, and was organized and mustered as a three-year regiment on July 26, 1861, under the command of Colonel Eleazer Paine.[240]

The men of the Ninth volunteered from Madison, St. Clair, and Mercer counties (Cos. A, B, C, D, E, and F), Alexander County (Co. G), Montgomery County (Co. H -*Hillsboro Volunteers*), Madison County (Co I – *Edwardsville Volunteers*) and Pulaski County (Co. K).

In September 1861, General Grant took possession of Paducah with the Ninth and Twelfth Illinois Infantry and remained there until the end of the year. In 1862 the regiment operated against Forts Henry and Heiman and fought at Fort Donelson[241],[242] on February 12-16. After moving to Pittsburg Landing, Tennessee, in March, the Regiment fought in the battle of Shiloh on April 6 and 7. From here, they marched southward and were at the siege of Corinth, Mississippi, from April 29 to May 30. They remained on duty at Corinth until March 1863, and during that time, they fought in the battle of Corinth on October 3 and 4 and were on an expedition to intercept Confederate General Forest in early January. On March 15, 1863, the regiment was mounted and remained so for the rest of the War. The One Hundred and Twenty-Eighth Illinois Infantry was consolidated with the Ninth Illinois Infantry on April 1, 1863, with the designation of the combined regiments being the Ninth Illinois Infantry. On August 3, 1863, 105 deserters from many different regiments were sent from Fort Pickering at Memphis, where they had been held as prisoners and added to the Ninth. The remainder of 1863 saw the Ninth operating in Alabama, Tennessee, and Mississippi. In May 1864, the Ninth Infantry joined Sherman's Atlanta campaign. Their activities included guarding the trains of the 15th and 16th Army Corps from Huntsville, Alabama, to Chattanooga, Tennessee, and scouting and skirmishing on the flanks of the contesting armies during the Atlanta campaign. The Regiment was engaged in battles at Buzzard Roost, Snake Creek Gap, Resaca, Dallas, New Hope Church, Allatoona Hills, Rome Cross Roads, Kenesaw Mountain, Peach Tree Creek, Decatur, Atlanta, Jonesboro, and Lovejoy Station. On September 9, 1864, 40 veterans

---

[239] Belleville, Illinois, *Belleville Democrat*, June 8, 1861, excerpt from a resolution of the "ladies of the town of Mascoutah" by Alwine Postal, Pres't, upon presentation of a flag to Company D of 9th Illinois.

[240] Fifty five year old Ohio born Paine graduated in 1839 from the United States Military Academy and served briefly in the Seminole Indian War in Florida; he passed the bar exam in Ohio in 1843. In 1848 he moved to Monmouth, Illinois where he practiced law and became friends with Abraham Lincoln. In 1864 a military commission found Paine guilty of extortion, unjust taxation,fencing stolen goods and immorality. (Wikipedia)

[241] Private John Rank, Company F,is credited with capturing an enemy flag at Ft. Donelson . The flag is thought to be from an Alabama unit but is without attributed identification. The flag is silk with silk fringe and measures 24" on the hoist by 33" on the fly with eight alternationg stripes—from top to bottom—white, blue, white, red, white, blue, white, red. It is edged on three sides with a pink fringe; the canton is solid dark blue. (Personal communication from Greg Biggs.)

[242] The Firelands Historical Society in Norwalk, Ohio has a national flag that was alledegedly awarded to Captain Webb for actions of the Ninth at Fort Donelson.

of the Ninth Illinois Infantry, along with 105 deserters that had been added to the regiment in August 1863, plus veterans and recruits of the Twenty-Seventh Illinois Infantry, formed the Ninth Illinois Infantry Consolidated Regiment. Following the surrender of Atlanta, they pursued Hood into north Georgia and Alabama then accompanied Sherman on the march to the sea. Following the siege of Savannah in December 1864, the Ninth moved north and fought at Whippy Swamp Creek, Salkehatchie Swamp, at the Edisto River bridges, Columbia and Florence, South Carolina, and in North Carolina at Bentonville. The Ninth was at the surrender of Johnston's army near Raleigh. The Regiment then marched via Richmond to Washington, DC, and marched in the Grand Review on May 24. The Regiment was mustered out on July 9, 1865, at Louisville, Kentucky. [243]

The Adjutant General's records show that the regiment originally retired a National and Regimental flag along with eight guidons. At the time of the most recent flag inventory, there were no guidons noted to be from the 9th Infantry; however, there were eight unidentified guidons. No other flags carried by this regiment are known to survive.

The shield-shaped star pattern in the above national flag is not typical of flags distributed by the Quartermaster Corps, nor is it typical of flags from other known flag manufacturers; therefore, this flag was probably homemade. It measures 49" on the hoist by 62" on the fly; there are no inscriptions on the flag.

The Belleville Weekly Advocate[244] reports the delivery of a flag by a "committee of gentlemen of our town" to Camp Defiance in Cairo. "The generous and truly patriotic ladies of Mascoutah, having made a splendid silk flag for the 9th Regiment Illinois Volunteer Militia, appointed the above-named committee to execute the very pleasant task of presenting the flag.....On the 24th, at the dress parade, which came off at 6 o'clock P. M. the committee, in behalf of the ladies of the town of Mascoutah, presented a regimental flag to the Ninth Regimental Illinois Volunteer Militia, Col. Paine commanding. The presentation and remarks by H. C. Fike were most patriotically replied to by private Wm. Orr, Esq., and Lieut. J. A. Cos. At the suggestion of Col. Paine, three rousing cheers were given for the stars and stripes, three for the ladies of Mascoutah, and three for the committee of gentlemen...." Perhaps this is the flag shown here with the shield-shaped star pattern.

---

243 AGR, Vol 1, p452-5.

244 Belleville, Illinois. *Belleville Weekly Advocate,* June 14, 1861

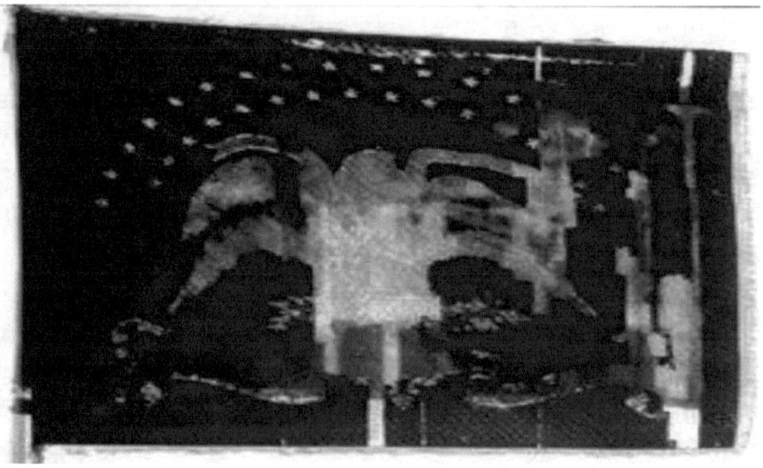

The Regimental flag measures 47" on the hoist x 69" on the fly and is probably of Quartermaster issue. This flag is not the presentation flag described below, since the flag below is described as having 13 stars above the eagle.

In June 1861, a flag made by the ladies of Belleville was presented to the 9th Illinois Infantry in Cairo, Illinois. In the Belleville Weekly Advocate,[245] the details of the flag and its presentation are given. "The patriotic ladies of Belleville have been engaged for some time past in preparing a flag for the 9th Regiment Illinois Volunteers. On Monday, an opportunity was given to the citizens to see it at the house of Sharon Tyndale Esq., and on Tuesday, it was duly forwarded to Cairo. The Flag is composed of a field of rich blue silk in the centre of which is embroidered the armorial device which constitutes the obverse of the great seal of the United States, consisting of a shield of thirteen stripes upon the breast of the American Eagle, displayed holding in his beak a scroll of white satin, inscribed with the motto "E PLURIBUS UNUM;" in his dexter talons an olive branch and in his sinster a cluster of thirteen arrows. Over the head of the eagle is a cluster of thirteen stars in the form of a semi-circle, and in his talon is also held a scroll on the obverse of which appears the motto, "Illinois U. S. Volunteers", and on the reverse, "Return Victorious." The eagle is embroidered in gray silk, and the stars and mottoes are worked in gold, all in the most skillful manner. It cost $1,10 and is the work of Mrs. Elliott of this city and not of St. Louis, as stated in the *Dispatch*." In the Sept. 9, 1864, issue of the Belleville Advocate, there is a summary of the history of the 9th as written by Rev. Marion Morrison. At the end of the history is the statement, "As a remarkable fact, we see it stated that Mr. Heimberger, the Color Bearer, carried the flag through every battle, and came off uninjured, and has now returned it to the kind and patriotic ladies who presented it to the regiment."

From this, I presume that the flag was returned to the ladies of Belleville. Its whereabouts today is unknown. There is another description of a flag in the Belleville Democrat[246], June 29, 1861. "We yesterday visited the house of Mr. Tyndale for the purpose of seeing the Flag which the ladies of this city have been preparing for a presentation to Col. Paine's Regiment. The Flag is the handiwork of a lady from St. Louis (at present residing here) named Mrs. Elliot and reflects great credit upon that lady's artistic skill in embroidery. It is composed of a field of heavy blue silk, in the centre of which a large grey Eagle is embroidered--in natural colors, with silk--holding in its beak a scroll of white satin upon which is engraved the national motto, while it grasps in its talons the insignia prescribed by the

[245] Belleville, Illinois, *Belleville Weekly Advocate,* June 28, 1861

[246] Belleville, Illinois *Belleville Democrat,* June 29,1861

regulations of the United States for regimental flags, another scroll streaming from these talons displays the words "Illinois U. S. Volunteers" on one side, and on the obverse "Return Victorious." We have said that the Eagle is embroidered in natural colors, with silk, and what is not a little remarkable, one side is an exact fac-simile of the other, while the outline of that noble bird is drawn with such accuracy, as the artist alone can command. The mottoes are embroidered in gold in fine style, while the arch of stars, which surmounts the whole, is a perfect chef-oeuvre of art....."

The “three-month” regiment was presented with a national flag in April 1861. “The following letter written by Captain Tucker of the Alton City Guards has been handed to us by a friend. The letter was prompted by the reception of a flag from some of our patriotic young ladies. May these ladies all soon find themselves in union with some good, loyal man and ever have the glorious Stars and Stripes to wave over them.”[247] There is a following letter of acceptance signed by B.W. Tucker, Capt. Alton City Guard. Capt. Tucker was from Company G of the Ninth, but he and most of the Company did not re-enlist. However, Company D of the Tenth Illinois Infantry “three-year” regiment, also came from Alton, Illinois, and in the Illinois Flag collection is a flag from the Tenth inscribed from the “Ladies of Alton.” Although the flag of the Tenth is preserved, the flag presented to the Ninth is apparently lost. No other flags carried by the Ninth Illinois Infantry are known to survive.

---

[247] Alton, Illinois. *Alton Telegraph,* May 3, 1861

10th Illinois Infantry

"Cheer up, cheer up; the night is past,
The skies with light are glowing.
Our ship moves proudly on, my boys,
And favoring gales are blowing.
Her flag is at the peak, my boys,
To meet the traitorous faction.
We'll hasten to our several posts
An immediately prepare for action.

CHORUS
Old Rosey is our man (repeat)
We'll show our deeds where 'er he leads,
Old Rosey is our man."[248]

The Tenth Illinois Volunteer Infantry was one of the first six Regiments formed in response to the Illinois Governor's order of April 16, 1861. This early "three-month" regiment was formed from the first four companies that reached Springfield--from Adams, Morgan, Knox, and Madison counties. Subsequently added were companies from Henderson, Clark, and De Kalb counties and three artillery batteries (the *Ottawa Artillery or Houghtaling's Battery* (Company F, Light Artillery), the *Springfield Artillery* or *Hopkins' Battery* (Company I, Light Artillery) and *McAllister's Battery* or the *Plainfield Artillery* (Company K, Light Artillery). On July 29, 1861, when the regiment mustered in for three years, the artillery companies were replaced. The three-year regiment was commanded by Colonel James D. Morgan[249] and was composed of men from Morgan County for Companies A and B -*Hardin Light Guards*; volunteers from Adams and Knox Counties filled Company C; troops from Madison County filled Company D; Henderson County supplied the men for Company E; troops from St. Louis filled Company F; Clark, Henderson and Lee Counties supplied the men for Company G; Company H – *Sandwich Guards*[250] came from DeKalb and Clark Counties; volunteers from Chester in Randolph County filled Company I and volunteers from Madison County comprised Company K. The troops were stationed in Mound City, Illinois in 1861 and moved to Bird's Point, Missouri in January 1862. The men saw their first action against Jeff Thompson's troopers near Sikeston, Missouri, in March. Later that month, the Tenth was engaged in the siege of New Madrid, Island No. 10, and Tiptonville. In April, the Tenth was ordered to Hamburg Landing, Tennessee, and engaged in the siege of Corinth, Mississippi April 29 to May 30. Until mid-June, the Regiment pursued the retreating enemy to Booneville, Mississippi; then were ordered to Nashville and fought in the siege of Nashville in September, October, and November. The Tenth Illinois Infantry remained at Nashville until July 1863. After assisting in

---

[248] Song popularizing General Rosecrans and sung by the Tenth Illinois Infantry and printed in the regimental history. Wilson, Captain Ephraim A., *Memoirs of the War,* W.M. Bayne Printing Company, Cleveland, (1893), p 174

[249] Fifty one year old Colonel James Dada Morgan was a Boston born merchant seaman. In 1834 he and his family moved to Quincy, Illinois where he opened a cooper shop and in 1837 joined the *Quincy Grays.* As a member of the *Quincy Grays,* he led a company of riflemen in the Illinois Mormon War. During the Mexican-American War, the *Quincy Grays* became Company G of the 1st Regiment of Illinois Volunteers and Morgan fought with that unit.

[250] Known as the "*Sandwich Guard.*"

driving Wheeler's Cavalry from the Sequatchie Valley, Tennessee, they crossed the Tennessee River (November 24) on pontoons to support Sherman's attack at Tunnel Hill and Mission Ridge. On January 1, 1864, the Regiment reenlisted as veterans. Following a 30-day furlough and with 200 recruits added, they moved to Rossville, Georgia, and accompanied Sherman's army in the Atlanta campaign. The Tenth fought at Rocky Faced Ridge, Buzzard's Roost Gap, Resaca, Rome, Dallas, New Hope Church, and Allatoona Hills in May; at Kennesaw Mountain, Pine Mountain, and Lost Mountain in June; and at Chattahoochee River, Peach Tree Creek, and Atlanta in July. In a flanking movement around Atlanta, the Regiment fought at Jonesboro and Lovejoy Station, then pursued Hood into Alabama in October. Finally, they marched with Sherman "to the sea," first fighting at Savannah, Georgia, then Pocotaligo, Edisto Rivers' Bridges, Banneker's Bridge, Cheraw, Fayetteville, and lastly, Bentonville, North Carolina. After Johnston's army surrendered, the Tenth marched through Richmond and Fredericksburg, Virginia, to Washington, DC, for the Grand Review on May 24, 1865. The Tenth mustered out on July 4, 1865, at Louisville, Kentucky.[251]

According to the records of the Adjutant General of the State of Illinois, the Tenth originally returned one National and one Regimental flag[252]. Currently, the collection contains only a national flag and a guidon. The National flag measures 68 x 77" and was presumably purchased and presented to the Tenth Regiment by the Ladies of Alton, Illinois

This flag is inscribed: "Presented to the 10th Ill. Vet. Inf. by the Ladies of Alton". There are no battle honors on this flag. A record of the presentation of this flag was not found.

[251] AGR, Vol. 1, p502-4.

[252] AGR, Vol. I, p159.

This swallowtail guidon is of the "stars and stripes" pattern and measures 25 ½" on the hoist by 36" on the fly.

There is a National flag at the Abraham Lincoln Presidential Library and Museum in Springfield, Illinois.[253]

On a tag is labeled "American Flag Made by the Union Ladies of Chester". The flag is "overall: 52 in x 80 in Cloth. The flag has 34 white stars on the blue field and six red and six white alternating horizontal stripes. Made by Union Ladies of Chester, Ill. and presented to Captain Morton McAtee of Co. I, 10th Illinois Infantry

No other flags carried by the 10th Illinois Infantry are known to survive.

---

[253] The photo of this flag and the information is supplied courtesy of the Abraham Presidental Library and Museum.

11th Illinois Infantry

It came at last; "Now, lads, the steel!"
The rushing hosts deploy;
"Charge, boys!—the broken traitors reel,
Huzza for Illinois!"
In vain, the rampart, Donelson,
The living torrent bars,
It leaps the wall; the fort is won,
Up goes the Stripes and Stars!
The proudest mother's eyelids fill,
As dares her gallant boy,
And Plymouth Rock and Bunker Hill
Yearn to thee—Illinois.[254]

The Eleventh Illinois Infantry was mustered for three months of service at Springfield Illinois on April 30, 1861. Like the other six Illinois "three months" regiments, it was remustered, July 13, 1861, for three years under the command of Colonel W. H. L. Wallace. [255] Soldiers for the Eleventh volunteered from the following counties-Stephenson county (Company A); Marshall county (Company B -*Marshall County Guards* and part of Company I); Marion county (Company C); Winnebago county (Company D—*Rockford Blues*); Effingham county (Companies E and G); Jasper county (Company F) and La Salle county (Companies H, and K -*Highland Rifle Company*- and part of Company I). The "three-month" regiment also had men from Fayette and Grundy Counties.

After initial garrison duties at Cairo, Illinois, the Regiment skirmished against Confederates at Charleston and Sikeston, Missouri. In February 1862, the Regiment moved against Fort Henry and, after its fall, took part in the siege against Fort Donelson. At Fort Donelson, the Regiment lost 329 men killed, wounded, or missing out of the 500 that had fought.[256],[257] In March, the men moved to Pittsburg Landing,

---

[254] Parker, H. I. *Proceedings of the First Reunion of the Eleventh Regiment Illinois Volunteer Infantry.* Ottawa, Illinois, Osman & Hapeman, 1875. p4. a poem entitled "The Capture of Fort Donelson" quoted from the Boston Daily Advertiser.

[255] Colonel William H Wallace was born in Ohio in 1821; in 1840, his family moved to Mount Morris, Illinois. AlthoughWallace had planned to study law with Abraham Lincoln he joined Theophilus Dickey's's law practice in Ottawa, Illinois and became licensed in law in 1846; that same year he joined the 1st Illinois Infantry and participated in the Mexicsn American War. In 1850 Wallace married Martha Ann Dickey, the daughter of Theolphilus Dickey who was to become the Colonel of the 4th Illinois Cavalry.

[256] In a letter dated February 17, 1862 at Fort Donelson, from L.D.Waddell, Capt., Company E, to his father, he writes, "My company was the color company, at which the rebels took particular aim; as fast as one man who carried it would be shot another would take his place, but the flag was brought through. Only one hundred and sixteen remains in the 11th regiment uninjured." *Knoxville Daily Register,* March 14, 1862

[257] "The colors of the regiment are riddled with shot and the staff was struck twice with bullets breaking off the spear at the top. The Color-Sergeant, McCaleb, was shot down and Color-Corporal Armstrong of Company H seized it and bore it from the field. Tell Helen Fisher that the pledge I made to her when she in behalf of the Ottawa ladies presented the flag to Co. I had been redeemed. The Eleventh lost nearly everything but their flag and their honor. It was the first flag planted on the inner fortification of Fort Donelson." Quoted from a letter written by W.H.L. Wallace to his wife from Fort Donelson on February 17, 1862. Wallace, Isabel, *Life and Letters of General W. H. L. Wallace*, Carbondale, Southern Illinois University Press, (reprint of 1909 edition) (2000) p.164-5.

Tennessee, and on April 6 and 7, fought in the battle at Shiloh.[258] From April 29 to May 30, the Eleventh participated in the siege of Corinth, then marched to Jackson, Tennessee, where they remained until August 1862. After further recruiting at Cairo, Illinois, the Regiment moved to Paducah, Kentucky, and spent the remainder of 1862 on Grant's Central Mississippi Campaign. 1863 saw the Regiment in Louisiana at Young's Point, Lake Providence, American Bend, Grand Gulf, the battle of Port Gibson, battles of Raymond and Jackson, Mississippi, battle at Champion's Hill, and the siege of Vicksburg. On April 23, 1863, the men of the One Hundred and Ninth Illinois Infantry Regiment were transferred to the Eleventh Illinois Infantry adding 589 men to the Regiment. Following the fall of Vicksburg, the Regiment moved downriver and occupied Natchez, Mississippi.[259] They remained there until mid-August, when they returned to Vicksburg and where they remained on guard duty until July 1864. During that time, the Eleventh was in action at Yazoo City, Black River Bridge, Benton, and Vaughan, Mississippi. In July 1864, the Eleventh was ordered on an expedition to Pearl River, then to Jackson, Mississippi and Clinton, Louisiana, on to the White River, and then in October to Memphis. In January 1865, the Regiment moved to Kenner, Louisiana, and was transported to Dauphin Island at the mouth of Mobile Bay, Alabama. There the Regiment participated in the siege of Spanish Fort in March and April, the assault and capture of Fort Blakely on April 9, and occupied Mobile Alabama on April 12. From Mobile, the Eleventh moved back to New Orleans, then to Alexandria, and finally Baton Rouge, Louisiana, to be mustered out July 14, 1865.[260]

Second Lieutenant James D. Vernay of the Eleventh Illinois Infantry Regiment, Company B, received the Medal of Honor. He had been detached as a volunteer to the steamer *Horizon* during the Vicksburg campaign. The medal was issued for "Served gallantly as a volunteer with the crew of the steamer *Horizon* that, under a heavy fire, passed the Confederate batteries."[261]

There are two flags of the Eleventh Illinois Infantry in the collection at the State of Illinois Military Museum.

This silk national flag measures 68" on the hoist by 74" on the fly. It has a square canton and 5 horizontal rows of stars. It was probably manufactured in New York for the New York depot.

---

258 Brigadier General Wallace was killed at the "Hornet's Nest" during the battle and command of the Regiment was passed to Thomas Ransom.

259 The commander of the Eleventh, Colonel Garrett Nevius, was killed near Vicksburg May 23, 1863.

260 AGR, Vol. 1, 542-3.

261 *Civil War Medal of Honor Winners.* published by the Civil War Centennial Commission of Illinois, July 1962. p5

This regimental flag measures 71" on the hoist by 74" on the fly. The fact that the stars above the eagle extend past the edges of the upper scroll and that the center of the lower scroll is raised with adornments at the end of the lower scroll is consistent with this flag being manufactured for the New York depot.

The national flag of the Eleventh mentioned above, as the first flag planted on the captured Fort Donelson, had been presented to Company I by the ladies of Ottawa. It was "a beautiful large silk one…and taken as the regimental flag, was used after the battle of Donelson at the burial of sixty-eight men in one grave. It was there presented to Colonel Wallace as a personal gift, which he sent home to his wife to be cared for by her. She considered it as one of her most valued possessions. It is now in the library at "The Oaks"—General Wallace's late home."[262] When the flag arrived in Ottawa, the staff was pitted by a rifle shot, the tip of the metal sheared, and the banner pockmarked with bullet holes.[263]Money was raised in Ottawa for the purchase of a new flag to replace the flag sent home to General Wallace's wife, Ann. In a letter to her husband on March 12, Ann describes the new flag, "The new flag is a beautiful blue silk one, but not the Stars and Stripes."[264] The description matches that of a blue silk Regimental flag. The Regimental flag pictured above may be the flag described by Mrs. Wallace. After the death of Isabel Wallace, the adopted daughter, the Wallace home—The Oaks—was purchased by the State of Illinois as a Museum. Unfortunately, in the early 1950s the State sold the mansion to private developers, and the furniture, Wallace papers, and flag were placed in "storage." Today, the whereabouts of any of this Wallace material, including the flag, is unknown.

No other flags carried by the 11th Illinois Infantry Regiment are known to survive.

---

[262] Ibid, Wallace, page 166

[263] Huffstodt, Jim, *Hard Dying Men, The Story of General W.H.L. Wallace, General T.E.G. Ransom, and Their "Old Eleventh" Illinois Infantry in the American Civil War (1861-1865),* Bowie, Md., Heritage Books, (1991), p84.

[264] Ibid Wallace, page 166

12th Illinois Infantry -- "*First Scotch Regiment*"

"From graves wherein they lie,
Shall one day burst a victory song,
The Union banner greets the sky,
And treason's flag shall trail ere long."[265]

The Twelfth Illinois Infantry was also known as the *First Scotch Regiment* and was organized at Springfield, Illinois. The "three-month" regiment was mustered on May 2, 1861, and the three-year regiment mustered on August 1, 1861, under the command of Colonel John MacArthur.[266] The men enlisted from Cook County (Companies A and parts of D and K); La Salle County (Company B); Vermilion County (Company C); Rock Island County (part of Company D); Edgar County (Company E); Jo Daviess County --*Galena Volunteers* and *Jo Daviess Guards*[267] (Company F); Perry County (Company G); Bureau County (Part of Company I) and Kankakee County (part of Company K). The "three-month" regiment also had men from Sangamon County.

On May 10, 1861, the Twelfth Illinois Infantry Regiment left Springfield for Cairo, Illinois, where it remained until September. From Cairo, the Regiment was ordered to Paducah, Kentucky, and remained there until February 1862, when they moved to and occupied Forts Henry and Heiman. The Twelfth saw its first fighting on February 12-16 at Fort Donelson. The capitulation of the Confederate forces at Donelson was the first Union victory in the West. In late February, the Twelfth was ordered to occupy Fort Sevier in Clarksville, Tennessee. From Clarksville, Tennessee, on March 1, the regiment embarked for Pittsburg Landing to fight in the battle of Shiloh on April 6 and 7. From April 29 to May 30, the Twelfth was in forces that advanced on and laid siege to Corinth, Mississippi. Following the fall of Corinth, they pursued the retreating enemy to Booneville, Mississippi, then returned to Corinth where they remained until September. Following skirmishes at Iuka and Burnsville in September the Regiment returned to fight in the battle of Corinth on October 3 and 4. During that battle, they drove the enemy from their works and captured a stand of enemy colors.[268] The Twelfth remained in Corinth until June 1863 when they moved to Pocahontas, Tennessee, to guard the *Chewalla to Corinth Railroad.* November 25, the Twelfth marched to Richland Station to guard the *Nashville & Decatur Railroad,* and on January 16, 1864, the Regiment was veteranized and sent home on furlough until March. Beginning in late spring, the Twelfth joined Sherman's Atlanta campaign and saw action at Resaca, Ley's Ferry, Rome Crossroads, Dallas, New Hope Church, Allatoona Hills, Marietta, Kennesaw Mountain, Nickajack Creek, Chattahoochee River, Decatur, the battle of Atlanta, Jonesboro, and Lovejoy Station. After the

---

[265] *Louisville Daily Journal,* November 17, 1861-from speech of Robert Harvey at the presentation of flags to the 12th Regiment

[266] Colonel John MacArthur was born Scotland in 1826, In 1849 he and his wife emmigrated to Chicago where he began working as a boilermaker. He joined the *Chicago Highland Guards* and at the outbreak of the war recruited men from the local Scottish community to serve in the 12th Illinois Infantry.

[267] Company F in the 3-month Regiment was also known as the *Jo Daviess Guards* and served as the color-company of the Regiment.

[268] "Private Daniel Osby, of Company K, captured a flag belonging to a Louisiana regiment and took the flag-bearer prisoner." O.R. Series 1, Volume 17, part 1, page 285.

fall of Atlanta, the Regiment marched to Rome, Georgia, in September and fought at Allatoona, Georgia, in October. Beginning in November, the Twelfth, Illinois "marched to the sea" and campaigned and fought at Savannah, Georgia, in 1864; moving northward, they fought at Salkehatchie Swamp, South Edisto River, North Edisto River, and Columbia, South Carolina. In North Carolina, in March, they fought at Bentonville, occupied Raleigh, and were at Bennett's House when Johnston surrendered. They marched in the Grand Review in Washington, DC, on May 24, then moved to Louisville, Kentucky, where they mustered out July 10, 1865.[269],[270]

The collection in Springfield at the State of Illinois Military Museum contains three flags belonging to the Twelfth Illinois—two national flags and one regimental flag.

This is one of the two National flags in Springfield from the 12th Illinois Infantry. At first glance, the star pattern is random, but closer inspection shows a St. Andrew's Cross with four stars between each arm of the cross. This star pattern is not typical of Federal manufacture. In his history of this Regiment, Gen. Augustus Chetlain writes: "Hon. E. A. Small, in behalf of the patriotic ladies of Galena, presented to the company a beautiful silk flag. His presentation address was eloquent and appropriate. I, as captain, accepted it in a brief speech in behalf of the company. This silk flag had been made by the ladies of Galena, headed by Annie Campbell (afterward Mrs. Gen. O. E. Babcock), who bought the material and who, on application to Captain U. S. Grant, was furnished with a design giving the dimension, etc."[271] Perhaps this flag is of the design furnished by Grant and presented by the ladies of Galena.

---

[269] AGR. Vol I, p574-6

[270] Dyer, Frederick H.. *A Compendium of the War of the Rebellion. Vol III*, p1049, New York, Thomas Yoseloff, 1959.

[271] Chetlain, Gen. A. L., *Recollections of Seventy Years*, The Gazette Publishing Company, Galena, 1899, pp 72-3. The presentation of the flag was originally reported in the *Weekly Northwestern Gazette*, April 29, 1861 and in the *Galena Daily Courier,* April 26, 1861.

The second national flag of this Regiment is pictured above. The stars are gold and arranged in six horizontal rows typical of flags manufactured by Gilbert and Hubbard of Chicago.

This may be one of the flags presented to the Regiment as described as follows. From the Louisville Daily Journal[272:] "When the orator of the occasion, Mr. Robert Harvey, mounted the stand the flags were for the first time unfurled since leaving Chicago and consisted of a national flag of the finest quality of silk, a regimental flag and a complimentary flag to the Colonel and a majority of the officers of the regiment who are of Scotland's noble sons. The regimental colors is composed of blue silk, regulation size, finest texture, fringed with yellow in the center of which was a gilt spread-eagle from the beak of which floated a ribbon upon which were inscribed the words, "E Pluribus Unum;" the eagle was encircled with thirteen stars over which in a half circle appeared a rich halo and underneath the eagle in gilt letters were the words "12th Reg. Ill. Volunteers." The other flag (complimentary) was of white silk with red border and yellow fringe ornamented in the center with a representation of the Scotch thistle, over and beneath which were inscribed the following familiar words (with the exception of McArthur's name being substituted for that of the hero of Lucknow) "Dinna ye hear the slogan? "Tis McArthur and his men!" The national and regimental flags mentioned above may be the flags now in the Illinois Military Museum; however, the white silk flag with the Scotch thistle in not in the Museum and its whereabouts is unknown.[273]

The description of the regimental flag above is characteristic of flags manufactured by Gilbert Hubbard & Co. of Chicago. Since the news article mentions that the flags were brought from Chicago, it is probable that Gilbert Hubbard & Co made all three flags, and the surviving national and regimental flags in the Illinois collection are the flags described.

---

[272] *Louisville Daily Journal,* Wednesday November 27,1861.

[273] *The Chicago Evening Journal,* November 25, 1861, describes the "complimentary" flag further: The last was a small white silk banner, richly mounted and emblazoned and decorated with a fine painting of the Thistle of Scotland. Above the thistle were the words, "DINNA YE HEAR THE SLOGAN?" and below "'TIS McARTHUR AND HIS MEN." In fine gold letters, while the apex of the beautiful hickory staff was surmounted with a solid god miniature of the battle-axe of Scotland.

This Regimental flag of the 12th Illinois Infantry Regiment is in tatters.

No other flags carried by the 12th Illinois Infantry Regiment are known to survive.

13th Illinois Infantry – "*Freemont's Grey Hounds*"

"Flag of the free hearts' only home,
By woman's hands to valor given,
Thy stars have lit fair Freedom's dome,
For all thy hues were born in heaven.
Forever float, thou standard-sheet,
Where breathes the foe but quakes before us,
With Freedom's soil beneath our feet,
And Freedom's banner waving o'er us."[274]

The Thirteenth Illinois Infantry was composed of volunteers from Lee County (Companies A-*Dixon Volunteers*--and Company C -*Amboy Volunteers*); Whiteside County (Company B and part of Company G); Rock Island County (Company D); De Kalb County (Companies E and F); Stephenson County (part of Company G); Kane County (Company H—*Aurora Rifles*); Cook County (Company I) and Du Page County (Company K – *DuPage County Rifles*). The Regiment was mustered in for three years of service on April 21, 1861, at Camp Dement, Dixon, Illinois under the command of Colonel John B Wyman[275]. In July, it was sent to Rolla, Missouri, where it remained until spring 1862. While stationed in Rolla, the Thirteenth was engaged in guarding supply trains and was part of General Fremont's force. The Regiment was given the name "*Fremont's Grey Hounds"* by Fremont himself due to the Regiment's exemplary march of 42 miles in a single day. In late 1862 and early 1863, the Thirteenth was part of Sherman's Yazoo Expedition. In the fight at Chickasaw Bayou on December 28, 1862, the Commander of the Regiment, Colonel John B. Wyman, was killed. The following day in a charge on the enemy's works at "Bloody Triangle," the Regiment lost 170 men killed, wounded, or missing and lost the Regiment's colors. In January 1863, the Thirteenth was part of an operation against Fort Hindman or Arkansas Post. Following the fall of the Fort, the Regiment moved downriver to Milliken's Bend and marched across Young's Point below Vicksburg. The Regiment joined in the actions against Vicksburg in May, joining the remainder of the army at Jackson, Mississippi, and fighting at Jackson and in the siege and assaults on Vicksburg. Following the fall of Vicksburg, the Thirteenth advanced on Jackson, Mississippi, and was in the siege on Jackson until the capture of the city on July 17. They remained in camp at Big Black until September when the Thirteenth accompanied Sherman to Chattanooga -- enroute, they skirmished at Corinth and Tuscumbia. In the Chattanooga-Ringgold campaign, the Regiment fought at Lookout Mountain and at Missionary Ridge, where on November 25, they captured the flag of the 18th Alabama Infantry. First Lieutenant Simeon Joselyn of Amboy was awarded the Medal of Honor for his actions at Missionary Ridge – "While commanding his company, deployed as skirmishers, came upon a large body of the enemy, taking a number of them prisoners. Lt. Joselyn himself shot their color bearer seized the colors of the Eighteenth Alabama Infantry,

---

274 *Amboy Times,* (Lee County), May 10, 1861, from speech of Rev. Mr. Day at the presentation of Colors to Company D of the 13th Infantry.

275 Colonel Wyman was born in Massachusetts and was 45 years old at the outbreak of the war. He had no education past age 14, and worked at a variety of odd jobs—worked in a mercantile store, as a clerk at a railroad manufacturing firm and conductor on the *New York & New Haven Railrord.* After moving to Chicago, in 1854, he organized the *Chicago Light Guard.* He landed a job on the *Illinois Central RR* and moved to Amboy, Illinois. He became mayor of Amboy just before the war.

and brought them back to his regiment.”[276][277],[278] The Regiment moved to Ringgold Gap, and it was at Ringgold Gap that Color Bearer Riley was killed (see below). In 1864 the Thirteenth was stationed at Madison Station, Alabama, and was mustered out on June 18, 1864. [279]

There are two national flags but no regimental flags of the Thirteenth Illinois Infantry in the collection at the State of Illinois Military Museum.

The Report of the Adjutant General of the State of Illinois[280] documents the above flag as having been the first flag flown over Richmond. “This flag was transmitted to the State with the following letters: “COMMONWEALTH OF MASSACHUSETTS, Executive Department, Boston, May 10, 1865. To His Excellency, Gov. Oglesby, Springfield, Ill.: GOVERNOR: I enclose with this a copy of a letter I received yesterday from Rev. Dr. Lothrop, a clergyman of the Brattle street church in this city, which tells its own story. Locke, who is a fine young fellow in appearance, brought it to me in person and brought with it the silk flag, the first Union flag displayed in Richmond on the day of its capture. Upon examination, this flag appears to belong to an Illinois Regiment, numbered the 13th: but of what arm of the service, whether infantry or cavalry does not appear. It was probably hanging in Turner's office as a rebel trophy. It belongs of right, therefore, to your State, and I hold it subject to your order, content in yielding it to you, to remember, as symbolical of the common patriotism of the whole country, that the first Union flag raised in Richmond was an Illinois flag by a Massachusetts soldier, I am, Governor, faithfully, Your friend and servant, JOHN A. ANDREWS, Governor of Massachusetts. *”*

Following this letter is another letter detailing the flying of this flag over Richmond. Both letters and the details of the capture of this flag at the Battle of Chickasaw Bayou (Dec. 29, 1862) are also

---

[276] *Civil War Medal of Honor Winners from Illinois, published* by the Civil War Centennial Commission of Illinois, July 1962, p8.

[277] Beyer, WF, and Keydel, OF, eds. *Deeds of Valor How America’s Civil War Heroes Won the Congressional Medal of Honor,* reprint by Longmeadow Press, (1993). p285.

[278] O.R. Series 1—Volume XXXI/2 [Series #55]

[279] Military History and Reminiscences of the Thirteenth Regiment of Illinois Volunteer Infantry in the Civil War in the United States, 1861-1865, Women’s Temperance Publishing Association, Chicago, (1892)

[280] AGR, Vol. 1, p159.

detailed in the unit's regimental history.[281] From the Confederate perspective, there is also a mention of the capture of this flag in Galveston Weekly News[282].. "Vicksburg, Miss., Dec. 29, 1862, Eds. News.-I forwarded you to-day an account of the battle of Vicksburg yesterday, but they have done, according to accounts, some of the tallest fighting to-day that has taken place since the battle of Fredericksburg…Amongst the trophies brought in are four beautiful regimental flags, one of the 48th Ohio, another the 29th Missouri Volunteers….Then there is the 31st Missouri, a beautiful flag. It is pretty badly shot; and the 13th Illinois is literally torn to pieces with bullets. Some of our boys recognized it as an old acquaintance they had met at Fort Donelson…S.M."

The flag was captured and taken to Richmond. Private John F. Locke of Company E, Thirty-ninth Regiment Massachusetts Volunteers, who had been captured earlier, was in Richmond for exchange at the time the city fell and, seeing the Federal cavalry enter the city, looked for a flag to raise over the city. He went to Major Turner's office at Libby prison, where he found two captured Union flags--one silk and one bunting. He hung the silk flag out of a third-floor window of the prison, thus raising the first Union flag over Richmond. The flag he raised was the flag of the Thirteenth that had been captured at Chickasaw Bayou.

This is the second of two National flags belonging to the 13th Illinois Infantry, which is at "rest" in Springfield.

The flag shown here, according to the Regimental History[283-] "was borne on the field of Ringgold Gap, Georgia, by Color-Sergeant Patrick Riley, of Company K, who was there shot through the breast and fell in such a manner as to be enveloped by the flag---and still has the stains of Sergeant Riley's life-blood." The stains seen on this flag are presumably those of Sgt. Riley. The battle honors on this flag cannot all be clearly read because of the deterioration of the silk and loss of paint. Originally the flag was inscribed: "Actually first at Chickasaw Bayou and Assault of 29th, JACKSON 14 MAY 63, VICKSBURG AND ASSAULT 22 MAY 63, JACKSON 10 JULY 63, TUSCUMBIA 26-27 OCT 63, LOOKOUT MOUNTAIN NOV 24th 63, MISSION RIDGE NOV 25 1863, RINGOLD GA NOV 27th

[281] Military History and Reminiscences of the Thirteenth Regiment of Illinois Volunteer Infantry in the Civil War in the United States, 1861-1865 (Woman's Temperance Publishing Association, Chicago, 1892) pp 281-282.

[282] *Galveston Weekly News*, January 28, 1863.

[283] ibid. p 281.

1863." The term "Actually first" was placed on the flag since there was a controversy with a Missouri regiment as to which was "actually the first" to gain the enemy's works at the battle of Chickasaw Bayou. Not appearing on this flag is the battle honor—"ARKANSAS POST," which is listed in the Official Army Register as an additional battle in which this regiment "bore an honorable part."[284]

In 1886 or 1887, the publishers of a book by Mary Livermore[285] sent a photographer and artist to photograph and paint important CW military flags. One of the flags that was photographed and painted was the flag of the 13th Illinois. It appears in plate III and is described as: "This flag is stained by the blood of Patrick Riley, color-sergeant, who was killed at Ringgold Gap, November 27, 1863. He was shot through the breast and fell in such manner as to be rolled up in the flag.[286] Although the flag depicted in Livermore's book is the same flag as shown here, in 1887 the battle honors did not appear on the flag. There is an unconfirmed story that the honors were applied at a reunion of the Regiment in 1890.

There is record of other flags having been presented to companies of the Thirteenth. Company A--the *Dixon Volunteers*-- were presented a unique flag sewn by the ladies of Dixon. The flag was presented to the Company by Miss Mary Williams. "A memento, unusual on the flag of our county, has been placed upon this flag by the fair hands of the ladies. It is the letter "**B**" in scarlet cloth cut from one of the flags that waved over the Fort on Bunker Hill. The ensign carrying this flag in the memorable battle of Bunker Hill was struck by a musket ball, and as he fell, Col. Maxwell, then a private caught it and bore it through the remainder of the fight. Having been riddled by the bullets of the enemy, its battered remains were voted to Col. Maxwell. As a token of the past, its pieces have been religiously preserved by his descendants, and it is through devotion to the same cause—the preservation of the liberties and nationality of the people of this country, that Mrs. M. Jerome, who is a great-grand-daughter of Col. Maxwell, has furnished the expressive "**B**" of such honored material to be placed on the flag."[287] There is no further history regarding this flag and its whereabouts today is unknown.

Company E of the 13th came from De Kalb County, Illinois, and was given a national flag. It is unlikely that either of the flags shown above are the same flag. "With a vague idea that each company required a banner, a beautiful silk flag was prepared and presented to our company by one of the young ladies before an immense crowd gathered to witness the novel scene."[288] Currently, the location of the flag of Company E is unknown.

According to the Amboy Times[289], a "Regimental" flag was presented to the *Amboy Volunteers,* who later became Co. C of Thirteenth Illinois Infantry. "The ladies of this city have got up a very fine banner for our volunteers, which they presented to them yesterday. The banner is of blue silk on which is a golden eagle holding the stars and stripes in his claws and is inscribed with the words, "We march to conquer." ...Rev. Mr. Day, on behalf of the ladies, delivered the following address: … I present you, Capt. Messenger, Officers, and Soldiers of the *Amboy Volunteers*, with this banner saying to you in the language of the brave dying Lawrence, "Don't give up the Flag" The whereabouts of this flag today is unknown.

---

[284] *Official Army Register of the Volunteer Force of the United States Army for the Years 1861, '62, '63, '64, '65.* Part VI, p232(Reprinted by Van Sickle Military Books, Gaithersburg, MD, 1987)

[285] Livermore, Mary. *My Story of the War: A Woman's Narrative of Four Year Personal Experience as Nurse in the Union Army, and Relief Work at Home, in Hospitals, Camps, and at the Front, during the War of the Rebellion.* A.D.Worthington and Company, 1887.

[286] Ibid., p 34.

[287] *Republican and Telegraph (*Dixon, Lee County), May 2, 1861

[288] Boies, Henry. *History of De Kalb County, Illinois* (O.P. Bassett, Printer, Chicago, 1868) page 294

[289] *Amboy Times,* May 10, 1861.

Company D also received a company flag[290]. "Rock Island, April 25—The Rock Island volunteer company, Capt. Williams, consisting of 120 men, left this forenoon for Springfield. A large concourse of citizens attended them to the depot, where a beautiful flag was presented to them, on behalf of the ladies, by Col. Danforth. A handsome sword was also presented to Lieut. McNeil by a lady of this city." This company flag is probably not one of the national flags in the State of Illinois collection.

As noted above, the Regimental history of the 13th mentions that a stand of their colors was captured at the Battle of Chickasaw Bayou. The National flag is accounted for and was the first flag to fly over Richmond. The fate of the Regimental flag is unknown, and no other flags carried by this regiment are known to have survived.

[290] *Chicago Tribune,* April 26, 1861.

14th Illinois Infantry

"The colors, which came out of this bloody conflict with forty-two bullet holes through them, fully attest to the gallantry of the command in that memorable struggle."[291]

The companies forming the Fourteenth Illinois Infantry Regiment were from Cass County (Company A), Shelby County (Company B), Macoupin County (Company C), Greene County (Company D), Menard County (part of Company E), Jersey County (Company F – *Jersey Guards*), Sangamon County (Company G), Christian County (Company H), Morgan County (Company I and part of E) and Scott County (Company K). After organization at Camp Duncan, Jacksonville, Illinois, on May 25, 1861, the regiment was mustered for three years of service[292] under the command of Colonel John M Palmer.[293] On June 19, the regiment was ordered to Quincy, Illinois and through the remainder of 1861, fought secessionist forces in Missouri, including a campaign against Springfield, Missouri, under Fremont. In February 1862, the regiment was ordered to Fort Donelson and arrived the day following surrender; then proceeded to Fort Henry and up the Tennessee River to Pittsburg Landing. At the ensuing Battle of Shiloh on April 6 and 7, half of the men of the Fourteenth who entered the battle were subsequently killed, wounded, or missing. Following the battle, the Fourteenth advanced southward to lay siege to Corinth, Mississippi, from April 29 to May 30. In June, the men marched to Memphis and remained on duty there until September 6. Following a march to Bolivar, Tennessee, on October 5, 1862, they fought the rebels retreating from Corinth at Metamora or the battle of Hatchie. As part of Grant's Central Mississippi campaign, the regiment guarded the *Mississippi Central Railroad* from Bolivar to Coffeeville, Mississippi, from October 31, 1862, until January 10, 1863, and went into winter quarters at Lafayette, Tennessee, on January 18. The regiment was ordered to Vicksburg, Mississippi, and from May 22 until July 4, was in the siege and capture of that city. Next, they pursued the retreating enemy toward Jackson, Mississippi and fought in the siege and capture of Jackson. In August, they marched to Harrisonburg, Louisiana, on the Ouachita River and captured Fort Beauregard; after that, the regiment returned to Vicksburg and remained on duty there until February 1864. Following veteran re-enlistment, the men were sent home on furlough. On return from furlough, they were ordered to Clifton, Tennessee, then marched to Rome, Georgia. From June through September, the Fourteenth did garrison duty at Allatoona Pass, Acworth, Big Shanty and Marietta, Georgia. July 1, 1864, the Fourteenth and Fifteenth Illinois Infantry were consolidated into the "Veteran Battalion of the Fourteenth and Fifteenth Illinois Infantry." In October, many the Battalion were either killed or taken prisoner by Hood and sent to Andersonville. Those not captured were mounted and served as scouts on Sherman's "march to the sea" and saw duty at Savannah, Pocotaligo, Columbia, Cheraw, Bentonville, and Raleigh. In the spring of 1865, with the arrival of new recruits, the Battalion was split into separate regiments, and following the surrender of Johnston, the Fourteenth marched in the Grand Review at Washington DC on May 24, 1865.

---

[291] AGR, Vol 1, p.626.(refers to the Battle of Shiloh)

[292] The Regimental Surgeon was Dr. Benjamin F. Stephenson who founded the Grand Army of the Republic after the War.

[293] Colonel Palmer was born in Kentucky in 1817.He was unable to receive eduation in Kentucky and this father moved the family to Illinois in 1831, In 1834 John Palmer entered a newly opened college in Alton, Illinois where he studied for the next 18 months. In 1839, after one year of study Palmer was admitted to the Illinois bar and opened an office in Carlinville. He became a Probate Judge in 1847and was elected to the State Senate in 1852. At the outbreak of the war he was one of the first to offfer his services and became the Coonel of the 14th Illinois Infantry. He was elected governor of Illinois in 1868.

The Fourteenth then proceeded to Fort Leavenworth, Kansas, where it mustered out on September 18, 1865.[294]

This is one of two national flags belonging to the 14th Illinois Infantry now at the State of Illinois Military Museum. Both of the national flags are in poor condition due to the deterioration of the silk and battle damage. This flag measures 68 x 72" and has a rectangular canton with the stars arranged in 6 horizontal rows. The stars and unit designation on the center stripe are gold. Based on this configuration, the flag was probably produced by Gilbert and Hubbard of Chicago. The battle honors are difficult to read but do include the following: 14th REGT ILL VOL, SHIELOH (sp?), SIEGE OF CORINTH, METAMORA (sp?), SIEGE OF VICKSBURG, JACKSON, CHAMPION HILLS, ATLANTA, SAVANNAH Sallketchie (sp?), COLUMBIA, FAYETTEVILLE, CHERAW, BENTONVILLE (sp?). *The various types of lettering for the battle honors suggest that different individuals added the honors at different times.*

This national flag of the Fourteenth does not bear any inscription. It appears that the flag originally had a square canton and has painted gold stars It measures 70 x 72" and is probably of New York manufacture.

---

294 AGR. Vol 1, p. 625-8.

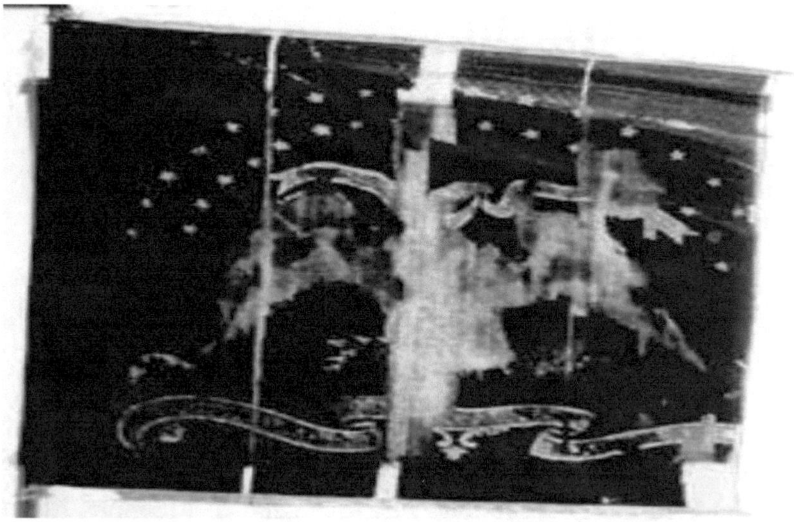

The Regimental flag also shows considerable damage and, at this time, measures only 70 x 62" (originally would have measured 72 x 78"). The double row of stars extending past the ends of the upper scroll, the raised center section of the lower scroll and the embellished ends of the lower scroll all suggest that this flag was manufactured in New York.

No other flags carried by this regiment are known to survive.

15th Illinois Infantry

"Our banner is a galaxy of glorious silver stars,
Freedom's history is written on its white and crimson bars,
In the face of Southern foes, we will flaunt our dear old flag,
And it never shall be lowered to a vile Secession rag.
It shall wave, forever free,
Forever, ever free,
Or a winding sheet shall be
For us and Liberty."[295]

The Fifteenth Illinois Infantry Regiment was organized in Freeport, Illinois, with men raised from the First Congressional District--Company A (*Woodstock Light Guards*) from Mc Henry County, Company B (*Boone Rifles*) from Boone County, Company C (*Ellis Rifles*) from Winnebago County, Company D (*Wayne Rifles*) from McHenry County, Company E (*Jo Daviess Guards*) from Jo Daviess County, Company F (*Algonquin Rifles*) from McHenry County, Company G (*Lena Rifles*) from Stephenson County, Company H (*Ogle Guards*) from Ogle County, Company I (*Lake County Union Guards* or *Union Rifles*) from Lake County and Company K (*Carroll Rifles*) from Carroll County. The regiment mustered into service May 24, 1861, for three years under the command of Colonel Thomas Turner[296]. In July, the Fifteenth was ordered to Mexico, Missouri, and early in the war, campaigned in Missouri and assisted in the capture of 1,300 rebels a few miles south of Sedalia. The regiment's winter quarters were at Otterville, Missouri. In February 1862, the regiment was ordered to Fort Donelson, Tennessee and arrived in time to take part in the Fort's surrender. The Fifteenth then marched to Fort Henry and there embarked for transport to Pittsburg Landing. On April 6 and 7, in the battle of Shiloh, the regiment was heavily engaged and suffered significant losses at the "Hornet's Nest."[297] In April and May, the Fifteenth participated in the Siege of Corinth and, following the evacuation of Corinth, marched to Memphis, and remained on duty there until September 6. During September, the regiment joined in the march to Bolivar and the Hatchie River, and on October 4, 1862, they met the rebels and defeated them at Metamora. From November through January, the Regiment guarded the *Mississippi Central RR* from Bolivar to Coffeeville, Tennessee and then returned to Memphis. In May 1863, the regiment was ordered to Vicksburg and participated in the assault of May 22 and in the siege and fall of Vicksburg on July 4. Later in July, they advanced on Jackson, Mississippi and were in the fighting at Jackson on July 12. During the time the regiment was camped at Natchez, Mississippi, the Fifteenth supported the capture of Fort Beauregard near Harrisonburg, Louisiana. From November until February 1864, the Fifteenth was in winter quarters at Vicksburg.[298] As part of Sherman's Meridian Campaign, the regiment fought

---

[295] *Chicago Evening Journal,* June 25, 1861. A portion of a poem signed by "AGNES"

[296] ColonelThomas Johnston Turner was born in Ohio in 1815. In 1838 he moved with his family to Freeport, Illinois, studied law and was admitted to the bar in 1840. He supported the Democratic party and served as judge of the probate court, member of the state house of representatives, established the first local newspaper (*Prairie Democrat*) and was the first elected mayor of Freeport, Illinois. With the outbreak of the war, Turner joined the Union army and was elected Colonel of the 15th Illinois Infantry.

[297] The Fifteenth Illinois Infantry lost 250 men killed and wounded and have more men buried in the National Cemetery at Pittsburg Landing than any other regiment.

[298] During this time the Regiment reenlisted as veterans allowing the veterans a furlough home.

at Champion's Hill, Mississippi, on February 4 and then was ordered to Georgia for Sherman's Atlanta campaign. While on guard duty around Acworth, Big Shanty and Marietta, on July 1, 1864, the Fifteenth was consolidated with the Fourteenth Infantry into the "Veteran Battalion Fourteenth and Fifteenth Infantry Volunteers." General Hood attacked the Battalion at Big Shanty and Ackworth, and many of the men were either killed or taken prisoner. The survivors were mounted and acted as scouts on Sherman's march to the sea and were at the fall of Savannah in December 1864. In 1865 the Battalion moved north in the Campaign of the Carolinas and fought at Beaufort and Columbia, S.C. and Fayetteville, Bentonville, Goldsboro and Raleigh, N.C. On April 28, enough new recruits were added at Raleigh, North Carolina, to divide the Battalion back into separate regiments. After Johnston's surrender, the Fifteenth Illinois Infantry marched in the Grand Review in Washington on May 24 and then was sent to Forts Kearney and Leavenworth, Kansas, where it mustered out on September 1, 1865.[299]

There are two National flags and one Regimental flag of the Fifteenth Regiment Illinois Infantry in the collection of the State of Illinois. The first flag carried by the regiment was a flag presented to the *Ogle Guards,* which became Company H of this Regiment. "The Company was presented with a flag in April 1861 by some of the women of Polo; my recollection (sp) is that Mrs. Phrocine Frisebee, who I think was formerly a Miss Whiteside, made the presentation speech. M. D. Swift, in his closing remarks, said, 'Here is to the Women of Polo, long may they wave.' This Company went (sp) to Freeport where with nine other companies, it was mustered into the United States service as Company H, 15th Illinois Vol. Infantry on the 24th day of May 1861, to serve three years unless sooner discharged. The regiment did not have a flag, so Company H's flag was used as a Regimental flag until April 6th and 7th, 1862, where at the battle of Shiloh it was tattered and torn with bullets and shells, then it was sent home, being in Capt. Swift's possession until his death, when it came into my possession and is in my possession at the present time."[300] This flag is described elsewhere as being made of silk and is depicted as being a national flag.[301]

The star pattern on this silk flag is not typical of flags manufactured for the Union Quartermaster. It measures 70 x 75" and may be the flag originally presented to Company H.

---

[299] AGR. Vol 1, p 671-4.

[300] From a written document dated Polo, Illinois, Dec. 13, 1905, and entitled *Polo Historical Society.* The document is signed, "Ex Sargt. Peter R. Cover" and is in the Logan collection of Illinois State Historical Library.

[301] An unidentified newspaper clipping in the "John W. Clinton Papers" in the Illinois State Historical Library.

This silk flag measures 77 x 69" with a square canton and gold stars, it is probably of New York manufacture.

The battle honors inscribed on the two National flags are: DONELSON, SHILOH APRIL 6th, SIEGE OF CORINTH, METAMORA, SIEGE OF VICKSBURG, JACKSON, CHAMPION HILLS, ATLANTA, SAVANNAH, SALKAHAHATCHIE (sp?), COLUMBIA, FAYETTEVILLE, CHERAW, BENTONVILLE

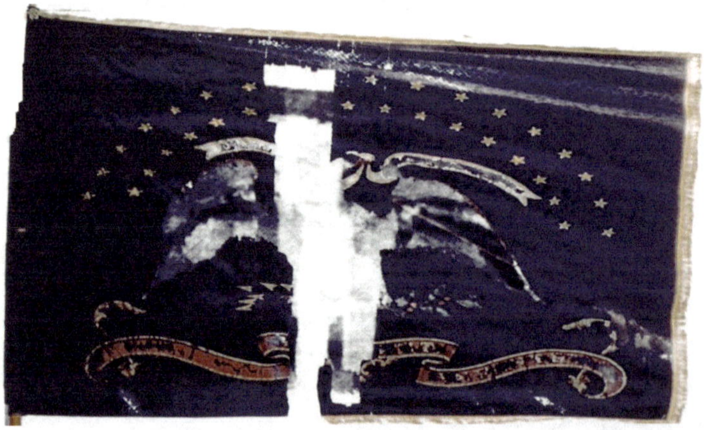

The silk regimental flag has stars extending past the ends of the upper scroll held by the eagle and an embellished lower red scroll with the center section of the scroll raised. This is typical of flags manufactured in New York. The flag measures 68 x 74".

Another existing flag from this regiment is framed and on display in a school in Lena, Illinois, in Stephenson County. The presentation of the flag to Company G is reported in the *Freeport Weekly Journal,* "Just before the Lena company left for this city, on Saturday morning last, the ladies of Lena presented them with a fine flag. The presentation was made by Mrs. Jesse Johnson, with the following very appropriate remarks: 'I present this flag to the company on behalf of the ladies of Lena. We ask you to accept it as an expression of gratitude for your willingness to engage in its defence. We pray the God of Nations blesses you and gives you success. We hope that you will always bear the stars and stripes and that they will continue to wave over the home of America, as long as we shall have a place among the nations of the earth'"[302] It is a 34-star national flag and has the accompanying information:

"This flag was made from pieces of ribbon stitched together by Lena women working in a millinery shop located in about the middle of what now is the first block of East Main Street. The *Lena Rifle Men* made up much of Company G of the 15th Illinois Regiment, which left Freeport for Alton, Illinois, on June 18, 1861. Alexander McEathron, sergeant and later Orderly sergeant, was the custodian of the flag. This flag was with Company G through the battle of Shiloh, on guard at a hospital and convalescent camp, at battles in Tennessee and Missouri, and at sieges at Vicksburg in July and August of 1863. Alexander McEathron brought the flag home when he mustered out on May 24, 1864. The use of the flag during the war had been limited to flying over encampments and at times when the company was guarding hospitals and convalescent camps. After the war, the flag served to drape the coffins of Civil War Veterans. At G.A.R. reunions and Memorial Day, it was proudly carried in the parade. Fred McEathorn, son of Alexander McEathron, presented the flag to the Sprague-Inman Post 577 of the American Legion Post at Lena after the death of his mother."[303]

Company D came from the area of Marengo in McHenry County. After the war, a member of the company wrote a history of Company D. In his book, he states:

"The 11th of May soon came around, and the usually quiet streets of Marengo were thronged with spectators, friends and relatives of the soldiers who had come to witness their departure. A few hours before leaving, the ladies of Marengo presented the Company with a fine flag, and 'neath its folds, we took a solemn vow never to disgrace it or bring it back until our flag could wave in triumph over all our land."[304] It is not known if this flag is still in existence.

Company E, from Warren in Jo Daviess County, also received a flag. "On the morning of their departure for Freeport, they were escorted to the depot by a large concourse of citizens, where, on behalf of the ladies of Warren, Rev. Mr. Probst presented the company with a very handsome flag, accompanied by an appropriate address."[305] The whereabouts of this flag today is unknown.

The first company recruited from Carroll County was presented a flag prior to their departure from Mount Carroll, the county seat. "While the men were busy polling up this company, the ladies of Mount Carroll were not idle, but their deft fingers had fashioned a handsome flag, which on Monday evening, April 29, was presented to the company, with a proper ceremony at the court house....Others of the volunteers also pledged themselves and comrades to stand by, defend and return the flag to the fair hands from which they received it."[306] This company became Company K of the Fifteenth Illinois Infantry. The further history of this flag is unknown. No other flags carried by the 15th Illinois Infantry Regiment are known to survive.

---

[302] *Freeport Weekly Journal,* May 15, 1861.

[303] Personal communication, Gerald E. Monigold.

[304] Barber, Lucius W. *Army Memoirs of Lucius W. Barber, Company "D," 15th Illinois Volunteer Infantry. Chicago*, J.M.W. Jones Stationery and Printing Co., 1894, p11.

[305] *The History of Jo Daviess County Illinois,* H.F.Kett & Co. Chicago, (1878) page383; *Galena Daily Courier,* May 14, 1861.

[306] *History of Carroll County Illinois,* H. R. Kett & Co., Times Building, 1878, page 281

16th Illinois Infantry

Yes there were the Union men who wept with joyful tears
When they saw the honored flag they had not seen for years
Hardly could they be restrained from breaking forth in cheers,
While we were marching through Georgia.[307]

The Sixteenth Regiment of Illinois Infantry was mustered into the Union army at Quincy, Illinois May 26, 1861, under command of Colonel Robert Frederik Smith[308]. The men volunteered from the following counties, Company A from Adams and McDonough Counties, Company B from McDonough County, Company C from Adams and McDonough Counties, Company D (*Union Rifle Company)* from Hancock County, Company E from Brown County, Company F from Henderson County, Company G from Adams and Schuyler Counties, Company H from Adams County, Company I from Adams and Hancock Counties and Company K from Pike County. In June, the Regiment moved to Grand River, Missouri, to guard the *Hannibal & St. Joseph R.R.* and saw their first action when mounted rebels attacked them at Monroe Station. It was near here that the Sixteenth captured a Confederate battle flag.[309] (See section on Captured Confederate flags). The remainder of the year was spent foraging and skirmishing in Missouri. In January 1862, the regiment was ordered to Bird's Point, Missouri[310] , and in March, was ordered further downriver to New Madrid, Missouri. The Tenth and Sixteenth Illinois Infantry aided in the capture of New Madrid on March 12-14 and then followed the retreating enemy and captured Island No. 10 on April 6 and Tiptonville, Tennessee, on April 8.[311] The regiment sailed further downriver to Osceola, Arkansas, on April 13 and then fought at Fort Pillow April 17-24. After the fall of Fort Pillow, the Regiment was transported to Hamburg Landing, Tennessee and led in the advance and siege of Corinth, Mississippi April 29 to May 30. After the evacuation of Corinth, the Sixteenth pursued the retreating enemy to Booneville, Mississippi, skirmishing with them at Tuscumbia Creek in late May. In July 1862, the regiment marched to Tuscumbia, Alabama, crossed the Tennessee River at Florence, Alabama, skirmished at Columbia, Tennessee and fought in the siege at Nashville until mid-November. They remained on guard duty at Edgefield, Tennessee, near Nashville until the following year--July 1863-- when the Sixteenth marched to Murfreesboro, Tennessee. One month later, the regiment marched to Columbia, Tennessee and then on to Athens, Huntsville and Stevenson, Alabama. In early October, the Sixteenth marched to Anderson's Gap and campaigned against Wheeler in the Sequatchie Valley. During November and December, the Regiment was at Kelly's Ford on the Tennessee River, unloading from boats provisions for Thomas' army at Chattanooga. In late December,

[307] *Marching Through Georgia,* Henry Clay Work, 1865

[308] Colonel Smith was born in Philadelphia in 1806 and died in 1893 and was buried in (Oakwood Cemetery) Hancock Couty, Illinois

[309] The flag was captured by Don Carlos Salisbury of Company C. "When the train reached Utica, on the Grand River, the rebels fired on the Union soldiers from a blockhouse. Here Don C Salisbury led the charge that drove them out and Don took their flag, as he was first to enter the fort-a banner having a rattlesnake grarding (sp) a palm tree opposed by a bunch of spears representing the Union. The banner was placed in the flag museum at Springfield, Illinois, with a label crediting Don with its capture". This is a personal communication from Kyle Walker who shared this quote from a book entitled "Western Adventure of Don Carlos Salisbury" page 1. (ed) Please turn to the chapter "Captured Confederate Flags" later in this book for a photo of the flag.

[310] Bird's Point was across the Mississippi River from Cairo, Illinois and on the *Cairo & Fulton RR.*

[311] At Tiptonville, the Union forces captured 5,000 prisoners and a large amount of artillery and small arms.

the Regiment re-enlisted as veterans, and on January 1, 1864, the regiment left for home on 30 days of furlough. On return, they rejoined the army at Rossville, Georgia. In May, they fought with Sherman's army in the Atlanta campaign and battled at Rocky Faced Ridge, Tunnel Hill, Buzzard's Roost Gap, Resaca, Rome, Dallas, New Hope Church and Allatoona Hills. In June, the Regiment fought at Pine Hill, Lost Mountain, and the assault on Kennesaw Mountain and in July at Ruff's Station, the Chattahoochee River, and Peach Tree Creek. The Sixteenth was on the front line during the siege of Atlanta in August and, in the flanking movement around Atlanta, was heavily involved in the battle at Jonesboro.[312] Following the pursuit of Hood into north Georgia and Alabama in October, the Sixteenth returned to Atlanta to join Sherman's "March to the Sea." The regiment had the honor of being the first regiment to take formal possession of Savannah. During February and March of 1865, the regiment marched north through the Carolinas and fought at Fayetteville, Averysboro and finally Bentonville, North Carolina—the last battle of the war for the regiment. After a month's rest, they marched on to Goldsboro, Raleigh, and Bennett's House, where Johnston surrendered to Sherman's army. On May 24, 1865, the Sixteenth marched in the Grand Review in Washington, DC, then proceeded to Louisville and was mustered out on May 24, 1865.[313]

There are six national flags and one regimental flag of the Sixteenth Illinois Infantry in the holdings of the State of Illinois Military Museum.

This is one of the National flags belonging to the Sixteenth Illinois Infantry in the State of Illinois collection.

This national flag is inscribed with numerous battle honors. Appearing on the flag are the following: NEW MADRID, ISLAND No 10, SEIGE OF CORINTH, NASHVILLE, CHICKAMAUGA, MISSION RIDGE, KNOXVILLE, ROCKY FACE, RESACA (sp?), 16th REGT ILLINOIS VET VOL, ROME, DALLAS, KENESAW MOUNTAIN, PEACHTREE CREEK, JONESBORO, ATLANTA, SAVANNAH, CHARLESTON, AVERASBORO, BENTONVILLE, RALEIGH.

[312] On September 1, 1864, the Regiment captured the flag of Swett's Mississippi Battery at Jonesboro.

[313] *Report of the Adjutant General of the State of Illinois,* H. W. Rokker, State Printer and Binder, Springfield, Illinois (1886) Vol. 2, pages 35-7.

This national flag is inscribed #16th REGT. ILLINOIS VOL. IN." The arrangement of the stars in the canton is typical of flags manufactured by Horstmann Bros. & Co of Philadelphia.

The above four national flags belonging to the Sixteenth Illinois are in very poor condition. They do not appear to be of Federal issue and are probably Company flags. In this regard I share the following[314]

"Speech by Miss Almarine Campbell

Presenting the flag to Co.- 16 Regiment.

Sam Hayes of Pittsfield, Illinois, Captain.

Friends of the 16th Regiment on behalf of our town, I welcome you home, for there is not a patriot nay not a Christian in our town whose heart does not leap for joy at the sight of your blue coats. This flag and your faces, recall forcibly to my mind a similar occasion when nearly three years ago on behalf of the ladies of Pittsfield I presented this flag to company K through your gallant fallen hero Major Sam Hayes[315]. Every man of the Company received at the same time a pocket Bible. Thus, equipped with those two emblems of man's first and holiest affections, love to his God and love of his country, you set out to meet the enemy. Could you have been better prepared? Captain Stuart[316], I take pleasure in again presenting this flag to your company-hoping that you may follow the footsteps of your honored predecessor upward in military rank and finally still upward."

---

314 Personal communication

315 Major Sam Hayes had died of "camp fever' August 6,1862

316 This is probably George W Stewart who was Captain of Company K

Perhaps one of the four national flags shown above was the flag that was "presented" to Company K of the 16th Illinois Infantry.

This Regimental flag of the Sixteenth Illinois Infantry measures 70" on the staff by 75" on the fly. Based on the position of the stars, the head of the eagle and the lower scroll, this flag was most likely manufactured by Evans & Hassall of Philadelphia.

No other flags carried by the Regiment are known to survive.

17th Illinois Infantry

"Up! Up, with the Banner!
On! On, to the Fight!
Throw its folds to the breeze,
And strike home for the Right!
Foul treason is rampant—
Has sworn that no more,
Our flag o'er the Union
Shall wave as of yore."[317]

The Seventeenth Illinois Infantry Regiment was mustered into service at Peoria, Illinois, on May 24, 1861 under the command of Colonel Leonard Fulton Ross.[318] The men volunteered from the following: Company A (*Peoria Blues* – also known as *Pope's Body Guard)*—Peoria County; Company B—Peoria County; Company C *(Fairview Independent Rifles)*—Henry County; Company D—Henry County; Company E (*Galesburg Invincibles)*—Knox County; Company F (*Monmouth Union Guards*)—Warren County; Company G (*Woodford Rifles)*—Fulton and Woodford Counties; Company H (*Fulton Blues)*—Fulton County; Company I (*Mercer Blues)*—Mercer County and Company K(*Mason Regulators)*—Mason County. After further training at Alton, Illinois, the regiment was sent to Bird's Point, Missouri, for garrison duty and then in pursuit of General Jeff Thompson. In September, they moved to Kentucky to aid in the construction of Fort Holt[319]. After returning to Cape Girardeau, they, with other regiments, were again sent in pursuit of General Thompson and defeated him at Fredericktown, Missouri, on October 21, 1861. The regiment returned to Cape Girardeau for provost duty until February 1862, when it was ordered south and battled at Fort Henry and then Fort Donelson. In April, the Seventeenth was ordered to Pittsburg Landing and fought in the Battle of Shiloh, April 6 and 7, where the regiment lost 130 men killed and wounded. Next, the regiment advanced with Union forces to the siege of Corinth, Mississippi (April 29-May 30) and, after the evacuation of Corinth, marched to Jackson, Tennessee. In July, the Regiment was sent for provost duty to Bolivar, Tennessee, where they remained until November 1862. In 1863 the Seventeenth moved to Memphis, Tennessee, then Lake Providence, Louisiana, saw action at Richmond, Louisiana, on January 29-30 and moved to Milliken's Bend in April. In Grant's Vicksburg campaign, the Seventeenth fought at Port Gibson on May 1, Bayou Pierre on May 2, Raymond on May 12, Jackson on May 14, Champion's Hill on May 16, and Big Black River on May 17. From May until July, they were in the siege at Vicksburg, and after Vicksburg's surrender, the regiment remained on garrison duty there until May 1864. The term of the regiment expired on May 24, 1864; an insufficient number of the men re-enlisted to continue the

---

317 Portion of a poem by Wm. Bounaeville, entitled "Up, Up With the Banner!" dedicated to Col. L.F. Ross, 17th Illinois, originally in the *Peoria Union,* but quoted in *Fulton County Ledger,* June 4, 1861.

318 Colonel Ross was born in 1823 in Fulton County Illinois and in 1829 the family moved to the city of Kansas on the Illinois river. After his father's death in 1837 the family moved to Canton in Fulton County and in 1841 Ross enrolled in Illinois College ; following graduation he studied law in Canton , was admitted to the bar in 1845 and opened a law practice in Canton. At the outbreak of the war with Mexico Ross and 2 of his brothers fought with the 4th Regiment of Illinois Volunteers. In the latter part of April 1861 Ross organized Company H of the 17th Illinois Infantry and was appointed its Colonel. Floyd, Frederick, General Leonard f. Ross. *Iowas Historical Record* IV (Iowa City 1888) 145-83.

319 Fort Holt on the Ohio River in Kentucky, directly across the river from Cairo, Illinois.

Seventeenth Illinois Infantry Regiment intact; therefore, those re-enlisting were consolidated with the Eighth Illinois Infantry and mustered out with that Regiment on May 4, 1866. The men whose terms of enlistment expired and who did not re-enlist were mustered out on June 4, 1864.[320]

There are two national and one regimental flag from the Seventeenth Illinois Infantry in the State of Illinois Military Museum collection. Both national flags are in very poor condition.

Despite its condition, the Regimental designation is still visible on the center stripe: 17TH REGT ILL VOL. *The flag measures approximately 73" on the hoist x 75"on the fly.*

Only fragments of this flag remain, held together by dark silk netting that was applied to the flag in the 1920s. This flag measures approximately 68" on the hoist x 75" on the fly.

[320] *Report of the Adjutant General of the State of Illinois,* Vol. II, p 61-3.

Based on the configuration of the eagle, the upper and lower scrolls and the arrangement of the stars, this regimental flag was manufactured by Horstmann Bros. & Co. of Philadelphia. This flag measures 70" on the hoist by 76" on the fly.

The band of the Seventeenth Illinois was presented a flag in July 1861:[321] "The ladies of Alton presented a splendid flag to our Regimental Band last evening, on dress parade. Hip! Hip! Hurrah! Who wouldn't be a soldier and belong to the Seventeenth?" No further information regarding this flag is known.

The "Fulton Blues" or Company H of the Seventeenth was presented a flag on departure from Lewistown, Illinois. "On Friday morning, the Fulton Blues left Lewistown enroute for their encampment at Peoria. The public square was filled with citizens—men, women and children of all classes—to witness the flag presentation and the departure of the company…The chairman called the immense crowd to order when Dr. R. R. McDowell, on behalf of the ladies of Lewistown, presented to the company, in an appropriate little address, a magnificent flag. Capt. Ross received the colors with the pledges that they should not be disgraced and handed them to Ensign Woolfolk, who received them in the name of the company with a short and appropriate address. The brass band had meanwhile played the 'Star Spangled Banner' and 'Yankee Doodle.' Capt. Ross proposed three cheers for the ladies—mothers, wives, sisters and friends of the company—who had so kindly provided for them in their departure."[322] There is no further history of this flag, and its whereabouts today is unknown.

No other flags known to have been carried by the 17th Illinois Infantry are known to survive.

[321] *Fulton County Ledger,* July 1861.

[322] *Fulton County Ledger,* May 14, 1861

18th Illinois Infantry[323]

Flag of our nation's stripes and stars,
The glorious standard of the free,
The banner borne in freedom's wars,
The hallowed gem of Liberty.
Fling out our country's banner wide,
Our emblematic stratagem,
Our Union never shall divide,
While floats that silken diadem.
Year after year, the brilliant stars
Shall indicate the strength of all;
Let all beware of civil wars,
That curse of monarchs, or freedom, falls.[324]

On May 19, 1861, at Anna, Union County, Illinois, Ulysses S. Grant, then the State Mustering Officer, mustered the Eighteenth Illinois Infantry Regiment from the Ninth Congressional District into State service for thirty days. After fulfilling its thirty days obligation, the regiment was mustered into the Union Army for three years of service under the command of Colonel Michael K. Lawler[325] on May 28, 1861. The men came from the following counties: Perry County (Company A and part of Company I); Gallatin County (Company B); Jackson County (Companies C and K); Alexander and Pulaski Counties (Companies D, E, and H), Jefferson and Wayne Counties(Companies F and G --*Wayne Guards*) and Jackson County (Company K). On June 24, the regiment moved to Bird's Point, Missouri and remained there on guard and construction duty until the end of August, when it moved to Mound City, Illinois, to guard the gunboat yards. In November, the Eighteenth routed Jeff Thompson in Missouri and returned to Cairo, Illinois. In January 1862, they were on reconnaissance near Columbus, Kentucky, under Grant and, in February, steamed up the Tennessee River to be one of the first at Fort Henry. February 13-16, in the battle at Fort Donelson, the regiment lost 200 men killed and wounded. On March 11, the Eighteenth embarked for Pittsburg Landing, Tennessee. At the subsequent Battle of Shiloh on April 6 and 7, the regiment lost three color bearers during the first day's fighting. Later, in April and May, the Eighteenth fought in the siege of Corinth, Mississippi. After the evacuation of Corinth, the Regiment was stationed at Bethel, Tennessee, in early June and marched on to Jackson, Tennessee, two weeks later. In September, the Regiment received two companies, "C" and "H", recruited at Carbondale, Illinois. (These were to replace the original companies "C" and "H" that had been consolidated with other companies in the Eighteenth following the loss of men at the Battles at Donelson and Shiloh.) Of the original 930 men mustered into the Eighteenth, only 556 remained prior to the arrival of these new

---

323 This Regiment has also been called the *Bloody Eighteenth".*

324 Portion of a poem, author anonymous, printed in the Wayne County newspaper-- *The Fairfield Gazette*, January 22, 1858. Men of the Eighteenth Illinois served from Wayne County.

325 Michael Lawler was born in County Kildare, Ireland in November 1814. Three years later the Lawler family emmigrated to the United States and settled in Gallatin County, Illinois. At the time of the Mexican-American war , 1846, Lawler joined the 3rd Illinois Volunteer Infantry with the rank of captain. After the war, he returned to his farm in Illinois, studied law, passed the bar exam and practiced law in Galatin County. In May, 1861 he recruited the 18th Regiment Illinois Infantry and was commissioned Colonel of the Regiment

troops. In December, Company K left the Regiment for Memphis and the Marine Brigade (Ram Fleet). May 18, 1863, the Eighteenth was ordered to Memphis and transported down the Mississippi River to Haines' Bluff on the Yazoo River; however, the Eighteenth saw little action at the siege and fall of Vicksburg. In July and August, the Eighteenth moved to Helena, Arkansas, followed by the battle and capture of Little Rock in September. They remained on duty at Little Rock, Pine Bluff and Duvall's Bluff through December. In 1864 the Eighteenth fought at Mount Elba, and in May, the three-year term of service expired for many of the men, and they returned to Springfield. The Eighteenth was reorganized with two companies of Veterans, one company of three-year recruits and seven companies of one-year recruits. The reorganized regiment remained at Little Rock, Arkansas and was mustered out there on December 16, 1865.[326]

There are no national flags of Union Quartermaster issue at the State of Illinois Military Museum. The national flag shown here in obverse and the reverse view is "homemade" and perhaps came with the replacement Companies "C" and "H" from Carbondale.

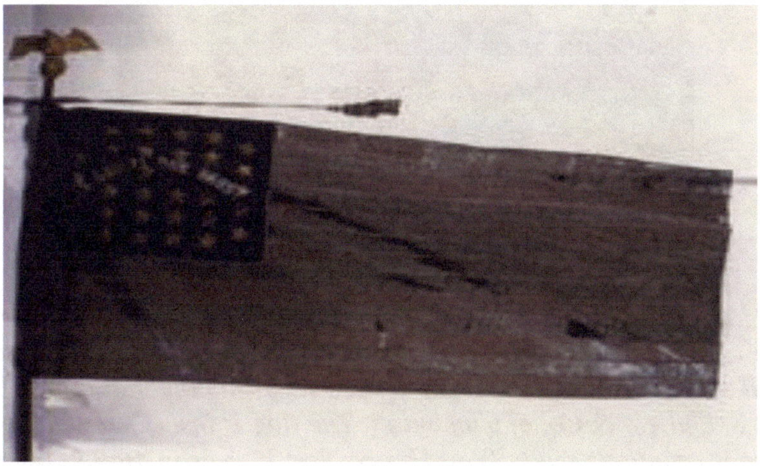

*This flag measures 42" x 90" and tassels are silver instead of the regulation blue and white and the stars are painted gold. The inscription in the canton reads "*Carbondale C. No.1.*"*

*This is the reverse of the flag. It is inscribed: "*In God is Our Trust.*"*

[326] *AGR,* Vol. II, p91-7.

According to *The Prairie Pioneer,* a national flag was presented to Company G, the *Wayne Guards*,[327] on May 14, 1861. "BANNER PRESENTATION by Miss Amelia Clemson, on behalf of the Ladies of Fairfield. Husbands, brothers and friends, on behalf of the patriotic ladies of Fairfield, I present you with a National Banner, our emblem of liberty and the sacred rights of humanity. This is the result of our toil—the work of our own hands—and in presenting it to you, we feel that we are confiding it to a sacred trust—that while the lifeblood beats warm in the heart, that flag will never be insulted; that ensign, made sacred by the glorious memories of the past, shall never trail in the dust in disgrace." There is no further information on the whereabouts of this flag.

A large center portion of the regimental flag was removed, probably for souvenirs. The flag now measures 68" x 68". Although the flag has deteriorated, it appears that there were originally rays above the upper scroll and stars. Flags with this feature were manufactured by Gilbert Hubbard & Co. of Chicago and John S. Shilleto & Co. of Cincinnati. The flag is inscribed with gold paint in the upper left-hand corner. The inscriptions read: "Fort Henry, Fort Donelson, Shiloh, Siege of Corinth, Siege of Vicksburg, Little Rock." *The sword, pistol and regimental flag of Michael Kelly Lawler, Colonel of the Regiment were "deposited in Springfield."*[328] *This is presumably that regimental flag.*

No other n carried by this regiment are known to survive.

[327] *The Prairie Pioneer*, Friday May 17, 1861. This newspaper was printed in Fairfield, Wayne County, Illinois.

[328] Dorris, Jonathan T, "Michael Kelly Lawler, Mexican and Civil War Officer". *Journal of the Illinois State Historical Society,* Volume XLVIII, No. 4, (Winter 1955), pages 166-401.

19th Illinois Infantry – "*Chicago Light Infantry*"[329]

"Forever float that Standard sheet!
Where breathes the foe but falls before us,
With freedom's soil beneath our feet,
And Freedom's Banner streaming o'er us!"[330]

The Nineteenth Illinois Infantry has a unique history in that four of its companies were organized before the fall of Fort Sumter and Lincoln's call for troops. The *Chicago Highland Guards* (Company E of the 19th) was an organized company in State service since 1855. After Lincoln's election in 1860, a company of "Wide-Awakes" formed the *Chicago Zouaves* with James Hayden as its Captain (he had been First sergeant of the original Ellsworth Zouaves Company). The *Chicago Zouaves* later became Company A of the 19th, captained by Hayden and Company K of the 19th, captained by John Clybourne. The *Chicago Light Infantry* commanded by Captain Frederick Harding was already organized as the fourth Chicago militia before Lincoln's call and became Company D of the 19th.

Since this regiment was to be filled from the State at large, the other six companies were: Company B from Stark County (*Elmira Rifles* or *Osceola Rifles*); Company C from Cook and Lake County (*Chicago Zouaves* or *Palatine Volunteers*); Company D (*Chicago Infantry* or *Washington Infantry)* from Cook County; Company F from Cass County (*Cass County Guards*); Company G from Cook County (this company was detached in January 1863 as *Bridge's Battery*); Company H from Rock Island County *(Moline Rifles*); and Company I from Cook County (*Anti-Beauregards*), Company K from Cook County (*Chicago Zouaves*). The full regiment was mustered into the United States service on June 17, 1861, under the command Colonel John B. Turchin[331]. The regiment was sent to Missouri in July and August but, in September, was ordered to Washington, DC. Enroute on the *Ohio & Mississippi Railroad*, the train broke through a bridge, and 24 men were killed and 105 injured. The transfer to Washington, DC, was canceled, and the regiment went into camp at Louisville and then Elizabethtown, Kentucky, for the remainder of the year. In 1862 the Nineteenth occupied Bowling Green until mid-February, when they advanced to Nashville, Tennessee and remained there until mid-March. April 9, the Nineteenth attacked Huntsville, Alabama and captured 170 prisoners along with 17 locomotives and 150 freight cars. Two days later, the regiment captured Decatur and Tuscumbia, Alabama and returned to Huntsville. In May, the Regiment was ordered to Fayetteville, Tennessee, and on June 2, joined the Expedition to Chattanooga. Mid-June, the regiment returned to Fayetteville, were on expeditions to Larkinsville and Stevenson, Alabama and guarded the railroad and bridges from Huntsville to Decatur until August. When the rebel army left Chattanooga to invade Kentucky, the Brigade, including the Nineteenth, was withdrawn to Nashville. In late December, the Nineteenth marched south and arrived at Murfreesboro on December 29. In the ensuing battle at Stone River, the Regimental commander Colonel

---

[329] Also known as *Washington Independent Regiment*

[330] Haynie, Henry, *The Nineteenth Illinois, A Memoir of a Regiment of Volunteer Infantry Famous in the Civil War of Fifty Years ago for its Drill, Bravery, and Distinguished Services.* Chicago: M.A.Donohue & Co., 1912. p383. (poem written by Joseph Drake, Co. K of 19th Illinois Infantry).

[331] Born Ivan Vasilovitch Turchininoff in Russia in 1822, his father served in the Russian military. Due his father's influence, Ivan attended a premier Russian military school and after graduation was assigned to Imperial Guards in St. Petersburg.In 1856 he and his wife left Russia and settled in Chicago and he changed his name to John Basil Turchin. In June 1861, he resigned his position with the *Illinois Central RR* and accepted a commission as Colonel of the 19th Illinois Infantry.

Scott was killed, and Colonel Alexander Raffen assumed command. A private in Company D captured a rebel flag.[332] The Nineteenth remained at Murfreesboro until June, then patrolled middle Tennessee until August. In September, the Nineteenth crossed the Tennessee River in the Chattanooga campaign; on September 11, they met the rebels at Davis's Cross Roads, Georgia and on September 18-21, fought at the great battle at Chickamauga. The Union forces retreated into Chattanooga and were besieged by the rebel forces from September 24 to November 23. When the Union took the offense in late November, at the battle of Missionary Ridge, the flag of the Nineteenth was the first flag planted on the summit[333]. In 1864 the Nineteenth joined Sherman's Atlanta campaign and fought at Buzzard's Roost Gap, Rocky Faced Ridge, Resaca, Kingston, Dallas, New Hope Church, and Allatoona Hills, Georgia. On June 9, the regiment marched to Ackworth, Georgia, moved to Chicago and was mustered out on July 9, 1864.[334],[335]

Both a National and Regimental flag from Nineteenth Illinois is in the State of Illinois Military Museum.

On the top stripe of this flag is inscribed:" STONE RIVER, Dec. 31, 1862, Jan. 2, 1863." *On the center red stripe in gold paint is inscribed:* "19th REGT. ILL VOLL." *The flag measures 69" on the staff by 71" on the fly. The canton is rectangular with gold stars characteristic of flags manufactured by Gilbert and Hubbard Company of Chicago.*

---

332 Ibid, Haynie, p 191. "A private in Company D had the good fortune to capture a rebel flag in that wonderful dash beyond Stone River; and it was afterward exhibited at the great Sanitary Fair in New York which Mrs. Mary Livermore had gotten up for the benefit of sick and wounded soldiers." This may refer to the flag of Wright's Tennessee Battery

333 Ibid. pp260-281;374;392.

334 *Report of the Adjutant General of the State of Illinois*, Vo. II, p91-7

335 ibid, Haynie

This Regimental flag is inscribed: " Presented by the Citizens of Chicago.*" It measures 70" on the staff by 72" on the fly. The pattern of the flag is not typical of flags manufactured for the Union Quartermaster. This is probably a specially designed flag made in Chicago.*

There are several other flags that were carried by this regiment.

"Immediately after the battle of Stone River, the people of Chicago determined to present our regiment with a new Stand of Colors. Our flags, already worn by usage, whipped into holes by the winds, riddled by bullets in numerous skirmishes or heavier engagements, were almost entirely done for in those two days near Murfreesboro, and this fact having come to the knowledge of our friends at home, in due course, Flags bright and beautiful, as rich and costly a set as ever gladdened any regiment, were in our hands; and for a matter of one year and five months, although in many a combat, never were those colors lowered in defeat."[336]

Unfortunately, these colors no longer exist, as described in the following:

"The Chicago Historical Society's rooms are full of historic trophies and histories, although some of its greatest treasures were destroyed in the great conflagration of 1871, including the colors of the Nineteenth given after the battle of Stone River, together with the Flags of Batteries A and B, given them after the battle of Fort Donelson;".[337]

There is a national flag of the 19th Illinois at the DuPage County Historical Museum in Wheaton, Illinois. This flag was folded and placed in an old lap desk when it was donated to the Museum. A card accompanying the flag reads, "First Union Flag flown at battle of Missionary Ridge, given by Col. Alexander Raffen to his daughter Mrs. C. Eadie of GE (Glen Ellyn)."

[336] Ibid, p 383-4.

[337] Ibid. p 62.

This silk flag, in the Museum in Wheaton, is not of a national issue and has 33 gold-painted stars outlined in red arranged in a star-shaped pattern on the canton. There is a blue border on the hoist end and a 2 ½" yellow-gold fringe on the other three sides. The flag measures 50" x 76". When I photographed the flag, I could not unfold it more than one-half because of the deterioration of the silk.[338]

A company flag was given to Company D: "At the theatre, on Saturday evening J.H. McVicker presented to Captain Fred Harding's Company an elegant silk flag bearing the motto, 'Retaliation—no mercy to traitors!'"[339] The location of this flag today is unknown.

The *Highland Guards* Company E-was presented a flag prior to its departure to Cairo as part of the Cairo Expedition. "The Highland Guard of this city, just before leaving for Cairo, were presented with a handsome flag, which was made by Miss Gabriel, whose father is First Lieutenant of the Company."[340] The further history of this flag is unknown.

Mrs. E. V. Carney, Mrs. Carrie Roche, and Miss Charlotte Carney presented a set of guide colors or guidons to the Regiment on November 11, 1861, prior to the regiment's departure from Chicago.[341]

The *Zouave Gazette,* published by the members of the Nineteenth Illinois Regiment, noted the following in the issue of November 15, 1861. "Lieut. Bremmer of company E arrived in camp on Wednesday morning, bringing with him a splendid set of guide colors for the Nineteenth Regiment, a present from Mrs. and the Misses Carney of Chicago. They are of heavy blue silk, on which is inscribed in gilt letters, '19th Reg't Ill. Vols.' The staff is of black walnut surmounted by an eagle, from whose beak depends on a pair of silken tassels. They are a magnificent present, and the Nineteenth boys appreciate them." Today's location of these two guide colors is unknown.[342]

"The Chicago Nineteenth—Our readers will doubtless recollect that a few days ago, allusion was made to the fact of the colors of this regiment being so soiled and battle-worn as to be unfit for further use. Messers. Chambers & Regan, sign and banner painters at 116 Clark Street, have just completed a most magnificent set for the regiment, the gift of the patriotic friends of this favorite regiment. Mr. Chambers is an artist of no mean spirit, and the canvas under his brush becomes instinct with life. The set of colors referred to, are really superior and will doubtless be gallantly borne in the front on many a hard-fought field."[343] There is no further record re this stand of colors.

No other flags carried by this Regiment are known to survive.

---

338 Personal observation, author.

339 *Chicago Tribune,* April 22, 1861.

340 *Chicago Tribune,* April 25, 1861

341 *Chicago Evening Journal,* November 27, 1861

342 From the *Zouave Gazette,* November 15, 1861 as quoted in the *Chicago Times,* November 22, 1861

343 *Chicago Daily Tribune;* February 17, 1863, p.4, c.3.

20th Illinois Infantry

"Come Tigers, let us raise our flag,
And stand beneath its stars,
With hands upraised and eyes fast fixed
Upon its crimson bars;
And let us swear by all that's right,
By all that's good and true,
That in the din and heat of fight
We'll keep its folds in view.
Then raise a rousing shout my boys
For the Girls of Old Champaign,
Whose fingers worked our Stars and Stripes,
For the bloody battle plain.[344]

The men of the 20th volunteered from the following counties: Champaign County (Company A-*Champaign Tigers or Champaign County Volunteers);* Will County *(*Company B and part of C)*;* McLean County(Company *C*); Livingston County (Company D); DeWitt County (Company E); Bureau County (Company F); Kankakee County (Company G); La Salle and Putnam Counties (Company H); Iroquois County (*Iroquois Light Guards*-Company I) and Kendall County (Company K). The Regiment mustered in for three years of service at Joliet on June 13, 1861, under the command of Colonel Charles Carroll Marsh.[345] They were transferred to Camp Pope at Alton, Illinois and then to Cape Girardeau, Missouri[346]. In October the Twentieth fought their first battle at Fredericktown, Missouri, then went into quarters at Birds Point, Missouri for the remainder of 1861. In February 1862 the Regiment sailed to Fort Halleck, Tennessee then moved to occupy Fort Henry. Later that month they were in heavy fighting at Fort Donelson and suffered 132 casualties. On April 6, Johnston's Confederate forces attacked the Twentieth at Pittsburg Landing (Shiloh). One hundred and thirty-six men were lost in the two days of fighting and John Wade McDonald, Company E, was awarded the Medal of Honor.[347] After the battle the Twentieth participated in the siege of Corinth, Mississippi (April 29-May 30) and spent the summer in pursuit of the enemy finally going into camp in Oostanaula, Tennessee in August. November 1862 until February 1863 the Regiment marched and counter marched and skirmished in Tennessee and Mississippi. In April 1863 the Twentieth joined Grant's forces in the Vicksburg campaign and fought at Port Gibson, Mississippi (May 1), Raymond (May 12), Champion's Hill (May 16), Big Black River (May 17) and finally in the assaults and siege of Vicksburg. After the fall of Vicksburg on July 4 the Regiment remained in Vicksburg until November 6

---

[344] Eugene Fauntleroy, "The Girls of Old Champaign", *Central Illinois Gazette,* June 26, 1861. (Company A was known as the "*Champaign Tigers*.")

[345] Colonel Marsh was 32 years old, a native of New York and after he moved to Chicago had become Captain of the *Chicago Light Guard.* With the outbreak of the Civil War, Marsh was called to Springfield and given command of 20th Illinois Infantry.

[346] *Iroquois (County)Republican*, August 8,1861. In a letter to the newspaper, Ezekiel B. Sleeth (Sgt of Company I) mentions that his unit captured a 'traitorous flag" off of a "tall liberty pole" after entering Dallas, Missouri on July 24, 1861.The flag was subsequently torn to pieces and "souveniered."

[347] *Civil War Medal of Honor Winners from Illinois,* published by the Civil War Centennial Commission of Illinois, July 1962. "Was severely wounded while endeavoring, at the risk of his life, to carry to a place of safety a wounded and helpless comrade."

when they went into camp at Big Black River, Mississippi. In February 1864 the Twentieth marched with General Sherman's expedition to Meridian, Mississippi. Shortly afterward the three-year term of the Regiment expired—March 1864—and those who re-enlisted left for home on veteran furlough. In May, the Veteran Twentieth Infantry Regiment marched through Decatur and Warrenton, Alabama, and then Rome and Kingston, Georgia to Acworth, Georgia and in June joined Sherman's Atlanta campaign. They fought at Kennesaw Mountain and skirmished at Nick-jack Creek, Marietta, and the Chattahoochee River. The Regiment went into battle at Atlanta with only 160 men; the Regiment was surrounded on July 22 and many men were captured. The Twentieth came out of the battle with only 16 men left.[348] Within a few days some of the men reappeared bringing the strength of the unit back to 35 men. The remnant of the 20th became the mounted scouts and bodyguard of General Leggett. They were engaged in actions at Jonesboro and Lovejoy Station, Georgia in September and in the pursuit of Hood into North Georgia and Alabama. They "marched with Sherman to the sea" and fought at Savannah and Beaufort, Georgia, Whippy Swamp, South and North Edisto River, and Columbia, South Carolina and Bentonville, Goldsborough and Raleigh, North Carolina. The Regiment occupied Raleigh on April 14 and were present at Bennett's House on April 26 for the surrender of Johnston and his army. The Twentieth marched in the Grand Review in Washington, DC on May 24, 1865, and were mustered out at Louisville, Ky. on July 16, 1865.[349], [350]

There are three National and two Regimental flags of the 20th Illinois Infantry in the State of Illinois collection.

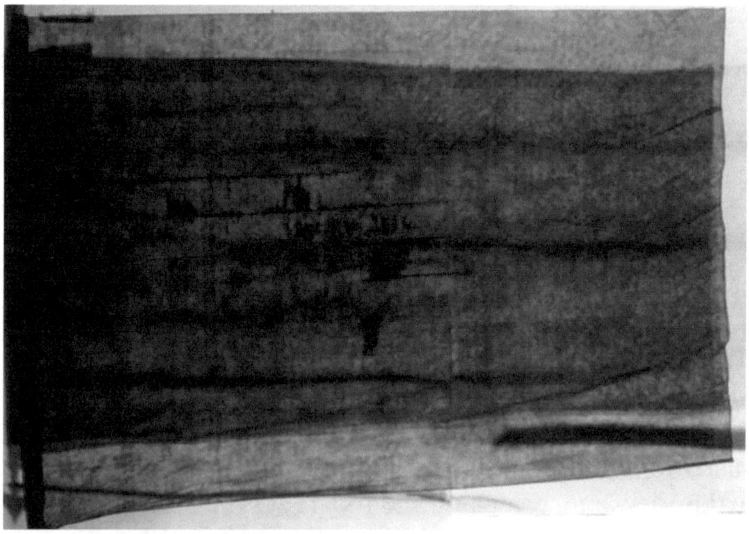

This National flag is contained in black silk netting but only a few silk shreds of the original flag remain.

[348] *History of Champaign County,* Brink, McDonough & Co., Philadelphia, 1879, p153. See below regarding loss of flag in this action.

[349] Behrens, Robert H., *From Salt Fork to Chickamauga,* The Urbana Free Library, Urbana, Illinois (1988). P29-37.

[350] Woodruff, George. *Fifteen Years Ago: or the Patriotism of Will County,* Joliet Republican Book and Job Steam Printing House, Joliet, 1876. Pp80-142

*The inscription on this flag is difficult to read:"*20th REGIMENT INFANTRY ? ? ? at Fredericktown, Donelson, Shiloh & *?" This flag appears to have been manufactured by the Gilbert Hubbard Company of Chicago.*

This National flag measures 70" on the staff by 75" on the fly. No inscriptions are seen.

This Regimental flag measures 71" on the hoist by 76" on the fly. The flag appears to be of New York manufacture.

A National flag inscribed with a motto was presented to Company A of the 20th Illinois Infantry. The current location of this flag is unknown, but perhaps it is the flag referred to in the poem above. A Champaign County newspaper[351] reports, "Two good hearted little girls, Miss Cady and Miss Nutting, by the aid of the citizens had procured a very handsome flag to present to the volunteers. Mr. Stoughton made a very fine presentation address which elicited great enthusiasm. Capt. Wolfe on receiving the flag in behalf of the company, made many telling appropriate remarks. The motto-- In hoc signe vinces -- By this Sign We Conquer --was heartily responded to by the Captain, who pledged his company, with one shout of enthusiasm, to preserve untarnished those Stars and Stripes and to bear aloft the old ensign that has waved triumphantly over many a field of battle and to return it without the loss of a single star or stripe. Three beautiful swords had been procured by the citizens and was presented in a speech by Mr. Oliver. One to Capt. Wolfe, one to 1st Lieutenant D. Bradley, and one to 2nd Lieutenant Kennard. .... Our citizens determined to uniform the company. They dispatched a committee to Chicago which purchased 500 yards of cloth and nearly all the new brass buttons in the city -- Col. Reed was appointed superintendent and with one accord the ladies of Champaign and Urbana turned out to aid in making up the uniforms for the company and with the help of several sewing machines finished them in time to send today. Surely the patriotic spirit which moved our Revolutionary mothers still lives."

---

[351] *Central Illinois Gazette,* May 13, 1861.

The flag above is a Regimental flag of the Twentieth. Most of the detail of the eagle in this flag is lost. However, rays above the eagle are found on flags manufactured by Gilbert Hubbard & Co of Chicago. This flag measures 69" on the staff by 76" on the fly. It is inscribed " Presented by the Citizens of Chicago for Gallant Conduct on the fields of Fredericktown & Donelson*". Although none of the Companies in the 20th came from Chicago, the Colonel of the Regiment, C. Carroll Marsh, was from Chicago. Therefore, after the fall of Fort Donelson the Twentieth was presented a stand of colors from Marsh's friends in Chicago.*[352,353]

There is record of another flag having been presented to this Regiment. "On the 17th of June (1861), another flag presentation occurred at the camp, when the ladies of Joliet presented Co. F, Capt. Hildebrant, with a beautiful flag. The speech on this occasion was made in behalf of the ladies by J. E. Streeter, Esq., and the reply was made by the gallant captain of the company."[354] There is no further description or history of this flag.

A flag was given to the Regiment, again by the ladies of Joliet, in 1863, "Wednesday March 11, 63 Elder Button returned from Joliet, Illinois with a flag for the Regt. He presented it tonight in a speech on behalf of the home people who sent it."[355] No further mention is made of this flag in Geer's diary and the location of this flag today is unknown.

Company E of DeWitt County was presented a flag in 1863. "AN ELEGANT FLAG—Captain North has taken back to company E of the Twentieth, an elegant silk flag to replace a flag worn out in service. The flag was presented by Colonel C. C. Marsh, late commander of that regiment. It is an elegant and costly affair, and reflects great credit upon the brave and generous donor."[356] The whereabouts of this flag today is unknown.

On August 1, 1861, it was reported that "a very handsome silk flag was presented to the Color

---

[352] *Chicago Daily Tribune, February 28, 1862; March 6, 1862.*

[353] Anderson, Mary Ann, ed. *The Civil War Diary of Allen Morgan Geer,* Cosmos Press, New York, 1977, p23.

[354] *Fifteen Years Ago: or the Patriotism of Will County, p29*

[355] *The Civil War Diary of Allen Morgan Geer,* p81. Elder Button probably refers to the chaplain of the Regiment, Charles Button from Joliet.

[356] *Clinton Union,* Saturday, October 10, 1863.

Guards." The flag was presented at Onarga, IL by Mr. Wm. B. Pierson in a neat, appropriate speech...which was responded to by Capt. Andrewson a brief, characteristic address." There is no description of the flag.[357]

None of the flags in the Illinois collection are identified as having been captured. However, apparently at least one flag of the Twentieth was captured during fighting before Atlanta. "Friday, January 10, 1890. THE OLD BATTLE FLAG. Mr. Frank Converse, of Decatur, called at the Public office the other day with the remnants of the old battle flag of the Twentieth Illinois Infantry. When Martin Mohrle was at home on furlough in 1864 he had a few shreds of the flag with him, part of which he gave his wife and she divided with Mrs. Frank Converse. Martin Mohrle was the color bearer of the Twentieth, and at the battle of Atlanta, on the 21st of July, 1864, he fell mortally wounded on the battle field and the flag was captured by the rebels. The shreds that Frank Converse had was all that was left of the flag under which the gallant Twentieth marched and fought from its first fight at Fredericktown, Mo., on the 21st of October 1861."[358]

No other flags carried by the 20th Regiment of Illinois Infantry are known to survive.

[357] *Iroquois Republican,* August 1, 1861. This presumably refers to the Company I of the 20th Illinois Infantry–*Iroquois Light Guards.*
[358] *Clinton Public,* Friday, January 10, 1890.

21st Illinois Infantry – "*Alexander's Infantry*"[359]

We're fighting for our Union,
We're fighting for our trust,
We're fighting for that happy land
Where sleeps our father's dust.
It cannot be dissevered
Though it costs us bloody wars;
We never can give up the land
Where floats the stripes and stars.
Hurrah! Hurrah!
For equal rights Hurrah!
Hurrah for the brave old flag
That bears the stripes and stars.[360]

The Twenty-first Illinois Infantry Regiment, from the Seventh Congressional District, was mustered into service for thirty days on May 15 at Mattoon, Coles County, under the "Ten Regiment Bill." On June 28, 1861 the Regiment was mustered in for three years of service under command of Colonel Ulysses S. Grant.[361] It was composed of companies from the following counties: Macon County (Company A), Cumberland County (Company B), Piatt County(Company C), Douglas County(Company D), Moultrie County(Company E), Edgar County(Company F), Clay County(Company G), Clark County(Company H), Crawford County(Company I) and Jasper County(Company K). On June 28, 1861, the Regiment proceeded to Ironton, Missouri for reconnaissance and instruction in company and battalion drill. In August Grant was promoted to Brigadier General and John Alexander was elected Colonel of the Regiment. In October the Regiment fought at Fredericktown, Missouri then returned to Ironton. 1862 saw the men in Missouri and Arkansas and they marched to Hamburg Landing, Tennessee in May and arrived during the last days of the siege of Corinth. In August the Regiment marched through Mississippi, Alabama, and Tennessee to arrive at Louisville, Kentucky in late September. The Twenty-first marched from Louisville on October 1 in pursuit of Bragg and met the rebels at Perryville on October 8 and at Stanford on October 14; the pursuit reached as far as Crab Orchard, Kentucky, then the Regiment returned to Nashville. In late December the Twenty-first moved southward to Murfreesboro, Tennessee and were engaged at the battle of Stone River December 30-31, 1862, and January 1-2, 1863. At that Battle the regiment lost more men than any other regiment engaged.[362] In 1863 the Twenty-first marched with General Rosecrans' army from Murfreesboro to Chattanooga, fought at Liberty Gap on June 24-27 and went into camp at Winchester, Tennessee. In August the Regiment crossed the Cumberland Mountains to Stevenson, Alabama, then crossed the Tennessee River and in September crossed Lookout Mountain and camped 50 miles south

---

[359] Also known as "*Seventh Congressional District Regiment*" and *"Mattoon Regiment"*

[360] *Decatur Weekly Gazette* (Macon County) May 13, 1863

[361] Grant was to become Commander in Chief of the Union Army and finally President of the United States.

[362] First Lieutenant, Nineveh S. McKeen of Company H was awarded the Medal of Honor for action at Stone River- "Conspicuous in the charge, where he was three times wounded," and at Liberty Gap where he "captured the colors of the 8th Arkansas Infantry, CSA." *Civil War Medal of Honor Winners from Illinois,* Published by the Civil War Centennial Commission of Illinois, July1862.

of Chattanooga. Later that month they marched north and fought at Dug Gap in the Pigeon Mountains and then in the battle of Chickamauga September 19 and 20. At Chickamauga the Regiment lost 238 officers and men; among the officers killed was the colonel of the Regiment, John W.S. Alexander. The remainder of that year the Twenty-first remained in camp at Bridgeport, Alabama. In March 1864 the term of the Regiment expired and after re-enlistment and a month's furlough the veterans rejoined the Regiment at Kennesaw Mountain to take part in the Atlanta campaign. The Regiment saw action at Buzzard' Roost Gap, Resaca, Dallas, New Hope Church and Allatoona Hills, Kennesaw Mountain, Chattahoochee River, Peach Tree Creek, and the siege of Atlanta. In a flanking movement around Atlanta the Twenty-first fought at Jonesboro and Lovejoy Station. In October they marched in pursuit of Hood to north Georgia and into Alabama. After a few days' rest they marched to Chattanooga and acted as escort for the wagon train of the Fourth Corps that passed over the Cumberland Mountains and arrived at Pulaski Tennessee in mid-November. The Twenty-first skirmished with the enemy at Columbia, Tennessee in late November. On November 30 the Twenty-first fought at Franklin, Tennessee, and on December 15-16 fought at Nashville. The Regiment pursued the enemy to the Tennessee River then marched to Huntsville, Alabama where they remained until March 13, 1865. Finally, they were sent to Texas by way of New Orleans. At New Orleans they embarked on a vessel to Matagorda Bay, disembarked for Victoria, Texas and marched to San Antonio where they were mustered out December 16, 1865.[363]

There are two National flags and one Regimental flag of the Twenty-first Illinois Infantry in the flag collection of the State of Illinois.

This National flag measures 70" on the hoist by 76" on the fly. Still visible on this flag are the following battle honors: FREDERICKTOWN; CHAPLIN HILL; STONE RIVER; LIBERTY GAP; CHICKAMAUGA; KENESAW; 21st ILL"S, V.V.I.; ATLANTA; JONESBORO; FRANKLIN; NASHVILLE . *The Official Army Register lists additional battle honors for this regiment which do not appear on the flag pictured above, the other honors are* "CORINTH, SMYRNA STATION, PERRYVILLE, KNOB GAP, STONE RIVER, LOVEJOY'S STATION, CHICKAMAUGA, PINE TOP."[364]

[363] *Report of the Adjutant General of the State of Illinois,*.H.W. Rokker, State Printer and Binder, Springfield, IL, 1886, Vol II, pp214-7

[364] *Official Army Register of the Volunteer Force of the United States Army for the Years 1861,"62, '63, '64, '65.* Part VI, p 245. (Reprinted by Ron Van Sickle Military Books, Gaithersburg, MD. 1987)

On the second national flag of the Regiment, note the Regimental designation on the center stripe, "21st REG ILL VOL INF". *Based on the arrangement of the stars, this flag was manufactured in Chicago by Gilbert Hubbard & Co. The measurements of this flag are not available.*

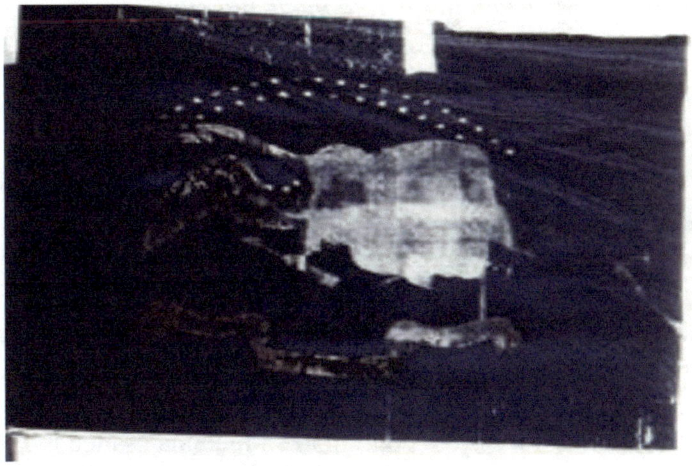

This Regimental flag is in poor repair with a major portion of the eagle missing.

The flag measures 70" on the hoist by 73" on the fly.

A silk flag was made by the "ladies at Paris. Ill., and presented to his company (Company F) when they marched away to be sworn into service. It was kept after the War by the flagbearer of the company, and at his death, the remaining members of the company voted to give it to Mr. Davis to keep. The writer saw the remains of what had evidently been a very beautiful flag, now rotted by time, but cherished for the sake of the past, where Mr. Davis keeps it in a special box, from which it cannot longer be taken as it would fall into tatters were it unfolded."[365] Paris is in Edgar County and the soldier referred to is Perry H. Davis who served in company F of the Twenty First Illinois Infantry. The whereabouts of this flag today is unknown, however, based on its condition in 1926, the flag has probably completely disintegrated.

There are no other flags carried by this Regiment that are known to survive.

---

[365] *Lynn Haven Free Press,* Lynn Haven, Florida, October 9, 1926

22nd Illinois Infantry

Unfurl the flag—our country's flag-
Upheld by gallant hearts and true—
A Star for every sovereign State
Emblazoned on its field of blue.
No mutilated banner ours
Of dear-bought honors basely shorn,
With half its glory-beaming stars
In frenzy from the Union torn.
Then be it ours to fence from harm
The glorious flag baptized in flame:
To keep its stars and stripes intact
The patriot's pride—the traitor's shame.
Still shall that banner stainless float
As when in joy and pride unfurled
Millions of hearts its color-guard
The hope of Freedom through the world.[366]

The Twenty Second Illinois Infantry was organized at Belleville, Illinois with men from the following counties: Clinton County (Company A); Madison County (Company B); St. Clair County (Companies C and K); Bond County (Companies D and E); Monroe County (Company F); Marion County (Company G) and Randolph County (Company H- *Union Guards*- and Company K). They were mustered in for three years at Caseyville, Illinois on June 25, 1861, under the command of Colonel Henry Dougherty.[367] In July they moved to Bird's Point, Missouri and on November 7 under General Grant were engaged at the battle of Belmont, Missouri where the Regiment lost 141 killed, wounded and missing. In January 1862, again under Grant, the Twenty Second made a reconnaissance into Kentucky to the rear of Columbus. This was followed by operations at New Madrid, Island No. 10 and Sikeston Missouri. In April the Regiment captured Tiptonville and marched on to Fort Pillow and Hamburg Landing Tennessee where they were involved in the siege of Corinth, Mississippi—April 29-May 30. In August the Twenty-Second marched to Nashville, Tennessee and fought in the siege of that city until mid-November. On December 26 they advanced on Murfreesboro, Tennessee and fought at the Battle of Stone's River on December 30 and 31 and January 1 and 2, 1863. Here the Regiment lost 199 of the 312 men who went into action. They remained at Murfreesboro until June. In August the Twenty-second crossed the Cumberland Mountains and the Tennessee River and September 19 and 20 the Regiment fought at Chickamauga where they lost 96 men in less than ten minutes. During October and November, the Regiment fought in the siege and battle of Chattanooga, fighting at Orchard Knob on November 23 and 24

---

[366] *Randolph County Democrat* (Chester), a portion of a poem by Francis A. Durivage. April 27, 1861. Co. H and I of the 22nd Illinois Infantry came from Randolph County.

[367] Colonel Henry Dougherty was born in North Carolina in 1827. His family moved to Carlyle, Clinton County, Illinois where his parents died, leaving him an orphan at age 8. He worked on a farm until age 16, Following a year with a Rocky Mountain Fur Company he enlisted as a private in *First United States Dragoons* in St. Louis. He fought in the Mexican War with the same military unit. After the close of the Mexican War, he joined Colonel Sumner's command and fought against the Navajo and Apache Indians. In 1852 he returned to his farm in Carlyle. At the outbreak of the Civil War, he was elected Colonel of the 22nd Illinois Infantry. *Civil War Harper's Weekly,* November 23, 1861

and at Mission Ridge November 25. In early 1864 the few remaining men fit for duty were in operations in the mountains of Eastern Tennessee. After Veterans furlough and recruitment in February and March 1864 the Regiment joined the Atlanta campaign and fought at Rocky Faced Ridge, Buzzard's Roost Gap, Resaca, Adairsville, Dallas, New Hope Church, Allatoona Hills and Kennesaw Mountain, Pine Hill, and Lost Mountain. The Regiment was mustered out on July 7, 1864. Those Veterans who reenlisted and recruits whose terms had not expired were transferred to the 42nd Illinois Infantry.[368]

There are currently five flags of the Twenty Second in the collection of the State of Illinois Military Museum in Springfield.

In 1863 a stand of colors from the Twenty Second was retired to Springfield in exchange for a new stand. As written by an officer of the regiment "We have just received from Springfield, Ill., a very fine stand of colors, for which we exchanged our old one for. They are now at Springfield, Ill., honorably discharged from the service for wounds received in action."[369]

Apparently "stand of colors" refers to a single flag since the Centralia Sentinel states: "Adjutant General Fuller received on Friday last, the old colors of the 22nd regiment, (Col. Dougherty's) Illinois Volunteers, accompanied by the following note; "This flag was presented to the 22d regiment at the organization at Belleville, May, 1861 by Judge Koerner, and was borne through the following conflicts: "Charleston, August 19, 1861; Belmont, November 7, 1861; New Madrid and Island No. 10, March and April, 1862; Farmington, 3d and 4th May; siege of Corinth, May 1862; Stone River, 30th and 31st December, 1862 and 1st 2d and 3d January 1863."[370]

The presentation of the above flag to the Regiment is reported in the June 22, 1861, issue of the Belleville Democrat. "Last evening (the 19th) the 8th[371] or the Col. Dougherty's Regiment left the Camp and marching by way of Illinois Street, preceded by their band, arrived in front of the Court House, where they formed by companies immediately opposite the facade of that structure. The band then ascended the steps and played a piece of popular music, when Judge Koerner presented a remarkably neat and rich silk flag to the regiment and accompanied its presentation by a little speech, in which many well-chosen and appropriate sentiments were uttered. We were agreeably disappointed by the well-timed and temperate remarks of the Judge, and when he adjured the soldiers of the 8th Congressional District to remember that they were going to war with their brethren, and in view of that fact, to so temper their conduct that no unnecessary cruelty should be laid to their charge,.....the flag was delivered to its future custodian--the standard bearer of Capt. Holcomb's company....Captain Abbott replied on the part of the regiment."[372] Note the reference to the 8th Regiment. This was the early name of the Twenty Second Regiment since it was raised in the 8th Congressional district.

The Adjutant General's Report published in 1886[373] lists a flag in the collection with the following honors: Charleston, Belmont, New Madrid, Island 10, Farmington, Siege of Corinth, Stone River. Presumably the above records all refer to the same flag. However, that flag is no longer in the collection and its whereabouts today is unknown.

---

[368] *Report of the Adjutant General of the State of Illinois,* Vol II, 1886, pp241-43

[369] *The Union Banner,*(published in Carlyle, Clinton County) July 2, 1863. This information is contained in a letter to the Editor from Major Samuel Johnson of the 22nd Ill., Volunteer Infantry.

[370] *Centralia Sentinel,* June 4,1863

[371] Refers to this regiment being the 8th Congressional District Regiment

[372] *Belleville Democrat,* June 22, 1861.

[373] *AGR,*Vol I, p 160.

The above National flag measures 58" on the hoist by 77" on the fly and is characteristic of flags manufactured by Hugh Wilkins of Louisville—gold stars in vertical rows. Flags from this manufacturer typically came on a staff topped by a sheet metal finial painted red. From this photo the finial details cannot be seen.

This national flag measures 73" on the hoist by 69" on the fly. No inscriptions appear on the flag. Horizontal rows of gold stars on a square canton with rows of 7, 7, 6, 7, 7 stars is typical of national flags manufactured in New York.

The above national flag is in poor condition and original size is unknown.

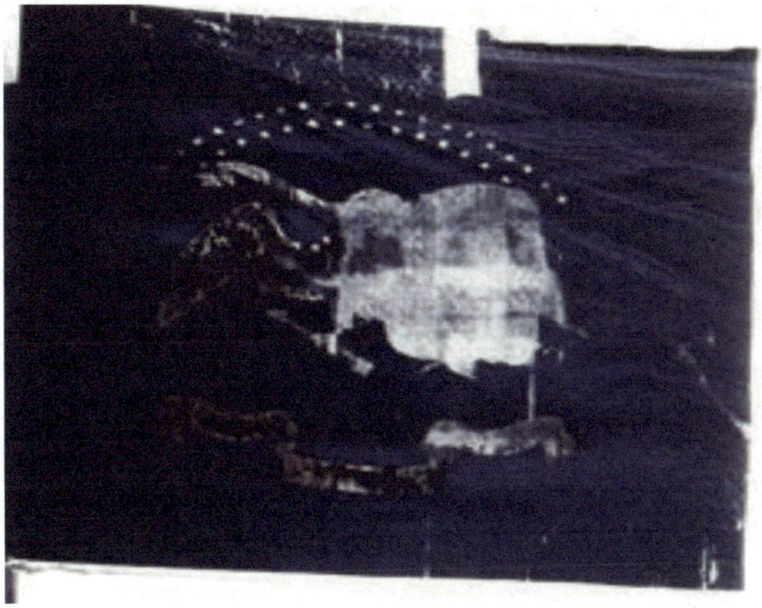

:

On the Regimental flag the major portion of the eagle is not present because of deterioration of the silk, however, the position of the stars and the banner under the eagle is similar to regimental flags manufactured in New York.

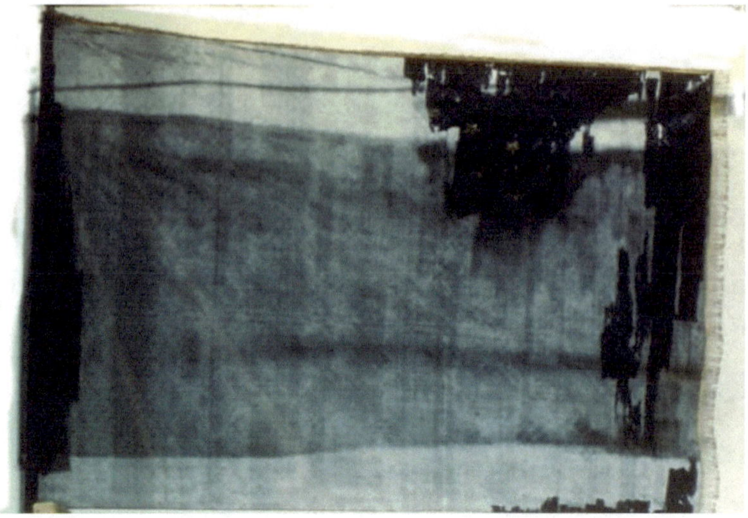

Only shreds remain of this regimental flag. No further information regarding this flag is known.

The Twenty Second Illinois captured at least one Confederate flag. "Our Regiment has been out on a little expedition through Mo., and I must say they chased old Jeff. Thompson as close as he had ever been driven yet; taking it all in all, it was a very interesting little scout. .... They captured a flag and took the caissons of one of Jeff's small guns, besides many other little trophies of their march....Yours as Ever, Henry Dougherty, Col. 22nd Ills. Vols."[374] This flag is not identified as being in the Illinois collection and no further description of the flag or from what Confederate regiment it was captured is known.

A unique patriotic gift was given by the ladies of Mount Vernon –"have recently sent to Cairo a beautiful comforter, for the sick bed of Col. Dougherty, who recently arrived there from Columbus. It is quite a magnificent affair with the thirteen stripes and thirty-four stars, and is a very appropriate present for one still confirmed to his bed by the wounds received in battling for his country's flag" [375] The "comforter" mentioned appears to actually be a flag.

No other flags carried by this regiment are known to survive.

[374] *The Advocate* (Belleville) March 14, 1862.
[375] *Chicago Tribune; Dec. 24, 1861.*

23rd Illinois Infantry – "*Irish Brigade*"

Ye sons of the Green Erin assemble,
And join in the battle array;
The usurpers and traitors shall tremble,
When they see the Brigade in the fray.
Go march to the battlefield proudly,
Let the foe at your might be dismayed,
And the trumpet of fame shall sound loudly,
The praise of the Irish Brigade.
In view of the guilt and the treason,
The Goddess of Liberty sighs;
Let us up and defend her in season,
And bring back the joy of her eyes.
Bear the stars and stripes proudly o'er you,
Nor e'er let your march be delayed,
'Til the foe flies in terror before you,
When charged by the Irish Brigade.[376]

Irish companies formed some of the early militia of Chicago. When Fort Sumter fell in 1861 these companies joined together with other volunteers to form a regiment of Irish volunteers –the "*Irish Brigade,*" or "*1st Irish Regiment*" or the "Twenty-third Illinois Infantry" commanded by James Mulligan. Companies included *Detroit Jackson Guard* -primarily from Detroit Michigan that formed Company A; the *Montgomery Guards*, an original Chicago military organization that was formed in 1842 became Company B; the *Chicago Jackson Guards* organized in 1853 was mustered as Company C; men from LaSalle County – the E*arl Rifles* became Company D; the men from Morris and Grundy Counties were the *Ogden Rifles* – Company E; the *Douglas Guards* from LaSalle County formed Company F; the *O'Mahoney Rifles* from Cook County became Company G; The *Ottawa City Guards* from LaSalle County were Company H and men from Cook County as *Shields Guards A* and *Shields Guards B* formed Companies I and K.[377] The Regiment mustered in for three years on June 15, 1861, under command of Colonel James Mulligan,[378]while in barracks at Kane's Brewery on West Polk Street in Chicago. In July they moved to Quincy by rail then on to the U.S. Government Arsenal at St. Louis. In August the Regiment was at Jefferson City, Missouri when it was ordered to Lexington, Missouri, a town on the south bank of the Missouri River. At Lexington the entrenched Regiment was attacked by Sterling Price and after three weeks of siege by the Confederates the Twenty-third surrendered on September 20. "Amongst the men of the 23rd that night the Regimental flag was torn into shreds. Each

---

[376] *Daily Chicago Times,* April 29, 1861. This is a portion of a poem "Dedicated to the Brigade now formed in Chicago by an Irishman, who, in heart and soul, sympathizes with the defenders of the Union, the Constitution and the Law. The author is identified only as "P.S.H"

[377] Von Allendorfer, Frederick, "The Western Irish Brigade, *Irish Sword II* (Dublin 1955), 177-83.and *Chicago Tribune, July 16, 1861.*

[378] Colonel Mulligan was born June 30, 1830, in Utica, New York of Irish descent. After his father's death, his mother remarried, and the family moved to Chicago. Mulligan read law and was admitted to the Illinois bar in 1856 and was commissioned a second Lieutenant in the *Chicago Shield Guards.* At the outbreak of the Civil War, Mulligan was instrumental in raising the 23rd Illinois Infantry also known as the *Irish Brigade. Wikipedia, the Free Encyclopedia,* 2023

received a small green patch, thus saving the flag from being taken as a trophy of war by the enemy."[379] The officers were held as prisoners while the men were released to find their way home. The Regiment was mustered out of service in October 1861. However, Colonel Mulligan, who had been exchanged, appealed to General McClellan who directed that the organization of the Twenty-third be retained and be considered as being continuously in service from the date of its original muster. The men of the Twenty-third reassembled in Chicago and guarded prisoners at Camp Douglas until June 1862 when they were ordered to Harper's Ferry, Virginia, then to New Creek, Clarksburg and Parkersburg, Virginia. In 1863 they remained in Virginia and fought at New Creek and Philippi and later were on Lee's flank during his retreat from Gettysburg. In April 1864 the Regiment was reorganized and after a month's furlough returned to Virginia. In July the Regiment fought at Leetown, Virginia, Maryland Heights, Maryland, Snicker's Gap, Virginia and Kernstown, Virginia where Colonel Mulligan was killed. "After he was twice wounded, his faithful followers raised him up and were about carrying him from the field. The rebels made a rush to capture the flag of the Irish brigade. Mulligan, perceiving it said--"Lay me down and save your flag." The command had to be repeated a second time before his comrades would desist from their labor of love. But they did obey; the flag was saved, and Mulligan fell into the hands of the enemy. He is a prisoner no longer. He has fought his last battle, and naught can awake him to glory again. ``[380] From August 1864 to December 1864 the Twenty-third fought under Sherman's command in the Shenandoah Valley and fought at Cedar Creek, Winchester, Charlestown, Berryville, Opequon Creek, Fisher's Hill and Harrisonburg. It was at Fisher's Hill that Private John Creed of Company C captured a Confederate flag for which he received the Medal of Honor[381]. In January 1865 the Regiment was transferred to Army of the James and was in front of Richmond and the defenses of Bermuda Hundred. In March and April, they were in the assault and capture of Fort Gregg[382] in front of Petersburg and afterward pursued Lee until he surrendered at Appomattox Court House. Sergeant Thomas McGraw of Company B received the Medal of Honor for "One of three soldiers most conspicuous for gallantry in the final assault at Petersburg, Virginia, April 2, 1865."[383] The Twenty-third was mustered out at Richmond, Virginia July 24 and discharged at Chicago, July 30, 1865. [384]

There is only a national flag and a guidon from the Twenty-third Illinois Infantry Regiment in the collection of the State of Illinois.

---

[379] Ibid.

[380] *Aledo Weekly Record,* August 3, 1864.

[381] *Civil War Medal of Honor Winners from Illinois,* Published by the Civil War Centennial Commission of Illinois, July 1962.

[382] Irish born, Corporal Patrick Highland, Company D, was awarded the Medal of Honor for "Conspicuous gallantry as color bearer in the assault on fort Gregg, Petersburg, Virginia, April 2, 1865." *Civil War Medal of Honor Winners from Illinois,* Published by the Civil War Centennial Commission of Illinois, July 1962.

[383] *Civil War Medal of Honor Winners from Illinois, Published* by the Civil War Centennial Commission of Illinois, July 1962.

[384] *Report of the Adjutant General of the State of Illinois,* Vol II, p276-80.

This National flag identified as a flag of the 23rd Illinois Infantry does not have any Regimental designation or battle honors.

There is no Regimental flag for the 23rd held by the Illinois Military Museum. However, as noted above, apparently the Regiment carried a green regimental flag early in the war, but the men destroyed this flag to avoid its capture. According to a member of the Regiment this regimental flag was known as the "Green of Erin."[385]

The Regimental guidon or flank marker reflects the Irish influence—a green background with a gold Irish harp. This color measures approximately 30" x 20".

There is record of three flag presentations (a total of five flags) to the Irish Brigade. The first was presented on April 27, 1861, at the McVicker's Theatre. The flag was presented by J. J. Sands and was described as "The flag is of silk, differing but little from the usual American colors, and having upon

[385] Crowell, Wm C.; "With Mulligan at Lexington, Mo." *The National Tribune Scrapbook,* No.2, p44.

a blue ground the usual number of stars—thirty-four in number—surrounded by the motto, "Our adopted country; we will defend it!" In the centre in gold is a fine representation of the American eagle. The ornamental work was done by Mr. DuBois and is an honor to him." Colonel Mulligan accepted the flag with an appropriate response. "At the conclusion of this, the scene opened in flat and exposed a gorgeous Temple of Liberty, wreathed with American flags and surrounded by members of the company dressed in the national colors, producing an impressive tableau which literally brought down the house."[386]

The second presentation was May 29, 1861. "This evening will take place the ceremonies of presentation of the beautiful stand of colors prepared by the patriotic ladies of Chicago for the Irish Brigade. The hour is fixed at 6 p.m., the place the new barracks of the Brigade, in the Kane Brewery, on West Polk Street."[387] The details of the presentation reflect on the importance of this presentation. "The committee of ladies were already seated upon the platform with the colors, and without further ceremony Miss Anne Lane, who was to make the first presentation was introduced to the audience. She was beautifully attired in an emblematic costume representing the Stars and Stripes. This costume, which would have attracted attention under any circumstances, was composed of a waist of blue, low in its neck, looped over the shoulders and falling gracefully to the hips. It was studded with silver stars, which gave a brilliant appearance. The skirt was composed of two flounces, the upper being of broad red and white stripes, descending half way to the feet.... She held in her hand a beautiful silk flag, in the style of the old Continental colors of the original Thirteen. The blue field was embellished with the eagle, surrounded by stars...Miss Emma A. Lantry was then introduced to the audience. Her presentation consisted of two magnificent banners of heavy silk, finished and ornamented in the most elaborate style. One of these was the American flag, and the other the flag of Ireland in all its original beauty.... The American flag was a magnificent specimen of workmanship, of heavy silk, surrounded with yellow fringe of the same material. The Irish colors were of the same material, of green ground with the Harp of Erin on one side and the American Eagle on the other." [388]

On June 14, 1861, a single flag was presented to the Regiment. "They were presented yesterday afternoon with a beautiful flag, the gift of young ladies. Daniel McElroy, Esq., made the presentation speech and Major Mulligan responded on behalf of the Brigade."[389] This flag is not further described.

The *Douglas Guards* from LaSalle are reported as arriving in Chicago with a stand of colors. "The *Douglas Guards* from LaSalle arrived in this city from LaSalle and gone into barracks with the Irish Brigade to which they are attached .... They carry two beautiful banners, 'the American Flag' and an Irish Flag. 'Long may they wave.'" [390]

It is possible that the flags carried by the *Douglas Brigade* are two of the flags in the State of Illinois flag collection.

No other flags carried by this Regiment are known to survive.

---

[386] *Chicago Evening Journal* April 29, 1861
[387] *Chicago Daily Tribune,* May 29, 1861.
[388] *Daily Chicago Times,* May 30, 1861
[389] *Daily Chicago Times,* June 15,1861.
[390] *Chicago Tribune;* May 31.1861.

24th Illinois Infantry – "*First Hecker Regiment*"

"Capt. Gunther's horse had been wounded by three balls, and many of our bravest men lay bleeding on the ground. But our centre, grouped around the regimental colors, refused to give way. "I received the flag to carry it on to victory," cried Joseph Brosch, the color-bearer, ready to die at his post, "never an enemy shall see my back." But immediately afterwards he too sank down holding the flag staff—the flag had already been shot to tatters—bravely aloft."[391]

Soon after the fall of Fort Sumter, a committee of German citizens from Chicago requested Julian Kune,[392] to organize a regiment composed of German Americans and other foreign nationals. Camp Robert Blum was established in Chicago, and men were recruited from Cook, McLean and LaSalle Counties and the State at large. Two companies that had previously been sent to Cairo to secure that part of the State for the Union, the "*Turner Union Cadets*" (a German athletic club became Company A) and the "*Lincoln Rifles*" (Company I) were released by General George McClellan from their Cairo service and joined this Regiment known as the *First Hecker Regiment* or the Twenty-fourth Illinois Infantry. Since the Illinois quota for recruits was already filled, it took a personal appeal to Lincoln to have this Regiment accepted and it was mustered into three years of service June 8, 1861 under the command of Colonel Frederick Hecker.[393] The remaining companies Company B (from Bloomington in McClean County); Co. C (*Illinois Rifles* or *Union Rifles);* Co. D (*Grenadier Company)*; Company F (*Lincoln Rifles)*; Co. G (*Turner Union Cadets)*; Co. H (*Pioneer Company);* Company K (Out of state - primarily from Wisconsin). After an initial move to Alton, Illinois the Regiment was ordered to Missouri to fight secession troops. In September they were ordered to Washington DC, but after arriving in Cincinnati they received counter orders and were sent to Louisville then south to Muldraugh Hill, Kentucky. In November the Twenty-fourth went into camp at Elizabethtown, Kentucky where they remained until February 1862. Receiving word of the fall of Fort Henry, the Regiment marched to Bowling Green, but when they arrived the town was in flames and they continued to Edgefield, Tennessee opposite Nashville. Nashville surrendered the following day; the Union troops entered the city and remained there until mid-March when they marched southward toward Murfreesboro, Tennessee and Huntsville, Alabama. April 4-11 the Twenty-fourth fought in actions, which captured Huntsville, Decatur and Tuscumbia, Alabama. At Huntsville, the Union troops not only captured the city but also all rebel soldiers garrisoned there, seventeen locomotives, one hundred and fifty railcars and military supplies. May 1 the Regiment marched to Athens, Alabama and remained there until the end of the month. In June they marched to Chattanooga, but due to lack of supplies the Regiment withdrew and did not attempt to occupy Chattanooga. The Regiment moved west from Chattanooga and met the rebels at Jasper, Tennessee and in July moved to Tullahoma, Tennessee to guard the *Nashville & Chattanooga Railroad.* In September, in response to Bragg's march into Kentucky, the Twenty-fourth moved to Nashville then marched to Louisville and met the rebels in the battle at Chaplin Hills near Perryville, Kentucky on October 8. The Regiment remained in camp at Mitchellsville, Kentucky until

---

[391] *Chicago Daily Tribune,* October 18, 1862

[392] Julian Kune was a man of Hungarian descent, born on the Hungarian-German border, who had been an officer in the Hungarian army, emigrated to the United States and had become the clerk of the Illinois Senate. Kune, Julian, *Reminiscences of an Octogenarian Hungarian Exile,* Chicago, Published by the Author, 1911.

[393] Colonel Frederic Hecker was born in Germany in 1811. He was an attorney and active in German politics prior to coming to the United States in 1848. He settled on a farm in Summerfield, Illinois, became a member of the Republican party and a friend of Abraham Lincoln. At the outbreak of the civil war, he was active in raising a German American regiment and was appointed Colonel of the 24th Illinois Infantry.

December 7 then moved toward Nashville and on December 26th they marched towards Murfreesboro. December 30 and 31 and January 1 and 2, 1863 the Twenty-fourth fought in the battle at Stones River. Following the battle, they remained in Murfreesboro until June 1863. In July they were part of a campaign into Middle Tennessee and saw action at Hoover's Gap. In September the Regiment crossed the Tennessee River and September 19-21 fought at the Battle of Chickamauga then fell back toward Chattanooga. During the siege of Chattanooga on November 25 the Regiment fought at Mission Ridge, pursued the enemy as far as Steven's Gap and returned to camp at Chattanooga. February 24, 1864, the Twenty-fourth fought at Dalton, Georgia. That evening, the commander of the Regiment, Colonel Mihalotzy was shot; he died three weeks later. The Regiment next saw action at Tunnel Hill, Buzzard's Roost, Rocky Faced Ridge, then in the Atlanta campaign at Resaca, Dallas, New Hope Church, Allatoona Hills and Kennesaw Mountain. In June the Twenty fourth was sent to the rear and mustered out August 6, 1864, at Camp Butler.[394]

There are two flags belonging to Twenty-fourth Illinois in the holdings of the State of Illinois Military Museum—one national and one regimental flag.

This national flag of the 24th Illinois Infantry is inscribed with the Regimental designation on the center stripe: "24th REGT ILL INF". *There are no battle honors on the flag. This flag is in poor condition and the stars can no longer be seen. However, there is a spade-shaped red painted finial on the flagstaff typical of flags manufactured by Hugh Wilkens of Louisville.*

The Surgeon of the Regiment, Dr. Wagner, describes the presentation of a flag to the Regiment.[395] "On the afternoon of the 18th day of June,1861, the old Hecker Regiment marched in dress parade at Camp Robert Blum, Chicago, shortly before its departure for the front, to receive a magnificent banner, the gift of Messars. Ludwig and Mueller to the regiment. We well remember the enthusiastic

[394] *Report of the Adjutant General of the State of Illinois,* Vol II, p323-7.

[395] Wagner, William. *History of the 24th Illinois Volunteer Infantry Regiment (Old Hecker Regiment).* Chicago, 1864, translated from the original German text, July 1911. p 5.

words with which Col. Frederick Hecker, on behalf of the regiment, accepted this flag and promised to keep it sacred. This flag waved at Perryville in the hottest rain of bullets, its silken stripes were shot to shreds and its bearer, the youthful Brosch, convulsively clutched it in his hands when he fell mortally wounded." There is no further information re this flag.

The presentation of another flag was also reported in the Chicago Tribune.[396] "The flag presented was a superb silk stand of colors, exquisitely mounted, the whole costing $800. The address of presentation was made by Otto Ludwig, Esq., in German. He said: "Soldiers of the Hecker Jaeger Regiment you behold the glorious Stars and Stripes, the banner of the Union. We give it into your charge, knowing that you will defend it." Colonel Hecker, on taking the flag, manifested much emotion." It is possible that the National flag pictured above is the same flag that was presented to the Regiment in Chicago. Dr. Wagner gives a more vivid description of this flag at the Battle of Perryville in a letter to the Chicago Tribune: "Capt. Gunther's horse had been wounded by three balls, and many of our bravest men lay bleeding on the ground. But our centre, grouped around the regimental colors, refused to give way. "I received the flag to carry it on to victory," cried Joseph Brosch, the color-bearer, ready to die at his post, "never an enemy shall see my back." But immediately afterwards he too sank down holding the flag staff—the flag had already been shot to tatters—bravely aloft. Quick as thought a rebel officer sprang forward from the column of the enemy, which was only a few paces from ours, in order to conquer our palladium, but a ball from Corporal Vogelberg's rifle laid him low-- at the same moment however, the gallant corporal, too, was struck down by the deadly lead. A terrible hand to hand conflict with butt and bayonet now ensued around the remnants of the flag. The enemy's flag was only twenty paces from ours, and twice the bearers of it were shot down by our riflemen; but from our heroic color guard one by one sank to the dust. Here our Kirchner received his death wound, and now Holden fell to the ground with shattered arms at the moment when he again leveled his rifle against an officer who had laid hands upon our flagstaff. Paul Kemmler, already bleeding from a wound in the arm, cried not to forsake the flag, but he, too, was felled by a second shot in the leg. During the melee the flagstaff was broken—it never should fall in the hands of the enemy. We have collected the remnants as a sacred memory to our dead brothers, and we shall send them to the citizens of Chicago, who once presented the flag to us."[397] It is possible that this is the same flag as pictured above. However, Dr. Wagner clearly states that the flag in question was returned to Chicago. It may be that the flag was burned in the Great Chicago fire.

---

[396] *Chicago Tribune,* June 19, 1861.

[397] *Chicago Daily Tribune,* October 18, 1862

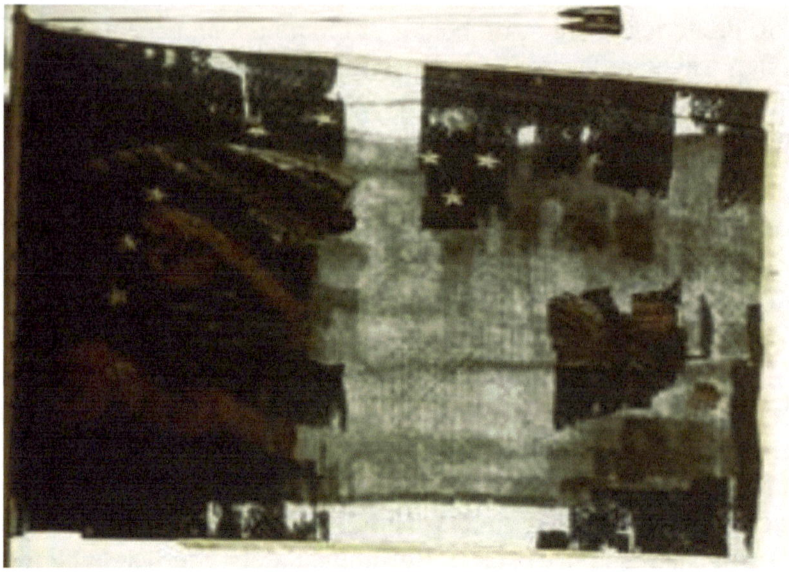

Much of this regimental flag has been lost due to deterioration of the silk and loss of paint. The pattern that remains is typical of flags made by Wilkins in Louisville.

A flag was presented to the *Turner Union Cadets* or *German Turners*. This was given while this company was on the Cairo Expedition (April 15, 1861, to April 29, 1861) and before becoming a company of the Twenty-fourth Infantry Regiment. "At their hall yesterday afternoon a superb American flag was presented to the gallant German Turnverein by the German ladies of Chicago. The splendid flag is made of the finest silk in the glorious "red, white and blue," and bears thirty-four stars. The staff is richly mounted in silver and bears a gold carved eagle. The flag is bordered with silver fringe, with costly cord and tassels, and has as streamers from the top of the staff the German tricolor in yellow, red and black ribbons."[398] There is no further history on this flag.

No other flags carried by this regiment are known to survive.

---

[398] *Chicago Daily Tribune,* April 26, 1861

25th Illinois Infantry – "*Coler Regiment*"

"Boys, we will plant the flag here, and rally the 25th around it, and here we will die!"[399]

William N. Coler, a Democratic leader and lawyer from Urbana in Champaign County organized the 25th Illinois Volunteer Infantry Regiment in the summer of 1861. Since he had served in the Mexican War, was a prominent citizen and had organized the Regiment he was commissioned its colonel. Companies A and B were organized in Vermilion County primarily from the towns of Ridge Farm, Georgetown, Bloomfield and Danville; Company C was organized in Champaign County with virtually all of the men coming from the town of Homer; Company D from Vermilion County was filled with men from Dallas and Fairmount; Company E volunteered from Coles County; Company F from Iroquois County; Company G was raised in Iroquois and Champaign counties; Company H was organized in Douglas County with most men from the town of Tuscola; Company I volunteered from Champaign County and was filled primarily by men from Middletown (now Mahomet); all the men of Company K were from Urbana, the county seat of Champaign County. The Regiment was mustered into service for three years at the U.S. Arsenal in St. Louis on August 4, 1861. They were transported to Jefferson City, Missouri in August then on to Sedalia and Otterville. They spent the winter of 1861 at Rolla, Missouri where twenty-three men died of disease. In February 1862 the Twenty-fifth was transported to Bentonville, Arkansas and fought in the battle of Pea Ridge. Following the battle the men marched to Batesville, Arkansas then on to Cape Girardeau, Missouri where they were loaded on steamers and transported to Pittsburg Landing, Tennessee (Shiloh). The Regiment was in the siege of Corinth, Mississippi in May and pursued the rebel forces to Boonesville, Mississippi in June. The summer of 1862 was spent marching and skirmishing in Tennessee, Mississippi, Kentucky, and Alabama finally arriving at Louisville, Kentucky in late September. In October the Regiment moved to Perryville but did not fight in the battle and then spent November and early December of 1862 scouting and foraging in the Nashville area. On December 22nd the Twenty-fifth marched to Murfreesboro, Tennessee and fought in the battle of Stones River December 30 through January 3, 1863. The Regiment remained at Murfreesboro until June 1863 when they marched toward Chattanooga crossing the Tennessee River under fire then crossing Sand Mountain, Lookout Mountain and Pigeon Mountain and going into camp south of Chattanooga. September 19 and 20, 1863 they fought in the battle of Chickamauga where they suffered 205 casualties out of the 337 who fought. The Twenty-fifth retreated to Chattanooga but fought again in late November in the assault on Missionary Ridge. The time for reenlistment came in January 1864 but fewer than 75% of the men reenlisted, therefore, the Twenty-fifth was not designated as a veteran volunteer regiment. In June they were sent south guarding a supply train for the Atlanta campaign. They fought at Kennesaw Mountain and Peach Tree Creek but early in the siege of Atlanta their term of service expired, and they were mustered out at Camp Butler on September 5, 1864.[400]

There is one national and one regimental flag from the 25th Illinois Infantry at the State of Illinois Military Museum.

---

[399] Eddy, Thomas Mears, *The Patriotism of Illinois,* Chicago; Clarke publishers, 1865, Vol I, p361. These are the words of Colonel Thomas Williams at the battle of Stone's River as he gathered the colors from the dead color bearer. Williams was soon shot and killed as he carried the colors.

[400] Behrens, Robert, *From Salt Fork to Chickamauga, Champaign County Soldiers in the Civil War,* Urbana Free Library, Urbana, 1988, p 45-78.

The National flag was manufactured in Cincinnati and is inscribed with battle honors. The battle honors on this flag are PEA RIDGE, ARK.; SIEGE OF CORINTH, MISS.; PERRYVILLE, KY.; NOLENSVILLE, TENN.; STONE RIVER; CHICKAMAUGA, GA.; MISSION RIDGE, TENN.; KENESAW MOUNTAIN, GA.; PEACH TREE CREEK; SIEGE OF ATLANTA, GA. *The Official Army Register also includes:* "DANDRIDGE, ROCKY FACE RIDGE, RESACA, CASSVILLE, LIBERTY GAP, PICKETT'S MILLS" *as battle honors in addition to the honors on the national flag pictured above.*[401]

This national flag is presumably one of the flags carried by the Regiment at Stones River. As described below, Colonel Williams of the Regiment was killed as he lifted the flag from the Regiment's fallen Color-bearer. On December 22nd the Twenty-fifth marched to Murfreesboro, Tennessee and fought in the battle of Stones River December 30 through January 3, 1863. The battle is described as follows: "The 35th Illinois, and the 25th Illinois were in the 14th Corps on the right. On the 30th of December, two companies of these regiments acted as skirmishers, under the command of Major McIlvain, of the 35th. They remained in position until three o'clock P. M., when Major McIlvain sent for another company, and commenced pressing back the enemy's skirmishers to a belt of timber. The rebel advance, however, rapidly caused our skirmishers to fall back. Col. Williams of the 25th, detached another company from his command, and it went forward, deploying as skirmishers while the brigade moved up to their support. In the fighting of the succeeding days these regiments bore themselves with determined bravery and heroism. General Woodruff in his report of the battle says, "Amid the glorious results of a battle won, it gives me pain to record the names of the gallant men who offered up their lives on the altar of their country. But we must drop the tear of sorrow over their resting places and offer our heartfelt sympathies to their relatives and friends, trusting that God will care for them and soothe their afflictions. And while we remember the noble dead, let us pay a tribute of respect to the gallant Colonel Thomas Williams, 25th Illinois regiment, who died in the performance of his duty. He fell with his regimental colors in his hands, exclaiming: "We will plant it here boys, and rally the old 25th around it, and if need there we will die." The story of the flag(s) carried at Stones River was further expanded well

---

401 *Official Army Register of the Volunteer Force of the United States for the Years, 1861, '62, '63, '64, '65,* Part VI, 1865 (reprinted by Ron R. Van Sickle Military Books, Gaithersburg, MD, 1987) p252.

after the close of the War. In a 1913 newspaper article from Homer, Illinois[402] the following appeared: "H. B. Hall, a veteran of the Civil War residing here has in his possession a valuable relic of the great struggle and one that he prizes very highly. It is the battle flag of Company C 25th Illinois volunteer Infantry and with the exception of a few jagged bullet holes is in a fair state of preservation. The flag was made by the women of Homer just before the boys of Company C left for the front in 1861....During the arduous summer of '61 it was carried in a knapsack until at the battle of Stone river the regimental standard of the 25th having been torn to shreds in the marches through the thick underbrush and timber, Col. Williams asked to borrow the company flag to be used as the regimental flag. The fighting was so fierce that the color bearer was shot dead and Col. Williams who had rescued the flag when the standard bearer fell was mortally wounded a few minutes later his life blood staining the flag. The flag was carried in the battles of Stone River, Chickamauga, Peach Tree Creek, Pea Ridge, on the 100-day campaign to Atlanta, Mission Ridge and Chattanooga and a number of other engagements." After the War the members of Company C in Homer, Illinois, kept the flag. The whereabouts of the flag today is unknown.

This Regimental flag of the 25th Illinois Infantry was produced in Philadelphia. It carries no battle honors.

No other flags carried by the 25th Illinois Infantry are known to survive.

[402] *Danville Commercial News,* March 21, 1913

26th Illinois Infantry

And the flag, the brave old flag,
No more with them will take sir.
They've buried that, and in its place,
They've got a pesky snake sir!
But we're the boys to keep the flag,
Our Washington fought under;
Long wave that flag, that good old flag,
The Stars and Stripes, by thunder![403]

The 26th Illinois Volunteer Infantry, was mustered for three years of service on October 31,1861 under the command of Colonel John Mason Loomis[404] The Regiment consisted of men from the following counties scattered throughout the state: Company A-Effingham County, Company B-Stephenson County, Company C-Bond County, Company D-Sangamon County, Company E-LaSalle County, Company F-Champaign and Vermilion Counties, Company G-White County, Company H-State at large, Company I-Champaign County and Company K-McLean County. Before the regiment had been fully filled with ten companies, seven companies were sent--unarmed—to Quincy, Illinois to provide protection from Confederates Price and Green. The men of the Twenty-sixth were armed with only hickory clubs. Later in 1861 the Regiment was filled to a total of ten companies, was armed with Tower muskets and ordered to Hannibal, Missouri to guard the *Hannibal & St. Joseph Railroad.* In February 1862 the Twenty-sixth Illinois marched to New Madrid, Missouri and engaged the rebels for the first time. On March 6 the Regiment exchanged fire with Confederate gunboats and then intercepted a retreating enemy from Island No. 10 capturing numerous prisoners. Transported up the Ohio and Tennessee Rivers the Twenty-sixth took part in the Siege of Corinth and fought May 8 and 9 at Farmington, Mississippi. June, July, and August were spent in northern Mississippi and Alabama but in September the men marched to Iuka, Mississippi and fought the enemy there on September 19. The Regiment joined in the pursuit of the retreating Confederate troops to Corinth and fought in the battle of Corinth on October 3 and 4, 1862. The remainder of 1862 was spent on guard duty in northern Mississippi. In 1863 the base at Holly Springs was broken up and the Twenty-sixth fell back with the army to Lagrange, Tennessee and assigned duty as provost guard. In March the Regiment marched to Collierville, Tennessee, remained three months and then as part of the Vicksburg Campaign was ordered to Haines' Bluff on the Yazoo River above Vicksburg. June 17-July 4 the Twenty-sixth fought in the siege of Vicksburg and after the fall of Vicksburg joined in the pursuit of the retreating enemy forces and engaged them at Jackson, Mississippi July 10-17. With the fall of Jackson, the Regiment returned to Vicksburg but in September 1863 moved to Memphis, Tennessee. In October the Twenty-sixth marched toward Chattanooga and on November 24 and 25 fought in the battle for Missionary Ridge; the

---

[403] From a poem entitled "The New Yankee Doodle" and published in the *Freeport Bulletin,* (Stephenson County) February 27, 1862. The author is not given.

[404] Colonel Loomis was born January 5, 1825, in Connecticut. In 1848 Loomis took over an existing lumber company in Milwaukee Wisconsin and in 1852 he moved the company to Chicago and formed a partnership with James Ludington. At the outbreak of the civil war, Loomis was a 1st lieutenant in the *Chicago Light Guard* with the reputation of being an exemplary military leader. Based on this reputation, Governor Yates urged Loomis to accept appointment as Colonel of the 26th Illinois Infantry.

Regiment pursued the retreating rebels to Ringgold, Georgia but were recalled and ordered to Knoxville. The Confederate siege of Knoxville was relieved before Union forces arrived and the Twenty-sixth marched back to Chattanooga and to winter quarters at Scottsboro, Alabama on the Tennessee River. In January 1864 four hundred and sixty-three men out of five hundred and fifteen men in the Regiment reenlisted and the Regiment went home on veteran furlough. On May 1, 1864, the Regiment joined Sherman's Atlanta campaign. The Twenty-sixth fought at Resaca, Kingston, Dallas, New Hope Church and Allatoona Hills. In June they were in the assault on Kennesaw Mountain and in July fought at Nickajack Creek, Chattahoochee River and the Battle and Siege of Atlanta.[405] In August and September the Regiment fought at Jonesboro and Lovejoy Station. The Twenty-sixth joined in the pursuit of Hood into north Georgia and Alabama during October; in November the Regiment linked with the Union forces in Sherman's "March to the Sea" and the campaign of the Carolinas. The Twenty-sixth Illinois Infantry was one of the first units to enter Columbia, South Carolina in February and fought in the Battle of Bentonville, North Carolina on March 20-21. Following the surrender of Johnston's army in April the Twenty-sixth marched to Richmond, then on to Washington, D.C. and marched in the Grand Review on May 24. The Regiment was mustered out on July 20, 1865, at Louisville, Kentucky.[406]

There are two Regimental flags, but no National flags carried by the Twenty-sixth Illinois Infantry in the State of Illinois Military Museum flag collection. By orders of the commanding General this Regiment was permitted to place the following honors on their colors: "New Madrid, Island No. 10, Farmington, Siege of Corinth, Iuka, Corinth 3d and 4th October 1862, Holly Springs, Vicksburg, Jackson, Miss., Mission Ridge, Resaca, Kennesaw, Ezra Church, Atlanta, Jonesboro, Griswoldville, McAllister, Savannah, Columbia, Bentonville."[407] Neither of the regimental flags carries battle honors.

---

[405] An incident relating to the Twenty-sixth's action at Atlanta is described –"The rebels rushed right upon our works, so many of them as lived to get there and sometimes were in the very act of vaulting over when killed. A rebel color-bearer ran up with his colors and with the spear head which was affixed to the top of the flag-staff run (sic) through a member of the 26th Illinois who was just on the other side of the works and killed him. Instantly another member of the regiment shot the daring rebel through the head, and leaping over the works captured his colors in safety." This was printed in the Lafayette, Indiana *Daily Journal,* August 6, 1864, and was reprinted from a letter sent to *Cincinnati Commercial,* July 29, 1864, that was entitled "Two Miles West of Atlanta".

[406] *Report of the Adjutant General of the State of Illinois,* Vol 11, p386-8.

[407] Ibid, 388.

On this regimental flag, the position of the stars and the raised center section of the scroll under the eagle with flourishes on each end of the scroll are typical of flags of New York manufacture.

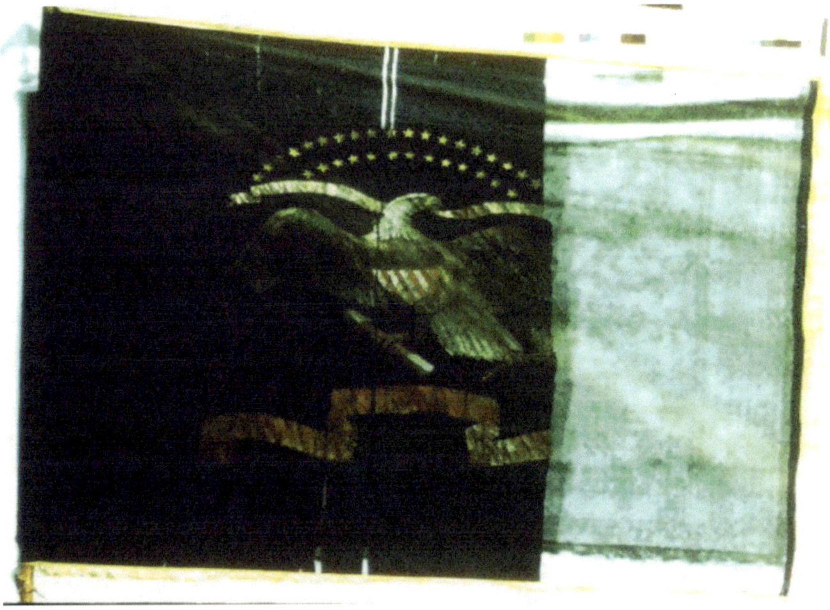

The position of this eagle and scroll identify the manufacturer of this Regimental flag as Horstmann Brothers & Co. of Philadelphia.

No other flags carried by this regiment are known to survive.

27th Illinois Infantry

"The Regiment went into the assault with seventeen officers, coming out with only seven, and the loss of the men was in about the same proportion. Color Sergeant Delaney, a brave soldier of Company K, from Jacksonville, Ill., planted the colors of the Twenty-seventh on the rebel earthworks, had a bayonet run through his breast and fell backward, while the Regimental colors fell inside the works. Delaney died the next day, mourning the loss of his colors."[408]

The Twenty-seventh Illinois Infantry was organized with seven companies at Camp Butler and mustered into Federal service on August 10, 1861, under command of Colonel Fazilo A. Harrington.[409] The men volunteered from the following counties—Company A-Adams County; Company B-Scott County; Company C-Pike County; Company D-Jersey and Madison Counties; Company E-Mason County (known at *Hutchens' Cavalry Company*); Company F-Macoupin County; Company G- Mercer County. On September 1, the men were ordered to Cairo where Companies H, I and K were added with volunteers from Jackson, Henry and Morgan Counties respectively. The men saw their first action at the Battle of Belmont, Missouri on November 7,1861. In 1862, together with other forces, the Twenty-seventh formed the "Mississippi Flotilla," and participated in the siege of Island No. 10, then disembarked and marched under command of Colonel N.B. Buford to attack Rebel forces at Union City, Tennessee. In April, the Regiment was the first to land on Island No. 10 and took the charge of prisoners, siege guns and equipment. Later that month they moved down the Mississippi River to attack Fort Pillow but were ordered back to Pittsburg Landing, Tennessee from where they marched to Corinth, Mississippi and besieged that city from April 29-1 May 30. During that time, they fought at Farmington, Mississippi and in June, July and August did foraging and guard duties in Corinth and Iuka, Mississippi. In September they crossed the Tennessee River and marched to Nashville where they remained until the end of December. While on march they skirmished at LaVergne, Tennessee and repulsed Forrest's attack on Nashville. On December 26, 1862, the Twenty-seventh marched with the "Army of the Cumberland " toward Murfreesboro,[410] Tennessee; they first met the enemy on the Nolensville Pike but pushed forward and engaged in the Battle of Stone's River. Following that battle the Regiment remained in Murfreesboro until the end of June 1863, when it moved with the Army against Shelbyville, Tullahoma and Stevenson, Tennessee and Bridgeport, Alabama. In September the Twenty-seventh crossed the Tennessee River toward Rome, Georgia and joined Rosecrans' forces at Chickamauga to fight in the battle there. After retreating to Chattanooga, the Twenty-seventh fought at Mission Ridge then marched to the relief of Burnside at Knoxville. From there they marched to Louden, Tennessee where they remained until April 1864. In May the Regiment fought in the Army of the Cumberland in the Atlanta campaign at Rocky-faced Ridge, Resaca, Calhoun, Adairsville, Dallas, Pine Top Mountain, Muddy Creek, Kennesaw Mountain (where the Regiment lost their colors- see description below), Chattahoochee River, Peach Tree Creek and the early siege of Atlanta. The Regiment was mustered out September 20, 1864, at Springfield. The Veterans of the Twenty-Seventh were consolidated with the Ninth Illinois Volunteer Infantry.[411]

---

[408] *Report of the Adjutant General of the State of Illinois,* H.W.Rokker, State Printer and Binder, Springfield, Illinois (1886) Volume II, page 416. Account of the fighting of the Regiment at the battle of Kennesaw Mountain.

[409] Colonel Harrington was born in Medina, New York, November 22, 1831. He entered West Point Academy on July 1, 1850, but resigned after three years. He was employed in Civil Engineering until the onset of the civil war. He offered his services to Governor Yates and was commissioned Colonel of the 27th Illinois Infantry.

[410] Colonel Harrington died on January 1, 1863, at Murfreesboro, Tennessee at the age of 31.

[411] *Report of the Adjutant General of the State of Illinois,* Vol. II, p413-16.

27TH ILLINOIS INFANTRY

There are three national flags in the State of Illinois flag collection. The original inventory of the flags also lists a regimental flag. However, a regimental flag from the Twenty-seventh is no longer in the collection.

The above National flag is in poor condition and does not bear any inscriptions.

This national flag bears the regimental designation on the center stripe: "27th ILLINOIS VOL INF."

This national flag of the Twenty-seventh is typical of flags manufactured in Philadelphia. Note the oval shaped pattern of the stars on the canton. It is inscribed with the regimental designation as well as battle honors. Inscribed on this flag are: BELMONT; UNION CITY; ISLAND 10; SIEGE OF CORINTH; LAVERGNE; STONE RIVER; CHICKAMAUGA; MISSION RIDGE; ROCKY FACE; RESACA; ADAIRSVILLE; DALLAS; MUD CREEK; PEACH TREE CREEK; KENESAW(sp) MOUNTAIN; ATLANTA.

There was another flag for the 27th, which was presented to the Regiment February 27, 1863. “Brid. Gen N. B. Buford presents a beautiful flag to the 27th and it appears on Dress parade this evening for the first time. In beautiful letters may be read. “Belmont”, “Island No. 10”, and Union City”.[412] Confederate forces captured this flag June 27, 1864, at Kennesaw Mountain. Its whereabouts today is unknown. In a letter home, Orderly Sergeant, Co. G, 27th Illinois Volunteer Infantry, S. B. Atwater wrote: "We afterwards moved farther to the right and kept up the skirmish til the 27th of June. ...On the right of our division (Newton's) was Davis' division, 14th corps, and still farther to the right was Geary's division, 20th corps; on our left was Stanley's division; our corps (the 4th). When the word was given we jumped over our works and with bayonets fixed and guns uncapped we went as fast as we could down one hill and up another.... some of our boys jumped over the rebel breastworks and never got back. Our flag, on which was inscribed "Presented by Brig. Gen. N. B. Buford to the 27th Reg't Ill. Vol. Inftry--Belmont, Union City, Island No. 10, Stone River" was planted on the works when the color Sergeant was twice shot and bayoneted; the flag was then taken by a Corporal, who mounted the works, and was shot, the flag falling among the rebels while he fell back dead on his comrades. It is needless to say we were repulsed." [413]

The Regiment apparently also lost a stand of colors at the Battle of Chickamauga.[414] “Soon after

---

[412] The Diary of Edward W. Crippin, Private 27th Illinois Volunteers, War of the Rebellion, August 7, 1861, to September 19, 1863. *Transactions of the Illinois State Historical Society,* 1909. p 262-3

[413] *Aledo Weekly Record,* July 27, 1864

[414] Eddy, T.M., *The Patriotism of Illinois.* Volume II, Clarke & Co., Publishers, Chicago, 1866, page 55.

Colonel Buford's promotion to Brigadier-General, he presented the regiment a magnificent stand of colors, which the men said should never fall into the hands of the enemy. At the battle of Chickamauga, the regiment made a charge upon a body of rebels protected by a stone wall. Two color bearers were shot down, when a third sprang on the wall, grasping the colors, when he, too, was killed, falling on the rebel side of the wall, where his body was seized by the enemy, and borne off with the colors, which the rebels retained as a trophy." No further information is known about these colors.

The Twenty-seventh captured Confederate colors near Lavergne, Tennessee on October 7, 1862.[415] The Regiment from whom the colors were captured is not known nor is the disposition of the colors known.

No other flags carried by this regiment are known to survive.

---

[415] The Diary of Edward W. Crippin, Private 27th Illinois Volunteers, War of the Rebellion, August 7, 1861, to September 19, 1863. *Transactions of the Illinois State Historical Society,* 1909, p251. "October 7th Monday we went fifteen miles to the town of lverme and routed a camp of rebles (rebels) cild (killed) thirty and wounded eighty took three hundred prisners (prisoners) their camp and equipage fell into our hands and some provision which we stood greatly in need of we got their colors and one piece of artilery (artillery) we had two cild (killed) and several wounded."

28th Illinois Infantry

"We're fighting for our Union, we're fighting for our trust, We're fighting for that happy land where sleeps our fathers' dust; It cannot be dissevered, tho' it cost us bloody wars-- We never can give up the land where float the Stripes and Stars.

Hurrah! Hurrah! For equal rights hurrah! Hurrah for the brave old flag that bears the Stripes and Stars."[416]

The Twenty-eighth Illinois Volunteer Infantry was composed of men from Mason County-Company A; Pike County-Company B; Scott County-Company C; McDonough County-Company D; Pike County-Company E; Menard County-Company F; Schuyler County- Company G; Fulton County-Company H; Pike County-Company I; Logan and Menard Counties-Company K. It was mustered in for three years of service at Camp Butler on August 15, 1861, under the command of Colonel Amory Johnson. In August the Regiment moved to St. Louis, was armed, and then moved to Thebes and on to Bird's Point, Missouri. In October the 28th marched to Fort Holt, Kentucky and remained there until January 1862 when they marched to Paducah. In February they moved up the Tennessee River and landed below Forts Heiman and Henry. The Rebels abandoned Fort Heiman and the Twenty-eighth was the first to enter the Fort. On February 13 a detachment of the 28th met Claiborne's Rebel Cavalry at Little Bethel Church and took 2 prisoners. In March the regiment was transported to Pittsburg Landing and on April 6 and 7 fought in the Battle of Shiloh, losing Lieutenant Colonel Kilpatrick along with 239 men killed, wounded and missing. Private Ellwood Williams received the Medal of Honor for his actions at Shiloh.[417] The Twenty-eighth was engaged in the siege of Corinth in May, and then marched to Memphis, arriving July 21, and staying there until September. En route back to Corinth in October the Regiment met the enemy at the Hatchie River, near Metamora, Tennessee, capturing a Confederate battery of six guns, caissons and one flag.[418] The Twenty-eighth moved on to Bolivar, Tennessee and then Lagrange, Tennessee in November. Beginning in December they guarded the railroad from Holly Springs to Waterford, Mississippi. In May 1863 the Regiment moved to Vicksburg, Mississippi and was in siege until the city's fall on July 4, 1863. On July 12, near Jackson, Mississippi the Regiment was ordered to charge the enemy across an open cornfield. The eight companies of the Twenty-eighth charged; of the 128 men who charged, 73 were killed or wounded and 16 were taken prisoner. In August the regiment moved to Natchez, Mississippi and remained there until October 1864. Early in 1864 the regiment re-enlisted as a veteran regiment and in May proceeded to Illinois for a veteran furlough. On return from furlough the men were in expeditions to Black Bayou, Sicily Island, Louisiana and Homochitto River, Mississippi. In November, while in Memphis, the regiment received 200 recruits

---

[416] Col. J. Geddes, *The Bonnie Flag with the Stripes and Stars,* Canton Weekly Register, July 20, 1863.

[417] Beyer, WF, Keydel, OF., eds. *Deeds of Valor*, reprint edition, Longmeadow Press, Stamford, CT (1993) p. 16. In the words of Private Williams, the recipient of the Medal of Honor: "After two hours of hard fighting at the battle of Shiloh, we were ordered to beat a retreat, with the Thirty-second Alabama Infantry close at our heels. In the heat of the action a box containing a thousand rounds of ammunition was inadvertently left between the lines. The box lay about one thousand yards from our lines, and in order to reach it under the heavy fire of both sides, we were compelled to crawl on our hands and knees. We reached the box in safety, but while we were returning with it Price fell mortally wounded. I then shouldered the box and started off for our lines. Having safely delivered the box, it was now my duty to look after my wounded comrade, and I immediately returned and brought him within the lines. I turned him over to the care of the hospital corps, where twelve days later he died from the effects of his wounds."

[418] There is no further information regarding this flag. There is no captured Confederate flag in the State of Illinois collection attributed to this regiment.

who were organized into two companies—the Regiment's eight companies had previously been consolidated into four companies. In 1865 the Twenty-eighth was sent to Kunerville, Mobile Point, New Orleans and Lake Pontchartrain, Louisiana. On February 17 the Regiment embarked for Mobile Bay, Alabama and in March they fought at Spanish Fort and after its evacuation marched to Fort Blakely. During this time the Regiment received four more companies of recruits from Camp Butler. In May the Regiment moved to Mobile, Alabama, then Brazos Santiago, Clarksville and Brownsville, Texas where they remained until March 1866 when they mustered out March 15, 1866.[419]

This national flag of the 28th Illinois Infantry measures 53" on the hoist by 88" on the fly. It appears to be home made and it is inscribed with battle honors:" FORT HENRY; FORT HEIMAN; LITTLE BETHEL FEB 13; PITTSBURG LANDING, TENN. APRIL 6 AND 7; HATCHIE OCT 5. The Official Army Register lists in addition to the above honors "SIEGE OF VICKSBURG, JACKSON, SIEGE OF CORINTH, FORT BEAUREGARD, SPANISH FORT*" as "battles in which this regiment bore an honorable part."*[420]

Samuel Howard, Company H of the 28th Illinois relates the story of a flag, perhaps this flag, in a letter to his niece who had seen the flags of the 28th in Springfield. "Suffice it to say that one of those two flags formed the original colors of our regiment at the time we left Camp Butler for the war in August, 1861; that this flag was successfully carried onward by our regiment at Belmont, General Grant's first battle of the war; at the capture of Forts Heiman, Henry, and Donelson; and at Shiloh, April 6-7, 1862, the deadliest great pitched battle of all time, where our division met with the greatest loss of any division; that we successfully upheld this flag at "Hell on the Hatchie River," in the following October, where our regiment lost more than 17 percent of all we had left in killed and wounded alone. During this battle one of the enemy's shells exploded in this flag and tore it all to tatters, leaving it substantially in the same condition you saw it at Springfield........After burial of the dead of both forces,

[419] *Report of the Adjutant General of the State of Illinois,* Vol. II, pp 441-44.
[420] *Official Army Register of the Volunteer Force of the United States Army for the Years 1861, '62, '63, '64, '65.* Part VI, p 257 (Reprinted by Ron Van Sickle Military Books, Gaithersburg, MD. 1987)

we carried this flag back to Bolivar, where, October 13th, we were reviewed by Jesse K. Dubois, then auditor of your State, representing Governor Yates. During this review, Colonel Johnson, of the 28th, through General Hurlbut presented the remnant of this flag to the State of Illinois, delivering it to the Auditor, who returned with it to Springfield, where it has been sacredly cared for ever since by the State of Illinois.[421]"

On this regimental flag, the position of the stars above the eagle and the raised center section of the lower scroll together with the embellishments on the ends of the scroll are typical of those flags manufactured in New York. This Regimental flag has an attached note which states, "Captured 12 July 1863." Based on the history of the Regiment this flag would have been captured during assault at Pearl River, south of Jackson, Mississippi[422]

A flag of the 28th was captured at Jackson, Mississippi but in the letter of Samuel Howard quoted above, his letter to his niece, goes on to describe the origin of this captured flag and the circumstances of its capture. "Upon learning of the destruction of our flag at the Battle on the Hatchie, the ladies of Vermont, your county, long before your birth, went to work and with their own hands made the silk flag which you saw in the State-house, and presented the same to our regiment in the early part of May, 1863 just before we left for Vicksburg. Thereafter we carried this silk flag through the entire siege and capture of Vicksburg, which was by far the greatest capture of all the wars, and fell into our hands July 4, 1863; and on the same day we started for Jackson, the capital of Mississippi....With a deep yell learned at Shiloh, which still rings in the ears of all survivors, these four brave regiments charged onward at double-

[421] Howard, Samuel Meek, *The Illustrated Comprehensive History of the Great Battle of Shiloh,* Franklin Hudson Publishing Col., Kansas City, Mo.1921.p 148-50. This information originally appeared in a letter to the *National Tribune*, 1914 written by the author.

[422] The capture of this flag is mentioned in the history of the Forty-first Infantry, "Several flags were captured, including those of the Twenty-eighth and Fifty-third, and the regimental flag of the Forty-first. *Report of the Adjutant General of the State of Illinois, Vol 3, p200.* The flag was captured by "Private Williams, Company F, of the 47th Georgia Volunteers." *Savannah Daily Morning News,* July 20, 1863.

quick for the Rebel earthworks—for destruction and death. The 28th then and there, and for the last time followed this same silk flag which you saw in the State-house at Springfield, the brave boys falling like grass before the sickle. They had proceeded only a short distance when Sergeant Thomas J. Metzler, carrying the colors, was pierced through and through by Minie balls and fell to the earth beneath its folds. Right then and there, Private John Hollingsworth, of Company H, grabbed this flag, hoisted it aloft, and shouted: "Come on, boys! I am with you now." He plunged onward and carried this silk flag clear over the main earthwork of the enemy, where he was taken prisoner, still unharmed. Several of our boys followed this flag clear to the deep ditch, and then saved their lives by leaping into the trench. And this is just the way our regiment lost this silk flag. "

Since the regimental flag identified above as having been captured and returned is clearly of New York manufacture, it cannot be the flag made by the ladies of Vermont as described by Howard. It is possible that the National flag of the 28th shown above with battle honors, may be the captured flag and the first flag of the regiment that was returned in tatters is lost and is no longer in the collection.

Finally, this regimental flag, also of New York manufacture, is attributed to the 28th. There is no other information on this flag.

In the Barry Museum in Barry, Pike County, Illinois are remnants of Barry's history. These include items from the McTucker GAR Post No. 154 and John McTucker Women's Relief Corps No. 91. Hanging on the walls of the Museum are framed GAR medals and reunion ribbons along with a silk flag. The flag has 13 stripes and 33 stars and is encased in fine black silk netting similar to the netting used in the early 1920s to "preserve" silk flags. The flag measures 6 feet on the staff and 8 feet, six inches on the fly. Printed on the flag is BARRY ILL, Co. I. 28th Reg..

According to local tradition, Henry Wing Brown of Barry enlisted in Company I of the Twenty-eighth Illinois Infantry on August 27, 1861, and carried the flag. Henry Brown or perhaps another member of Company I brought the flag back to Barry at the end of the Civil War.

On May 2, 1917, Henry Brown died and the following appeared in his obituary: "In June, 1863 when a part of Co. I was at home on a veteran furlough, the women of Barry made and presented to Co. I a flag, which came into Mr. Brown's possession. This flag he cared for a nearly sacred thing and this flag today rests on his casket." [423]

Based on local lore, the flag was next found among the possessions of Henry Brown's nephew, Louis Gerald Brown, when the nephew died in 1947. The flag then came into the possession of Louis Brown's daughter Jessie Mahala Brown.

In 1963 Norton Walther opened a general store/museum in Barry. He sold antiques in an "Old Country Store " flavor but he also collected historical material related to the history of Barry. Jessie Brown gave or sold the flag of Company I to Norton Walther. After 21 years Walther sold his business and in 1984, he received permission from the city of Barry to renovate the unused second floor of the Barry Public Library as a Museum. This flag now hangs in the Barry Museum located on the second floor of the Barry Library.

No other flags carried by the 28th Illinois Infantry are known to exist.

---

[423] *Barry Adage*, May 9,1917

29th Illinois Infantry

On the ninth of April sixty-five
Long remembered be the day.
In range of Blakely's batteries,
General Canby's forces lay.
They had waded sloughs and marshes
Marched over concealed torpedoes,
This proximity to gain.

Onward rushing to the muzzles,
Of Huge death-dealing guns,
Each vieing to be foremost,
And cheering as they run.
Mounting the rebel ramparts,
With shouts that rend the air,
And plant the "Emblem of the Free,"
Our glorious colors there.[424]

The Twenty-ninth Illinois Infantry Regiment was organized at Camp Butler near Springfield, Illinois during the summer of 1861. The men of the Twenty-ninth volunteered for Company A from Hardin County; for Company B from Alexander and White counties, for Company C from Gallatin County, for Company D from Gallatin and White counties, for Company E from Saline County, for Company F from Pope County, for Company G from White County, for Company H from Edgar County, for Company I from Sangamon County and for Company K from Massac County.[425] The regiment mustered for three years of service at Camp Butler on August 19, 1861 under the command of Illinois born Colonel James S Rearden, a merchant from Shawneetown, Illinois. With base at Cairo the 29th was part of expeditions to Bloomfield, Missouri in October and Columbus, Kentucky in January. In a Brigade commanded by Colonel R. J. Oglesby the regiment fought at Fort Henry February 2-6 and at Fort Donelson February 12-16. At Fort Donelson the Twenty-ninth was heavily involved and lost 100 men killed and wounded. In March the Regiment proceeded to Savannah, Tennessee and then to Pittsburg Landing. It was here on April 6 and 7 that they fought in the Battle of Shiloh where they lost 100 men out of an effective strength of 400. In April and May, they fought in the Siege of Corinth, Mississippi and in June they marched to Jackson, Tennessee and from there had excursions to Bolivar and Brownsville, Tennessee in July and August. On October 1, 1862, the Twenty-ninth was sent to reinforce Union forces at Corinth; they arrived too late for the battle but did pursue Rebel forces to Ripley, Mississippi then returned to Jackson. When they went into camp at Holly Springs in December, Companies D and K were sent north to guard Jackson, Tennessee. On December 20, the Union forces at Holly Springs surrendered to Confederate General Van Dorn and eight companies of the Twenty-ninth

---

[424] This is an excerpt from a poem entitled "Blakely" that was written by Thomas Washington Danner, Co. G, 29th Illinois Infantry and shared by his great-great granddaughter LaVonda Smith.

[425] Mustered in with and attached to the 29th Infantry Regiment as Cavalry Company B was *(Egleton) Carmichael's Cavalry* Company afterwards Company B of Stewart's Independent Cavalry Battalion and finally Company B of the 15th Illinois Cavalry

were paroled, sent to Benton Barracks and their books and records destroyed.[426] Companies D and K, who had been sent to Jackson, were assigned to the Western Navy, Mississippi River Squadron, and attached to gunboats *Tuscumbia*, *Tyler,* and *Petre*l. In July 1863 the eight paroled companies of the Twenty-ninth were exchanged, returned to duty, and ordered to Vicksburg, Mississippi. While at Vicksburg, on October 16, the One Hundred and Thirty-first Illinois Volunteer Infantry was consolidated with the Twenty-ninth Illinois Infantry. On December 1, 1863, the consolidated Twenty-ninth Illinois Infantry Regiment moved to Natchez for garrison duty and in January 1864 the men of the Regiment reenlisted and were mustered as veterans. After veteran furlough home in July 1864 the men returned to Natchez and remained there until October 1864 when they were ordered to Memphis and then Paducah. The Twenty-ninth was in the area of Memphis and Moscow, Tennessee until January 1, 1865, at which time they embarked for New Orleans and camped near that city at Kenner. In March and April, the Regiment campaigned against Mobile, Spanish Fort and Fort Blakely in Alabama. In June they sailed by steamship to Galveston, Texas, for duty guarding the Texas Railroad at Millican, Hempstead, Brenham and Beaumont, Texas. The Regiment mustered out November 6, 1865. [427,428]

The following flags were all identified as flags of the 29th Illinois Infantry when photographed. However, several of the flags may be those of the 29th USCT. Nevertheless, two of the flags can definitely be identified as being flags of the 29th Illinois Infantry because of appropriate battle honors.

*This national flag is inscribed as follows: "29th Ill. Vo....*Fort Henry, Fort Donaldson *(sp),* SHILOH, CORINTH, SIEGE of VICKSBURG, Blakely, Spanish Fort. *Since portions of the flag are missing, there were probably additional inscriptions. The fact that the inscriptions are of various types of lettering suggests that the inscriptions were not all placed on the flag at the same time.*

---

[426] It is unknown if the Regiment's colors were captured at this time.

[427] *Report of the Adjutant General of the State of Illinois,* Vol. II, p 501-2

[428] Dyer, Frederick H., *A Compendium of the War of the Rebellion,* Volume III, p 1057-8.

This national flag is of Philadelphia manufacture and bears no inscriptions. It may be the flag of the 29th USCT.

Although difficult to see, this flag is inscribed " 29th REG….., DONALD…., SHILOH. *Therefore, this is also a flag of the Twenty-ninth Illinois Infantry.*

This Regimental flag is inscribed in lower scroll as a flag of the 29th Volunteer Regiment Illinois Infantry.

*This Regimental flag has "29th ILLS VOL. REGIMENT INFANTRY in the lower scroll and is inscribed with "*Donaldson*" and "*Shiloh*" -- battles fought by the Twenty-ninth Illinois Infantry.*

29TH ILLINOIS INFANTRY

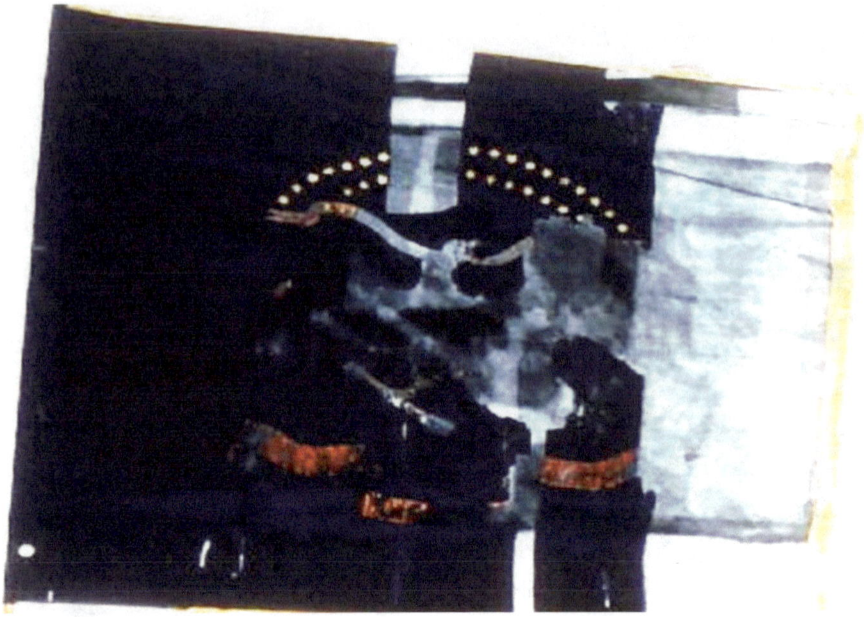

No inscriptions remain on this regimental flag. The pattern of the flag is typical of those flags manufactured by Longley & Brother of Cincinnati. This may a flag of the 29th USCT.

No other flags carried by the 29th Illinois Infantry Regiment are known to exist.

29th United States Colored Infantry

Hundreds on hundreds fell;
But they are resting well;
Scourges and shackles strong
Never shall do them wrong.
O, to the living few,
Soldiers, be just and true!
Hail them as comrades tried;
Fight with them side by side;
Never, in field or tent,
Scorn the black regiment![429]

"Come, now, colored men! Step forward and fill up the ranks of the 1st Illinois Colored Regiment. Your services are wanted in the field. You owe them to the country which makes you free. —You will receive the *one hundred dollars* bounty allowed by Cook county, and Congress will put you on the same footing with white soldiers as to Government bounty and pay, before the close of the present session. Rally to fill up the ranks."[430]

Illinois, particularly the southern portion of the State where the settlers had come from the South, was intolerant to blacks. Laws forbad blacks the right to vote or hold office, required them to register at the county seat as proof of their "freedom," permitted flogging of blacks, denied the right of education, and did not allow blacks to serve in the militia. In the northern part of the state there was a greater tolerance of blacks and an Underground Railroad helped slaves move from cities along the Mississippi River to northern cities and then to Canada. However as late as 1862, a black in Hancock County was arrested, found guilty and fined for being in the State more than ten days with intention to remain permanently; on appeal to the State Supreme Court the verdict was upheld.[431] The first significant change at the Federal level was the Militia Act of 17 July 1862, which authorized the President "to receive into the service of the United States...persons of African descent." This law did not mean that blacks would be organized into military regiments to bear arms, but rather to serve only as a labor force. The role of blacks in the military changed with the implementation of the Emancipation Proclamation on January 1, 1863, which stated that blacks "will be received into the armed service of the United States to garrison forts, position, stations...." Subsequently, the Bureau for Colored Troops was established under the Adjutant General and boards were created to examine white officers for competency to serve in black regiments. Black men were to be recruited directly into the service of the United States and not into the State Regiments.

An important distinction, however, was that the black soldiers who volunteered into the United States Colored Troops (USCT) would be credited toward the quota assigned to the State in which they enlisted.

---

429 *The Black Regiment,* portion of a poem written by George Henry Boker and downloaded from www.civilwarpoetry.org

430 *The Weekly Union Banner* (Carlyle, Illinois), January 14, 1864. Quoting from the *Chicago Journal.* The editor of the *Weekly Union Banner* adds the editorial comment, "The above is from the Chicago Journal; we insert it here to show how the world moves onward. What would have been thought of this had it appeared three years ago?"

431 Allen, John W., "Slavery and Negro Servitude in Pope County, Illinois, *Journal of the Illinois State Historical Society,* Volume 42, 1949, pages 411-23

In the first year of the Civil War there were more volunteers for the Illinois regiments than there were positions to fill. However, as the War progressed it became apparent that this would not be a three- or six-month war and that more troops would be needed. By 1863 Illinois, with a diminishing number of volunteers to fill its quota, was facing a potential draft to fill its regiments. The realization that a full regiment of 1000 blacks raised in Illinois would significantly offset Illinois' quota resulted in Governor Yates requesting and receiving authorization to raise the Twenty-Ninth US Colored Infantry. The Regiment was organized at Quincy, Illinois as the First Regiment Illinois Volunteers and the first enlistments occurred on November 1, 1863. Recruiting for a Colored Infantry in Illinois was difficult for several reasons, 1) despite the assurances-- there were pay and bounty inequities between the white and black recruits, 2) prior recruiting in Illinois from other states had taken many of the black candidates, 3) despite the prohibition, out-of-state recruiters continued to recruit in Illinois in competition with Illinois recruiters, 4) The Chicago newspapers, particularly the *Chicago Times*, did not support and even opposed the enrollments and 5) the citizens of Illinois were lukewarm to the idea.

Captain John Bross, Company A of the Eighty-eighth Illinois Infantry, was appointed by Governor Yates as the Colonel of the Colored regiment and was charged with recruiting.[432] As companies were filled, they organized at Quincy and the recruits were supplied with clothing and equipment but no arms. On April 25, 1864, Companies A, B, C, D and E, were mustered into service and the First Regiment Illinois Volunteers (Colored) became the Twenty-ninth United States Colored Infantry.[433] With minimal training and no weapons the men departed for Chicago on April 28 then on to Pittsburgh, Baltimore and finally Washington, D.C. On May 2 the Regiment marched to a training camp, *Camp Casey,* where they were issued Springfield Rifles. For the next month they trained with another new regiment, the Twenty-eighth U.S. Colored Infantry from Indiana. May 30 the Twenty-ninth left *Camp Casey* and joined the Army of the Potomac, Ninth Corps, at White House Landing on the Pamunkey River. June 12 the Twenty-ninth moved toward the James River as rear guard of the Ninth Corps. During this time the Union Army was fighting at nearby Cold Harbor. By the end of June, the Union army was entrenched before Petersburg and the Ninth Corps was part of a plan to breach the Confederate lines and occupy Petersburg. A tunnel was dug under the Confederate fortifications and packed with four tons of explosives. At 4:40 AM on July 30, 1864, the mine was exploded. Over the next several hours, instead of the Union forces breaking through the Confederate lines and seizing Petersburg, the Union forces including the Twenty-ninth Colored Infantry poured into the crater created by the explosion and became trapped. The disaster was a result of a combination of problems including a late change of plans, misinterpretation of orders, lack of carrying out orders and lack of presence of the necessary commanders on the field. The Twenty-ninth entered the crater with colors flying. Color bearers Isaac Stevens and Frederick Bailey were killed. Three other color bearers—Joseph Brown, John Maxon and James Brockway were wounded. As Brockway fell, Colonel Bross seized the colors and said, "The man who saves these colors shall be promoted." A Confederate soldier later reported that a colonel "seized his colors and sprang over the protective ditch before he was shot down."[434] Despite these efforts the colors of the Twenty-ninth were lost at the Battle of the Crater. The Confederate counter-attack on the men in the crater resulted in heavy loss of life and there were charges that the Confederates killed black soldiers and white officers of the black regiments after they had surrendered. The losses for the Regiment are poorly documented, but overall, it appears that in addition to Colonel Bross, the Regiment lost eight other officers and more than 200 men killed, wounded or missing.

---

[432] John Armstrong Bross was a Chicago attorney and a Douglas Democrat. His brother, William Bross was a Republican and co-owner of the *Chicago Times.*

[433] Company F joined the Regiment on July 9.

[434] Miller, Edward A. Jr., *The Black Civil War Soldiers of Illinois,* The University of South Carolina Press, Columbia, 1998, page 71.

Colonel Bross was buried on the battlefield with his men. When the bodies were subsequently moved to the Popular Grove National Cemetery, his body was not identified. The black soldiers had performed well in their first battle; they had proven themselves a valuable addition to the Union cause. Before the battle the Twenty-ninth had only six of the usual ten companies and many of the companies were at minimal strength. Following the battle, they were even more depleted of both officers and men. In July 1864 new procedures for recruitment of black soldiers went into effect—recruiting officers would no longer be paid per recruit, recruits would be ordered to designated assembly points for distribution to regiments in the field and pay and bounties would be the same for white and black recruits. For the Twenty-ninth this meant that new recruits and new companies would be filled with men from other states, not only Illinois. In the months following, the Twenty-ninth acted primarily in support and guard duties. During September and October recruits were added to existing companies from Illinois, Maryland and Virginia and new companies G and H were formed with men from Maryland and Virginia. During this same period new officers were being added--unfortunately some of them had very little combat experience. In December and January 1865 Companies I and K were mustered in with men from Delaware and Maryland. Thus in 1865, for the first time, the Twenty-ninth had a full quota of ten companies. In mid-November 1864 the Regiment had been assigned to the Army of the James and effective December 4, 1864, a new Corps, the Twenty-fifth Corps, was created. This was a new all black corps commanded by Major General Godfrey Weitzel. For the first three months of 1865 the Twenty-ninth was in winter camp at Chaffin's Farm north of the James River in Virginia. In late March the Regiment crossed the James River and marched south to join the Army of the Potomac. They followed Lee's retreating army to Appomattox Court House and at least Company H witnessed Lee ride his gray horse to the surrender. In May the Twenty-fifth Corps was ordered to Texas. On board steamers they finally reached Brazos in south Texas on June 22. June 24 the Twenty-ninth marched to White's Ranch four miles from the mouth of the Rio Grande River and in July they moved to Rio Grande City. November 6, 1865 the Regiment mustered out at Brownsville, Texas and departed for Springfield, Illinois.[435]

There is a national and regimental flag in the flag collection of the State of Illinois identified as flags of the Twenty-ninth USCT.

The national flag does not have a regimental designation and there are no battle honors inscribed. The oval pattern of the stars is typical of flags manufactured in Philadelphia.

[435] Miller, Edward A. Jr., *The Black Civil War Soldiers of Illinois.* This history of the Twenty-ninth U.S. Colored Infantry is an excellent and detailed source of information of not only of the Regiment but also of the backgrounds and subsequent history of the men who served in the Regiment.

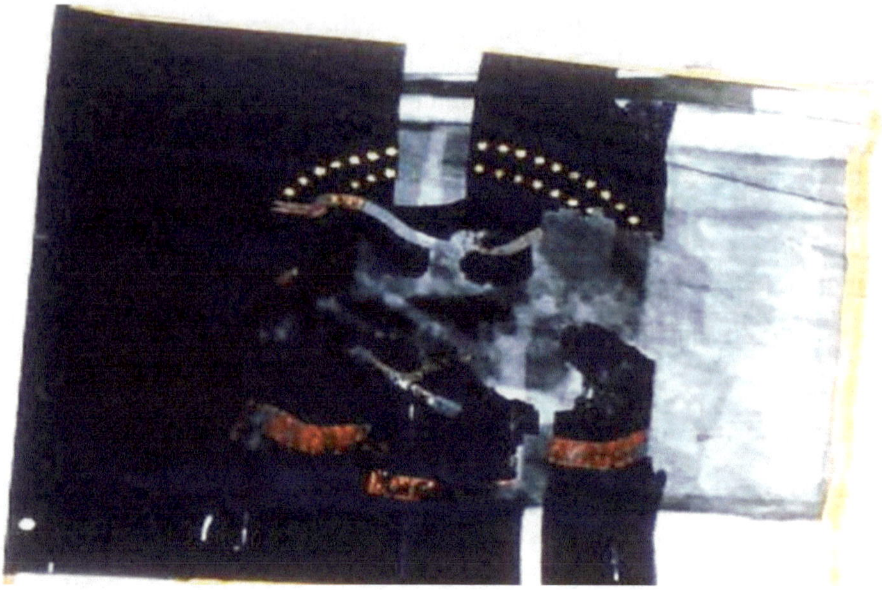

The regimental flag is in poor condition and does have any readable inscriptions.

No other flags carried by the 29th USCT are known to survive.

30th Illinois Infantry

"Brave Henry in the battle fell,
While in his hands he bore,
The emblem of his country dear,
Though stained with battle gore.
So nobly did he bear our flag,
His comrades all did say,
Never was brave Henry known,
To falter in a fray."[436]

The Thirtieth Illinois Volunteer Infantry was organized at Camp Butler and mustered in for three years of service August 28, 1861, under the command of Colonel Phillip B. Fouke [437] of Belleville, Illinois. Companies A and G came from Mercer County, Company B from Sangamon County, Companies C and E from Randolph County, Company D (also known as *O' Harnett's Cavalry Company*) from Crawford County, Company F from Clark and Edgar Counties, Company H from Clinton and Macoupin Counties, Company I from Bond and Clinton Counties and Company K from Clinton County. The Regiment was ordered to Cairo, Illinois in September and in October were ordered on scout into Kentucky near Columbus. On November 7 they fought at Belmont and captured "*Watson's New Orleans Battery*."[438] In 1862 the Thirtieth moved from Cairo up the Tennessee River and took part in the capture of Forts Henry and Heiman (February 6) and Fort Donelson (February 12-16). After traveling to Pittsburg Landing, Tennessee in April 29-30 they fought at the siege of Corinth, Mississippi. On June 4 and 5 the Regiment marched to Bethel, Tennessee and occupied Jackson, Tennessee on June 8. The Thirtieth marched to Estenaula and Denmark, Tennessee in August and on September 1, 1862, the regiment marched toward Medon Station on the *Mississippi Central Railroad.* There they met the rebel cavalry and defeated them in four hours of fighting. The next day they marched into Medon then on to Jackson, Tennessee. In November the Thirtieth marched from Jackson to LaGrange, Tennessee then to Water Valley, Mississippi arriving there December 19. In December the Regiment moved toward Memphis and arrived January 19, 1863. On February 22, 1863, the Thirtieth traveled to Lake Providence, Louisiana (above Vicksburg) and to Milliken's Bend, Louisiana in April. After joining Grant's army, the Thirtieth moved below Vicksburg and crossed the Mississippi River near Bruinsburg. On May 1 the Regiment fought at Thompson's Plantation, Port Gibson, on May 12 near Raymond, on May 14 at Jackson and on May 16 at Champion's Hill, Mississippi. They crossed the Black River May 17 and participated in the siege of Vicksburg and the fighting at Jackson, Mississippi during June and July. In October the Thirtieth moved toward Canton, Mississippi and fought at Bogue Chitto Creek.

---

[436] Portion of a poem written by Miss Ann Steele of Sparta, Randolph County and quoted by G.B. McDonald in his history of the Regiment in the Sparta News, 1916. G.B. McDonald was the brother of Henry McDonald the color bearer who died defending the flag. There are three soldiers in the 30th Illinois named Steele who were from Randolph County. Presumably Ann Steele is the sister of one or all of these soldiers. McDonald, Granville B. *A History of the 30th Illinois Volunteer Regiment of Infantry*, printed by the Sparta News, Sparta, Illinois. (1916) p 31.

[437] Colonel Fouke established and published the *Belleville Advocate,* was admitted to the Illinois Bar in 1845, served as a member of the State of Illinois House in 1851 and was elected as a Democrat to the Thirty-sixth and Thirty-seventh Congress.

[438] ibid p 6.McDonald, J.B. Even though the Battery was captured, "but to thick timber, it had to be abandoned and left on the field."

January 1864 the regiment reenlisted as a veteran organization and saw action in the Meridian Campaign. After furlough in March the Thirtieth left Cairo with the "Tennessee River Expedition" and proceeded to Clifton, Tennessee then marched to Pulaski, Tennessee, to Athens, Alabama and to Huntsville, Alabama. From there they marched to Ackworth, Georgia where they joined Sherman's Army in the Atlanta Campaign. In that Campaign they saw action at Kennesaw Mountain, Brushy Mountain, Nickajack Creek, Howell's Ferry, Chattahoochee River, and the Siege and Battle of Atlanta. In the flanking movement around Atlanta the Thirtieth fought at Jonesboro in August and at Lovejoy Station in September. Following the pursuit of Hood into Alabama in October the Regiment joined Sherman on his "March to the Sea." The Thirtieth took part in the siege of Savannah and fought at Pocotaligo, Salkehatchie Swamps, Barker's Mills and Columbia in South Carolina, and then Bentonville and Goldsboro in North Carolina. They occupied Raleigh in April then marched via Richmond to Washington DC for the Grand Review on May 24, 1865. The Thirtieth moved to Louisville for Provost duty and was mustered out July 16, 1865.[439]

There is one national and two regimental flags of the Thirtieth Illinois Infantry in the collection of the State of Illinois Military Museum.

The pattern of this National flag is not typical of the Quartermaster issue and is probably a homemade flag. Note the attached Regimental designation-"30th REG ILL VOL" *(See the history in the information below.)*

A tag on this flag states that the flag was captured at the Battle of Britton's Creek in September 1862. However, this information cannot be verified.

The story behind the history of the national flag probably begins at the mustering of the Regiment: "The Thirtieth Infantry was mustered into the service of the United States at Camp Butler, near Springfield, August 28, 1861, and here it was that the battle flag was presented to the Regiment by

[439] *Report of the Adjutant General of the State of Illinois,* Vol. II, p 537-8.

the women of Sangamon and surrounding counties. They were particularly interested in this regiment for it was recruited largely from this vicinity." [440]

There is confirmation of the capture of this 30th Infantry "battle flag" before Atlanta, Georgia on July 22, 1864. "This was a cool, daring, brave charge and well executed. It was in this charge that our flag bearer, Henry McDonald, was killed. He was standing on top of the breats (sic) works, waving the flag. We lost our flag, but there was not enough of the boys left to defend it. The flag bearer was the brother of the writer, but I must say there never was a braver soldier ever marched under the stars and stripes.... The flag was captured by J.C.Leird, a member of Co. "A" 27th Tennessee Regiment, C.S.A. After the battle, J.C. Leird turned the flag over to Gen. Hardee."[441]

When Confederate General Hardee died in 1873, the captured flag was found in his effects by his daughter. She read a note pinned to the flag detailing by whom, where and when it was captured. Hardee's daughter eventually located Mr. Leird in Ripley, Tennessee and turned the flag over to him. Mr. Leird thought that the flag should be returned to Illinois and contacted Col. Poilson, commander of Confederate Post No. 8 in Chicago. He in turn contacted the Columbia GAR Post to help find an officer of the 30th Illinois Infantry. Captain Elijah David, Co. A, the only surviving officer of the Regiment was found in Toledo, Illinois and arrangements were made to return the flag to him. In a ceremony in Chicago on November 28, 1911, the flag was returned to Captain David: "The transfer of the flag from Confederate to Union hands is one of the saddest and most pathetic stories in recent history. Elaborate preparations had been made for turning over the flag to Capt. E. B. David, of Aledo, the sole surviving officer of the gallant regiment. He received it. He attempted to speak, but in the midst of his remarks the aged, gray-haired veteran broke down entirely. He was so completely unnerved by the occasion that he could not continue. After vainly striving to speak he tottered to where the flag lay—the flag that meant so much to him, and beneath which he had fought so many times, had seen his friends and fellow soldiers mowed down—and buried his face in it and wept. He wept for grief as it vividly recalled to him those fateful days and for joy at a reunited country and the return of the old dear flag".[442]

---

[440] *J. of the Illinois State Historical Society:* Vol 4, January 1912, page 493-6; The Battle Flag of the Thirtieth Illinois Infantry, After half a Century, Tattered and Torn, Comes to Memorial Hall, Springfield".

[441] McDonald,GB, *A History of the 30th Illinois Volunteer Regiment,* pp 29-30; *Sparta News,* 1916

[442] J.Illinois State Historical Society, vol. 4, 1912; pp 493-6

The above two Regimental flags of the Thirtieth Illinois Infantry bear no inscriptions or battle honors.

Judge James M. Rogers, of Carlyle, purchased a flag for the Thirtieth in recognition of the Regiment's service at the battle of Belmont. The flag was presented to Colonel Philip B. Fouke while the Regiment was at Camp McClernand, Cairo.[443] The flag is not described, and it is possible that it is one of the flags shown above.

No other flags carried by this regiment are known to exist.

[443] From the *Springfield Journal,* November 26, 1861 as quoted in the *Chicago Times,* November 28, 1861

31st Illinois Infantry – "*Dirty First*"

"O Flag of our Union, the flag of our pride,
Though trampled by traitors and basely defied
Fling out to the glad winds your red, white and blue,
For the heart of the North land is beating for you.
And her strong arm is serving to strike with a will,
Till the foe and his boastings are humbled and still.[444]

The first Colonel of this regiment was John "Black Jack" Logan, a lawyer and a Democrat from "Little Egypt" (Carlinville) who became one of the Union's premier civilian generals. He fought at Bull Run as a volunteer in a Michigan regiment but resigned to return home to "Little Egypt" to recruit an Illinois regiment. "Little Egypt" is the area of Southern Illinois settled by inhabitants from Southern states who supported the Confederacy.[445] The regiment was organized at Cairo, Illinois with volunteers from the southern counties of Illinois; Perry County-Company A (*Dresser's Battery*); Saline County - Company B; Williamson County -Company C (*Dollins' Cavalry Company*); Johnson and Pope Counties- Company D; Union County-Company E; Johnson, Scott, Williamson and Pulaski Counties- Company F; Saline County-Company G; Alexander, Jackson and Williamson Counties-Company H; Franklin and Tazewell Counties- Company I and Marion County-Company K. The Thirty-first mustered in for three years of service, September 18, 1861, and less than two months later fought at Belmont, Missouri. In 1862 they fought at Forts Henry (February 7) and Donelson (February 15), Tennessee where the Regiment lost 250 men killed and wounded. The Thirty-first remained at Fort Donelson until late April when they were transported to fight in the siege of Corinth, Mississippi, but arrived only to find the city evacuated. From Corinth the Thirty-first marched to Jackson, Tennessee and spent the summer scouting in that area. In the fall they returned to Corinth and followed the retreating enemy to Ripley, Mississippi with skirmishes at Chewalla and Big Hill, Tennessee and along the Tuscumbia River. In January 1863 the men marched to LaGrange and then to Memphis, Tennessee and on to Lake Providence, Louisiana, arriving there on March 10. April 30 the army crossed the Mississippi River below Grand Gulf and fought at Thompson's Plantation near Port Gibson on May 1, Ingraham's Heights on May 3, Raymond on May 12, Jackson on May 14, Champion's Hill on May 16, and Big Black River Bridge on May 17; they were in the siege and assault of Vicksburg from late May until the city's surrender July 4, 1863. During the siege the Regiment operated against Fort Hill north of Vicksburg. Irish born Corporal Thomas C. Murphy of Company I, received the Medal of Honor for his actions at Vicksburg.[446] The remainder of 1863 the Thirty-first saw action at Monroe, Louisiana and Canton,

---

[444] "The Flag of our Union", *Centralia Sentinel,* August 6, 1863 (the men of Co. K of the 31st came from Marion County with many from the Marion County town of Centralia.)

[445] At a meeting of the citizens of Williamson County on April 15, 1861, they had voted that in the event that the border states seceded "the interest of the citizens of southern Illinois imperatively demands at their hands a division of the State and we hereby pledge ourselves to all means in our power to effect the same and to attach ourselves to the southern Confederacy." *Cairo City Weekly Gazette,* April 25, 1861.

[446] Thomas Murphy from Pekin in Tazewell County received the Medal of Honor: "Voluntarily crossed the line of heavy fire of Union and Confederate forces carrying a message to stop the firing of one Union regiment on another." Vicksburg, Mississippi, May, 1863. *Civil War Medal of Honor Winners from Illinois,* published by the Civil War Centennial Commission of Illinois, July 1962. p 12.

Mississippi. In early 1864 the Thirty-first under Sherman was in the Meridian campaign. After reenlistment the men went on furlough in March and April and on return were sent to north Georgia for the Atlanta campaign. They skirmished at Big Shanty and Brush Mountain and were in the assault on Kennesaw Mountain. They fought at Nickajack Creek, Chattahoochee River and Battle of Atlanta. In August, in a flanking movement around Atlanta, they fought at Jonesboro and Lovejoy Station. Following their pursuit of Hood into north Georgia and Alabama they joined Sherman's "March to the Sea." The Regiment was at the siege of Savannah, Georgia in December. In 1865 the Thirty-first skirmished at Fort Pocotaligo, Salkehatchie, Binnaker's Bridge, Orangeburg, Cheraw and Columbia, South Carolina and in Fayetteville and at Bentonville, North Carolina. They then marched further to Goldsborough and Raleigh, North Carolina, through Richmond to Washington DC for the Grand Review on May 24. They then moved to Louisville, Kentucky on Provost duty, until July 19, 1865, when they were mustered out.[447,448]

In the collection of the State of Illinois Military Museum there are three flags of the Thirty-first Illinois Infantry Regiment—two national and one regimental.

This national flag with a square canton and the stars arranged in 5 horizontal rows is typical of flags manufactured by manufacturers in New York. There are no identifications or battle honors on this flag.

[447] *Report of the Adjutant General of the State of Illinois,* Vol II, p 574-7
[448] Dyer, Frederick H. *Compendium of the War of the Rebellion, Vol* III, p 1058-9.

The above National flag is in very poor condition. According the records of the Adjutant General it formerly contained the following inscriptions: BELMONT, FORT HENRY, DONELSON, SIEGE OF CORINTH, TUSCUMBIA RIVER, THOMSON HILLS, INGRAHAM HEIGHTS, RAYMOND, JACKSON, CHAMPION HILLS, BLACK RIVER BRIDGE, SIEGE OF VICKSBURG 22 MAY, FORT HILL, MONROE LA, MERIDIAN, BIG SHANTY, BUSH MOUNTAIN, KENESAW MOUNTAIN, SIEGE OF ATLANTA, ATLANTA JULY 21, 22 AND 23, LOVEJOY STATION, JONESBORO, SAVANNAH, POCOTALIGO, SALKAHATCHIE, CHARLESTON, ORANGEBURG, COLUMBIA, CHERAW, FAYETTEVILLE, BENTONVILLE, GOLDSBORO, RALEIGH

This Regimental flag of the Thirty-first, based on the position of the stars and the scroll with flourishes on the ends beneath the eagle is typical of flags manufactured in New York.

A report of one flag presentation to the Regiment has been found. After the battle of Belmont, a flag was presented to the Regiment. "FLAG PRESENTATION—On Tuesday afternoon the 31st regiment, Col. Logan, were presented with a splendid regimental flag by the ladies of Carbondale. The presentation speech was made by John A. Hull, editor of the *Carbondale Times,* in a brief and appropriate manner, which was replied to by Col. Logan in his usual effective style; after which Gen. McClernand addressed the regiment and complimented them highly for their gallant conduct at the battle of Belmont."[449] There is no further history re this flag and no other flags carried by this Regiment are known to exist.

---

[449] *Cairo City Weekly Gazette,* December 12, 1861.

32nd Illinois Infantry

"Not only did we fight at Shiloh,
But we fought at Hatchie too.
And Prices hordes were put to flight,
By the fighting Thirty Two.
T'was on the morning of the fifth,
Of October, Sixty Two.
The Bridge was crossed through shot & shell.
By the Illinois Thirty Two.
We drove them back to Mattamoras,
And there they made a stand,
But Gen. Hurlbut made a charge,
With his gallant little Band.
We took their Field Artillery,
When they were put to flight.
These with three hundred Prisoners,
Were the fruits of Hatchie Fight."[450]

The Thirty-second Illinois Infantry was organized at Camp Butler and mustered in for three years of service on December 31, 1861, under the command of Colonel John Logan.[451]. By special authority, the Regiment was initially comprised of ten infantry companies, one cavalry company (*Gilmore's Cavalry Company* from Greene County) and one artillery battery (*Rodgers' Battery*). Prior to taking the field, the cavalry company was transferred to Twelfth Illinois Cavalry Regiment, and the artillery battery to Second Illinois Light Artillery, Co. K. The ten infantry companies volunteered from the following counties: Lee County—Co. A; Hancock and Logan Counties—Co. B; Macoupin County—Co. C; Greene County—Co. D; Fayette, Shelby and Greene counties—Co. E; Madison County—Co. F; St. Clair County—Co. G; Morgan County—Co. H; Macoupin, Wabash and Peoria counties—Co. I and Jasper County filled Company K. On January 29, 1862, the Regiment arrived at Cairo, drew arms and moved to Bird's Point, Missouri. In February they traveled the Tennessee River to Fort Henry where Company A escorted *Rodgers' Battery* to Fort Donelson and participated in action there. Company A was then detached as sharpshooters on a gunboat in Eastport, Mississippi. Companies C and K embarked on gunboats *Tyler* and *Lexington* and fought against a rebel battery at Pittsburg Landing, Tennessee on March 1. The remainder of the Regiment moved from Fort Henry to Pittsburg Landing in mid- March and fought at the Battle of Shiloh on April 6 and 7, 1862 where more than fifty per cent of the men who went into action were killed or wounded. The commander of the Regiment, Col. John Logan was wounded, and Lt. Col. John Ross was killed. Next the Thirty-second fought at the siege of Corinth, Mississippi—April 29-May 30, and afterwards pursued the retreating Rebels to LaGrange and Grand

---

[450] Hough, Granville W., ed. "Diary of a Soldier in Grant's Rear Guard" *(1862-1863), J. of Mississippi History,* Volume 45, August 1983. Portion of a poem written by William Owns, Co. F. Thirty-second Illinois Infantry.

[451] Colonel John Logan was a 51-year-old physician from Carlinville, Illinois. He had fought in the Black Hawk War and was the cousin of General John A. Logan, the commander of the 31st Illinois Infantry. *Alton Evening Telegraph,* August 27, 1885, page 3.

Junction, Tennessee. October 1862 the Regiment marched from Bolivar, Tennessee to the Hatchie River and at Metamora on October 5 engaged the Rebel forces that had retreated from Corinth. In November the Thirty second captured 100 prisoners at Lamar, Mississippi then, due to the loss of the supply depot at Holly Springs they retreated to guard the *Memphis & Charleston RR* until March 1863. Next the Regiment garrisoned Memphis and on May 11 moved to Young's Point, Louisiana and on May 16 to Grand Gulf, Mississippi. The Regiment fought in the Siege of Vicksburg and after the fall of Vicksburg on July 4 they pursued the retreating forces of General Johnston to Jackson, Mississippi. During the remainder of 1863 the men saw action at Natchez Mississippi, Harrisonburg, Louisiana and Fayette, Mississippi. January 1864 the Regiment moved to the rear of Vicksburg, where they re-enlisted as Veteran Volunteers. In April the Volunteers, who had been on furlough, re-assembled at Camp Butler and moved to Bird's Point, Missouri, then in May they moved by boat down the Tennessee River to Clifton, Tennessee and marched through Alabama to join Sherman's Atlanta campaign at Ackworth Georgia. They fought at Kennesaw Mountain, Nickajack Creek, Big Shanty and the Chattahoochee River and Turner's Ferry. The Regiment was on guard duty at Marietta Georgia from July 18 until November 3. During November, those men whose term of service had expired were discharged and the remainder of the Regiment, under Sherman, began the "March to the Sea." Enroute to Savannah, the Thirty-Second skirmished at Oconee River crossed the King's Bridge on the Ogeechee River, assaulted a fort outside Savannah and entered Savannah in mid-December 1864.

January 1865 began the campaign in the Carolinas. The Thirty-second saw action at Pocotaligo, Salkehatchie Swamps, Orangeburg, Columbia[452] and Cheraw South Carolina, then Fayetteville and finally Bentonville, North Carolina. On April 13 the Regiment reached Raleigh and remained there until Johnston's surrender. From Raleigh they marched via Petersburg, Richmond and Alexandria to Washington, DC for the Grand Review on May 24. In June they moved to Louisville, Kentucky, to St. Louis, to Fort Kearney, Nebraska and finally Fort Leavenworth, Kansas. They mustered out at Fort Leavenworth on September 15, 1865.[453]

The original inventory of flags reported for the Thirty second Infantry by the Adjutant General of the State of Illinois listed one national and one regimental flag. There are now four flags in the collection—three national and one regimental.

---

[452] A flag of the Thirty Second Illinois Infantry was displayed on the courthouse at Columbia. "The first two flags displayed in Columbia were those of the 32nd Illinois Infantry (on the town hall by Adjutant Fenwick Hedley) and the 13th Iowa Infantry on the state house. The 32nd Illinois captured two Confederate colors. There is no report as to which Confederate Regiments the colors belonged or as to what happened to the captured colors." *Report of the Adjutant General of the State of Illinois,* Vol II, p. 611.

[453] Ibid p. 609-11

32ND ILLINOIS INFANTRY

This national flag bears no inscriptions. The oval star pattern is typical of flags manufactured in Philadelphia.

This national flag is inscribed: "Pittsburg Landing….., SHILOH April 6…..,32nd Regt Ills Infn….., SIEGE of C…., MATAMORA…" *Some of the inscription has been lost. The oval star arrangement is typical of flags manufactured in Philadelphia.*

The honors inscribed on this national flag are: FORT DONELSON, SHILOH. HATACHIE, SIEGE OF CORINTH, VICKSBURG, JACKSON, KENNESAW MOUNTAIN, NICKAJACK, ATLANTA, SAVANNAH, POCOTALIGO, COLUMBIA, SALKAHATCHIE, RALEIGH, BENTONVILLE. *This is a 35-star flag, therefore was probably manufactured late in the War. Note that the printing of the inscriptions is all done with similar lettering, suggesting that all the inscriptions were applied at the same time. This flag may have been carried in the Grand Review.*

This regimental flag measures 72" on the hoist and 73" on the fly. It was manufactured late in the War since there are 35 stars above the upper scroll. This flag, like the national flag above may have been carried in the Grand Review. The design of the flag is typical of flags of New York manufacture.

No other flags carried by this regiment are known to survive.

33rd Illinois Infantry – "*The Teachers' Regiment*"

"Fair is our flag— "Flag of the Free!"
But the truths it announces are fairer.
Rare is our flag in its beauty:
But its mystical teachings are rarer.
White are the stars in the union;
But the justice they stand for is whiter.
Bright are the red and white stripes;
But the glory they tell of is brighter.
Clear are its colors resplendent;
But the rights they assure are clearer.
Dear are our memories of it;
But our hopes for its future are dearer."[454]

Charles Hovey, an eminent educator, succeeded in creating a state institution for the training of teachers-Illinois State Normal University- at Bloomington, Illinois.[455] In Washington DC during the first Battle of Bull Run, Hovey asked President Lincoln for authority to raise an Illinois Regiment with students as the nucleus. Permission granted, Hovey returned to Illinois and within three weeks had raised nine companies for his regiment. The men of Company A (*Normal Rifles)*-were the graduating class of State Normal University and were presented a "silken banner" by the young ladies of the school.[456] Since Hovey wanted a regiment of students, teachers and other educators, men from 80 of the 102 Illinois counties volunteered for his regiment. However, Company K from Cass, Menard and Morgan counties enrolled as an already formed company-*Bateman Rifles*. After muster on September 3, 1861, the men moved to camp at Ironton, Missouri where they remained through the winter. During that time in October 1861, they saw their first action with skirmishes at Big River Bridge and at Fredericktown, Missouri. In March 1862 they marched to Helena, Arkansas and in July fought the Texas Rangers at Cache Creek near Cotton Plant, Arkansas.[457] In October the men boarded the transport *Des-Moines* and were transported north to Memphis, on to Cairo and debarked at Sulfur Springs for the march to Pilot Knob, Missouri. They remained in camp until November when they began a three-month campaign in southeast Missouri. On March 10, 1863, the Thirty-third broke camp, marched to St. Genevieve on the Mississippi River, boarded the transport, *Illinois,* and landed at Milliken's Bend Louisiana on the 26th. For the next three months they were engaged in the Vicksburg campaign and fought at Port Gibson—May 1,

[454] Way, Virgil G., History *of the Thirty-third regiment Illinois veteran volunteer infantry in the Civil war, 22nd August 1861, to 7th December, 1865,* Gibson City Press of the Gibson Courier, 1902, p 62. Poem written Robert Allen Campbell

[455] Colonel Hovey resigned as president of the University to recruit the men for the 33nd Infantry and to assume the command of the Infantry Regiment.

[456] *Illinois Civil War Sketches,* "The Schoolmasters' Regiment*"; Civil* War Centennial Commission, No. 4, June 1963.

[457] Way. Virgil, ibid, page 28. "Just at this time occurred the most thrilling incident of the day. The little cannon had been abandoned, and the foe was almost upon it….Capt. Potter called out, "Boys, save the gun." Hearing the call of his captain, Orderly Sergeant Edward A. Pike, of Company A, who was as colossal in courage as he was in stature rushed to the gun when the enemy was almost near enough to have struck him with their sabers, seized the trail (sic) and tore down the road with the cannon as if it had been a baby wagon. For this brave act Sergt. Pike received a medal of honor from the Secretary of War."

Champion's Hill—May 16, Black River Bridge—May 17 and the Siege of Vicksburg and fall of Vicksburg—April 19-July 4. On July 5, 1863, the Regiment pursued the retreating rebels east to Jackson, Mississippi and laid siege to Jackson. Among those wounded in the pursuit was "Private Joe" Fifer of Co. C.[458] In August they moved to New Orleans and were engaged at Bayou Teche in October. On November 15, the Thirty-third embarked on the sea-going steamer *Clinton*, sailed for Texas, landed on St. Joseph Island, and participated in capture of Fort Esperanza on the east end of Matagorda Island. The flag of the Thirty-third was the first raised over the Fort. They remained around Indianola, Texas, reenlisted as veteran volunteers in January 1864 and went home to Bloomington on furlough in March. The Regiment reorganized at Camp Butler then proceeded to New Orleans in April 1864. They were assigned guard duty at Bayou Boeuf, Bayou L'Ours, Tigerville, Chacahoula, Terre Bonne, Bayou Lafourche, Bayou Des Allemands and Boutte, Louisiana. In September the non-veterans were discharged home. In March 1865 the Thirty-third was assigned to the Mobile campaign and participated in battles at Spanish Fort and Fort Blakely and the surrender of Mobile. While encamped in Montgomery, Alabama they received word of the surrender of Lee and Johnston. In May they marched to Selma then went by rail to Meridian, Mississippi. Here men were transferred from the One Hundred and Seventeenth, One Hundred and Twenty-second and One Hundred and Twenty-fourth Infantry Regiments to fill the Thirty-third Infantry back to full strength. The men were mustered out November 24, 1865, at Vicksburg, Mississippi.[459]

The women of State Normal University gave Co A a handmade flag which was the first flag carried by this Regiment. This is probably the flag that now is at the McLean County Historical Society Museum in Bloomington Illinois. (No photo available.) According to Virgil Way in his regimental history,[460] "The flag used by the 33rd from its organization up to October 20th, 1861, was one presented by the women of Bloomington to the "Normal Rifles" and was carried by that company before its connection with the regiment. This flag was taken care of by Col. Charles E. Hovey, and at his death was presented to Capt. J. H. Burnham of Bloomington, who now has it in his possession."

Letters written between Mary and Charles Elder are in the possession of the McLean County Historical Museum. Charles Elder writes in one of his letters to Mary dated Oct 20th, 1861, from Ironton, Missouri, "A gentleman from Chicago in behalf of the public schools of that city presented our regiment with a beautiful set of colors; one an ordinary flag with stars and stripes surrounded with a wide fringe of yellow or orange silk with the inscription 33 reg't, Ill. Vol. in gold. The other flag of deep blue silk in size perhaps 8 x 10 feet surrounded like the other with orange fringe in the centre is an Eagle with a scroll with inscription "Presented by Teachers of Chicago to Normal Regt." Most an elegant thing. They cost $130.00 and I think are appreciated by everyone in the reg't." This presentation is also reported in the Chicago Tribune.[461]

There is a more complete description of these flags in the *Chicago Times*[462], "A splendid stand of colors, prepared for presentation to Col. Hovey's Normal Regiment (the Thirty-third), is now on exhibition at the rooms of the Board of Education, No. 119 Clark Street. The larger flag is of silk, has a field of blue, surrounded by an orange fringe, and is about six feet square. In the center are the United

---

[458] "Private Joe" became governor of Illinois and served 1889-93. Bennett, James O'Donnell, *"Private Joe" Fifer,* Pantagraph Printing & Stationery Co. Bloomington, Illinois, 1936.

[459] Way, Virgil G., *ibid*

[460] Way, ibid, p 62.

[461] *Chicago Tribune;* Oct. 29,1861, p.4

[462] Quoted from the *Chicago Times.*—original in the October 18, 1861 issue of the *Illinois Statesman.*

States coat of arms. Above the eagle are the words, “Presented by the Teachers of Chicago” beneath, “The Thirty-third Regiment of Illinois Volunteers.” The cost of the set of colors was $130. They were made my Messrs. Gilbert, Hubbard & Co., and the decorations were done by Dubois, the well-known artist. The funds to pay for the flags were raised wholly in the public schools of the city.”

Virgil Way mentions that Mr. Benjamin R. Cutler, a representative of the teachers of Chicago, presented the flags at Arcadia, Missouri and the regiment, until after the Battle of Vicksburg, carried these flags “when they had become much faded, worn and tattered. After the regiment had returned from Jackson and was in camp on the river, Mr. George W. Spofford, Principal of the Foster School and chairman of a committee of the Chicago teachers, came to Vicksburg and presented the regiment with a new stand of colors, taking the old ones with him on his return.”[463]

Based on a story from the *Chicago Times* in 1865, apparently this second set of colors was returned to the Chicago teachers. “ The principals of the public schools of the city, together with several of the High School teachers, met at 4:30 o’clock yesterday afternoon, in the rooms of the board of education, to welcome, Col. I. H. Elliott, of the Normal or Thirty-Third regiment, and to receive again the flag which had been presented to the regiment in 1863 by the teachers of the public schools…The flag yesterday returned showed upon its folds the marks of battle and the trace of long and arduous service….Mr. Cutler moved that the colors be returned to the State Normal University by Col. Elliott. The motion prevailed.” [464] Colonel Elliott states “In compliance with the request of that meeting, I took them direct to Normal and turned them over to President Hewitt. There was a meeting in the chapel of all the instructors and students to receive them.”[465]

A reunion of the regiment was held in Bloomington, Illinois in 1875 and the men returned to Normal University. “The old regimental flag of the 33rd, tattered and war-torn, was brought in by Prof. Hewitt, from the Normal Museum, where it was long since deposited as a precious relic.” They returned to the Ashley House where, “The old regimental flags, which were tattered and torn by rebel bullets, were displayed from the east awning of the hotel. One of the flags, being the first which was carried by the regiment during the war, had upon it the names of the different battles through which the regiment had passed.”[466] It was also these colors that were used at the “Seventh Reunion of the 33rd Illinois Veteran Association” held in Bloomington, Illinois, “The Star Spangled Banner was participated in by about a dozen young ladies in costume, assisted by several young men holding flags, among them the regiment’s old banner being most prominent.”[467] The McLean County Museum, in Bloomington, Illinois, has a flag identified as having been carried by the 33rd Illinois Infantry. The flag is in a box in shards. Most of the colors are deep purple. Possibly this is the regimental flag of the 33rd Illinois.

Virgil Way also states that at least one set of regimental colors was secured through the Quartermaster, and these were deposited in Springfield. The following national flag and regimental flag are in Springfield in the Illinois collection. There is a second national flag not shown here.

---

[463] Way, *Ibid,* p63

[464] *Chicago Times* December 14, 1865.

[465] Way, Virgil G , *Ibid,* p64

[466] Reunion of the 33d Illinois Regt. At Bloomington, Illinois, October 21st, 1875. Report of the Proceedings, p5

[467] Seventh Reunion 33 Illinois Veteran Association, held at Bloomington, Illinois, January 22-23, 1890, p 5.

This is the National flag of the 33rd Illinois measures 68" on the hoist by 73" on the flay. The flag originally bore inscriptions on both the red and white stripes. At this time only the top inscription can be seen and reads FREDERICKTOWN. *The center red stripe does read "33" but the remainder of the inscription is lost.*

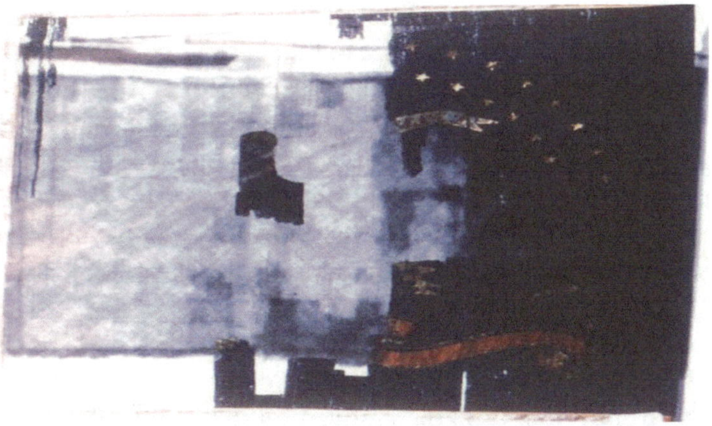

Very little remains of the regimental flag of the Thirty-third. Based on the position of the stars above the eagle the flag appears to be of New York manufacture.

The men of Company B came from DuPage, Stark and LaSalle counties. "B Company was given a flag by the "Ladies of DuPage County". It floated over Fort Morgan…On the assembling of the companies to assume operations as a regiment; this beautiful flag was placed into the keeping of Lieutenant Nelson G. Gill; by him it was sent north to be cared for… After the war was over it graced with its presence a flagstaff on a school building in Holly Springs, Miss., in which Mrs. Gill was conducting a school. The hand of a vandal who was too cowardly to face the flag in open battle set fire to the building. That flag, waving in its majesty, the emblem of peace and good will, even to a lately rebellious people was burned with the building."[468] No other flags carried by the 33rd Illinois Infantry are known to survive.

---

[468]Way, Virgil G., ibid p97

34th Illinois Infantry - "*Rock River Rifles*"

"About four o'clock in the afternoon, the pluckiest incident ever witnessed by the regiment occurred in front of the left wing. A steady, regular fire was in progress all the time, but it had no terrors for the tall, athletic artilleryman who sprang over the rifle-pits with a Confederate flag in his hand and leisurely trotted fifty or sixty yards to the front and drove his flagstaff into the ground, lifted his hat in salute, and as deliberately turned and trotted back into their lines. He was greeted with a chorus of cheers from our whole line, and a volley of musketry also…"[469]

The volunteers of this Regiment came from northwestern Illinois, the valley of the Rock River; thus, they mainly came from the counties of Whiteside County—Companies A and B; Lee County—Company C; Ogle County—Companies E, F and H; and Carroll County—Company I (*Hickory Rifle Guards)*; but some came from further south -- from Christian County—Company D; Morgan County and Randolph—Company G (*Liberty Home Guards)* and Edgar County—Company K. Colonel Edward N. Kirk, an attorney from Sterling in Whiteside County raised the regiment and became its first Colonel when the regiment mustered in for three years of service on September 7, 1861. On October 2 the Regiment was sent by rail through Illinois, Indiana, and Ohio down through Lexington and Louisville, Kentucky to Camp Nevin, which was situated south of Elizabethtown, Kentucky near the Nolin River. In December they marched further south to Munfordville, Kentucky and remained there until February 1862 when they marched north through Nashville to Pittsburg Landing, Tennessee. The regiment arrived in time to engage the enemy at the Battle of Shiloh on April 6 and 7. From Shiloh they moved south, fought in the siege of Corinth, Mississippi, and at the end of May pursued the retreating rebels to Booneville. The Regiment next marched to Iuka; Mississippi crossed the Tennessee River at Florence, Alabama and was in operations along the line of the *Memphis and Tennessee R. R.* in June, July, and August. In August they marched to Nashville then pursued Bragg into Kentucky in September and October resulting in actions at Floyd's Fork (October 1), near Clay Village (October 4) and at Perryville, Kentucky (October 8). They returned to the Nashville area and remained there until December 26, 1862, when they advanced to Murfreesboro, Tennessee and fought in the Battle of Stone's River December 30 through January 2, 1863. It was at this battle that the colors of the Regiment were captured.[470] They remained at Murfreesboro until June 1863 then moved south toward Liberty Gap, fighting there June 24-27. On July 1 the Thirty-fourth marched to Tullahoma, Tennessee, in August to Bellefonte, Alabama and in October to Anderson's Cross Roads. In November they were at Chattanooga and fought in the Battle of Chattanooga, at Tunnel Hill, Missionary Ridge, and Chickamauga Station. December 22, 1863, the Thirty-fourth re-enlisted as a veteran regiment and in January 1864 started home on veteran furlough. Upon return, in March, they went into camp at Rossville, Georgia and in the Atlanta campaign fought at Rocky Face Ridge, Buzzard's Roost Gap, Resaca, Rome, Dallas, New Hope Church, Allatoona Hills, Big Shanty, Kennesaw Mountain, Pine Hill, Lost Mountain, Chattahoochee River, Peach Tree Creek, and the siege of Atlanta. In the flanking movement around Atlanta, in August they led a charge on the rebel works at Jonesboro, in September fought at Lovejoy Station then pursued Forrest across Tennessee to Florence, Alabama. The Thirty-fourth returned with the army to Atlanta and joined Sherman's "March to the Sea." This campaign included the siege of Savannah, actions at Averysboro, Bushy

---

[469] Payne, Edwin, *History of the Thirty-Fourth Regiment of Illinois Volunteer Infantry.* Allen Printing Company, Clinton, Iowa, (1902) pp109-110. The author is referring to an incident at the battle of Resaca on May 14, 1864. Bravery in carrying the flag was cheered and rewarded by both sides.

[470] ibid, Payne, pp 44 and 56-7

Swamp, and Bentonville, North Carolina, and occupation of Raleigh. Following Johnston's surrender the Regiment marched to Washington, DC via Richmond for the Grand Review on May 24, 1865. They moved to Louisville, Kentucky and were mustered out July 12, 1865.[471]

The flag collection at the State of Illinois Military Museum contains three national and one regimental flag.

This national flag is inscribed "34th REGt. ILLs. VOLs." *There is a tag on this flag stating that it was captured. In the Regimental History the author, recounting the Battle of Stone's River, details two separate skirmishes in which the Regiment lost a flag. On December 31, Payne relates the following:*[472] *"The Thirty-fourth soon became engaged in an almost hand-to-hand combat. Already twelve men had been killed and sixty wounded, but the regiment heroicly sought to maintain its line. The strife over its colors was terrible and bloody. Five color-bearers fell in quick succession, but as fast as they fell, the flag was raised aloft and flaunted in the face of the foe! The entire color-guard being killed or wounded, its colors over so much precious blood had spilled were trailed in the mud and borne off the field by the hands of the enemy."*

Payne describes the loss of a second set of colors that took place at Stones River on January 2, 1863.[473] "The lines of the enemy, advancing rapidly by a "left oblique" first struck the position of the Thirty-Fourth Illinois, which opened fire at the first opportunity and stood their ground until they became mixed up with the foe, and a battle for the regimental colors was soon decided in their loss to the enemy, but not without the loss of life and blood on both sides in that special melee. Every color-guard and bearer was either killed or wounded, Santee, of Company C; Wright, of Company F; William Wendle of Company D; W.H.Steele of Company F, who received seven wounds, and now carries three of the bullets in his body. He handed the colors to Lieut. John M. Smith, of Company H, saying, "Trail them, until you get away;" but Lieut. Smith was immediately riddled with five shots and instantly killed.

---

[471] Payne, Edwin, *History of the Thirty-Fourth Regiment of Illinois Volunteer Infantry.*

[472] Ibid. page 44

[473] Ibid. pp 56-7

Someone whose name has not been obtained, seized the colors from Lieut. Smith, and ran a few paces and handed them to one of Edgarton's artillerymen, who was mounted, to bear to the rear, but he, too, was shot, and the enemy having captured a large number of the regiment, the colors fell in the hands of the Second Arkansas infantry, of Gen. Cleburne's Division, and were carried off the field by the soldiers who guarded the captured men of the regiment to the rear as prisoners of war." It is unclear which of the above two scenarios resulted in the loss of the Thirty-fourth's National flag that is shown above.

This National flag is inscribed as the flag of Company C of the 34th Illinois whose men came from Lee County. There is no history regarding origin of this company flag. The flag is inscribed as follows: SHILOH Company C 34 REGIMENT ILLINOIS INFANTRY 1861 CORINTH *and has 34 stars. The flag is homemade, and the letters and stars are sewn on the flag. It measures 69" on the hoist by 76" on the fly.*

This national flag contains 35 stars; therefore, it was probably issued late in the War. There are no inscriptions or regimental designations on the flag. Based on the shape of the canton and arrangement of the stars, this flag was produced in New York.

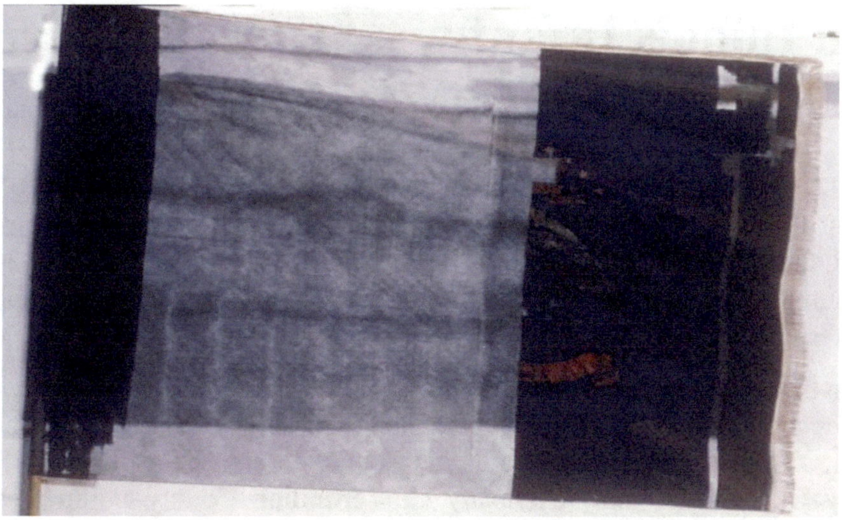

Little remains of the Regimental flag. The two portions of the original flag are held in place by silk netting. The flag appears to be of standard Quartermaster Dept. issue.

A "banner" of the Thirty-fourth Illinois Infantry was on display at the Great North-Western Fair in Chicago in 1865. A letter from the commander of the Regiment, Colonel Peter Ege, states, "I have the honor, herewith, to transmit the remnant of the banner carried by my Regiment in the campaign against Atlanta, the Georgia campaign, and the campaign in North and South Carolina. This banner has been literally cut to pieces in our action at the charge of Jonesboro, Ga., twelve bullets pierced it and one cut the flagstaff off." Either the 35-star national flag or the remnant of the regimental flag shown above may be the "banner" displayed at the Chicago Sanitary Fair

Company G, *Liberty Home Guards*, was presented a flag by the ladies of Liberty, Illinois. As reported in the Weekly Randolph Democrat[474] "After the election of officers, the company, numbering ninety men, was drawn up in order to receive a splendid flag, presented by the ladies of Liberty. The flag was presented by Mrs. Dr. Vance, with the following remarks: 'In behalf of the Union ladies of Liberty, I present this flag. This union of the stars and stripes is emblematic of our sentiments. This flag, that has so long waved over us in triumph, we exhort you to stand by and protect, and never to permit it to be dishonored!'"

There is no further record of this flag and its whereabouts today is unknown.

No other flags carried by this regiment are known to survive.

[474] *Weekly Randolph County Democrat,* May 18, 1861.

35th Illinois Infantry – "*Gustavus Smith's Independent Regiment*"[475]

"Lieutenant-Colonel Chandler, carrying his regimental colors after 7 color sergeants had been killed or wounded, the colors receiving more than thirty bullet holes, planted them on the works, where they were soon joined by those of the Sixty-eighth Indiana, Eighty-sixth Indiana, and Fifty-ninth Ohio (the two latter of General Beatty's brigade)."[476]

The Thirty-fifth Illinois Infantry was organized in Decatur, Macon County and mustered in for three years' service at Jefferson Barracks near St. Louis on August 28, 1861, under the command of Gustavus Smith[477]. The men volunteered from Piatt County-Company A; Christian, Shelby and Fayette counties-Company B; Shelby County-Company C; Vermillion County-Companies D, E, F, and I; Fayette County-Companies G and H and Fayette and Effingham counties-Company K. In September the Regiment moved to Jefferson City and on to Otterville, Missouri, then marched to Sedalia to join in the advance on Springfield, Missouri. For the remainder of 1861 and early 1862 the Thirty-fifth was stationed and skirmished in southern Missouri and northern Arkansas. May 6-8 the Thirty-fifth was heavily engaged against rebel Missouri troops in the Battle of Pea Ridge (Elkhorn Tavern), Arkansas where the Regiment lost 115 men. In late May the Regiment arrived at Hamburg Landing, Tennessee, marched to Farmington, Mississippi and took part in the siege of Corinth, Mississippi until the city was evacuated on May 30. After pursuing the fleeing enemy as far as Booneville, Mississippi they returned to the area of Holly Springs, Mississippi in June, marched to Nashville in August and moved into Kentucky in October. On October 8 they skirmished with Bragg's forces at Perryville (Chaplin Hills) and returned to Louisville. December 26 the Thirty-fifth marched to Murfreesboro, Tennessee and December 30 and 31 and January 1 and 2, 1863 fought at Stones River. In that battle the Regiment lost 86 men killed, wounded, missing, or taken prisoner. They remained at Murfreesboro until June 1863 and in August crossed the Cumberland Mountains and the Tennessee River in the Chickamauga campaign. At the battle of Chickamauga in September the Thirty-fifth was in the midst of fighting and of 299 men who entered the battle, 160 men were casualties—killed, wounded or missing. November 25, 1863, the Regiment was in the assault on Missionary Ridge and lost 54 of the 212 who led in the charge up the mountain. "At Missionary Ridge Lt. Col. W. P. Chandler, commanding the regiment, was the first to plant the Stars and Stripes on the enemy stronghold, the entire color guard being killed or wounded."[478] March 1864 the Regiment marched to the relief of Knoxville, Tennessee and in May joined Sherman's Atlanta campaign fighting at Rocky Faced Ridge, Buzzard's Roost Gap, Resaca, Adairsville, Dallas, Pumpkin Vine Creek, New Hope Church, Allatoona Hills, Pickett's Mills, Kennesaw Mountain, Ruff's Station, Chattahoochee River, Peach Tree Creek and the Siege of Atlanta. Following the fall of Atlanta, the Regiment marched to Chattanooga. The Regiment did not re-enlist, therefore on September

---

475 The Thirty-fifth Illinois Infantry was originally known as "Col. G. S. Smith's Independent Regiment" and was organized in Decatur, Illinois by G. A. Smith a buggy manufacturer.

476 O.R.-- SERIES I--VOLUME XXXI/2 [S# 55] NOVEMBER 23-27, 1863. --The Chattanooga-Ringgold Campaign. No. 62. --Report of Brig. Gen. August Willich, U. S. Army, commanding First Brigade. Refers to Battle of Missionary Ridge

477 Colonel Smith was a 41 year old carriage maker in Decatur prior to the outset of the war.

478 Banton, OT, Editor, *History of Macon County 1976,* Printed by the Macon County Historical Society, Decatur, Ill. (1976), page181.

27, 1864, the Thirty-fifth Illinois Infantry mustered out at Springfield, Illinois.[479]

There is one national and one regimental flag of the Thirty-fifth Illinois Infantry in the collection at the State of Illinois Military Museum in Springfield, Illinois.

The inscriptions on this National flag read as follows: PEA RIDGE, PERRYVILLE, CHICKAMAUGA, MISSION RIDGE, ROCKY FACE, RESACCA (sp?) & DALLAS, 35th ILLs. INFANTRY, MUD CREEK, KENESAW, CHATTAHOOCHIE, ATLANTA. *The flag measures 69" on the hoist by 73" on the fly, appears to have either 33 or 34 stars and has a canton and star arrangement characteristic of flags produced in Cincinnati.*

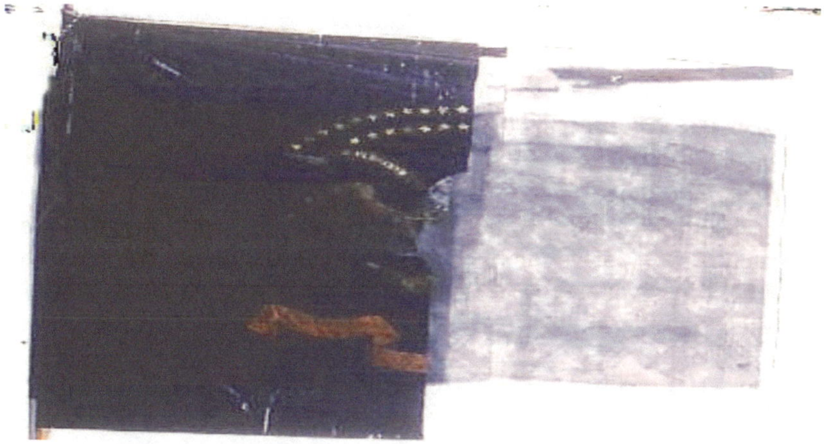

The Regimental flag of the Thirty-fifth Illinois Infantry is severely damaged and only 50% of the flag remains. No inscriptions are seen on the remaining portion of the flag.

No other flags carried by this regiment are known to survive.

[479] *Report of the Adjutant General of the State of Illinois,* HW Rokker, State Printer and Binder, Springfield, Illinois, (1886), pp713-15.

36th Illinois Infantry – "*Fox River Regiment*"

Come boys, a song for our own flag,
Our colors old and dear:
Each shred and tatter time hath made,
We every one revere.
Then fling it once more to the breeze—
Unfold its crimson bars,
And let our friends know we retain
Our galaxy of Stars.
Hurrah! Hurrah!
For freedom far and wide;
Hurrah for the dear old flag,
That fills our hearts with pride.[480]

The Thirty-Sixth Illinois Infantry was organized at Camp Hammond, near Aurora, Illinois during August 1861. Attached to the infantry regiment were two independent cavalry companies--Cavalry Company A from Kane County and McHenry County and Cavalry Company B[481] from Kane County. The men for the Thirty-Sixth Infantry Regiment volunteered as follows: Companies A and B[482]—from Kane County, Company C[483]—from Warren County, Companies D, E, F and I[484]--from Kendall County, Company G[485]—from Grundy County, Company H[486]—from McHenry County and Company K[487]—from Du Page County. The Thirty-Sixth was mustered in for three years' service September 23, 1861, under the command of Colonel Nicholas Greusel.[488] On September 24 the Regiment moved through Quincy, Illinois and St. Louis to Rolla, Missouri[489] where it remained until mid-January 1862; the men then marched to Springfield, Missouri and on to Cape Girardeau, Missouri, arriving May 11. En-route they

---

480 Bennett, L.G., Haigh, Wm. M., *History of the Thirty-Sixth Regiment Illinois Volunteers, During the War of the Rebellion,* Knickerbocker & Hodder, Printers and Binders, Aurora, Ill, 1876. P 554. First verse of a poem written by Corporal James L. Dryden, Co. C and sung to "*The Bonnie Blue Flag.*"

481 Cavalry Company A was also known as the *Aurora Dragoons, Hamilton's Escort, Jenks' Cavalry, Rosecrans' Escort or Willis' Cavalry* and Company B was also known as the *Plato Dragoons, Davis' Escort, Sherer's Cavalry, or Smith's Cavalry.* These two cavalry companies usually performed on detached service from the remainder of the regiment and after December 1862 were consolidated with the 15th Illinois Cavalry and later the 10th Illinois Cavalry; their history is included in the history of the 15th Illinois Cavalry.

482 The *Elgin Greys* (Company A) and the *Aurora Guards* (Company B)

483 The *Young America Guards* (Company C)

484 The *Lisbon Rifles* (Company D), *Bristol Light Infantry* (Company E), *Newark Rifles* (Company F), *Oswego Rifles* (Company I)

485 *Morris Guards* (Company G)

486 *Woodstock Rifles* (Company H)

487 *Wayne Rifles* (Company K)

488 Colonel Greusel was born in 1817 in the Kingdom of Bavaria. He emmigrated with his family to the United States and in the Mexican-American War he led a Michigan volunteer company. At the onset of the Civil War he joined a three-month Illinois regiment, rose to field officer rank and was apponted by Governor Yates as Colonel of the 36th Illinois Infantry.

489 At Rolla Cavalry Companies A and B were detached from the infantry regiment.

fought at Bentonville (March 6), Leetown (March 7) and Pea Ridge, Arkansas (March 8). The Thirty-Sixth left Cape Girardeau on May 23 on the steamer *Planet,* debarked at Hamburg Landing, Tennessee and marched to Corinth, Mississippi arriving on May 30. The Regiment then pursued the retreating rebel forces to Booneville, Mississippi and finally to Rienzi, Mississippi; the Thirty-Sixth remained at Rienzi until September 6. Ordered to Cincinnati the Regiment went into camp across the Ohio River from Cincinnati at Covington, Kentucky. Next the Thirty-Sixth was ordered to Louisville and under General Sheridan started in the pursuit of Confederate General Bragg in the Kentucky campaign. On October 8 the Regiment fought at Perryville, Kentucky following which they marched to Nashville and remained there until the end of December. On December 26 the Thirty Sixth advanced on Murfreesboro, Tennessee and fought at Stones River, Tennessee December 30 through January 3, 1863.[490] The Regiment remained at Stones River on the Shelbyville Pike until late June 1863. In the summer of 1863, the Thirty-Sixth marched in the Middle Tennessee campaign and in September marched through the Cumberland Mountains, crossed the Tennessee River, and arrived at Chickamauga, Georgia. They fought at Chickamauga—September 19-20, at the siege of Chattanooga—September 24-November 23, at Orchard Knob—November 23-24 and at the battle of Missionary Ridge on November 25. The Thirty- Sixth claimed their colors were among the first planted on Missionary Ridge and the actions of Sgt. Leverett Kelley resulted in his receiving the Medal of Honor.[491] Following this action the Regiment marched under Sheridan's command for the relief of Knoxville, Tennessee. January 1, 1864, the men re-enlisted as veterans and on January 6 they left for home on furlough. May 3, on return, the veterans and new recruits of the Thirty-Sixth Infantry joined the Atlanta campaign and fought at Rocky Face Ridge, Resaca, Adairsville, Dallas, New Hope Church, Kennesaw Mountain, Pine Hill, Lost Mountain, Chattahoochee River, Buckhead, Peach Tree Creek and in the siege of Atlanta. In the flanking movement around Atlanta, the Regiment fought at Jonesboro—August 31-September 1 and Lovejoy Station—September 2-6. On September 25 the men were ordered back to Chattanooga and then left there on October 2 marching in pursuit of Hood into Alabama. In the Nashville campaign of November and December the men of the Thirty-Sixth fought at Columbia, Duck River, Spring Hill, the disastrous Battle of Franklin and on December 15-16 at the Battle of Nashville.[492] The Regiment joined in the pursuit of the retreating forces of Hood through Franklin, Columbia and Pulaski, Tennessee to the Tennessee River. The Illinois troops went into winter quarters at Huntsville, Alabama on January 6, 1865. Leaving Huntsville by rail on March 28, they proceeded to Chattanooga and into East Tennessee, where at Blue Springs on April 10 they received news of Lee's surrender and on April 15 the news of Lincoln's assassination. They returned to Knoxville, and were ordered to New Orleans, Louisiana on June 23. The Thirty-Sixth was placed on special headquarters duty by General Sheridan and remained at New Orleans until October 8, 1865, when mustered out of service and returned to Springfield, Illinois.[493]

In the military flags collection at the State of Illinois Military Museum there are two national and one regimental flag carried by the Thirty Sixth Illinois Infantry Regiment.

---

[490] The Regiment lost 65 men killed, 151 wounded and 15 missing. This was the largest number of men killed in the Regiment during the War.

[491] Sgt. Leverett M. Kelley, Company A, from LaSalle County, received the Medal of Honor: "Sprang over the works just captured from the enemy and calling upon his comrades to follow, rushed forward in the face of deadly fire and was among the first over the works on the summit, where he compelled the surrender of a Confederate officer and received his sword." Missionary Ridge, Tennessee, November 25, 1863. *Civil War Medal of Honor Winners from Illinois,* Published by the Civil War Centennial Commission of Illinois, July 1962, p 8.

[492] "Corpl. Edgar S. Case, Company E, Thirty-sixth Illinois, captured a rebel flag at the battle of Nashville, December 15, 1864." O.R. Series 1—Volume XLV/1 [Series #93]

[493] *AGR,* Vol III, p35-7.

This National flag of the Regiment has a star pattern typical of flags manufactured in Philadelphia. The flag measures 69" on the hoist by 72" on the fly, is in excellent condition and therefore was presumably issued late in the War. The following battle honors are inscribed on the flag: PEA RIDGE Mar 6,7 & 8, 62 PERRYVILLE Octo. 8. 62 STONE RIVER, Dec. 31. 62 CHICKAMAUGA, Sept. 20. 63 MISSION RIDGE Nov. 25. 63 RESACA. May 14. 64 PLEASANT HILL. May 17. 64 KENESAW MOUNTn June 27. 64 PEACHTREE CREEK. July 20. 64 ATLANTA CAMPAIGN SPRING HILL. Nov. 29. 64 FRANKLIN. Nov 30. 64 NASHVILLE Dec 15 & 16. 64 [494]

The second National flag of the Regiment is in poor condition, however battle honors for Pea Ridge and Perryville are still visible and a portion of the inscription denoting that this is the flag of the 36th Illinois Infantry remains on the center stripe. The condition of the flag is consistent with battle damage and weathering and was probably issued early in the War.

---

[494] The official list of battles in which this regiment bore an honorable part are: "PEA RIDGE, CORINTH, PERRYVILLE, STONE RIVER, HOOVER'S GAP, CHICKAMAUGA, MISSION RIDGE, ROCKY FACE RIDGE, RESACA, ADAIRSVILLE, NEW HOPE CHURCH, KENESAW MOUNTAIN, PEACHTREE CREEK, ATLANTA, JONESBORO, SPRING HILL, FRANKLIN, NASHVILLE." *Official Army Register of the Volunteer Force of the United States Army for the Years 1861, '62, '62, '64, '65.* Part VI, Indiana-Illinois, Published by order of the Secretary of War, Adjutant General's Office, Washington, August 31, 1865, (reprinted by Ron R.Van Sickle Military Books, Gaithersburg, MD, 1987) page 272

The Regimental flag of the Thirty-sixth Infantry is typical of flags manufactured by Shilleto for the Cincinnati Depot and distributed by the Quartermaster of the Union Army. The flag shows very little wear.

There is a piece of a National flag from this Regiment in the DuPage County Historical Museum in Wheaton, Illinois. The flag bears an area where a shell allegedly passed through the flag. (No photo available.)

No other flags carried by the 36th Illinois Infantry are known to survive.

37th Illinois Infantry- "*Fremont Rifles*"

"Our bugles sang truce and the night-cloud had lowered,
And the sentinel stars kept their watch in the sky;
When thousands had sunk to the earth overpowered,
The weary to sleep, and the wounded to die."[495]

The *Fremont Rifles* (also known as the *Illinois Greyhounds*) were mustered into service on September 18, 1861 at Chicago, under the command of Colonel Julius White.[496] Companies A (*Rock Island Rifles)* and H (*Moline Guards)* were from Rock Island County, Companies D (*Cicero volunteers*) and G (*Turner Rifles)*from Cook County (a portion of D was from Manistee, Michigan- *Manierre Rifles*), Companies C (*Lake County Guards)* and F (*Lake County Rifles)* from Lake County, Company B (*Lafayette Rifles)* from Stark and Henry County, Company E (*Mendota Rifles)* from La Salle County, Company I (*Audubon Rifles)* from Boone County and Company K (*Vermillion Zouaves*[497]*)*from Vermilion County. Before leaving Chicago, they were presented colors from the Chicago Board of Trade (see details below) and proceeded to Booneville, Missouri where they went into camp. Late in October as part of Fremont's army the regiment advanced to capture Springfield, Missouri then returned north and camped on the Lamine River near Syracuse Missouri. In March 1862 the men marched by way of Cassville, Missouri to Sugar Creek, Arkansas and fought at Pea Ridge[498]. During heavy fighting the regiment captured a Confederate flag[499] and the actions of Captain William P. Black, Company K, resulted in his receiving the Medal of Honor.[500]Following the battle the regiment was stationed at Cassville, Missouri on outpost duty. During the summer of 1862 the Thirty-seventh defeated rebels at Neosho, Missouri but spent most of the summer guarding the frontier between Cassville and Springfield. In October they marched to Osage Springs, Missouri and drove the enemy from Newtonia[501] and marched on to Huntsville, Arkansas. October 28, they surprised the rebels at Fayetteville, Arkansas, took

---

495 The Battle of Prairie Grove, Arkansas was fought December 7, 1862. Brevet Brigadier General Eugene B. Payne (Payne had been a 1st Lt. in Co. E of the Thirty-seventh Illinois Infantry) wrote this poem about the evening at the end of the battle. *Prairie Grove,* MOLLUS-DC, War papers No. 52, read at the stated meeting of March 2, 1904.

496 Prior to the war 44 year Colonel Julius White had been a prominent Chicago insurance agent and president of the Chicago Board of Trade and a friend of Lincoln.

497 Chicago Evening Journal, September 20, 1861

498 Mullins, Michael A., *Fremont Rifles, A History of the 37th Illinois Veteran Volunteer Infantry.* Broadfoot Publishing Company, Wilmington, North Carolina, 1990 p68 "Color Sergeant Benjamin Manning .... displayed conspicuous bravery on fourth advance. Corporal Swap, a member of the color guard, recalled that the major had rebuked Manning for withdrawing and as the 37th went forward, the color sergeant furious at being censured, marched the color guard consisting now of only three men about two rods in advance. The small group knelt down so the regiment could fire over them, while Sergeant Manning, kneeling on his left knee and holding his flag staff in his left hand, drew his revolver with his right and proceeded to annihilate the Southern Confederacy."

499See section in this book "Captured Confederate Flags" for details of the flag and its capture.

500 Captain William P. Black, Company K, from Vermilion County "Singlehanded, confronted the enemy firing a rifle at them and thus checking their advance within 100 yards of the lines." Pea Ridge, Arkansas, March 7, 1862. *Civil War Medal of Honor Winners,* Published by the Civil War Centennial Commission of Illinois July 1962, p5

501 1st Lt. Wells H. Blodgett, Company D, from Cook County, received the Medal of Honor for his actions: "With a single orderly, captured an armed picket of eight men and marched them in as prisoners." Newtonia, Missouri, September 30, 1862. *Ibid-Civil War Medal of Honor Winners,* p 3.

prisoners and returned to Osage Springs. The remainder of 1862 was spent in pursuit of the enemy in southwest Missouri and northwest Arkansas until they went into camp at Camp Lyon, Missouri on December 1. December 3 orders were received to relieve General Blunt, then besieged at Sugar Hill, Arkansas. On the morning of December 7, they engaged the enemy at Prairie Grove, Arkansas and in a day long battle lost one seventh of their number either killed or wounded. Lt. Col. John C. Black received the Medal of Honor for his leadership that day.[502] After pursuit of the enemy as far as Fort Smith, Arkansas, they returned to Prairie Grove and wintered there skirmishing with the rebels. April 24, 1863, the Thirty-seventh marched to St. Louis and on to Cape Girardeau, where they engaged the enemy and drove them to the St. Francis River at the Battle of Chalk Bluffs. Next, the Regiment was sent to Vicksburg where they took an active part in the siege and entered the fallen city July 4, 1863. Later in July they captured Yazoo City taking many prisoners, then pursued the enemy to the Big Black River. After return to Vicksburg, they were sent to Port Hudson and then on to New Orleans where they went into camp at Carrollton, Louisiana in August. In September the Thirty-seventh met and defeated the enemy near Morgan's Bend. In October the Regiment took possession of Brownsville, Texas and guarded the Rio Grande River until February 1864 when they re-enlisted as a veteran regiment. After a thirty days' furlough the men proceeded to Memphis, Tennessee and after pursing General Forrest in Tennessee returned to Memphis and were sent to Atchafalaya Bayou, Louisiana. From May through September 1864 the Thirty-seventh was at Morganza, Louisiana, St. Charles, Arkansas, and Duvall's Bluff, Arkansas where they went into winter quarters. January 1865, they received orders to return to New Orleans and from there were transported to Barrancas, Florida. In March they marched to Pensacola, Florida and across the Perdido and Escambia Rivers to Pollard, Florida, skirmishing en route. In April the Regiment participated in the storming and capture of Fort Blakely, Alabama. For actions in this battle 1st Lt. Thomas Payne received the Medal of Honor.[503] On April 20th they boarded steamers and proceeded up the Alabama River to the rebel prison at Cahawba (or Cahaba), Alabama and took on board the Union prisoners confined there. On April 29 they reached Montgomery, Alabama, and on May 15 Mobile. June 28, they embarked on a steamer and arrived at Galveston, Texas July 1. On July 17 they arrived at Houston, Texas and were stationed guarding the railroads surrounding Houston. On May 15, 1866, the Thirty-seventh Illinois Infantry was mustered out at Houston and reached Springfield, Illinois May 31, 1866.[504,505]

---

[502] Lt. Col. John C. Black from Vermilion County, "Gallantly charged the position of the enemy at the head of his regiment, after two other regiments had been repulsed and driven down the hill and captured a battery; was severely wounded." Prairie Grove, Arkansas, December 7, 1862. ibid. *Civil War Medal of Honor Winners,* p 5

[503] 1st Lt. Thomas H. L. Payne received the Medal of Honor "While acting regimental quartermaster, learning of an expected assault, requested assignment to a company that had no commissioned officers present; was so assigned, and was one of the first to lead his men in the enemy's works." Fort Blakely, Alabama, April 9, 1865. *Civil War Medal of Honor Winners,* ibid p 9.

[504] Herr, George W. *Episodes of the Civil War, Nine Campaigns in Nine States.* The Bancroft Company, San Francisco, 1890, p 398-405

[505] Ibid Mullins, p 326. 0

37TH ILLINOIS INFANTRY

At the State of Illinois Military Museum there are three flags belonging to the Thirty-seventh Illinois Infantry-one National and two Regimental

There has been loss of some of the original inscription but those remaining read as follows: 37th ILLINOIS V.V.INFANTRY Pea Ridge, Prairie Grove, Chalk Bluffs, ------?[506] *The presentation of this flag is detailed in the Regiment's History*[507]*-"At half past twelve on that day the regiment marched from Wrights Grove to the front of the Board of Trade building-----Then it was presented with the national flag which displayed "37th Illinois" in letters of gold on one of the white stripes."*

The presentation of this flag is also mentioned in the Regimental history by Henry Ketzle of Company A, "till on the 19th, in columns of platoons, after a march to and thru the city, we drilled in front of The Board of Trade Building—formed in open square and after several spirited and patriotic speeches by Judge Joe Knox and others, the Regiment was presented with its colors and flag from members of above-named Board of Trade." [508] The Ketzle history goes on to state, "The 21st (of September) marched in and through St. Louis and halted in front of General J. C. Fremont's Hdqtrs., where we were highly complimented on our soldiery appearance and to make words more sure Mrs. Jessie Fremont tied red, white and blue ribbons to our flag and colors, as our colors had imprinted in its folds a picture of J. C. Fremont scaling the rocky mountains." There is not further explanation regarding the origin or disposition of this flag.

There is a report of another National flag presentation. "The citizens of Lyons, Cook County, had a patriotic time last Monday evening, in presenting a beautiful flag of the Stars and Stripes, costing fifty dollars, to Capt. Messer's Company of the Fremont Rifles, composed of their sons and brothers."[509]

---

[506] The official list of battles in which this regiment bore an honorable part: "PEA RIDGE, SIEGE OF VICKSBURG, PRAIRIE GROVE, BLAKELY, CHALK BLUFF."" *Official Army Register of the Volunteer Force of the United States Army for the Years 1861, '62, '62, '64, '65.* Part VI, Indiana-Illinois, Published by order of the Secretary of War, Adjutant General's Office, Washington, August 31, 1865, (reprinted by Ron R.Van Sickle Military Books, Gaithersburg, MD, 1987) page 274.

[507] Ibid. Mullins, page 9

[508] Ketzle, Henry, *Illinois Greyhounds,* a private history held by the Ketzle family and published on the Internet.

[509] *Chicago Tribune;* Sept. 11, 1861, p4

This Regimental flag of the Thirty-seventh Illinois Infantry is characteristic of flags manufactured in New York.

The above regimental flag is in poor condition.

No other flags carried by the 37th Illinois Infantry are known to survive.

38th Illinois Infantry

"Our regimental colors were all borne off the field flying, though four color-bearers in succession, of the Twenty-first Illinois Volunteers, were shot down, and two of the color-guard of the Thirty-eighth Illinois Volunteers, three of the color-guard of the Fifteenth Wisconsin Volunteers, and four of the color-guard of the One hundred and first Ohio Volunteers fell."[510]

West Point graduate Colonel William P. Carlin[511] of Carrollton, Illinois commanded the men of the 38th Illinois Infantry. They were volunteers from Sangamon County, (Company A), Effingham County (Company B), Champaign and Fayette Counties (Company C), Crawford County (Company D), Coles County (Company E), Logan County (Company F), Mason and Tazewell Counties (Company G), Jasper County (Company H and K), and Edwards County (Company I). After muster on August 15, 1861, the men were ordered to Pilot Knob, Missouri. On October 20, 1861, they fought at the battle of Fredericktown, Missouri where they defeated Confederate General Jeff Thompson who commanded the Missouri State Guard and the Eleventh Missouri Infantry. The Regiment then returned to Pilot Knob for winter quarters. In March 1862 the Thirty-eighth marched to Reeves Station on the Black River and on to Pocahontas, Arkansas; en route they skirmished at Putnam's Ferry, Missouri. The men marched to Jacksonport Arkansas in April and to Cape Girardeau, Missouri in May. The Regiment moved by steamer to Hamburg Landing, Tennessee where they participated in the Siege of Corinth, Mississippi—April 26-May 30. The Thirty-eighth pursued the retreating enemy to Booneville, Mississippi until June 12. During the summer of 1862 they marched through Tennessee, Mississippi, Alabama, and Kentucky and arrived in Louisville, Kentucky in late September. In pursuit of Bragg the Regiment moved into Kentucky in early October and on October 8, 1862, at the battle of Perryville, Kentucky, (Chaplin Hills) the Regiment led an assault and captured an ammunition train, two caisson, and a hundred Confederate prisoners. In November, the Thirty-eighth Illinois Infantry along with the Fifteenth Wisconsin Infantry captured a Confederate wagon train together with a hundred horses and mules at Harpeth Shoals, Tennessee. At Knob Gap on December 26 the Regiment engaged the Confederates and captured two rebel cannons. December 30, 31 and January 1 and 2 the 38th fought at Stone's River; the men were involved in some of the worst of the fighting and sustained a total of 177 casualties. Company F from Logan County lost 31 of its 62 men. The Thirty-eighth Infantry remained at Murfreesboro, Tennessee until June 1863 when it moved toward Tullahoma and fought at Liberty Gap and there captured the colors of the Second Arkansas Infantry.[512] After marching across the Cumberland Mountains and crossing the Tennessee River the men encamped fifty miles south of Chattanooga in September 1863. On September 13 the Thirty-eighth marched over Lookout Mountain to Dug Gap but did not fight in that battle. However, on September 19 they began fighting along Chickamauga Creek and were in the heaviest of fighting at the battle of Chickamauga the following day. Of the 301 men who entered the battle, 180 were killed, wounded or missing. Following the battle, the Union Army retreated into Chattanooga and the Thirty-eighth went into winter quarters south of Chattanooga on the Tennessee River. In late January 1864 the Regiment marched toward Dalton, Georgia and made a successful

---

[510] O.R.-- SERIES I--VOLUME XX/1 [S# 29] Report of Col. William P. Carlin, Thirty-eighth Illinois Infantry, Commanding Second Brigade. December 26, 1862-January 5, 1863. --The Stone's River or Murfreesboro, Tenn., Campaign.

[511] 31year old Colonel William Carlin was born in Greene County Illinois. He received an appointment to the United States Military Acaddemy at Westpoint and graduated in the class of 1850. Prior to the civil war he had served in the regular US Army fighting in the Indian Wars. He commanded Fort Bragg in California in 1859-1860..

[512] O.R. Series 1—volume XXIII/2 [Series #35]

surprise raid on the Rebel outpost at Dalton. Having completed their three-year tour of duty, the majority of men in the Regiment re-enlisted and left for home on veteran furlough on March 28, 1864. They rejoined their brigade at Ackworth, Georgia June 9. In the Atlanta campaign the Regiment fought at Resaca, near Kingston, Dallas, New Hope Church, Allatoona Hills, Pine Hill, Kennesaw Mountain, Chattahoochee River, Peach Tree Creek and in the lines before Atlanta. In the flanking movement around Atlanta the Thirty-eighth fought at Rough and Ready, Jonesboro and Lovejoy Station. They marched in pursuit of Hood to north Georgia and Alabama finally arriving in Chattanooga on October 30, 1864. In November the Thirty-eighth marched toward Franklin, Tennessee and on November 30 fought in the Battle of Franklin. That same night they marched to Nashville and on December 15 participated in the battle at Montgomery Hill. March 13, 1865, the Regiment moved by rail to east Tennessee then on to Asheville, North Carolina in April. In May they returned to Nashville and in June moved down the Mississippi River to New Orleans and then to Indianola, Texas. The Thirty-eighth Illinois was mustered out on December 31, 1865, at Victoria, Texas.[513,514]

Although the inventory of the flags held by the Adjutant General of the State of Illinois as originally reported in 1885,[515] listed two regimental and two national flags for the Thirty-eighth Illinois Infantry at this time there are only two national flags in the inventory.

This national flag of the Thirty-eighth Illinois Infantry is of typical Philadelphia manufacture and measures 73" on the hoist by 76" on the fly. Inscriptions on this flag read: SIEGE OF CORINTH, May, 1862; CHAPLIN HILLS, Oct 8, 1862; KNOB GAP, Decr 26th, 1862; STONE RIVER, Decr. 31st, 1862; 38th REGT ILL. V.V. INF; LIBERTY GAP, June 25th, 1863; CHICKAMAUGA, Sept. 20th, 1863; ATLANTA, 1864; FRANKLIN, Novr, 30th, 1864; NASHVILLE, Decr, 15th, 1864.

---

513 *Report of the Adjutant General of the State of Illinois*, H.W. Rokker State Printer and Binder, Springfield, Ill. (1886) Vol. 3, pp71-4.

514 Behrens, Robert H., *From Salt Fork to Chickamauga, Champaign County Soldiers in the Civil War,* The Urbana Free Library, Urbana, Ill., (1988), pp. 115-20.

515 Ibid, Vol. 1, p. 160.

38TH ILLINOIS INFANTRY

This national flag belonging to the Thirty-eighth Illinois Infantry appears to have portions cut from it as well as significant deterioration of the material. Inscriptions on this flag read: FREDERICKTOWN; CORINTH; CHAPLIN HILLS; 38 REG. ILL.INF.; KNOB GAP; STONE RIVER. *The flag measures 68" on the hoist by 74" on the fly.*

No other flags carried by the 38th Illinois Infantry are known to survive.

39th Illinois Infantry— "*Yates Phalanx*"

"Hark! There's a shout!
Raise me up, comrades! We have conquered, I know!
Up, up, on my feet, with my face to the foe!
Oh! There flies the flag, with its star spangles bright,
The promise of glory, the symbol of right!
Well may they shout!"[516]

As an honor to the Governor of the State of Illinois, Richard Yates, who used his influence in Washington to bring this Regiment into service, the men of the Regiment named themselves the *Yates Phalanx*. The Regiment was organized in Chicago with volunteers from Will County (Companies A and E – *Florence Rifles*- formed from the residents of Florence Township); McLean County (Companies B, H and part of I); Livingston County (Company C); Ogle County (Company D); Cook County (part of Companies F and G -*Northern Rifles*); Marion County (part of Company F); DeWitt County (part of Company I) and La Salle County (Company K). The Regiment mustered in for three years' service on October 11, 1861, under the command of Colonel Austin Light[517] and was sent to Camp Benton at St. Louis, Missouri. The Thirty-ninth was one of the few Illinois Regiments to fight primarily in the Eastern theatre and on October 29 was sent to Williamsport, Maryland for arms and equipment. In December it crossed the Potomac River to guard the *Baltimore and Ohio Railroad.* A rebel force, commanded by Stonewall Jackson, attacked three companies of the Thirty-ninth on January 3, 1862, near Bath, Virginia forcing withdrawal of the Regiment back across the Potomac River to Maryland. In January and February, the Thirty-ninth was first deployed to guard Cumberland, Maryland but was next sent to guard the bridge at New Creek, Virginia; then to protect the *Baltimore and Ohio Railroad.* On March 23, 1862, the Thirty-ninth fought and defeated Stonewall Jackson at the Battle of Winchester, Virginia. In May 1862 the Regiment marched to Fredericksburg, Maryland where they learned the next day of the defeat of Union General Banks in the Shenandoah Valley. Therefore, the Regiment was ordered back to the Shenandoah Valley and after a few days' rest marched on to Alexandria, Virginia where the men were loaded on transports and sailed to Harrison's Landing on the James River. The Regiment remained at Harrison's Landing until August 16 when ordered to Fortress Monroe, Virginia. September 1, the Thirty-ninth was sent to Suffolk, Virginia and remained there until January 23, 1863, when they reported to General Foster at Newbern, South Carolina. It was here that a flag was presented to the Regiment from Governor Yates.[518] During February and March the Thirty-ninth camped on St. Helena Island, South Carolina. April 1, 1863, the Regiment took part in the expedition against Charleston, South Carolina and landed first on Folly Island, then was ordered to Morris Island and the assault on Fort Wagner. The Regimental colors of the Thirty-ninth were the first to fly over Fort Wagner. January 1, 1864, the Thirty-ninth left for home on veteran furlough. By March, due to added recruits, the strength of the Regiment

---

[516] Clark, Charles M., MD, *The History of the Thirty-Ninth Regiment Illinois Volunteer Veteran Infantry in the War of the Rebellion, 1861-1865,* published under the auspices of the Veteran association of the Regiment, Chicago, 1889. p. 215-Verse from a poem written by Rev. William E. Miller as comment on the fighting at Deep Run.

[517] Colonel Light was a Chicago merchant who was dismissed from Service by the War Department one month after his muster. I have ben unable to find the cause of his dismissal. (ED) He was replaced by Colonel Thomas Osborn

[518] Ibid: p 103 While at Newbern, the regiment received an elegant flag, from Governor Richard Yates of Illinois, bearing his portrait, and at the hour of 'dress parade' it was placed in the hands of the 'color guard 'with appropriate comments.

had increased from 450 to 750 men. On May 5, 1864, the Thirty-ninth joined General Butler's expedition up the James River; on reaching Bermuda Hundred they advanced into the interior and May 16 met the enemy at Drewry's Bluff, Virginia. The Union Army was driven back, and the Regiment lost 200 men killed, wounded or missing. On May 20 the Regiment lost another 40 men in action at Ware Bottom Church. In June the Thirty-ninth fought Longstreet's Corps near the Petersburg and Richmond Pike and lost 35 men. August 14 the Regiment crossed to the north side of the James River and led a charge on the enemy's works at Deep Run. During this action Henry Hardenburgh captured the colors of the 10th Arkansas Infantry for which he was awarded the Medal of Honor.[519] At the Battle of Deep Run the Regiment lost an additional 104 men. In late August the Regiment was ordered to the trenches in front of Petersburg and fought at Chapin's farm and against the enemy near Darlington Road,[520] seven miles south of Richmond. The winter of 1864 the Regiment remained in the works on the north side of the James River and had frequent skirmishes with the enemy. In March 1865 the Thirty-ninth received 100 recruits and later that month took part in the actions that resulted in the fall of Petersburg and Richmond. As part of that action the Regiment charged Fort Gregg and was the first regiment to plant its flag on the Fort.[521,522] The Regiment witnessed the surrender of Lee at Appomattox Court House and mustered out December 5, 1865.[523,524]

On December 16, 1865, the Thirty-ninth Illinois Infantry gathered in Springfield to receive their final pay and surrender their flags to Adjutant-General Haynie. In the chapel at Camp Butler, the regimental historian, Charles Clark relates that three flags were surrendered. "No. 1 was a "prize flag," awarded by the State Agricultural Society, as a premium for superior drill and discipline, and was called "The Agricultural Flag." The brazen eagle, ball and socket, attached to this flag, were presented by Major-General John Gibbons, commanding the Twenty-Fourth Corps, for gallant conduct in the assault on Fort Gregg, Petersburg, Va., April 2d, 1865."[525] The origin of this flag is also described by Clark, "In the early part of September the State Commissioners of Agriculture, while in session at Chicago, voted a flag to be presented to the best drilled company of soldiers in camp at Chicago, and the regimental officers decided to enter into competition for it; and Adjutant Frank B. Marshall, formerly a member of the Ellsworth Zouaves, was selected to organize and command a company made up of men of his

---

519 Clark, ibid, pp210. "A brave young private, Henry M. Hardenburgh, of Company G, captured one of the latter after a hand-to-hand fight with the color-sergeant of the Tenth Alabama, whom he left dead on the field. General Birney, our corps commander, to whom he delivered the flag, complimented him very highly. Since coming here, while on duty in the trenches, he was mortally wounded by a piece of shell. A day or two after his death his appointment as First Lieutenant in the Thirty-Sixth United States Colored Troops was received at our headquarters from Major-General Butler for gallantry in the field, but it came too late. He is silent in the grave, all unmindful of earthly rewards."

520 Clark. Ibid. p 226, In regard to Color-Sergeant, George W. Yates of Company A, "He called his guard about him and told them that in all probability a charge would be made at that point and it would be a desperate affair; and 'Boys, I shall in all likelihood fall. When the order is given to charge, let not one of you desert those colors. Save them, whether I am lost or not. 'When the order was given to charge, Sergeant Yates sprang forward with the colors like a deer. But no sooner did he come in sight of the rebel works than he became the target of a terrible volley from their guns and fell pierced with four balls."

521 *Civil War Medal of Honor Winners From Illinois,* Published by the Civil War Centennial Commission of Illinois, July 1962. Abner P. Allen, of Company K was awarded the Medal of Honor for "Gallantry as color bearer in the assault on Fort Gregg, Petersburg, Virginia, April 2, 1865.

522 A brazen eagle, ball and socket for the regimental flag was presented to the Regiment for their bravery at Ft. Greg. (see below)

523 Clark, ibid.

524 *Report of the Adjutant General of the State of Illinois,* H.W.Rokker, State Printer and Binder, Springfield, Ill, 1886, Vol. 3, pp. 140-5.

525 Clark, Ibid, p.282.

choosing from any material in the regiment….and on the day appointed for the contest and the adjudication they won an easy victory and were awarded the flag."[526] The eagle bears the inscription "Presented to the Thirty-Ninth Illinois Veteran Volunteers by Major-General John Gibbons, Commanding Twenty-fourth Army Corps, For Gallantry in the Assault upon Fort Gregg, Petersburg, Va., April 2d, 1865."

This is the flag awarded to the Regiment by the State Commissioners of Agriculture who were meeting in Chicago in September 1861. The flag is heavily damaged but the remaining inscription reads, "PRESENTED 39th____ BY ILL. ST. AGL. SOCIETY, SE________. *The reverse lettering seen on the flag reads,* "____t Wagner, Fort Gregg."

Clark goes on to say, "No. 2 had been presented by Governor Yates, and contained a life-size picture of his Excellency. This flag, General Mann said, had been consigned to the care of a Sergeant of the same name, who had been shot down in one of the engagements, while unfurling it to the breeze, and when his lifeless body was about to be removed for burial, it was discovered that his right hand still grasped a shred of the flag, and that its folds were saturated with his life-blood. The name of Sergeant Yates should therefore be associated with that old flag."[527] There is no remnant of that flag in the Museum's holdings.

Finally, Clark writes, "No. 3, in its tattered, riddled and ragged condition, bore unmistakable traces of original beauty, and was the gift of a young lady named Miss Arion, and had been tenderly and sacredly regarded for her sake." [528]

---

[526]Clark, ibid, p. 4.

[527]Clark, Ibid p. 282.

[528] Clark, ibid, p 282

This is possibly flag No. 3 described by Clark. Inscribed on this flag are the following: WINCHESTER, Va. March 23, 1862 MORRIS ISLAND SC, July__, 1863 WEIR BOTTOM CHURCH, May 20, 1864, DEEP RUN, August 16, 1864, PETERSBURG F________ DARBY TOWN ROAD, *Oct 13, 1864, APPOMATTOX C.H. RICHMOND.*

This national flag is of Philadelphia manufacture, based on the pattern of the stars on the canton. Inscribed on this flag are the following: WINCHESTER, Va. March 23, 1862; MORRIS ISLAND SC, July__, 1863; WEIR BOTTOM CHURCH, May 20, 1864; DEEP RUN, August 16, 1864; PETERSBURG F________; DARBY TOWN ROAD, Oct 13, 1864; APPOMATTOX C.H. ; RICHMOND.

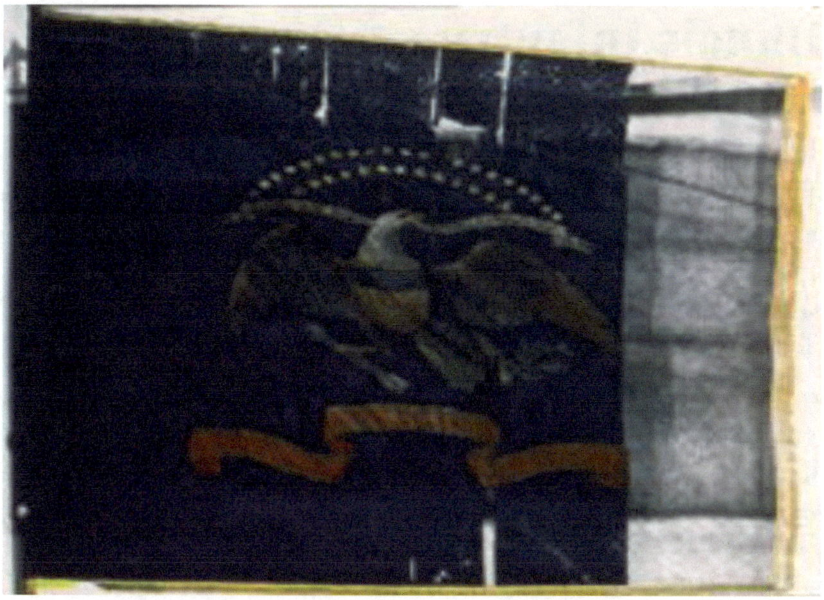

Finally also in the State of Illinois collection is a Regimental flag of the Thirty-ninth Illinois. The flag bears no inscriptions and is not specifically mentioned in the history of this Regiment.

No other flags carried by the 39th Illinois Infantry are known to survive.

40th Illinois Infantry – "*Egyptian Regiment*"

"*July 3d* (1862, Holly Springs, Mississippi) A member of company E, while examining the contents of a barrel sitting in one of the lower rooms of said building, discovered an old *Union Flag*, which appeared to have been concealed there from the beginning of the war. The flag was taken out and unfurled to the breeze, where it delightfully waved, to the joy of the Unions boys and the shame of the Rebel citizens. That flag was carried by Company E until the regiment arrived at Memphis, where it was hoisted, about forty feet high, in Fort Pickering, close to the bank of the Mississippi river."[529]

Georgia born, Stephen G. Hicks[530] was a practicing lawyer in Salem, Illinois when the Civil War broke out. By the end of May 1861, he had raised a regiment from the following counties of Southern Illinois: Hamilton County –Co. A (*Hamilton County Company)*; Fayette and Marion Counties- Co. B (*Kinmundy Company)*; White County-Co. C; Wayne County—Co. D (*Burnt Prairie Company)*; Wayne County -Co. E (*Fairfield Company)*; Franklin and Hamilton County-Co. F; Hamilton and Wayne County-Co. G (*New Baltimore Company)*; Fayette and Marion County-Co. H; Clay, Edwards and Wabash Counties-Co. I and Clay County-Co. K. The Regiment was mustered into service for three years on August 10, 1861, and proceeded to Jefferson Barracks, St. Louis. In September they moved down the Mississippi river to Bird's Point, Missouri, drew uniforms and moved on to Paducah, Kentucky where they remained for six months. On March 6, 1862, the order was given to move to Savannah, Tennessee and later to camp at Pittsburg Landing, Tennessee. April 6 and 7 the Fortieth fought in the battle of Shiloh where Colonel Hicks was wounded. Later that month the Regiment advanced on and sieged Corinth, Mississippi until May 30; following the siege the Fortieth marched to Memphis, Tennessee via LaGrange, Grand Junction and Holly Springs and were on duty at Memphis until late November. The Regiment then went into winter quarters at Davis' Mills in northern Mississippi. After a scouting expedition in northern Mississippi in April 1863,[531] the Regiment was detailed to guard the *Memphis and Charleston Railroad.* In June the Fortieth marched to Memphis, boarded the steamer *Crescent City* and moved down the Mississippi River and up the Yazoo River to Snyder's Bluff at the rear of Vicksburg. At Black River, Mississippi the Regiment confronted Johnston's troops until the fall of Vicksburg on July 4. The Fortieth joined the pursuit of Johnston's army to Jackson, Mississippi and fought there on July 16. After this battle the Fortieth was engaged in destroying railroads and bridges in the Jackson area, then returned to Vicksburg and encamped until late September. From Vicksburg the Regiment marched to Chattanooga arriving there in late November; there they fought in the battle at Tunnel Hill on November 23-24 and at Mission Ridge on November 25. After pursuit of the retreating enemy the Regiment was ordered to the relief of Burnside at Knoxville, Tennessee. On January 1, 1864, the Regiment mustered in as a Veteran Regiment and after furlough joined the Atlanta campaign in May. The Fortieth fought at Kennesaw Mountain and Brush Mountain in June, Nickajack Creek, the

---

[529] Hart, Sergeant E. J., *History of the Fortieth Illinois Inf., (volunteers),* H. S. Bosworth, Cincinnati, 1864, page 104. The Fortieth sacked the courthouse at Holly Springs and the flag was found in the lower level of the courthouse.

[530] Colonel Hicks, who was half Cherokee Indian, was born in in Jackson County, Georgia in 1809. He moved to Illinois to live with his uncle after the death of his father. He served in the Black Hawk War and Mexican-American War, took up the practice of Law and was elected to the Illinois General Assembly. In 1847 he organized the 6th Illinois Infantry and was appointed Lt. Colonel. At the outreak of the civil war he volunteered to raise an infantry regiment.

[531] Ibid, Hart, page 131. "Our good old banner was covered with its oil cloth case, but now, as there was a chance of seeing our traitorous foe, it was best to display our colors. The boys shouted to the color-bearer, 'Sam, for God's sake pull that oil cloth cover off the flag, and let the old star spangled banner wave!'" Apparently, the Regiment protected their colors very well.

Chattahoochee River and Atlanta in July. In the flanking movement south of Atlanta the Regiment fought at Jonesboro August 31-September 1 and Lovejoy Station September 2-6. Following the fall of Atlanta, the Fortieth pursued Hood's army toward Chattanooga and into northern Alabama and then returned to Atlanta. On November 16, 1864, the Fortieth joined the "march to the sea" and saw action at Griswoldville Georgia, and Savannah Georgia in November and December. In 1865 in the campaign through the Carolinas the Regiment was engaged at Salkehatchie Swamps, South and North Edisto River, at Columbia, South Carolina, and at Bentonville, Goldsboro and Raleigh, North Carolina. En-route to Washington the Regiment marched through Petersburg and Richmond; the Fortieth marched in the Grand Review at Washington D. C on May 24, 1865, and was mustered out of service at Louisville, Kentucky on July 24, 1865. [532,533,534]

The Fortieth Infantry retired four national and two regimental flags at the end of the War. These remain in the collection of the State of Illinois Military Museum.

The star pattern and shape of the canton of this flag are typical of national flags manufactured in Philadelphia. The inscriptions on this flag are now very difficult to read because of deterioration of the paint, however they originally read: SHILOH; SIEGE OF CORINTH; SIEGE OF VICKSBURG; JACKSON MISS; MISSIONARY RIDGE; NEW HOPE CHURCH; BLACK JACK KNOB; KENNESAW MT; ATLANTA JULY 22,___? EZRA CHAPEL; JONESBORO; LOVEJOY; GRISWOLDVILLE; SIEGE OF SAVANNAH; COLUMBINE CREEK; BENTONVILLE. *It measures 71" by 71".*

---

532 Eddy, T. M., *The Patriotism of Illinois,* Clarke & Co., Publishers, Chicago, 1865, Vol. 1, pp318-26.
533 *Report of the Adjutant General of the State of Illinois,* H.W. Rokker, Printer and Binder, Springfield, ILL, 1886, Vol 3, pp. 171-4.
534 Hart, Sgt. Ephraim J., *History of the Fortieth Illinois Inf., (Volunteer.),* H. S. Bosworth, Cincinnati, Ohio (1864)

This national flag is inscribed: U.S. VOLUNTEERS 40th Ills Inft; SHILOAH (sp?); SIEGE OF CORINTH; VICKSBURG; JACKSON MISS; CHATTANOOGA, NOV 24 & 25, 1863. *The flag is in very poor condition and the number of stars is uncertain.*

The star pattern on this national flag is unclear. There are no inscriptions.

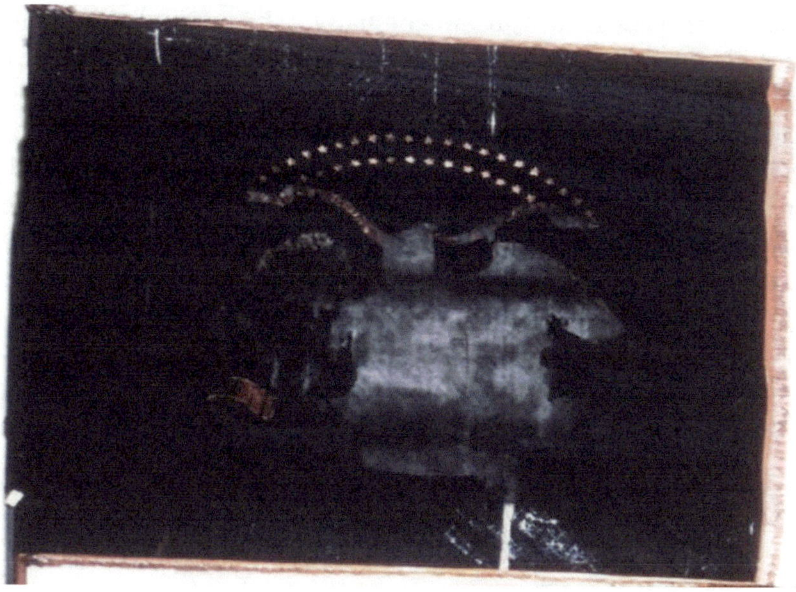

This regimental flag is in poor condition. There are no inscriptions on the flag.

Only a tattered remnant remains of this regimental flag.

In a letter written by J.J.B from Germantown, Tennessee on July 6, 1863, to the Editor of the Centralia Sentinel,[535] (It has been) "Sometime since the Regiment was honored by the presentation of a beautiful national and regimental flag by the citizens of Monroe County." Presumably two of the flags in the Illinois collection are those two flags, but they are not so identified.

A flag was presented to Company E in early August of 1861. "...I will relate an incident which occurred at Jeffersonville, Wayne County. The young ladies of that place made a neat flag, and, in the

[535] Centralia Sentinel, July 9, 1863.

presence of a large audience, presented it to the company, when one of their number, Miss Mary A. More, read a patriotic and interesting address to the soldier boys in which she made an earnest appeal to them 'never to bring disgrace on the dear old flag, but ever defend it against the cruel wiles of the treacherous foe, and to place it in triumph and honor on the battlements of every rebel stronghold which might chance to come in their way.'"[536] This flag was presumably "homemade" and probably is not one of the flags now held in the State of Illinois collection. There is no further mention of this flag in the regimental history and the further history of this flag is unknown.

No other flags carried by the Fortieth Illinois Infantry are known to survive.

[536] Ibid, Hart, page 12.

41st Illinois Infantry –"*Warner's Guards*"[537]

"Several flags were captured, including those of the Twenty-eighth and Fifty-third, and the regimental flag of the Forty-first. The latter was shot down five times, the color bearer being killed each time. Sergeant H. M. Strever, who carried the National colors of the Forty-first, was severely wounded, but he carried the flag, torn and tattered, in triumph off the field."[538]

The Forty-first Illinois Infantry was organized at Decatur, Illinois during July and August of 1861. The volunteers of the Forty-first came from the following counties: Company A (*Decatur Grays*) from Macon County; Company B from Moultrie County (with most of the men from the towns of Sullivan and Bethany); Companies C, F and K were from De Witt County (with the majority of the men from the town of Clinton-*Clinton Zouaves*); Company D from Coles County (with most men from Mattoon); Company E from Macon County (with most men from Blue Mound); Companies G and I from Christian County (the men from Co. G were from Taylorville and the men from Co. I from Mount Auburn and Illiopolis), and Company H from Shelby County (*Shelby Tigers)*. The Regiment was mustered in for three years on August 5, 1861, under the command of Colonel Isaac Pugh. [539] In September the Regiment moved with other troops to fortify Paducah, Kentucky. In February 1862 the Forty-first fought at both Fort Henry and Fort Donelson and on April 6 and 7 at the battle of Shiloh. In that battle the Regiment lost 200 men killed and wounded including Lieutenant Colonel Ansel Tupper. April 29 through May 30 they were engaged at the siege and fall of Corinth, Mississippi. From Corinth the Regiment marched through northern Mississippi and Tennessee to Memphis arriving there on July 21. In October, after a forced march, they met the enemy on the Hatchie River (Battle of Metamora) and pursued them in retreat. The Forty-first went into winter quarters at Moscow, Tennessee guarding the *Memphis and Charleston Railroad* until March 1863 when the men were ordered back to Memphis. On April 12 the Regiment met the enemy at Coldwater, Mississippi then returned to Memphis. On May 12 the troops boarded transports for Vicksburg and landed at Haines' Bluff on May 22 to take part over the next forty-seven days in the siege of Vicksburg. Following the fall of Vicksburg, the Regiment met the retreating enemy in a desperate battle at Pearl River and the regimental flag of the Forty-first was captured.[540] The remainder of 1863 was spent at Vicksburg and Natchez, Mississippi and the Regiment went into winter camp on the Big Black River near Vicksburg. In 1864 two hundred of the veterans re-enlisted and were give a thirty-day furlough. During this furlough the Regiment received word of a riot at Charleston, Illinois in which men of the Fifty-fourth Illinois Infantry had been shot by rebel sympathizers.[541] On arrival the Forty-first they found that although the citizens were aroused and under arms, the riot was over. During the veterans' furlough the non-Veterans were sent on the Red River Campaign and saw action at Pleasant Hill, Cloutiersville, Alexandria, Boyce's Plantation, Well's

---

[537] Also known as *Warner's Guards*

[538] Battle of Pearl River near Jackson, Mississippi on July 12, 1863. *Report of the Adjutant General of the State of Illinois,* Vol 3, p 200. The flag was captured by the 32nd Alabama as noted in the *Savannah Republican,* July 17, 1863, "The colors of the 41st and 53rd and 3d Illinois Regiment of Pugh's Brigade, taken at Jackson on the 12th, are claimed by the 32nd Alabama, 3d Florida, and the 47th Georgia Regiments, as their trophies." Note that Pugh's Brigade included the 3d Iowa not the 3d Illinois.

[539] Colonel Pugh was born in Christian County Kentucky in 1805. He subsequently moved to Decatur Illinois and served in both the Black Hawk War and the Mexican-American War.

[540] See footnote 1. In the battle the Regiment lost over 200 killed, wounded and taken prisoner.

[541] Coleman, Charles H., Spence, Paul H., "The Charleston Riot, March 28, 1864" *J. of the Illinois State Historical Society,* March 1940, Volume XXXIII, No. 1.

Plantation, Bayou Boeuf and Yellow Bayou, Louisiana and Lake Chicot, Arkansas. Following the return of the Veterans from furlough, the Regiment moved to Nashville and to Tunnel Hill to guard the railroad there and fought at Moon Station, Big Shanty, Marietta and Kennesaw Mountain. The term of those who did not re-enlist expired on July 21; on July 22 General James McPherson was killed near Atlanta and the Forty-first escorted his body home. After the fall of Atlanta, the Forty-first joined the "march to the sea" and participated in the siege of Savannah. On January 4, 1865, the Forty-first was consolidated with the Fifty-third Illinois Infantry, forming companies H and G of the Fifty-third.[542]

There are only two flags of the Forty-first Regiment Illinois Infantry in the Illinois collection of the State of Illinois Military Museum.

The oval pattern of the stars on this national flag is typical of flags manufactured in Philadelphia. The inscriptions on this flag are difficult to read because of deterioration of the silk and paint. They originally read: 41st Ills.___(?), FORT HENRY, FORT DONELSON, PITTSBURG LANDING, SIEGE OF CORINTH, METAMORA.

[542] *Report of the Adjutant General of the State of Illinois,* Vol 3, p 198-201.

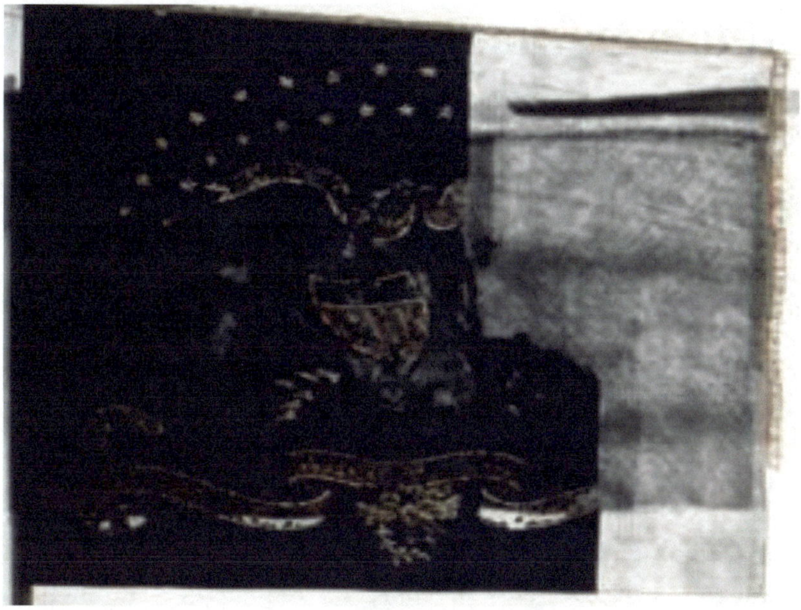

The general pattern of this flag is typical of those manufactured in New York. According to the records of the Adjutant General of the State of Illinois this was the Regimental flag captured at Pearl River and taken to Richmond. On March 26, 1886, the flag was returned to the State of Illinois.

No other flags carried by the 41st Illinois Infantry are known to survive.

42nd Illinois Infantry – "*Douglas Brigade, 1st Regiment*"

O God! Our banner save!
Make it for ages wave!
God save our flag!
Preserve its honor pure,
Unstained may it endure,
And keep our freedom sure!
God save our flag![543]

The Forty-second Illinois Infantry Regiment was organized in Chicago by Colonel William A. Webb and was also known as the *Douglas Brigade, 1st Regiment*. The men volunteered from the following counties: Company A (*Kewanee Volunteers*) from Henry County; Company B (*Robbins Rifles*) from Cook and Henry counties; Company C (*Oneida Guards)* from Knox County; Company D (*Kankakee Valley Guards*) from Kankakee County; Company E (*Cass County Guards*) from Cass County Michigan; Company F (*Drummond Guards*) from Hillsdale County, Michigan; Company G (*Roberts Guards or Mystic Guards*) from Cook County (also included men from Iowa, Michigan and Indiana); Company H (*Stuart Guards and Wolverine Guards*) from the State of Michigan plus Kankakee, St. Clair and Fulton counties in Illinois; Company I (*Kane County Guards, Webb Guards*) from Kane County; Company K (*DeKalb Rifles)* from De Kalb County. After muster on July 22, 1861, for three years under the command of Colonel Webb,[544] the Regiment joined General Fremont at Tipton, Missouri. That fall they went into winter quarters at Smithton, Missouri where they remained until February 1862. They marched via St. Charles, Missouri and arrived at Fort Holt, Kentucky in late February and occupied Columbus, Kentucky in early March. In mid-March the Forty-second moved to Island No. 10 where they were engaged until the Island's surrender on April 11, 1862. Later that month the Regiment moved to Hamburg Landing, Tennessee and then on to the siege of Corinth, Mississippi—April 29-May 30; during that time on May 9, 1862, the Regiment fought at Farmington, Tennessee and led the advance in pursuit of Beauregard's Army to Baldwin, Mississippi. The Forty-second occupied Cortland, Alabama from July 25 until September 3 when they left for Nashville, Tennessee going by way of Decatur, Alabama. En-route they battled at Columbia, Tennessee and arrived at Nashville on September 13. The Regiment remained in Nashville during the siege of Nashville, then left on December 10 in the Murfreesboro campaign. They engaged the enemy at the Battle of Stones River on December 30 and 31 through January 1-3, 1863, with loss of 22 killed, 116 wounded and 85 taken prisoner. The Regiment remained in Murfreesboro until March when they pursued Van Dorn to Columbia, Tennessee and returned to Murfreesboro. April through August the Forty-second campaigned in middle Tennessee. In September they marched to Georgia, crossed Lookout Mountain, and fought in the battle of

[543] *Chicago Daily Times,* May 10, 1861. Portion of a poem by Rev. Edward Hopper that originally appeared in the *Journal of Commerce*

[544] Colonel William Webb was born in Maine in 1830. He graduated in the Class of 1853 from the United States Military Academy at West Point. He took a special interest in the organization of the *Douglas Brigade* and was elected Colonel of the 42nd Illinois Infantry, the first regiment of the Douglas Brigade. He died six months after muster of his regiment on December 24, 1861 of "exposure and fatigue".

Chickamauga on September 19 and 20. Here they lost 28 killed, 128 wounded and 28 taken prisoner. After retreat to Chattanooga, on November 28, they fought in the battle of Mission Ridge; on December 27 they camped at Stone's Mill, Tennessee and in January the term of the Regiment expired, and they re-enlisted as a Veteran Volunteer Regiment. After veteran furlough, the men returned to Chattanooga April 27, 1864. They entered the Atlanta campaign and fought at Rocky Face Ridge, Resaca, Adairsville, Dallas, New Hope Church, Allatoona Hills, Pine Mountain, Kennesaw Mountain, Chattahoochee River, Peach Tree Creek and the siege of Atlanta. In the flanking movement around Atlanta, the Regiment fought at Jonesboro on August 31-September 1 and at Lovejoy Station on September 2-6. Following the fall of Atlanta, the Regiment saw further fighting at Spring Hill and Franklin, Tennessee[545] in November. December 15 and 16 they engaged the enemy in the battle of Nashville. December through April 1865 the Forty-second was on duty at Huntsville and Decatur, Alabama. In June the Regiment moved by rail to Johnsonville, Tennessee and then by water to New Orleans. On July 18 they proceeded to Port Lavaca, Texas and remained on post duty until December 16, 1865, when they mustered out.[546,547]

The State of Illinois Military Museum holds one national and one regimental flag of the Forty-second Illinois Infantry in its collection. Neither flag is inscribed.

The national flag has a double oval pattern of stars typical of flags of Philadelphia manufacture; it measures 70" by 70".

---

545 "Corpl. John H. Smith, Company E, and Private John Mollison, Company B, Forty-second Illinois, each captured a battle-flag at Franklin, November 30, 1864. Private E. Downs, Company K, Forty-second Illinois captured three battle-flags at Franklin, November 30, 1864." One of the colors was that of the Sixth Alabama. *O.R. Series 1—Volume XLV/1 [Series #93]*

546 Dyer, Frederick H., *A Compendium of the War of the Rebellion,* New York, Thomas Yoseloff, (reprinted by Sagamore Press, Inc, 1959) Vol. 3, p. 1064

547 *Report of the Adjutant General of the State of Illinois,* Vol 3, pp244-5

The regimental flag above has a pattern of 20 stars in the first arc with 15 stars in the second arc and the stars come to the ends of the ribbon held in the mouth of the eagle. This pattern is typical of flags manufactured by Shilleto of Cincinnati.

In the Battle of Spring Hill, Tennessee on November 29, 1864 one of two flags of the Regiment was captured by Confederate forces: "we were ordered to hold said line as long as possible, but having 350 entirely new recruits, who had no drill at all and never were under fire, I did not expect to hold such a line very long…and being advised twice by my superiors, the field officers of the Sixty-fourth Ohio, I finally ordered my regiment in retreat, and while doing so the colors of the regiment became separated and the sergeant and all the color guard with one of them were killed and the flag was captured by the enemy. We retreated about half a mile, when we reformed the regiment with only one flag, and the loss of some 110 killed, wounded, and missing…F.A.Atwater, *Major Forty-second Illinois, Commanding Regiment.*"[548] There if no description of the captured flag, therefore it is not clear whether the national or regimental flag was captured.

At least one flag was presented to the Regiment. "A beautiful flag was presented to the Douglas Brigade on Thursday afternoon by J. Logan Chipman, Esq., of the *Chicago Times,* on behalf of Miss Marian Stuart, the daughter of Lieut. Col. David Stuart…The flag is of regulation design and is the national colors. It is made of heavy silk and is one of the most beautiful emblems of the kind we have lately seen."[549] There is no further information about this flag.

No other flags carried by the Forty second Illinois Infantry are known to survive.

[548] O.R.--Series I—Volume SLV/1 [S# 93] page 275-6.

[549] *The Chicago Evening Journal,* September 7, 1861

43rd Illinois Infantry – "*Koerner Regiment*"

The Forty-third Illinois Infantry, under the command of Colonel Julius Raith,[550] was organized at Camp Butler and mustered into three years' service on October 12, 1861. Companies A and B were from St. Clair County; Company C from Knox, Sangamon, and Henry counties; Company D (*Sigol Rifles)* from Cook County; Company E from Rock Island County with some men from Davenport, Iowa across the Mississippi River; Company F was from Monroe County; Company G and H from St. Clair County; Company I from southern Ohio and Company K from St. Clair County. The Regiment remained in the St. Louis area until February 1862 when they sailed on the steamer *Memphis* to Fort Henry, on to Fort Donelson and finally to Savannah, Tennessee. In March they moved to Pittsburg Landing, Tennessee and fought in the Battle of Shiloh on April 6 and 7. At that battle the Regimental commander, Colonel Julius Raith, was mortally wounded and of the 500 men of the Forty-third who fought, 206 were lost including 49 left dead on the field. April 29-May 30 the Regiment advanced on and was in the siege of Corinth, Mississippi. In June they marched to Bethel then to Jackson, Tennessee then to Bolivar Tennessee where they remained until May 1863 constructing fortifications. At that time, they were in actions against Forest in late December and early January 1863. In May 1863 the Forty-third moved to Memphis, Tennessee and then to Vicksburg, Mississippi. Actions around Vicksburg included those on the Yazoo River, Mechanicsburg, Haines Bluff and the siege of Vicksburg. After the surrender of Vicksburg on July 4 the Forty-third moved to Big Black River, and Snyder's Bluff. In August they were at Helena, Arkansas and fought at Bayou Fourche and at the surrender of Little Rock, Arkansas on September 10, 1863. They remained at Little Rock until March 1864 when they bridged the Ouachita River and arrived at Arkadelphia, Arkansas on March 29. They remained in Arkansas and skirmished at Prairie D'Ane, Camden and Princeton returning to Little Rock on May 3, 1864; they remained there until muster out on November 30, 1865.[551,552]

There are three flags of the Forty-third Illinois Infantry in the collection of the State of Illinois Military Museum—two national flags and one regimental flag.

---

550 Colonel Raith was born in Germany in 1819, came to the United States with his family in 1836 and settled in St. Clair County, Illinois. He led the 2nd Illinois Volunteers in the Mexican-American War and at the outbreak of the Civil War, he aided Gustave Koerner in recruiting and organizing a German Regiment.

551 Dyer, Frederick H., *A Compendium of the War of the Rebellion,* Sagamore Press, New York, 1959, Vol. 3, p. 1065.

552 *Report of the Adjutant General of the State of Illinois,* Vol. 3, p. 276-9.

The inscription on this national flag reads: Liberty and Union/ Now and Forever. *The flag measures 62 x 77" and the 34 stars are sewn to the blue canton. The flag is homemade and not typical of flags supplied by the Union Quartermaster. The maker of this flag is unknown.*

This national flag has a square canton and arrangement of stars typical of flags manufactured for the New York depot. Inscribed on this flag are the following: FT DONELSON; SHILOH; SIEGE OF CORINTH; LITTLE ROCK; FORT DE RUSSEY; PLEASANT HILL; CLOUTEVILLE, MORRIS PLANTATION; AORY ELLS; PRAIRIE BAYOU; YELLOW CHICOT; TUPELS; FRANKLIN & NASHVILLE.

The regimental flag of the Forty-third is of typical New York manufacture and bears no inscriptions.

No further flags carried by this regiment are known to survive.

44th Illinois Infantry -- "*Northwestern Rifles*"

The Forty-fourth Illinois Infantry was also known as the "*Northwestern Rifles.*" (Also spelled as *North-Western Rifles*) The men were primarily of German descent. The men of Company A volunteered from La Salle and Tazewell Counties (primarily from the towns of Pekin and Peru); Company B from Branch County, Michigan; Company C from Livingston, Marshall and Putnam counties; Company D from Ohio—mainly Fulton County, Ohio; Company E from Cook and Logan counties; Company F from Jefferson and Washington counties; Company G from Winnebago County; Company H from southwestern Michigan; Company I from Jefferson County and Company K from Cook, Logan, La Salle and Tazewell counties. The Regiment was organized at Camp Ellsworth in Chicago and mustered into service for three years, under the command of Colonel Charles Knobelsdorff,[553] on September 13, 1861, and immediately ordered to St. Louis. After receiving arms from the St. Louis Arsenal, they moved to Jefferson City, Missouri, arriving there on September 29. In October the Regiment was in Springfield, Missouri, in November at Wilson's Creek then Rolla, Missouri where they went into winter quarters. In February 1862 the Regiment started for Springfield, Missouri to attack Price but he retreated, and the Forty-fourth pursued him culminating in the battle at Pea Ridge March 6 and 7. After the battle they marched to Batesville, Arkansas then to Cape Girardeau, Missouri for a trip downriver to Pittsburg Landing, Tennessee. On May 29 and 30 the Regiment was at the Siege of Corinth, Mississippi then in pursuit of the retreating enemy to Booneville, Mississippi. They went into camp at Rienzi, Mississippi where they remained until August 26. Because of enemy threats against Cincinnati and Covington, the Regiment was sent to Cincinnati, Ohio and crossed the Ohio River to Covington, Kentucky. After withdrawal of the enemy, the Regiment boarded cars for Louisville, Kentucky and arrived there on September 19. On October 1 the Forty-fourth left in pursuit of Bragg and fought at Perryville, Kentucky on October 8; they then marched to Nashville where they remained until November 27. In December they moved to Murfreesboro, Tennessee and fought at Stones River on December 30 and 31 and January 1 and 2, 1863. They remained at Murfreesboro until June when they marched to meet the enemy at Hoover's Gap, Shelbyville and Tullahoma, Tennessee. In August the Forty-fourth embarked on the Chickamauga campaign and fought in the battle of Chickamauga September 19 and 20 then fell back to Chattanooga where they remained until November 23. On November 25 the Regiment led a charge on Mission Ridge and was credited by Sheridan with being one of the first regiments to plant its flag on the rebel works. On November 27 the Regiment began a forced march to Knoxville to relieve Union forces under siege but arrived after the siege had been lifted. After campaigning in east Tennessee, they went into camp at Blaine's Cross Roads, Tennessee. During this time the men suffered from the cold and from hunger, existing on one-quarter rations. Despite this three-fourths of the men re-enlisted as veterans in January 1864. On January 16 and 17, 1864 the Regiment met the enemy at Dandridge, Tennessee but fell back to Knoxville and then to Kingston, Tennessee where they made a stand until January 30. They were ordered to Chattanooga and left on veteran furlough in February. On April 14, 1864, the Forty-fourth started back to the field to join the Atlanta campaign. They subsequently saw action at Rocky Face Ridge, Buzzard's Roost Gap, Dalton, Resaca, Adairsville, Dallas, Pumpkin Vine Creek, New Hope Church, Allatoona Hills, Kennesaw Mountain, Pine Hill, Lost Mountain, Smyrna, Chattahoochee River, Peach Tree Creek, and the siege of Atlanta. In the flanking movement south of Atlanta the Regiment fought at Jonesboro August 31 to September 1 and Lovejoy Station on September 2 –6. The Forty-fourth pursued Hood into North Georgia and Alabama in October. On November 30th the Regiment fought in a bloody battle at Franklin, Tennessee; the next day they reached

[553] Colonel Knobelsdorff was dismissed August 20,1862 for disability and replaced by Colonel Wallace Barrett.

Nashville and fought in the battle of Nashville on December 15 and 16. After pursuing the retreating rebels to the Tennessee River the Regiment went into camp at Huntsville, Alabama on January 5, 1865. March 28 the Regiment was ordered to Knoxville, and then on April 22 to Nashville and in June moved to New Orleans. In July the Forty-fourth was transported to Texas and they were mustered out September 25, 1865, while in camp on the La Placido River in Texas. [554,555]

There are two national and one regimental flag of the Forty-fourth Illinois Infantry in the State of Illinois Military Museum.

This national flag of the Forty-fourth is in poor condition. Originally it was inscribed as follows: 44th ILLs VOLs.; CHAPLIN HILLS; PEA RIDGE; STONE RIVER; ROCKY FACE; RESACA; CALHOUN; ADAIRSVILLE; DALLAS; LOST MOUNTAIN; KENESAW MOUNTAIN; PEACH TREE CREEK; ATLANTA; JONESBORO; LOVEJOY; SPRING HILL; FRANKLIN; NASHVILLE Dec. 15 & 16, 1864. *The flag measures 70" on the hoist by 68" on the fly.*

---

[554] Dyer, Frederick H. *A Compendium of the War of the Rebellion,* Sagamore Press, Vol. 3, pp 1065-6.
[555] *Report of the Adjutant General of the State of Illinois, Vol. 3, pp328-31.*

This national flag of the Forty-fourth is also in poor condition. The star pattern cannot be determined.

The regimental flag is also in poor condition. It appears to be of New York manufacture. There is no inscription on this flag

No other flags carried by the 44th Illinois Infantry are known to survive.

45th Illinois Infantry -- "*Washburne Lead Mine Regiment*"

"Shiloh was a terrible battle, and now after fifty years have slipped by, I sit in my easy chair and occasionally dream of the past. I seem to hear again as vividly as then, the booming of cannon, the rattle of musketry and the whiz of the minnie-ball, amid the cries and groans of my comrades who touched elbows with me, and I ask myself: "Can it be? Was I there, or is it a wild fancy of the brain?" The scenes come too vividly before my memory to doubt it, and I thank God that I was able with my comrades to bear a humble part in saving, to those who come after us, this grand nation, and in helping to perpetuate but one flag, the Stars and Stripes—the 'heaven-born banner'—to float over a reunited land and people."[556]

The popular name of this Regiment was the "*Washburne*[557] *Lead Mine Regiment*." John E. Smith[558] of Galena, Illinois who subsequently became its Colonel organized the Regiment. The lead mines in northwestern Illinois supplied the majority of the lead used in Civil War ammunition; five of the companies in this Regiment came from the area of the lead mines-Company A (*Mount Carroll Independent Rifles)* from Carroll and Jo Daviess counties; Company B (*Union Riflemen)* from Jo Daviess County; Companies C and D (*Galena Greys*) and E (*Goodwin Rifles)* from Jo Daviess and Carroll counties; Company F came from Boone and Winnebago counties; Company G (*Rockford Rifles)* from Winnebago County; Company H from Rock Island County; Company I from Mercer County and Company K from Knox County. The Regiment was mustered in for three years' service on Christmas Day, December 25, 1861, at Camp Douglas in Chicago. January 12, 1862, they left for Cairo and in February left Cairo for Fort Henry arriving after surrender of the Fort. The Forty-fifth saw its first action at Fort Donelson February 13 -15 and, "The colors of the 45th were the first to be planted on the ramparts of Fort Donelson."[559] The Regiment remained at Fort Donelson until March 4 when they took boats to Savannah, Tennessee. March 25, they moved to Pittsburg Landing, Tennessee and fought in the battle of Shiloh on April 6 and 7. On April 24 the Forty-fifth left camp at Shiloh and marched to the siege of Corinth, Mississippi. In June the Regiment was ordered to Jackson, Tennessee and remained there guarding the railroads until November. On August 31 Armstrong's rebel cavalry attacked the Union lines at Medon; sharp fighting occurred, and Company C of the Forty-fifth was captured. Late in November the Forty-fifth marched south as far as Springdale, Mississippi then returned north, crossed the Tallahatchie River, and arrived in Memphis February 1863. Later that month the Regiment boarded transports and May 1 found them on the east bank of the Mississippi below Vicksburg. The Forty-fifth took part in many of the battles at Vicksburg, but the most disastrous battle occurred June 25. A mine

---

556 Crummer, Wilbur F., *With Grant at Fort Donelson, Shiloh and Vicksburg,* E. C. Crummer & Co. Oak Park, Ill., 1915, page 80

557 The name "Washburne" was added to honor E. B. Washburne a member of Congress from the Galena district who was instrumental in raising the Regiment.

558 Colonel John Smith was born in Bern, Switzerland in 1816. His father had served under Napoleon Bonaparte and after the fall of Bonaparte, he emigrated with his family to Philadephia. While living in Philadelphia the future Colonel Smith was trained to be a jeweler and practiced his trade there for the next 20 years. He continued practicing the jewelry trade when he moved to Galena. Prior to the outbreak of the civil war he was an aide de camp to Illinois Governor Richard Yates. On December 26, 1861 he was appointed Colonel of the 45th Illinois Infantry. Wikipedia, 2023

559 Adair, John M., *Historical Sketch of the Forty-fifth Illinois Regiment, with a complete list of the Officers and Privates and an Individual Record of Each Man in the Regiment,* Lanark, Carroll County Gazette Print. 1869, p 4.

had been dug under the rebel lines at Fort Hill and charged with explosives; the Forty-fifth was selected to rush the enemy line after the explosion. However, when the Regiment rushed into the crater, they were faced with a secondary embankment protecting the enemy lines and were met with murderous fire. Sergeant Henry Taylor, a color-bearer, was the first to plant the Regiment's colors on the rebel works during the fighting and for his actions received the Medal of Honor.[560] Eighty-three of the Regiment's officers and men were either killed or wounded in this action. On July 4 the Union troops entered Vicksburg. "On account of gallant services previously rendered, this regiment was, by Gen. Grant's special order, given the place of honor in front and right of the column that first marched into Vicksburg after the surrender of Gen. Pemberton, and, consequently, their flag was the first of the victorious banners to plant itself before the eyes of the captured foe."[561] The Regiment remained at Vicksburg on provost duty until October 14. They skirmished with the rebels at Bogue Chitto on October 17, then went into camp at Black River east of Vicksburg. In December and January 1864, the men re-enlisted as Veterans and went home on furlough in March. When they returned the Forty-fifth joined the Atlanta Campaign and were involved in actions at Kennesaw Mountain, Nickajack Creek, Chattahoochee River, Leggett's Bald Hill, and the battle of Atlanta. In the flanking movement south of Atlanta the Regiment fought at the battle of Jonesboro and Lovejoy Station in August and September. After operations against Hood in North Georgia and Alabama they joined the "march to the sea." During the next five months the men of the Forty-fifth fought at Savannah, Georgia, Pocotaligo South Carolina; they marched through Orangeburg, Columbia and Cheraw, South Carolina and northward, through Fayetteville, Goldsboro, Raleigh and Greensboro North Carolina. In May the Forty-fifth marched through Richmond and on to Washington DC for the Grand Review on 24, 1865. The Regiment was mustered out of service at Louisville, Kentucky July 12, 1865, and left for home.[562,563, 564]

There are four flags of the Forty-fifth in the flag collection at the State of Illinois Military Museum; one is a national flag and three are regimental flags. Several flags were presented to the regiment and one or more of these presentation flags may be among those in the collection.

The ladies of Galena presented a stand of colors to the Regiment in November 1861. The colors are not described.[565] At the second annual reunion of the Regiment at Freeport, Illinois on September 28, 1869, Captain John M. Adair of the Regiment read the Regimental History. In it he said "Presently our curiosity was satisfied by the presentation in the name of the ladies of Galena, of two splendid banners. They were accepted in the name of the regiment by Col. Smith and in the course of his remarks promised that they should never trail in the dust or be tainted with dishonor and you sanctioned his word by cheer upon cheer. That was the first time we flung our banner to the breeze, and marched back to camp with our flag proudly floating."[566] Captain Adair does not further describe the flags, but they may be two of the flags described below.

---

[560] Beyer, WF and Keydel, OF, Eds. *Deeds of Valor*, reprint of 1903 edition by Longmeadow Press, Stamford, CT. 1993, p 215.

[561] Reed, Capt. HB, *The Back-Bone of Illinois in Front and Rear From 1861-1865* Springfield, Ill., H.W. Rokker, Printer and Binder, 1886, p.7. Refers to one of the flags of the Forty-fifth now in the collection of the State of Illinois

[562] *Report of the Adjutant General of the State of Illinois*, Springfield, Ill. HW Rokker State Printer and Binder, 1886, Vol. 3, pp. 363-66.

[563] Dyer, Frederick H., *A Compendium of the War of the Rebellion,* Sagamore Press, New York, 1959, Vol. 3, p. 1066.

[564] Adair, John M., ibid. pp. 1-16.

[565] *Weekly Northwestern Gazette,* November 26, 1861

[566] Adair, John M., *Historical Sketch of the Forty-fifth Illinois Regiment, with a complete list of the Officers and Privates and an Individual Record of Each Man in the Regiment,* Lanark, Carroll County Gazette Print. 1869, p2-3.

A flag presented to the Regiment is also described in a newspaper account in 1863.[567] "The ladies of Galena, alike distinguished for their patriotism and large-hearted liberality, may well feel proud of the history of the splendid flag presented by them to the WASHBURNE LEAD MINE REGIMENT before it left Camp Washburne in the fall of 1861…Its history has never been dishonored, but it is indissolubly entwined with the memories of so many brave men who have nobly fallen fighting under its folds…But it was a crowning glory to the Regiment that this brave old flag, the work of the fair hands of our Galena ladies, was selected from an army of a hundred thousand of the bravest men that ever trod a battle field, to first float from the towers of captured Vicksburg."

In October 1863, "the old battle-stained set of colors, which at the formation of the regiment…we presented to you as you sent forth to measure swords with the embattled hosts of tyranny on Southern soil," were returned to the Ladies of Galena. A "new set of colors, which we, the patriotic Ladies of Galena, have promised you," was then presented to the Regiment.[568] It is not possible to determine whether any of these flags are in the Illinois collection.

This national flag shows signs of significant wear. It may be the flag mentioned above as the first flag flown over Fort Donelson or the first flag to be flown over Vicksburg. The flag measures 71" on the hoist by 72" on the fly.

---

[567] *Weekly Northwestern Gazette,* July 27, 1863.

[568] *Weekly Northwestern Gazette,* October 6, 1863.

This Regimental flag is of standard Quartermaster issue and is typical of flags manufactured in Philadelphia. The flag shows very little signs of wear and was probably new near the end of the War. Inscriptions on the flag read: FORT HENRY, FORT DONELDSON (sp?), SHILOH, SIEGE OF CORINTH, MEDAN STATION, THOMPSONS MILLS, BAYOU PIERRE, RAYMOND, JACKSON, CHAMPION HILLS, VICKSBURG, FORT HILL, MERIDIAN, BOUGE CHITTO, CHUNKY STATION, ATLANTA, SAVANNAH, POCOTALIGO, CHERAW, ORANGEBURG, FAYETTEVILLE, COLUMBIA, CHARLESTON, BENTONVILLE.

This is the reverse view of the above flag.

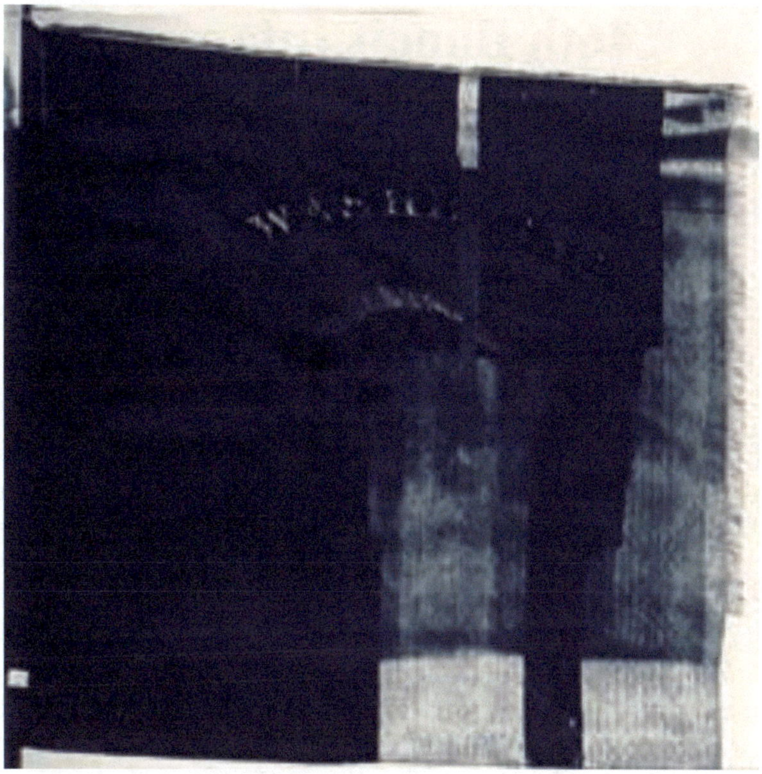

This Regimental flag is inscribed with the name "Washburne." It is not of typical Quartermaster issue and is well worn and measures 69" on the hoist by 67" on the fly. It is possibly one of the two flags presented to the Regiment in Galena as described above.

No other flags carried by this regiment are known to survive.

46th Illinois Infantry

Hip! Hip! Hip!
Hurrah! Hurrah! Hurrah!
Our colors are well worth sustaining,
From them we'll never fly;
But we'll conquer or we'll die
In defense of our Happy Land of Canaan.[569]

Company A (*Stephenson Farmers Light Infantry)*, Company B (*Stephenson County Yeomen)*, Company C, along with Companies G and K of the Forty-sixth were recruited in Stephenson County. The men of Company C (*Freeport German Rifles)* were of German descent and the officers of the Company had all been born in Germany. Twenty-four of the recruits in Company K were from Caledonia, Minnesota; they had left Minnesota to join a Missouri Regiment but were sent to Springfield and assigned to the Forty-sixth Illinois Infantry. Company D was from Lee County, Company E from Whiteside County, Company F from Clay County, Company G (*Stephenson County's Best* or *Young Lions)* from Stephenson County, Company H from De Witt and Lee counties and Company I from Will County. The Regiment was organized at Camp Butler, mustered on December 28,1861 under command of Colonel John A. Davis [570]and ordered to Cairo, Illinois in February 1862. From there they were immediately transported to Fort Donelson where they engaged the enemy on February 14 and 15. After the fall of the Fort they marched to Fort Henry and in March were transported to Pittsburg Landing, Tennessee. In the subsequent battle of Shiloh on April 6 and 7 the Forty-sixth suffered heavy losses in both officers and men. From April 29 to May 30, they were in the siege of Corinth, Mississippi and in June marched to Memphis, Tennessee where they remained until September. On October 5 the Regiment fought at the Battle of the Hatchie or Metamora and Colonel John A. Davis was fatally wounded. On November 3, 1862, the Forty-sixth marched to Lagrange, on November 28 to Holly Springs and on November 30 arrived at Waterford, Mississippi and went into winter quarters. However, only twelve days later the Regiment was ordered to Yacona Station in northern Mississippi and remained in northern Mississippi and southern Tennessee guarding railroads until March 1863. In March they marched to Memphis. In April they moved to Hernando, Mississippi and returned and in May embarked for Vicksburg. At Vicksburg, during the night of May 25 while detailed on picket duty, the enemy captured five companies of the Regiment. In total one hundred and four men and seven officers were captured but seventy escaped.[571] The remainder of the Regiment was actively engaged in the siege of Vicksburg until its fall July 4, 1863. On July 12th they arrived at Pearl River south of Jackson, Mississippi and engaged in the siege of Jackson; when the enemy retreated the Regiment returned to Vicksburg. January 4, 1864, the men of the Forty-sixth re-enlisted as a Veteran Regiment and were furloughed home.

---

[569] Jones, Thomas B., *Complete History of the 46th Regiment Illinois Volunteer Infantry, a Full and Authentic Account of the Participation of the Regiment in the Battles, Sieges, Skirmishes and Expeditions in which it was Engaged.* Freeport, Wm. H. Wagner & Sons, 1907, p14. Chorus of song sung by the Glee Club of the 64th Infantry at the presentation of the colors to the 46th Infantry.

[570] Colonel Davis was born in New York state in 1824 and at age 14 moved with his parents to Rock Run Township in Stephenson County, Illinois. Prior to the War he was a farmer.

[571] *Complete History of the 46th Illinois Veteran Volunteer Infantry,* Freeport, IL, Bailey & Ankeny, Printers, 1866, pp. 54-5.

Following recruitment, the Regiment returned to Vicksburg in March. In May and June, the Regiment was on expedition to Yazoo City and Benton, Mississippi and July skirmished with the enemy at Jackson, Mississippi. In August the Forty-sixth sailed downriver to Morganza Bend, Port Hudson and Clinton, Louisiana. On September 13 the non-veterans of Companies A, B and C mustered out of service; the remainder of the Regiment moved to Memphis. While in camp at Memphis the non-veterans of Companies E, F, H, K, and I mustered out. January 2, 1865, the Regiment proceeded to Louisiana then into camp at Dauphin's Island, Alabama. The remainder of the time of service the Forty-sixth saw action at Spanish Fort, Fort Blakely, Mobile and New Orleans. While at Baton Rouge January 20, 1866, the Regiment mustered out and embarked for Camp Butler.[572]

In the State of Illinois Military Museum flag collection, there are three national and two regimental flags identified with the Forty-sixth Illinois Infantry. The first Colors were presented to the Regiment while at Camp Butler. Mrs. M. M. Sheets in behalf of the ladies of Stephenson County stated in her presentation, "Into your hands we give it, and may a stripe never, never fade, nor a star dim in the sunlight of heaven."[573] This presumably refers to a national flag. However, she goes on to say, "The American Eagle which here represents us, will have his eyrie, and the young Eagles will perch upon our cliffs, and hill tops when these traitors and this rebellion have done their utmost and sunk to an infamous grave. Our country "E Pluribus Unum" still. …If Truth is Eternal and will survive the lapse of ages and the roll of time, then strike, and "On to Victory" be your guide." The mention of an eagle suggests a regimental flag. Perhaps the ladies of Stephenson County presented both a national and regimental flag.

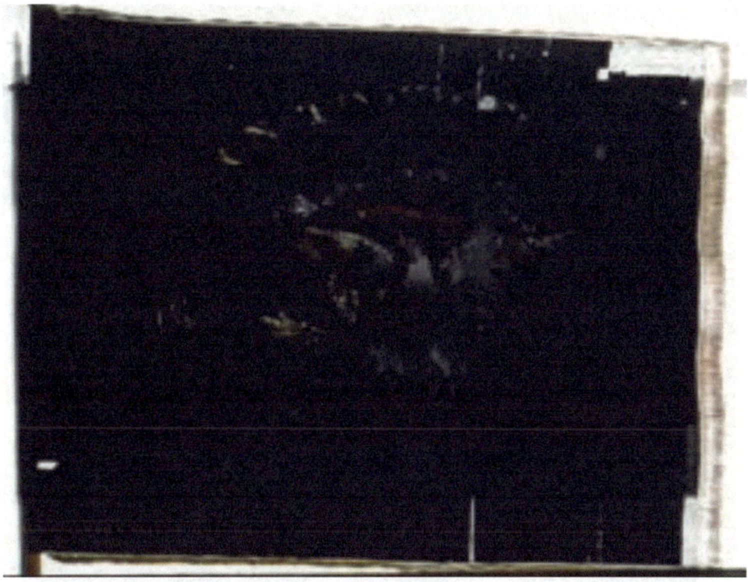

Based on the description above this is the Regimental color presented by the ladies of Stephenson County. It is inscribed "ON TO VICTORY." *This flag is typical of those manufactured by Gilbert Hubbard Co. of Chicago; it measures 69 "on the hoist by 68" on the fly.*

[572] *Report of the Adjutant General of the State of Illinois,* Vol. 3, pp. 405-8
[573] Jones, Thomas B. ibid., page 12-14.

Of the national flags in the collection this is certainly the flag that shows the most wear. Therefore, this may be the national flag presented at Camp Butler. The regimental history refers to "the old flag that had been shot away" and that Captain John Hughes as a souvenir brought a fragment of the flag home.[574] *It is not clear from the history whether only a portion of the flag remained and was brought home, or whether a portion of the flag that had been shot off of the flag was brought home and the remainder of the flag stayed with the Regiment. In any event, of the three national flags in the collection, this is most likely the national flag presented by the ladies of Stephenson County.*

This color is inscribed: 46th REG ILL INFANTRY

[574] Jones, Thomas B. ibid. pp 85-7.

This national color is inscribed: 46th ILL INFTY; DONELSON Feby 14th 15th 16th 62 ; SHILOH Apl. 6th & 7th 62; SIEGE OF CORINTH ___? ; MATAMORA Oct 5th 62. *Note: there are no battles recorded on the flag that were fought after 1862*

This regimental flag of the Forty-sixth is typical of flags manufactured in New York.

No other flags carried by the regiment are known to survive.

47th Illinois Infantry

"That was when we wanted to show our colors—when the enemy could see us. "Here we are!" the Flag shouts to the skirmish-line, feeling its way through the dense woods hunting for us; "Here we are! This is My Regiment, right under my folds! Train your guns this way! You'll find us more easily than you can lose us!"[575]

The Forty-seventh Illinois was part of the "Eagle Brigade" so named because it included the Eighth Wisconsin Infantry Regiment whose mascot was a young American Eagle named "Old Abe." Prior to the outbreak of the Civil War a militia organization-- the *National Blues* had formed in Peoria. By the outbreak of the War Peoria had three additional militia companies, the *Peoria Guards,* the *Emmett Guards,* and the *Peoria Rifles.* Many of the men from these companies enlisted in the Forty-seventh Illinois Infantry. The Regiment was mustered in for three years' service at Peoria on August 16, 1861under the command of Colonel John Bryner.[576] Peoria County furnished the men for Companies A, C, F and part of H; Tazewell County furnished Companies E and part of B (*Washington Rifle Company)*; Woodford County furnished Company I and part of B; Marshall County furnished Companies D and G; Stark County, Company K and Bureau County part of Company H. By the end of December 1861, the Regiment was in Otterville, Missouri where it remained until February 1862. In March they campaigned at New Madrid and Island No. 10. In April the Regiment moved by steamer to Hamburg Landing, Tennessee. They advanced in the direction of Corinth, Mississippi and on May 9 were engaged at Farmington, Mississippi where Lt. Colonel Daniel Miles of the Regiment was killed. On May 28 the Forty-seventh was engaged near Corinth then pursued the rebels as far as Booneville, Mississippi. September 19, they fought at Iuka, Mississippi, although the enemy forces under General Sterling Price were defeated, Major John Cromwell of the Forty-seventh was taken prisoner.[577] On October 3 and 4 the Regiment fought in the battle of Corinth and on October 3 Colonel William Thrush was killed while leading a charge against enemy forces. In November and December, the Regiment was in Grant's Central Mississippi Campaign with operations on the *Mississippi Central Railroad.* In January 1863 the Forty-seventh returned to Corinth then moved by rail to Ridgeway Station, Tennessee where they remained guarding the railroad until mid-March when they marched to Memphis. From here they moved by steamer to Duckport, Louisiana twelve miles above Vicksburg and marched down the west side of the Mississippi River, crossed at Grand Gulf and marched to Jackson, Mississippi. There on May 14, 1863, they engaged in the battle that resulted in the capture of Jackson. May 22, they participated in the charge on the enemy works at Vicksburg and on June 4 participated in the defeat of the enemy at Mechanicsville, Mississippi. After the fall of Vicksburg, the Regiment camped at Bear Creek, east of Vicksburg until mid-November when they moved to LaGrange, Tennessee to guard the *Memphis and Charleston Railroad.* In January 1864 the Regiment left LaGrange and went into camp at Black River Bridge near Vicksburg. In March, on the Red River Expedition, the Regiment participated in the capture of Fort DeRussy, Louisiana and fought at Henderson's Hill, Pleasant Hill, Cloutierville, Belle Prairie, and Bayou de Glaze, Louisiana. In June they moved upriver and were engaged at Lake Chicot, Arkansas. In July 1864 those veterans who had re-enlisted were sent home on furlough. In September, the returning veterans and new recruits to the Regiment left Memphis on the White

---

[575] Burdette, Robert J., *The Drums of the 47th,* Indianapolis, The Bobbs-Merrill Company, 1914, pp 65.

[576] Colonel Bryner was born in Pennsylvania in 1820. He moved to Peoria in Peoria County and beame a prominent citizen of the town. After the Civil War, one of the GAR posts in Peoria was named in his honor.

[577] Cromwell rejoined the Regiment and was promoted to Colonel of the Regiment but was killed at Jackson, Mississippi May 16, 1863.

River expedition arriving in Brownsville, Arkansas and from there marched north into Missouri in pursuit of General Price's army. In November a detachment of veterans was sent to Chicago to assist in quelling any problems that might arise on Election Day. This detachment was then sent to Camp Butler where it was joined by 200 new recruits. These men rejoined the remainder of the Regiment at East port, Mississippi in early 1865. The Forty-seventh fought at Spanish Fort and at Fort Blakely Alabama in March and April 1865. They were mustered out at Selma, Alabama January 21, 1866.[578,579]

Only two flags are known to exist from the Forty-Seventh Illinois Infantry, and both are in the collection of the Adjutant General in Springfield, Illinois. The National flag is in tatters and only remnants of a few stripes remain.

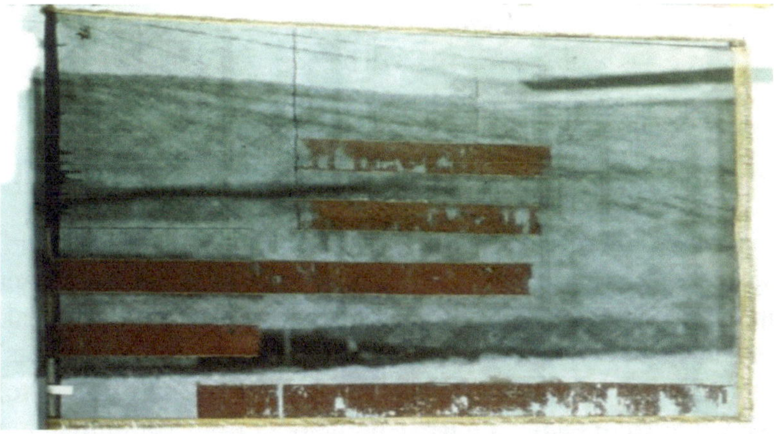

The Regimental flag is of typical New York manufacture and is in very good condition. There are no battle honors on the flag.

No other flags carried by the 47th Illinois Infantry are known to survive.

---

578 *Report of the Adjutant General of the State of Illinois,* Vol. 3, pp405-8.

579 Bryner, Cloyd, *Bugle Echoes, The Story of the Illinois 47th*, Phillips Bros., Printers and Binders, Springfield, Ill, 1905

48th Illinois Infantry

Come all you soldiers brave and true,
And let us show the rebel crew
What we brave soldier-boys can do
With rebel bands and traitors too.

CHORUS:
Roll on, roll on, sweet Union, roll on
And let this brave soldier go home, go home.
From Maine to Mexico we claim,
And Liberty shall be her name.
The Stripes and Stars shall wave again
When copperheads shall have no claim.[580]

Isham N. Haynie raised this Regiment in the southern part of the state and became its Colonel. Haynie was practicing law in Marion Illinois in 1846 when he volunteered for service in the Mexican War and served as First Lieutenant in the 6th Illinois Infantry. Prior to the outbreak of the Civil War, he had served in the Illinois Legislature and as Judge of the Court of Common Pleas at Cairo. In January 1865 he was appointed Adjutant General of Illinois. The men of Company A volunteered from Pope County; Companies B and K from Clay County; Company C from Hardin County; Company D from Washington County; Company E from Crittenden County Kentucky (across the Ohio River from Hardin County Illinois); Company F from Marion County; Company G from Wabash County; Company H from White County and Company I from Wayne County. After organization at Camp Butler in September 1861 the men departed for Cairo Illinois on November 11 and remained there until January 1862 when they moved under Grant's command down the Mississippi River to Fort Jefferson Kentucky. From there they moved to Fort Henry and entered the surrendered Fort on February 6. February13 through 16 they fought at Fort Donelson where Lt. Col. Thomas H. Smith (age 33) of the Forty-eighth was killed. Next, they sailed on the Tennessee River to Savannah, Tennessee and then on to Pittsburg Landing, Tennessee. At the subsequent Battle of Shiloh on April 6 and 7 the Regiment lost half of its numerical strength and Col. Haynie and Lt. Col. William Sandford were among the wounded. In May the Regiment was involved in siege of Corinth, Mississippi and in June was ordered to garrison duty at Bethel, Tennessee where they remained until May of 1863. In June 1863 the Regiment marched to Memphis and then to Vicksburg, Mississippi. The Forty-eighth fought in operations around Vicksburg until the town's surrender on July 4; then they advanced with Sherman and fought at Black River, Mississippi on July 5 and Jackson, Mississippi until July 16. In September the Forty-eighth Regiment arrived back in Memphis and in October Chattanooga, Tennessee. In November the Regiment fought at Tunnel Hill and Missionary Ridge; from here they marched to Knoxville, Tennessee to relieve Burnside. The regiment then went down the Tennessee River to winter camp in Bridgeport, Alabama. In January 1864 the three-year term of the Regiment having expired--over ninety percent of the men re-enlisted and in January and left for home on veteran furlough and to recruit new members; four hundred and fifty men left on furlough but eight hundred returned. Beginning in May 1864 the Forty-eighth was in the Atlanta

[580] Wm. Prosise, Co. F, 48th Reg. Ill. Vol. Infantry, *Centralia Sentinel,* November 5, 1863.

campaign and fought in Georgia at Resaca, Dallas, New Hope Church, Allatoona Hills, Kennesaw Mountain, Nickajack Creek, Chattahoochee River and the Battle of Atlanta. In a flanking movement around Atlanta the Forty-eighth fought at Jonesboro August 31-September 1 and Lovejoy Station September 2-6. In pursuit of rebel General Hood, they moved to Marietta and Resaca, Georgia then to Jacksonville, Alabama and back to Atlanta. In November 1864 they marched with Sherman in the "March to the Sea" and fought at Savannah, Fort McAllister in Georgia then in 1865 at Salkehatchie Swamp, South Edisto River and Columbia South Carolina and Bentonville, Goldsboro, and Raleigh North Carolina. After Johnston's surrender, they marched to Washington D. C. via Richmond and marched in the Grand Review on May 24, 1865. They moved to Louisville, Kentucky then to Little Rock Arkansas where they were mustered out August 15, 1865.[581,582]

There are three national and one regimental flag of the Forty-eighth Illinois Infantry in the collection of the State of Illinois Military Museum.

This national flag is inscribed --FORT HENRY, FORT DONALDSON, 48th REG ILL VOL, SHILOAH, CORINTH. *Although the arrangement of the stars on this flag is characteristic of New York manufactured flags the stars are unusually large and flag does not have a fringe. It is quite large, measuring 70" on the staff by 83" on the fly. Origin and manufacturer of this flag is uncertain, and the flag may be homemade.*

Since this flag is apparently not of Quartermaster issue, it may be the flag presented to Colonel Haynie as reported in the *Cairo City Weekly Gazette,* "We understand that the citizens of Cairo as an earnest of their high estimation of the officers and men of the 48th Illinois regiment, Col. I. N. Haynie, have decided to present to the regiment a handsome flag. Col. Haynie is a citizen of Cairo and such testimony of regard we think eminently proper and deserving."[583]

---

581 Eddy, T.M., *The Patriotism of Illinois,* Chicago, Clarke & Co., 1865, Vol I pp326-30, Vol II p. 231

582 *Report of the Adjutant General of the State of Illinois,* Vol 3, pp 494-5.

583 *The Cairo City Weekly Gazette,* December 19, 1861.

This tattered national flag has the oval star pattern characteristic of Philadelphia manufactured flags.

This thirty-five-star national flag has an attached white stripe inscribed "FORTY-EIGHTH." *The square canton with five horizontal rows of seven stars is typical of New York manufacture.*

This thirty-five-star regimental flag has an upper arc of 18 stars with a lower arc of 17 stars; the stars extend well past the upper scroll ends. The lower red scroll has a raised center section and gold flourishes at the ends of the scroll. These findings identify this flag as being of New York manufacture.

No other flags carried by this regiment are known to survive.

49th Illinois Infantry

"Oh, give us a flag,
All free without a slave;
We'll fight to defend it as our fathers did so brave;
The gallant Comp'ny "A",
Will make the rebels dance
And we'll stand by the Union if we only have a chance"[584]

The Forty-ninth Illinois Infantry was mustered in for three years' service at Camp Butler on December 31, 1861 under the command of Colonel William R. Morrison.[585] The volunteers for the Regiment came from: Companies A and H-Monroe County; Company B-Monroe and Randolph counties; Companies C and F-Washington County; Company D-Madison County; Company E-Macoupin and St. Clair counties; Company G-Madison and Wabash counties; Company I-Randolph and Washington counties and Company K-Jefferson County. On February 8, 1862, the men were ordered to Cairo then on to Fort Henry. On February 11 they engaged the enemy at Fort Donelson where they lost fourteen killed and thirty-seven wounded including Colonel William Morrison, Regimental commander. In March the Regiment moved to Pittsburg Landing, Tennessee and on April 6 and 7 was engaged at the Battle of Shiloh where the Forty-ninth sustained seventeen killed and ninety-nine wounded including then commander of the Regiment, Lieutenant Colonel Phineas Pease, and Major William W. Bishop. April 20 through May 30 the Forty-ninth was in the siege of Corinth, Mississippi and then stationed at Bethel, Tennessee guarding the railroad until March 1863. On March 10 the Forty-ninth marched to Germantown and White's Station Tennessee to guard the railroad. They remained there until August 10 when they moved to Memphis and on to Brownsville, Arkansas. In September they participated in the capture of Little Rock, Arkansas. While at Memphis in January 1864 three fourths of the men re-enlisted and were mustered as veteran volunteers. January 27 the Regiment moved to Vicksburg, joined Sherman's Meridian campaign, and returned to Vicksburg on March 3. March 10 as part of the Red River expedition they fought at Fort DeRussy, Louisiana and at Pleasant Hill, Louisiana, then returned to Memphis in June. June 24 the veterans were ordered home on furlough while the detachment of non-veterans remained under command of Captain John A. Logan and fought at the Battle of Tupelo in July. The regiment was re-united in August in time to participate in the expedition to Oxford, Mississippi followed by return to Jefferson Barracks, Missouri. They drove General Price from Franklin, Missouri and pursued the enemy through Missouri in September and October. In November the men moved to Nashville, Tennessee and fought in the battle of Nashville on December 15 and 16. December 24 they were ordered to Paducah, Kentucky and the non-veterans were mustered out. The veterans remained at Paducah on garrison duty until muster out on September 9, 1865.[586,587]

---

[584] *Hoist up the Flag;* by an anonymous poet.

[585] Colonel William Ralls Morrison was born in 1824 on a farm near Waterloo, Illinois. He attended the local schools and MCKendree College in Lebanon, Illinois. He served in the Mexican-American War, studied law and was admitted to the bar in 1855, served as a member of the State house of representatives, served as speaker of the house and was instrumental in the organization of the 49th Illinois Infantry.

[586] Eddy, T.M., *The Patriotism of Illinois,* Chicago, Clarke & Co. Publishers, 1866, Vol II, pp231-3.

[587] *Report of the Adjutant General of the State of Illinois,* H.W.Rokker, State Printer and Binder, Springfield, Ill, 1886, Vol 3, p 525.

49TH ILLINOIS INFANTRY

There are two national and one regimental flag of the Forty-ninth Illinois Infantry in the collection of the Adjutant General of the State of Illinois.

This national flag is inscribed: 49th REGT ILLINOIS. *Very little of the flag remains except for remnants of stripes and a fringe.*

This National flag contains 34 gold stars arranged in five horizontal rows of 7,7,6,7,7. This star pattern together with a square canton is characteristic of flags manufactured in New York. The flag measures 70" on the hoist and 70" on the fly.

This regimental flag displays an eagle with a shield and gold stars above the upper scroll. The center section of the lower scroll is lowered. This flag is typical of those supplied by Longley & Brothers of Cincinnati.

The *Centralia Sentinel* published a letter from Colonel Pease of the Forty-ninth sent to the *Waterloo Advocate* regarding a flag presentation. "Germantown, Tenn., June 16, 1863. Editor Waterloo Advocate: It becomes my pleasant duty, on behalf of the regiment to acknowledge the receipt, through Mr. Samuel Thing, of your county, of a beautiful regimental flag, gotten up and presented to the 49th, by the citizens of Monroe County."[588] The regimental flag pictured above may be the flag mentioned in this letter.

No other flags carried by the 49th Illinois Infantry are known to survive.

---

[588] *Centralia Sentinel,* August 27, 1863. Waterloo is a town in Monroe County, Illinois.

50th Illinois Infantry- "*Blind Half Hundred*"[589]

"Come all you noble patriots, and give a listening ear,
Concerning the Fiftieth Illinois, and of her past career.
We are from the Western Prairies, and from the sucker state,
And have met many adventures, which here I will relate.
At For Donelson our Regiment was engaged
Most bloody and desperately the battle strife did rage.
From ten to sixty pounders fell among our ranks like hail,
From redoubt, ridge and ravine the Rebel batteries blazed.
But after four days' fighting, with all their skill and strength,
To fight our fearless army they made no more attempt.
And dropped the Rebel colors, 'mid Rebel taunts and jeers,
The Stars and Stripes were hoisted 'mid three prolonged cheers.[590]

Companies A, B (*Clayton Guards*), Company C (*Quincy Cadets* and *Payson Guards)*, Companies D and E volunteered from Adams County. Nearby Hancock County provided men for Company F and part of Company G and K; adjacent Brown County filled Company H; Warren County sent men to Company I and Clinton and Fulton counties sent men to complete companies G and K. The Regiment was organized at Camp Wood near Quincy in Adams County and mustered for three years' service September 12,1861 under the command of Colonel Moses M Bane.[591] October 9 the Regiment left for Hannibal, Missouri and campaigned in that state until January 21,1862 when ordered to Cairo, Illinois. February 6 the Fiftieth marched into Fort Henry and February 13, 14 and 15 were in action against Fort Donelson, where the flag of the Fiftieth was the first planted within the Fort. On March 25 they were ordered to Pittsburg Landing, Tennessee and were engaged in the battle of Shiloh on April 6 and 7.[592] In May they participated in the siege of Corinth, Mississippi and in June pursued the enemy to Booneville, Mississippi and then returned to camp at Corinth. The Fiftieth fought at the battle of Corinth in October and pursued the retreating enemy as far as Ruckersville, Mississippi and then returned to their base at Corinth. In April 1863 the Regiment moved toward Tuscumbia, Alabama and skirmished at Bear Creek, Cherokee and Newsome's Farm Alabama then returned to Corinth. In November the men were ordered from Corinth eastward through Eastport, Mississippi on to Waterloo Alabama then north to

---

589 Hubert, Charles F., *History of the Fiftieth Regiment Illinois Volunteer Infantry in the War of the Union.* Western Veteran Publishing Co., Kansas City, Mo. (1894) p50. According to the author of the Regiment's history the name "Blind Half Hundred" came about because several of the men in the Regiment were "minus" and "and several were "squint and cross-eyed." Also, in a brawl with the 16th Illinois Infantry a brick was thrown striking a man from the Fiftieth in the eye leaving him "half blind."

590 Portion of a poem read at the 22nd Reunion of the 50th Illinois Infantry. *Annual Report of the 22nd Reunion of the 50th Illinois Infantry Reunion Association,* held at Quincy, Illinois, October 14 and 15, 1908.

591 Colonel Moses Milton Bane was born in 1826 in Athens County, Ohio. He entered college at age 21 and in 1844 he graduated from Sterling Medical College at Columbus, Ohio. Dr. Bane next moved to Payson in Adams County, Illinois and began a medical practice. At the outbreak of the civil war he was active in the recruitment of the 50th Illinois Infantry and was appointed Colonel of the regiment. In May 1865 he entered Harvard Law School; in 1878 he was appointed Governor of Utah by President Rutherford Hayes. Colonel / Dr Bane is buried in the Arlington National Cemetery.

592 At Shiloh, Commander of the Regiment, Colonel Moses Bane, a physician from Payson, Illinois lost his right arm and thus was unable to resume his medical practice following the War.

Pulaski and into camp at Lynnville, Tennessee. The men were mounted on November 17. January 1864 three quarters of the men re-enlisted and went home on veteran furlough. On March 15 the veterans returned to Lynnville and were ordered to Athens, Alabama. After a month in northern Alabama, the Regiment marched to Chattanooga then to Chickamauga Creek in early May. Later in May they fought at Resaca and Lays Ferry and arrived at Rome Georgia at the end of the month. September 29 the non-veterans of the Regiment were mustered out and returned home to Illinois. In early October the Fiftieth engaged Hood's army at Allatoona. Although the enemy was repulsed the Regiment lost 87 men killed, wounded or missing. The men again fought on Cave Spring Road near Rome and in November moved to Atlanta and in December met the enemy at Little Ogeechee River near Savannah and entered Savannah on December 21. In 1865 the Fiftieth moved north through South Carolina then in March fought the enemy at Bentonville, North Carolina. After Johnston's surrender the Fiftieth moved north and marched in the Grand Review on May 24. In June the men moved to Louisville, Kentucky and on July 13, 1865, were mustered out of service.[593,594]

There is a national and regimental flag of the Fiftieth Illinois Infantry in the collection of the Military Museum of the State of Illinois in Springfield and a National flag of the Fiftieth at Grant's tomb in New York.

At Chillicothe, Missouri Company K, under the command of Capt. Timothy D. McGillicuddy joined the Fiftieth. McGillicuddy had served in the *Marion County Battalion*-- Company B,[595] a three-month regiment with headquarters in Hannibal, Missouri during the summer of 1861. The ladies of Hannibal had presented a flag to Company B (McGuillicuddy's Company –the *Quincy Cadets*) of the Battalion and the flag had been returned to Battalion headquarters at the close of the three months service and disbandment of the company.

In September of 1861 McGillicuddy recruited a company for three years' service under the authority of General Fremont. It was this company (Company K) that joined the Fiftieth Illinois at Chillicothe. On December 26, while the Fiftieth was at St. Joseph, Missouri, a delegation of citizens from Hannibal presented this same flag to Company K. Captain McGillicuddy placed the flag in the hands of Sinclair Watts as the flag's color-bearer. This flag was used as the Regiment's color until a full set of colors could be provided to the Regiment. It was under this flag that the Regiment moved from Quincy to St. Joseph, Missouri.

When the Fiftieth returned from St. Joseph Missouri to Quincy in January 1862, colors for the Fiftieth still had not arrived. At the request of Colonel Bane, Captain McGillicuddy agreed that the flag of Company K could become the flag of the Regiment on the condition that Sinclair Watts be appointed color-bearer. Therefore, it was this flag that was first planted on the works at Fort Donelson. At the battle of Shiloh in April 1862 the flag was riddled with bullets and during the advance on Corinth the flag was further damaged by underbrush. July 17, 1863, Captain Cramer of Company A returned from Quincy with a pair of beautiful white silk guidons lettered in gold "50th Ills. Vols." These were a gift of Mrs. Col. Swarthout and her sister. According to Hubert, this made the battle flags look so shabby that a new stand of colors was requested and was received on August 8, 1863.[596]

The "original" Regimental flag was returned to Captain McGillicuddy who brought it home after the War and kept it for the next forty years. On July 18,1901 he brought the flag to Hannibal Missouri

---

[593] Hubert, ibid.

[594] *Report of the Adjutant General of the State of Illinois,* Vol 3, pp 558-9

[595] These men were from Marion County Missouri not from Marion County Illinois.

[596] Hubert. ibid, p 425.

to return it to the donors. He found that the donors were gone; therefore, he left the flag with the William T. Sherman GAR Post 43, Department of Missouri.

October 14, 1903, the Sherman Post proposed the return of the flag to the 50th Illinois Infantry Reunion Association and this was done. At the 17th Annual Reunion of the 50th Illinois Infantry, held in Quincy, Illinois in October 1904, General Greenville Dodge proffered a request for the Association to place this flag in General Grant's Tomb and the Association voted in agreement. The flag was forwarded to the Quartermaster's Office for placement in Grant's tomb March 29, 1905.[597] Personal inquiry was made to the National Park Service regarding the status of this flag. Unfortunately, when the flag was removed from display at Grant's Tomb in 1976 the flag had deteriorated -- it was wrapped around the staff and placed in storage. Reproduction flags have since been placed in Grant's tomb.

This second set of colors was carried until June 19, 1864, when the Regiment received a third set of colors. This third set of colors was carried by the Fiftieth through the end of the War and was carried in the Grand Review and deposited in Springfield. The disposition of the other set of colors is not mentioned in the Regimental History and their whereabouts today is unknown.

This is the National flag identified as belonging to the Fiftieth Illinois Infantry. The rectangular canton with the stars arranged in seven horizontal rows is typical of flags manufactured in Cincinnati. It measures 70" on the hoist by 73" on the fly.

[597] *Report of the 18th Annual Reunion of the 50th Illinois Infantry Reunion Association,* Quincy, Illinois, October 5th and 6th, 1904, pp 95-8.

Very little remains of the regimental flag of the Fiftieth. The manufacturer cannot be identified.

The Flag photo of the flag with Capt. Timothy McGillicuddy is from the Report of the 50th Illinois Infantry Reunion held at Quincy, Illinois in October of 1904.

Finally, the Fiftieth, in a drill competition at Louisville, Kentucky on July 3, 1865, was awarded a banner as the best drilled Regiment in the 4th Division of the 15th Army Corps. The banner or flag was called Excelsior and at the end of the War was placed in the hands of Colonel Hanna and taken to his home in Golden, Illinois.[598] The flag is now in Quincy, Illinois in the possession of the Historical Society of Quincy and Adams County.

Photo of the "Excelsior" Flag won by the Fiftieth Illinois in a drill competition on July 3, 1865. Note that the center of the flag has the cartridge case labeled "40 Rounds" the designation of the Fifteenth Army Corps.

No other flags carried by the Blind Half Hundred are known to survive.

---

[598] Hubert, ibid, pp 435-8.

51st Illinois Infantry – "*Chicago Legion, Ryan Life Guard*"

"THEIR NAME IS LEGION-The 51st Regiment Illinois volunteers, known as the *Chicago Legion*, have been in active service during the last three months, and are without their regimental colors. We hear that the friends of this fine corps, whose number is also "legion," have taken this matter in hand, and ordered a superb set to be presented at an early day. Any who wish to add their mite towards a well-deserved compliment to these, our brave Chicago boys can have an opportunity by calling on Messrs. H.C. Livingston & Co., 181 South Water Street or Munson, the stationer, at 140 Lake Street."[599]

The Fifty-first Illinois Infantry was organized at Camp Douglas with date of muster for three years -- December 24, 1861, under the command of Colonel Gilbert W. Cumming.[600] This Regiment was also known as the "*Ryan Life Guards*" or the "*Chicago Legion.*" Most of the officers were from Chicago; the volunteers for Company A (*Union Railway Guard)* were all from Cook County and men from Cook County were scattered in all of the other companies. Company B (*Tucker Light Guard)* was from Champaign and Du Page counties; Company C was from Iroquois County; Company D from McLean and Woodford counties; Company E (*Sturgis Light Guard)* from Champaign and Vermillion counties; Company F from Mason County; Company G (*Higgins Light Guard)* from Lake County; Company H from Rock Island County, Company I from Knox and Henry counties and Company K from Bureau and Knox counties. The Fifty-first also had a band under direction of William Tidmarsh. The members of the band mustered out in 1862 when the War Department ordered regimental bands disbanded. February 1862 the Regiment left Chicago and encamped at Camp Cullum on the Kentucky shore of the Mississippi River. In April they moved against New Madrid and Island No. 10 compelling the surrender of the Confederate forces. In April a steamer transported the men to Hamburg Landing, Tennessee. From there they engaged in battle at Farmington, Mississippi and the Siege of Corinth, Mississippi. In July the Regiment marched to Tuscumbia, Alabama and assumed railroad guard duty. In September the Fifty-first moved to Nashville, Tennessee and in late December marched down the Nolensville Pike to Murfreesboro, Tennessee to battle at Stones River December 30-January 3, 1863. During that battle the Regiment sustained fifty-seven casualties—with one officer and six enlisted men killed. Following the battle, the Regiment remained at Murfreesboro until June. In June 1863 the men moved toward Chattanooga, Tennessee to begin the battle for capture of that city. They crossed the Cumberland Mountains and reached Bridgeport, Alabama on July 30. In September the Fifty-first crossed the Tennessee River, and on September 19 engaged in the battle of Chickamauga where

[599] *Chicago Daily Tribune,* May 1, 1862

[600] Colonel Cumming was born in New York in 1817. After study of law he became a member of the New York bar. He also joined a New York militia and eventually became colonel of the regiment. He moved to Chicago in 1858 and continued his law practice. In 1861 he he was instrumental in the recruitment of men for the 51st Illinois Infantry and was appointed Colonel of the Regiment by Illinois Governor Richard Yates. *Find A Grave website.*

they captured the flag of the 24th Alabama.[601,602] There the Regiment lost 18 killed, 92 wounded and 18 missing; a total of 128 casualties out of the 209 men who fought—casualties of 61 percent—the highest rate of any Union regiment that fought at Chickamauga. In November the Fifty-first, at the battle of Missionary Ridge lost 30 of the 150 men engaged; they then marched toward Knoxville, Tennessee to relieve General Burnside. In January 1864 the Fifty-first returned to Chattanooga and in February mustered in as veteran volunteers and left for home on furlough. In March the men returned and became part of Sherman's Atlanta campaign. They subsequently fought at Rocky Face Ridge, Resaca, Adairsville, Dallas, New Hope Church, Allatoona Hills, Pine Hill, Lost Mountain, Kennesaw Mountain, Chattahoochee River, Peach Tree Creek, and the siege of Atlanta. In a flanking movement around Atlanta, they fought at Jonesboro August 31-September 1. In late September the Regiment marched back to Chattanooga and in late October 105 new recruits joined the Regiment. They were involved in the pursuit of Hood through Alabama to Spring Hill, Tennessee where Hood's forces attacked the Union troops. The Union forces retreated to Franklin, Tennessee where Hood attacked and the Fifty-first suffered heavy casualties.[603,604] On December 1, 1864, the Union army retreated to Nashville where it defeated the Confederate forces. The Regiment wintered at Huntsville, Alabama and in April 1865 while at Nashville, Tennessee, new recruits forming a new Company I, joined the Fifty-first. In July the Regiment was transported to New Orleans then to Port Lavaca, Texas, then to Camp Placido Texas. On September 25, 1865, the Fifty-first was mustered out at Camp Irwin, Texas.[605,606]

The State of Illinois Military Museum has three flags belonging to the Fifty-first Illinois Infantry-one national flag and two regimental flags. The Chicago Daily Tribune described the first set of colors issued to the Regiment. These colors were bought with donations made by citizens in Chicago. "COLORS FOR THE FIFTY-FIRST ILLINOIS—A magnificent stand of colors, comprising the Stars and Stripes and regimental colors of blue silk surmounted by the eagle, with "New Madrid and Island No.10" inscribed upon the folds, has been prepared for the 51st Regiment of Illinois volunteers, and will be on exhibition at Munson's, 140 Lake street, for a few days."[607] There are no inscriptions on the national flag in the Illinois collection however the two regimental flags appear to have had inscriptions. Unfortunately, only shreds

---

601 *OR 30, Part 1, Page 490.* This flag is probably now at the Chicago Historical Society (collection no. 1920.1653) and is probably misidentified as "Confederate Garrison flag that floated from the flagstaff over the Town Hall of Vicksburg, Miss." The flag is blue cashmere—36" on hoist by 54" on fly—with a white silk St. George's Cross-, 4" side. Personal communication Howard Michael Madaus. There is further documentation that the flag was brought back by the 51st –*Chicago Tribune, October 6, 1863* "CHICKAMAUGA FLAGS—What remains of the flag of the 51st regiment Illinois Volunteers will be on exhibition, for a few days at the store of Munson, Skinner & Co., 140 Lake Street; also, a rebel flag, captured from the 24th Alabama, at the same battle."

602 From the diary of Benjamin T. Smith, 51st Illinois, Illinois State Historical Library, "Boyd sprang up with his two hundred men, who sent in a volley at short range, and ordered a charge, the rebs were taken by surprise and scattered in all directions, down the hill we went, and so close to us were they, we captured a squad of prisoners, including their color guard and flag. They proved to be a part of the 24th Alabama regiment. Boyd tore the flag from its staff and folded it up, put it inside his blouse."

603 O.R. Series 1—Volume XLV/1 (Series #93) It was at Franklin that the colors of the 51st were captured but recaptured by Corporal. John Miller, Company B, 24thWisconsin Infantry and brought from the field.

604 "Sergt. M. Lewis, Company E, Fifty-first Illinois captured a flag at Franklin, November 30, 1864." *O.R. Series 1—Volume XLV/1 [Series #93}*

605 Behrens, Robert H., *From Salt Fork to Chickamauga,* The Urbana Free Library, Urbana, Illinois (1988), 123-35.

606 *Report of the Adjutant General of the State of Illinois,* Volume III, Springfield, Illinois, H.W.Rokker, State Printer and Binder, 1886, pp 590-92

607 *Chicago Daily Tribune,* May 6, 1862.

remain from the original regimental colors, and it is impossible to determine if any of those flags originally bore any inscriptions.

The star pattern on this national flag is uncertain, and there are no inscriptions.

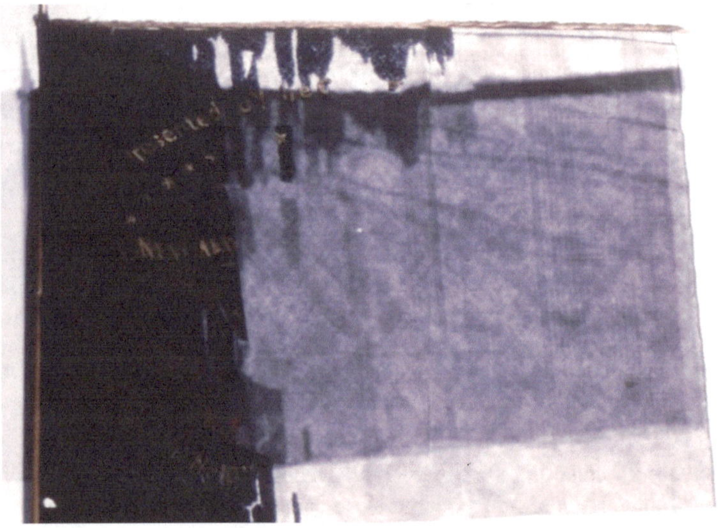

This Regimental flag, though in shreds still bears a portion of the original inscriptions. Visible on the flag are "Presented by the C…" and "New Madrid". *This is presumably the flag whose presentation is mentioned in the Chicago Tribune and probably was originally inscribed "Presented by the Citizens of Chicago". It is also presumably the flag described, as "What remains of the flag of the 51st regiment Illinois Volunteers will be on exhibition, for a few days, at the store of Munson, Skinner & co., 140 Lake Street."[608] Charles C. Merrick, 2nd Lieutenant of Company G, had brought the flag back to Chicago.[609]*

---

[608] *Chicago Daily Tribune*, October 6, 1863
[609] *Chicago Daily Tribune,* October 2, 1863

The above regimental flag was inscribed with battle honors. Only "Mission Ridge" is still visible. No other details are visible. The flag appears to be homemade and not a typical regimental flag supplied by the Quartermaster of the Union Army.

A note in the Chicago Tribune suggests that the Board of Trade purchased a "replacement" flag for the colors of the Fifty-first damaged at Chickamauga, "The honored battle-torn flag of the 51st Illinois regiment alluded to on Friday was yesterday exhibited ...appropriate remarks were made, which were well received, resulting in a decision by the members of the Board of Trade to contribute for the purchase of a new flag, to be presented to the gallant heroes of the 51st."[610] No other flags carried by the 51st Illinois Infantry are known to survive.

[610] *Chicago Daily Tribune,* October 30, 1863.

52nd Illinois Infantry --"*Lincoln Regiment*"[611]

Then up with our banner so glorious,
The star-spangled red-white-and-blue,
We'll fight till our flag is victorious,
For Lincoln and Liberty too![612]

The Fifty-second Illinois Infantry was organized at Geneva in Kane County, Illinois and mustered in for three years on November 19, 1861, under the command of Colonel Thomas William Sweeney. [613] The men volunteered from the following counties: Company A (*Kaneville Rifles*), Company D, Company G (*St. Charles Guards)*, Company H (*Lincoln Rifles)*, Company I *(Dundee Freemen)* and Company K (*Hunter Guards)* all volunteered from Kane County, while Company B (*Lamoille Guards)* came from Bureau County, Company C (*Sycamore Rifles)* from De Kalb County, Company E (*Winnebago Sharpshooters)* from Winnebago County and Company F (*Fulton Tigers)* from Whiteside County.[614] In November the Regiment was ordered to Missouri[615] and spent December and January guarding the *Hannibal and St. Joseph Railroad.* In late January 1862 they moved to Smithland, Kentucky and on February 16 moved by steamer to Fort Donelson, Tennessee where they arrived in time to be assigned the duty of guarding and transporting rebel prisoners to the newly opened prison at Camp Douglas in Chicago. The Regiment returned to St. Louis on March 7 and left to join the Army of the Tennessee arriving at Pittsburg Landing, Tennessee on March 18. On April 6 and 7 the Regiment was engaged in the battle of Shiloh where the Regiment lost 170 men killed, wounded, or missing out of 500 who entered the battle. Edward B. Spalding, Sergeant of Company E received the Medal of Honor for his actions on April 6, 1862.[616] In May the Fifty-second fought at the Siege of Corinth, Mississippi and pursued the retreating rebels to Booneville, Mississippi then returned to Corinth. October 3 and 4 the Regiment fought in the battle of Corinth and lost 70 men killed or wounded. On December 9 the Regiment accompanied an expedition into northern Alabama where they fought and drove the enemy from Little Bear Creek and returned to Corinth. In February 1863 the men left with wagons for sawmills at Burnsville, Mississippi, but

---

611 The Regiment also known as the "*Lincoln Rangers*"

612 *Lincoln and Liberty Too;* verse 5, campaign song for Lincoln's 1860 campaign written by Jesse Hutchinson, jr.

613 Colonel Sweeney was born in Cork, Ireland in 1820. Following the death of his father, he and his mother came to New York in 1836 where he joined a local militia, the *Baxter Blues.* With the outbreak of the Mexican-American War, the *Baxter Blues* became Company A of the 1st New York Volunteers. In the fighting at Churubusco Sweeney sustained a wound necessitating the amputation of his right arm. Due to his bravery, he was awarded a commission as lieutenant in the Second US Infantry. He was promoted to Captain in January 1861 and posted to the St. Louis arsenal. He was appointed Colonel of the 52nd Illinois Infantryin January 1862. Rory T. Cornish, Encyclopedia of the Civil War, WW Norton & Co, New York, page 1912

614 Attached to this Infantry Regiment was *Franklin Gilbert's Cavalry* Company (Cavalry Company A) from Kane County. Afterward the Company became Company H of the 12th Illinois Cavalry Regiment and then Company G of the 15th Illinois Cavalry Regiment

615 A "Secesh Flag" was taken from a house in Missouri and sent home by C. Linn of Company F. "The flag was about two and a half feet long and one foot wide, with two red stripes and one white, a pale blue ground in the corner surmounted by twelve stars. The stars were made each of six pieces sewed together; the red stripes were composed of four shades of color, and the white also of four shades, untastefully arranged and roughly sewed together." *Fulton Weekly Courier,* February 28, 1862.

616 *Civil War Medal of Honor Winners from Illinois,* Published by the Civil War Centennial commission of Illinois July 1962, p 4. "Although twice wounded, and thereby crippled for life, he remained fighting in open ground to the close of the battle."

because of the heavy rains they were able return to Corinth with only two hundred feet of lumber. In April the Regiment moved to northern Alabama. They marched through Burnsville, Mississippi through Iuka, crossed Bear Creek and engaged the rebels near Little Bear Creek, Alabama on April 23, 1863. On April 27 they met the enemy at Town Creek and drove them from the area, then marched back to Corinth. The Regiment was ordered to Germantown, Tennessee in August and remained on duty there guarding the railroad until the end of October. The Fifty-second then marched east, crossed the Tennessee River, and arrived at Pulaski, Tennessee for provost duty. January 1864 three-fourths of the Regiment reenlisted as veteran volunteers and left for Illinois on furlough. They returned to Pulaski in late February with the addition of one hundred and seventy-three recruits and in April moved south to Chattanooga. In May the Regiment became part of the Atlanta campaign and fought at Snake Creek Gap, Resaca, Lay's Ferry, Rome Cross Roads, Dallas, Allatoona Hills, Kennesaw Mountain, Nickajack Creek and Decatur. In July they fought before Atlanta and in August at Jonesboro. After the capture of Atlanta, the Regiment moved back to Rome, Georgia and then struck cross-country in Sherman's "march to the sea." Following the battle at Savannah, Georgia in December the men marched north for actions at Salkehatchie Swamps, South Edisto River, North Edisto River, Columbia and Cheraw, South Carolina; and in North Carolina they fought in the major battle at Bentonville. Following the surrender of Johnston and his army on April 26, 1865, the Regiment marched north toward Washington, D.C., via Richmond, Virginia. On May 24, 1865, they marched in the Grand Review in Washington. D. C. and were mustered out on July 5 at Louisville, Kentucky.[617,618]

There are two flags--one national and on regimental--of the Fifty-second Illinois Infantry in the collection at the State of Illinois Military Museum.

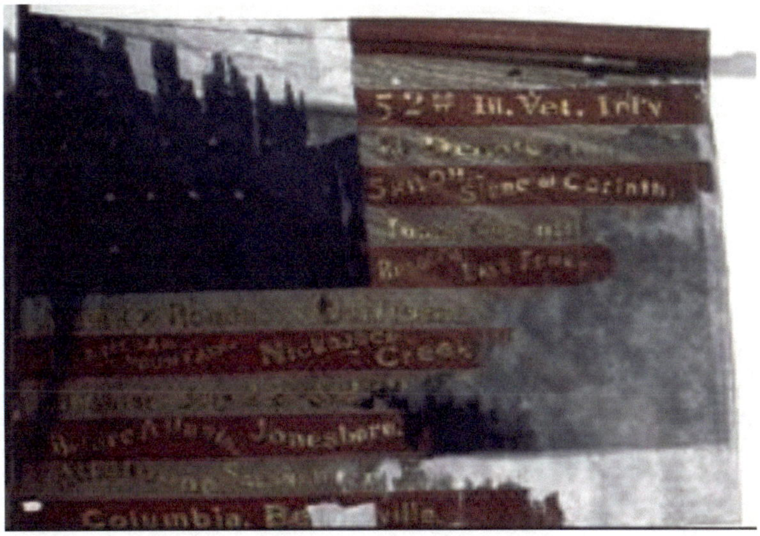

This National flag carries the Regiment's battle honors. Inscribed on the flag: FORT DONELSON, SHILOH, SIEGE OF CORINTH, IUKA, CORINTH, RESACA, LAYS FERRY, ROME CROSS ROADS, DALLAS, KENESAW MOUNTAIN, NICKAJACK CREEK, DECATUR, BEFORE ATLANTA, JONESBORO, ALLATOONA, SAVANNAH, COLUMBIA, BENTONVILLE. *The square canton and arrangement of stars is typical of flags manufactured in New York. The flag measures 69" on the hoist by 74" on the fly.*

617 Boies, Henry L., *History of De Kalb County, Illinois*, O.P. Bassett, Printer, Chicago, (1868) pp.328-39.
618 *Report of the Adjutant General of the State of Illinois,* Volume III, Springfield, Ill, H.W.Rokker, State Printer and Binder, (1886), pp. 621-2

The regimental flag is typical of New York manufacture and is not inscribed with battle honors. It measures 70" on the hoist by 72" on the fly.

No other flags carried by the 52nd Illinois Infantry are known to survive.

53rd Illinois Infantry

"All around in every direction, lay the bodies of those who had but recently been brave and valiant soldiers, some wearing the blue and some the gray, but who now were sleeping their last sleep, many of them unrecognized and unrecognizable, many without means of identification and who must be included in that plaintive list – 'missing,' always made up after a battle."[619]

The Fifty-third Illinois Infantry Regiment was organized at Ottawa, Illinois. It was originally organized with ten infantry companies, *Cogswell's Independent Battery Light Artillery*[620] became Company A and Ford's Cavalry Company (Cush*man's Brigade*)[621]was organized with the regiment but did not become part of the regiment. The men of infantry companies A, B, C, D (*Batavia Guards*), F and K volunteered from La Salle County; Company E from Kankakee County; Company G from Livingston County; Company H from Marshall County; and Company I from Mason County. The Regiment mustered in for three years' service in January 1861 under the command of Colonel William H. W. Cushman,[622] left Ottawa February 27, 1862, and traveled to Chicago to guard prisoners at Camp Douglas. In March the men traveled to Savannah, Tennessee and saw their first action at Shiloh, Tennessee on April 7, 1862. From April 29 until May 30 the Regiment participated in the siege of Corinth, Mississippi. After the evacuation of Corinth, the Fifty-third marched to Grand Junction, Tennessee, then on to LaGrange, Tennessee, south to Holly Springs, Mississippi, back north to LaGrange, Tennessee finally arriving in Memphis in late July 1862. On October 5 the men saw their next fighting at the Battle of Hatchie River, Tennessee (also known as the battle of Metamora). As part of Grant's Central Mississippi Campaign, the Fifty-third spent the remainder of 1862 marching and skirmishing in Tennessee and northern Mississippi. In May 1863 the Regiment embarked for Young's Point, Louisiana and were in the Vicksburg campaign until July 4 when the city surrendered. On July 12 the Regiment charged the Confederate works at Pearl River near Jackson, Mississippi; of the 250 men who entered the battle only 66 were not either killed, wounded, or captured. All the color-bearers were either killed or wounded and the Regiment's colors captured.[623] The Regiment camped in the Vicksburg area for the remainder of 1863. In 1864, the Regiment re-enlisted as veterans and in March returned to Ottawa on veteran furlough. While there, companies E and I were consolidated into company E and men for a new company I were recruited. The Fifty-third rejoined Sherman's army at Kingston, Georgia and in June and July 1864 were engaged in the siege and battle of Atlanta. In the flanking movement around Atlanta the Fifty-third fought at Jonesboro on August 31 and September 1. After pursuit of Hood's army into Alabama, the Fifty-third returned to Atlanta and in November joined the "march to

---

[619] J, O, Harris," Recollections of an Army Surgeon," *Blue and Gray III,* (1894): p 142. Harris wrote this after the battle of Shiloh.

[620] *Cogswell's Independent Battery of Light Artillery* was mustered in as Company "A" Artillery but was detached as a Battery March 17, 1862.

[621] *Ford's Independent Cavalry* Company was organized with the Fifty-third Infantry but was attached to District of Columbus, Kentucky. It did not become part of the 53rd Illinois.

[622] The citizens of Illinois petitioned Secretary of War Simeon Cameron for the appointment of Cushman as Colonel of a volunteer regiment. This petition was endorsed by Lincoln. Lincoln and Lincoln-related documents from the Illinois State Archives.

[623] *Report of the Adjutant General of the State of Illinois,* Vol 3, H.W.Rokker, State Printer and Binder, 1886, p659. The flag was captured by the 3d Florida as reported in the *Savannah Republican*, July 17, 1863, "The colors of the 41st and 53rd and 3d Illinois Regiment of Pugh's Brigade, taken at Jackson on the 12th, are claimed by the 32nd Alabama, the 3d Florida, and the 47th Georgia Regiments, as their trophies." Note that Pugh's Brigade includes the 3d Iowa not the 3d Illinois.

the sea." In December they fought at Savannah, Georgia and in January 1865, the remaining 222 men of the Forty-first Illinois Infantry consolidated with the Fifty-third. The consolidated Fifty-third saw further action at Salkehatchie Swamps, Salkehatchie River, South Edisto River, Orangeburg, Cheraw, and Columbia South Carolina then Fayetteville, Bentonville and Goldsboro, North Carolina, Following the Confederate surrender the Regiment marched to Washington DC via Richmond, Virginia for the Grand Review on May 24, then moved to Louisville, Kentucky where they mustered out July 22, 1865.[624]

There are two national flags belonging to the Fifty-third Illinois Infantry Regiment in the State of Illinois collection; neither is identified as having been captured at Pearl River. There are no regimental flags from this Regiment in the collection.

This national flag appears to have bloodstains and may be the flag captured at Pearl River (Battle of Jackson, Mississippi) and covered with the blood of color-bearer George Poundstone.[625] *"Another heroic color-bearer was George Poundstone, who carried the stars and stripes in front of the 53d Illinois Infantry, when regiment went into the fight at Jackson, Miss., on July 12th, 1863, with 366 men, and came out reduced to 72 men. Heroic Poundstone made desperate attempts to keep the enemy from capturing the flag. After being shot through the heart, he tore the flag from the staff and thrust it in his bosom; his heart's blood has written upon the flag he loved so well a record of his devotion and bravery that speaks more plainly than could epitaph on marble."*

---

[624] Ranstead, H.E., *A True Story and History of the Fifty-third Regiment Illinois Veteran Volunteer Infantry*, (1910)

[625] H.B.Reed, Capt. *The Back-bone of Illinois, Front and Rear, H.W.Rokker*, Printer and Binder, Springfield, Ill., p.7.

Very little remains of this national flag of the Fifty-third.

No other flags carried by the 53rd Illinois Infantry are known to survive.

54th Illinois Infantry

MURDER!

The 54th Regiment, Illinois Volunteers offer ONE THOUSAND DOLLARS REWARD for the apprehension of John H. O'Hair, Sheriff of Coles County, J. Elsberry Hanks, John Frazier, James W. Frazier, Henderson O'Hair, Jesse O'Hair, B. F. Toland, and B. F. Dukes. All of whom were engaged in the murder of Major York and four soldiers of the 54th Infantry and the wounding of several others in Charleston on Monday March 28th, 1864.[626]

The Fifty-fourth Illinois Infantry was organized at Camp DuBois in Anna, Illinois and was part of the "Kentucky Brigade."[627] The volunteers came from the following counties: Company A-Christian County; Company B-Douglas County; Company C-Coles County; Company D-Effingham County, Company E-Jasper County, Company F-Edgar and Shelby counties, Company G-Clark and Coles counties, Company H-Shelby County, Company I-Coles, Douglas and Effingham counties and Company K from Logan, Peoria, DeKalb, LaSalle, Cook and Tazwell Counties and Davenport, Iowa. The Regiment mustered in for three years on February 18,1862 under the command of Colonel Thomas Harris (a 40-year-old merchant from Paris, Illinois) and was ordered to Cairo, Illinois. In March the Regiment was ordered to Columbus, Kentucky, and that fall, three companies were ordered to Humboldt, Tennessee to guard the railroad. In December the Fifty-fourth was sent to Jackson, Tennessee, and Company B, plus all of the sick and the Regimental records were detached at Union City, Tennessee. On December 21, Forrest captured the detached men and all the records; the records were destroyed. In early 1863 the Regiment remained at Jackson with two companies at Medon Station and two at Toone's Station. In May the men were ordered to Haines's Bluff on the Yazoo River and remained there confronting Johnston's army until the fall of Vicksburg. On July 24 the Regiment moved to Helena, Arkansas as part of Steele's expedition against Little Rock; they swam the Arkansas River during the attack on Little Rock, entering the city on September 18, 1863. The Fifty-fourth remained at Little Rock and in early 1864 three quarters of the Regiment re-enlisted and were sent home on veteran furlough. During this furlough, on March 28, the Regiment was involved in a riot at the courthouse in Charleston, Illinois. Major Shubal York, surgeon of the Regiment, and four privates (Oliver Sallee, Alfred Swim, James Goodrich, John Neer) were killed; also killed was William G. Hart, deputy provost marshal and member of the Sixty-second Illinois Infantry. The men were in a "shoot out" with Copperheads;[628] this was the only Civil war battle to take place on Illinois soil. The Fifty-fourth returned to Little Rock on April 30 but left on May 18 to pursue rebel General Shelby. They met Shelby's forces June 26 near Clarendon, Arkansas then pursued the enemy across the Cache River, finally returning to Duvall's Bluff. In August the Regiment was sent to guard the *Memphis and Little Rock Railroad*; there were five railroad stations, and two companies were assigned to guard each station. Shelby attacked on August 24 and although six companies were concentrated at one station after five hours of fighting the six companies of the Fifty-fourth were driven out and captured; Companies E and H at a distant station were not attacked and escaped capture. The captured men were paroled at Jacksonport, Arkansas and reached

[626] Charles H. Coleman and Paul H. Spence, "The Charleston Riot, March 28, 1864," *Journal Illinois Historical Society,* XXXIII, (1940): 7-56.

[627] The Kentucky Brigade consisted of the 54th, 60th, 62nd and 63rd Illinois Infantry Regiments

[628] Coleman, Charles, Spence, Paul, "The Charleston Riot, March 28, 1864" *J. Illinois Historical Society,* Vol XXXIII, (1940) p 7-56.

Benton Barracks, Missouri on September 9. They were exchanged December 5, 1864, and sent back to their former station, Hickory Station, on the *Memphis and Little Rock Railroad.,* On January 1865 they rejoined the remainder of their Regiment. On June 5, 1865, the men were sent to Little Rock via Pine Bluff and Fort Smith. The Fifty-fourth mustered out at Little Rock on October 15, 1865.[629,630]

There are two national flags but no regimental flags of the Fifty-fourth Illinois Infantry in the Collection of the State of Illinois.

This national flag is inscribed with battle honors and the star pattern is typical of flags manufactured in Philadelphia. Inscribed on this flag: VICKSBURG, LITTLE ROCK, CLARENDON, GRAND PRAIRIE, 54 ILLs INFty.

Only remnants remain of this national flag of the Fifty-fourth.

No other flags carried by the 54th Illinois Infantry are known to survive.

---

[629] *Report of the Adjutant General of the State of Illinois,* Volume 3, H.W.Rokker, Printer and Binder, Springfield, IL 1886, page 690-1,

[630] Eddy, T.M. *The Patriotism of Illinois,* Clarke & Co. Publisher, Chicago, 1866, Volume 2, p245-7

55th Illinois Infantry - "*Douglas Brigade, 2nd Regiment*"

Hail to the flag with crimson bars!
Hail to thee, flag with golden stars!
With thy sky-like field—its thousand sparks—
Christened with blood from brave men's hearts.
Wave, precious emblem of freedmen, free;
Loosed from the bonds of slavery.
Then give ye three cheers for the field of bright stars,
Not one was dimmed in the carnage of war.
Our night was so dark, who feared not the day?
One star more was given to guide us to day.
Sign of God's care that shall never cease;
Emblem of freedom and perfect peace.[631]

The Fifty-fifth Illinois Infantry Regiment was organized at Camp Douglas in the fall of 1861 by Chicago attorney-Colonel David Stuart[632] and mustered in on October 31, 1861. Men from the northern portion of the state volunteered to fill this Regiment. The volunteers for Company A (*Canton Rifles* originally *Fulton Cadets)* were from Fulton County; Company B from Cook, Coles, Ogle and De Kalb counties; Company C from Winnebago County; Company D from Fulton County; Company E (*Washington Guards)* from DuPage and Kane counties; Company F (*Bushnell Light Guards)* from McDonough County; Company G (*Lyons Guards*) from La Salle and McDonough counties; Company H from Carroll and Ogle counties; Company I from Grundy and La Salle counties and Company K from Knox County. The Regiment left Camp Douglas December 9, 1861, for Benton Barracks, St. Louis and in January 1862 moved down river to Paducah, Kentucky. After rearming, the Regiment moved by boat and captured Columbus, Kentucky. In March 1862 the Fifty-fifth moved up the Tennessee River and debarked at Pittsburg Landing. April 6 and 7 the Regiment fought in its first battle, the battle of Shiloh. The Regiment had 512 men engaged but after the battle only 215 remained to answer the next roll call. En-route to Corinth, Mississippi the Fifty-fifth skirmished at Russell's House (near Corinth, Mississippi) and on the evacuation of Corinth was the first Regiment to enter the fortifications on May 30. The summer of 1862 was spent in western Tennessee and Mississippi and included the capture of Holly Springs, Mississippi on July 4. The Regiment spent the next four months in camp at Memphis, Tennessee. At the opening of the Vicksburg campaign the Regiment joined Sherman's command and first fought at Chickasaw Bayou, Mississippi in late December. January 1863 found the Regiment

[631] *Report of the Proceedings of the Association of the Fifty-fifth Illinois Veteran Volunteer Infantry at Their Second Reunion, Moline Illinois November 10 and 11, 1886,* Davenport, Iowa, H.L.Wagner, Job Printer, 1887, p42-3. Portion of a poem written and recited by May Roberts and dedicated to the surviving members of the 55th Illinois Vet. Vol. Infty.

[632] Colonel David Stuart was born in Brooklyn, New York in 1816. He attended Amherst and Oberlin Colleges, graduating in 1838. He studied law, and was admitted to the New York bar. In 1853 he was elected as a Democrat to the 33rd Congress. In 1854 Stuart moved to Chicago and became the lawyer for the *Illinois Central Railroad.* At the coutbreak of the civil war he enlisted in the Union Army, and was commissioned lieutenant colonel of the 42nd Illinois Infantry and then colonel of the 55th Illiinois Infantry Regiment on October 31, 1861.

engaged at Arkansas Post on the Arkansas River; the rebel force was captured, and Fifty-fifth moved back down river to Young's Point, Louisiana. The Regiment moved with Grant's army to the rear of Vicksburg and fought in the battle of Champion Hill on May 16. Since Vicksburg was surrounded, to avoid a prolonged siege Grant ordered an assault on the Confederate lines north of Vicksburg on May 19. The colors of the Fifty-fifth were advanced to within pistol shot of the rebel works and were riddled with bullets.[633] Among the boy musicians in the Fifty-fifth was Orion P. Howe, age fourteen. For his actions during the assault on May 19 he was awarded the Medal of Honor.[634,635] May 22 was selected for the second assault on rebel lines.[636] One hundred and fifty volunteers (since the likelihood of death was high only unmarried volunteers were accepted) were to lead the assault called the "forlorn hope." Of the one hundred and fifty volunteers, 19 were killed, and 34 wounded; 14 men volunteered from the Fifty- fifth and four of them were killed in the assault. Fifty-three of the survivors of the "forlorn hope" were awarded the Medal of Honor, including seven from the Fifty-fifth Illinois Infantry.[637] Following the fall of Vicksburg the Regiment pursued General Johnston to Jackson, Mississippi then went into camp on the Black River. In September they moved to Memphis then marched to Chattanooga arriving there November 21, 1863. En-route they saw action at Cherokee Station and Tuscumbia, Alabama. On November 25 the Regiment was in the assault on Mission Ridge and in December marched to relief of Knoxville, Tennessee. They went into camp at Bellefonte Station, Arkansas on December 26. January 6, 1864, the Fifty-fifth marched to Larkinsville, Alabama and remained there until April when the Regiment veteranized but only after exacting the right to elect officers. The existing field officers all failed election and quit the service. Following a furlough, the veterans rejoined the Union army at Big Shanty and fought in the Atlanta campaign at Resaca, Dallas, New Hope Church, Allatoona Hills, Kennesaw Mountain, Nickajack Creek, Chattahoochee River and the Battle of Atlanta. It was for his actions at Atlanta on July 22, 1864, that Milton L. Haney, the Regimental Chaplain, was awarded the Medal of Honor.[638] After the battles of Jonesboro and Lovejoy Station the Regiment pursued Hood into Alabama. They began the "March to the Sea" in November and saw action at Fort McAllister and Savannah, Georgia, Duck Creek, Salkehatchie Swamps and Columbia, South Carolina and Fayetteville, Bentonville and Goldsboro, North Carolina. Following surrender of the Confederate forces the Regiment marched northward for the Grand Review in Washington DC on May 24. They mustered out August 14, 1865, at Little Rock Arkansas.[639,640]

---

[633] *The Story of the Fifty-fifth Regiment Illinois Volunteer Infantry in the Civil War,* by a Committee of the Regiment, Clinton, Mass., 1887 p. 236.

[634] Ibid, p 237-41.

[635] *Civil War Medal of Honor Winners from Illinois,* Published by the Civil War Centennial Commission of Illinois, July 1962. Page 7. — "A drummer boy, fourteen years of age, and severely wounded and exposed to a heavy fire from the enemy, he persistently remained upon the field of battle until he had reported to General W.T. Sherman the necessity of supplying cartridges for the use of troops under command of Colonel Malmborg." Vicksburg, Mississippi, May 19, 1863 (After the War, Howe was appointed to USNA but did not graduate. He became a dentist and practiced and died in Springfield, Mo.)

[636] Bears, Edwin Cole, *The Campaign for Vicksburg,* Volume III, Morningside House Inc, Dayton, Ohio, 1986, pages 813-52.

[637] Ibid. *Civil War Medal of Honor Winners from Illinois.* Robert A. Lower, Jacob Sanford, John Warden, John H. Fisher. James Larrabee, Jerome Morford received the Medal for "Gallantry in the charge of the "Volunteer Storming Party' " and Robert M Cox was awarded the Medal for "Bravely defended the colors planted on the outward parapet of Fort Hill."

[638] Ibid., *Civil War Medal of Honor Winners from Illinois,* page 7, "Voluntarily carried a musket in the ranks of his regiment and rendered heroic service in retaking the Federal works which had been captured by the enemy." Atlanta, Georgia, July 22, 1864."

[639] Brown, John G. *Report of the Proceedings of the Association of the Fifty-fifth Illinois Veteran Volunteer Infantry at Their First Reunion.* Chicago: James Guilbert, Printer, 1885, pages 21-31

[640] *The Story of the Fifty-fifth Regiment Illinois Volunteer Infantry in the Civil War 1861-1865,* by a Committee of the Regiment, W.J.Coulter, Clinton, Mass., 1885.

There were three stands of colors presented to this Regiment. The first was presented by the Quartermaster at Camp Douglas in 1861 and was sent back to Chicago from Vicksburg in 1863 and deposited with the Chicago Historical Society. (Presumably this stand of colors was lost in the Chicago Fire.)

The second stand of colors was also presented by the Quartermaster and was captured July 22, 1864, by the Thirty-ninth Alabama Infantry at Atlanta, Georgia. The Adjutant of the Thirty-ninth Alabama was Henry B. Tompkins, who later became an Atlanta judge. Francis Fisher, a veteran of the Fifty-fifth Illinois, contacted Judge Tompkins in 1884 and inquired as to the whereabouts of the captured colors. Judge Tompkins stated that a blue silk flag (regimental flag?) had been captured by his regiment, but he made no mention of a national flag. Judge Tompkins went on to relate that he had given the flag to Major General H. D. Clayton of Clayton, Alabama who then lost the flag from his headquarters' wagon.

Mr. Fred P. Fisher, of Chicago, presented a third stand of colors to the Regiment. The Regiment received them in North Carolina, after the march to the sea and in time for the Grand Review in Washington, DC. "This flag had all the battles in which the regiment had participated inscribed upon it. It is not certainly known what became of this flag, but it is supposed that it was deposited in the rooms of the Chicago Historical Society, and if so, it must have been consumed, together with the first flags, in the great fire of October 1871."[641] It appears that this third "stand of colors" refers to only a national flag. In the Regimental History the author states, "While in camp about four miles north of Raleigh the Fifty-fifth received a beautiful national flag, upon which were inscribed the names of the battles in which the regiment had participated."[642]

At the 1886 reunion of the Regiment, the following statement was made, "let it be inserted here, that the remnant of one of our National and one of our Regt. Banners, carried by the 55th while in the field, (probably the first ones,) are in 'Memorial Hall," Springfield, Illinois."[643]

The initial inventory of the flags by the Adjutant General lists only one national and one regimental flag in keeping with the report given at the Regiment's 1886 reunion.

However, there are now three flags--one national and two regimental flags-- in the collection of the State of Illinois. If the information given at the reunion of the Regiment in 1884 is correct, then the Fifty-fifth had three stands of Quartermaster issued colors; based on information in the Regimental history the third stand of colors was only a national flag. It is clearly stated that the first stand of colors was returned to the Chicago Historical Society and was presumably destroyed in the Chicago fire. The second stand of colors was captured at Atlanta. Since there is a national flag identified as having been captured, the national flag in Springfield must be one of the two flags captured at Atlanta. However, there are also two regimental flags in the collection, and unless one or both are misidentified one of these regimental flags must be the flag captured before Atlanta and "lost" by Major General Clayton. The other regimental flag may be misidentified and belong to another regiment especially since the original inventory of the flags lists only one national and one regimental flag belonging to the Fifty-fifth. There is no flag of the 55th in the collection inscribed with battle honors, therefore perhaps only the inscribed national flag was deposited with the Chicago Historical Society.

---

641 *Report of the Proceedings of the Association of the Fifty-fifth Illinois Veteran Volunteer Infantry at Their First Reunion.* Chicago: James Guilbert, Printer, 1885, pages 47-48

642 Ibid., *The Story of the Fifty-fifth Regiment Illinois Volunteer Infantry in the Civil War 1861-1865,* , p430.

643 *Report of the Proceedings of the Association of the Fifty-Fifth Illinois Veteran Volunteer Infantry at Their Second Reunion held in Moline, Ill., Nov. 10 and 11, '86*, p11.

There is tag affixed to the staff of this flag stating that it was captured.

The Regimental flag is not inscribed with battle honors. It is typical of flags manufactured in New York.

This regimental flag is also identified as being a color of the Fifty-fifth Illinois Infantry Regiment. It is not inscribed with battle honors. .

In January 2012 a flag was sold by Sotheby's in New York. It is a national flag that is said to have been carried by William Lawrence Shallenberger of Co. D of the 55th Illinois Infantry at the Battle of Shiloh. (There are several spellings of the last name and the O.R. lists a William Shellenbergher as a member of Co. D of the 55th Illinois Infantry.) The details associated with the sale of the flag state that Shallenberger was the color-bearer of the company and died on April 6, 1862, the first day of the Battle of Shiloh from a wound to the chest. This personal flag was found in his pocket and was used to cover his body and then sent home to his family. The flag is described as a hand sewn silk flag with 34 stars **printed** on the canton and with blood and bullet holes in the body of the flag. The size of the flag is not given. The flag was purchased and donated to the Shiloh National Military Park, and it is now in their collection. (The flag and an image of Shallenberger are on display at the park and Shallenberger is buried in a marked grave in the cemetery)

This is the flag carried by William Shallenberger

No other flags carried by the 55th Illinois Infantry are known to survive.

56th Illinois Infantry – "*Mechanic Fusiliers*"[644]

The US Army Transport *General Lyon* sailed from Southport North Carolina on the morning of March 29, 1865, traveling north to Fort Monroe, Virginia. The ship carried the 56th Illinois Infantry. After two days at sea, on March 31, the vessel approached Cape Hatteras, North Carolina. The waves were mountainous and the winds at hurricane strength. At 10 a.m. a fire erupted in the engine room most likely followed by an explosion. It is estimated that over 550 men perished in the sinking of the *Lyon* with only 28 survivors.[645]

The Fifty-sixth Illinois Infantry[646] was mustered into service February 27, 1862, under the command of Colonel Robert Kirkham, a merchant from Shawneetown Illinois, at Camp Mather near Shawneetown in Gallatin County. The men volunteered from the most southern counties of Illinois—Company A-Massac County; Company B-White County; Company C (*Indiana Company – Mechanics Fusaliers)*-Hamilton and White counties; Company D-Gallatin, White and Saline counties; Company E-Saline and Williamson counties; Company F-Hamilton and Wabash counties; Company H-Gallatin and Saline Counties; Company I-Franklin and Williamson counties and Company K-Saline and Pope counties. General Grant immediately ordered the Regiment to garrison duty at Paducah, Kentucky. In April the Regiment moved by steamboat up the Tennessee River to Hamburg Landing and marched to join the siege of Corinth, Mississippi. Following the rebels' abandonment of Corinth, the Regiment joined in the pursuit of Beauregard's army to Booneville, Mississippi and remained in northern Mississippi during the summer of 1862. In October Confederate forces attacked Corinth and the Fifty-sixth with the Tenth Missouri drove the rebels from their works and repulsed rebel reinforcements. During the winter of 1862 the Regiment campaigned in central Mississippi and escorted wagons and supplies to Memphis in December. In the spring of 1863, the Fifty-sixth was ordered to Vicksburg going through the Yazoo Pass and joining Grant's army at Young's Point. They then marched south to occupy Grand Gulf, Mississippi in May and fought at Champion Hills and joined in the siege of Vicksburg. On June 25 the Regiment was ordered to reinforce General Logan's command in the assault on Fort Hill. Following the blowing up of the Fort the Fifty-sixth occupied the crater, a portion of the night, and lost heavily in both killed and wounded. After the fall of Vicksburg, they moved to Helena, Arkansas then to Memphis and finally to Chattanooga where they remained until the end of November. On November 23 the regiment crossed the Tennessee River and attacked the rebels right flank at Missionary Ridge; during the next three days fighting the Regiment sustained heavy losses. Following the battle at Missionary Ridge, the Fifty-sixth garrisoned Whitesburg, Alabama on the Tennessee River. May 1864 the Regiment joined Sherman's Atlanta campaign and through November guarded the line of communication and railroads at the rear of the army. During this time the Regiment was reinforced with a new Company I (composed of veterans from the Thirteenth Illinois Infantry) and a newly organized Company D. In mid-October Confederate General Hood attacked Resaca, Georgia and the Fifty-sixth with other regiments defended the city against five times their number. The Fifty-sixth Illinois joined

[644] The word "fusilier" has a variety of spellings in the records from the 1860s. I have used each spelling as in the original record.

[645] Extracted from *New York Times,* April 3, 1865

[646] There were two regiments numbered 56. The first was organized in Chicago between October 31, 1861, and February 1, 1862, and was called the "Mechanic Fusileers." With only five companies and because of turbulence, discord, and outright mutiny of some of the men, it was mustered out of service in accordance with orders from the War Department in February 1862. However, many of the men joined the ranks of the 57th Illinois Infantry. The second regiment given the number 56 is the subject of this report.

Sherman at Atlanta in November 1864 for the "March to the sea." The men saw action at Savannah, Georgia, Salkehatchie Swamp, South Edisto River, and Columbia South Carolina and Bentonville and Goldsboro, North Carolina. On March 25, 1865, two hundred and five men of the Regiment boarded the *US General Lyon* for the trip to Fortress Monroe. On March 27 in a storm off Cape Hatteras fire erupted in the engine room and only 29 survivors were rescued including just 5 men from the 56th Illinois Infantry.[647] Following the surrender of General Johnston the remnants of the Fifty-sixth marched to Washington DC for the Grand Review on May 24. The Regiment mustered out at Little Rock Arkansas on August 12, 1865.[648]

There are three flags--one national and two regimental-- identified as belonging to the Fifty-sixth Illinois Infantry in the State of Illinois Military Museum. The flag pictured here is the only intact flag. The other two flags are only shreds. It is not known if any of the Regiment's colors were lost in the sinking of the *US General Lyon* described above.

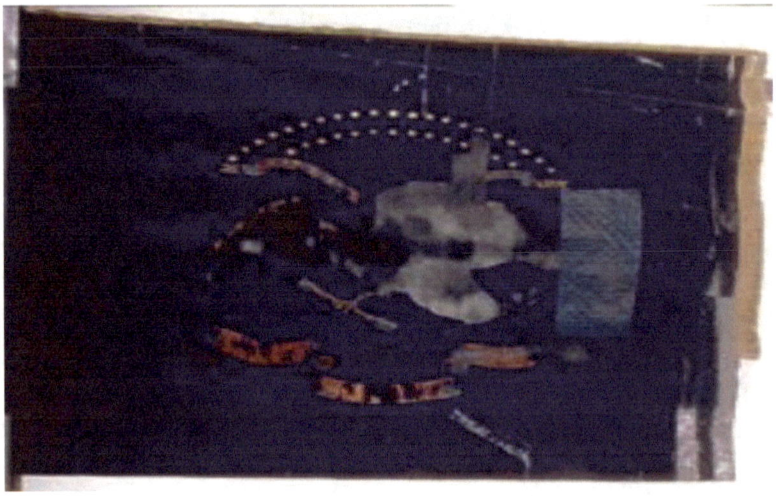

There is a silver plate on the staff of this regimental flag which reads:" Presented to the 56th Illinois by Mrs. Maria Raum, January 1863. Sub hoc signo vinces." Translated this means-With this sign – Conquer. The story behind this presentation is not known.

A stand of colors was presented to the Regiment on October 23, 1861. "The presentation of colors to Col. Wilson's Mechanic Fusileers Regiment occurred yesterday afternoon at Camp Webb, Wright's Grove. Hon. John M. Wilson presented the colors with an eloquent speech, which was most appropriately answered by Col. Wilson in behalf of the Regiment… The colors of this Regiment consist of the United States flag, elegantly mounted and a regimental flag of blue silk, with appropriate mechanical emblems wrought in the centre and fringed with yellow silk. Both these flags are sent direct from the War Department and are, therefore, the gift of the nation."[649] The further history of these colors in unknown.

No other flags carried by the 56th Regiment of Illinois Infantry are known to survive.

---

647 *New York Times*-- Monday April 3, 1865 and Friday April 14, 1865

648 *Report of the Adjutant General of the State of Illinois,* HW Rokker, State Publisher and Binder, 1886, Volume 4, pp 31a-31d.

649 *Chicago Tribune;* Oct.24,1861

57th Illinois Infantry- "*National Guards*"

"A notable feature of the regiment is the pioneer force consisting of twelve men equipped by the Colonel himself at his own expense in huge bear skins."[650]

The Fifty-seventh Illinois Infantry was organized at Camp Douglas, Chicago in late 1861 by Colonel Silas D. Baldwin[651] and mustered in for three years on December 26, 1861. The men of the Regiment came from the *Princeton Regiment*[652] and from those recruited for the Fifty-seventh Infantry. The Regiment was comprised of men from—Company A-LaSalle County; Companies B, F, H and K-Bureau County; Companies C, E and G-Cook County; Company D-Henry County (this company was from the Swedish settlement at Bishop Hill); Company H-in addition to those men from Bureau County there were also men from Henry and Knox counties and Company I-from the State at large. In February 1862 the Regiment moved by rail to Cairo, Illinois then by steamer to a point two miles below Fort Donelson on the Cumberland River. The Regiment was in support of artillery batteries and Fort Donelson surrendered before the Fifty-seventh was engaged. In March they moved up the Tennessee River to Crump's Landing and then on to Pittsburg Landing, Tennessee. On April 6 and 7, at the Battle of Shiloh, the Regiment lost 187 men and officers killed, wounded and missing. At the end of April, the Regiment advanced toward Corinth, Mississippi and took part in the siege until the end of May then pursued the retreating rebels to Booneville, Mississippi and returned to camp near Corinth. In October the Fifty-seventh as part of the Third Brigade repulsed a Confederate attack on Corinth and aided in capturing 2,000 prisoners then pursued the retreating enemy to the Hatchie River and returned to Corinth where they remained until April 1863. In April and May the Regiment marched to Tuscumbia and skirmished at Town Creek, Alabama and returned to Corinth. In mid-September the Fifty-seventh as part of the 2nd Division raided LaGrange, Tennessee, passing through Holly Springs and on to the Hatchie River in pursuit of Forrest. As Forrest retreated south the Division returned to Corinth. In November the men moved toward Iuka, Mississippi, crossing the Tennessee River and went into camp at Pulaski and then Lynnville, Tennessee. In January 1864 the entire Regiment except for Company C re-enlisted as veterans and left for home on furlough. The Regiment returned from furlough in March rejoining Company C at Athens, Alabama. They moved by train to Chattanooga, Tennessee then marched to Rossville, Georgia and went into camp on the banks of Chickamauga Creek. As part of the Atlanta campaign the Fifty-seventh saw action at Resaca, Ley's Ferry, Rome Cross Roads, Dallas, New Hope Church, Allatoona Hills, Kennesaw Mountain Chattahoochee River and battle of Atlanta. In September the Regiment fought at Jonesboro, Lovejoy Station and moved to Rome, Georgia. On October 4, Companies A and B were sent to reinforce the garrison at Allatoona Pass. The garrison was attacked by rebel General French but held out until reinforcements arrived. Company F and a portion of Company D were sent to Chattanooga with prisoners taken at Allatoona. On the Regiment's return, at Dalton, Georgia, the command, and the garrison at Dalton, along with a regiment of colored troops were taken prisoner by Hood's army without a fight. On November 10, 1864, the Regiment with 504 enlisted

[650] *Chicago Daily Tribune,* February 8,1862.

[651] Colonel Silas Baldwin was dismissed for cowardice from service on March 12, 1863.

[652] The 56th Illinois Infantry was organized at Princeton, Illinois (*Princeton Regiment*) but only five companies were recruited. Due to the inadequate number of companies plus problems with mutiny and discord on the part of some of the volunteers, the original 56th Infantry was disbanded and many of the men of the 56th were combined with five companies of the organizing 57th Infantry to form the 57th Illinois Infantry (*National Guards*). Subsequently another 56th Illinois Infantry Regiment was organized in southern Illinois and was mustered into service.

men and 17 officers departed on Sherman's "March to the sea." The Regiment first saw action December 10 through 21 at Savannah, Georgia; in late December all non-veterans in the ranks of the Regiment were mustered out of service, their term having expired. In January 1865 new recruits from Illinois joined the Regiment. In February the army crossed the Savannah River and skirmished at Branchville, Salkehatchie and the Edisto Rivers then entered Columbia, South Carolina. In March they fought at Bentonville, North Carolina, occupied Goldsboro, and Raleigh and after surrender of Johnston's army marched to Washington DC for the Grand Review on May 24. The Fifty-seventh mustered out at Louisville, Kentucky July 7, 1865.[653,654]

The State of Illinois Adjutant General's original inventory lists one national and three regimental flags for the Fifty-seventh Infantry. Only the fringe remains from one of the regimental flags in the collection. The other three flags are shown here.

This flag is inscribed: 57th REGT. ILL. VOLS. *The canton contains 33 stars, and the stars are arranged in an oval pattern. This is consistent with a flag produced in Philadelphia early in the War.*

[653] Cluett, William W., *History of the 57th Regiment Illinois Volunteer Infantry, from Muster in, Dec. 26, 1861 to Muster out, July 7, 1865,* T.P.Streeter, Lessee Republican Job Department, Princeton, 1886. (Cluett was a sixteen-year-old drum Major in the Regiment)

[654] *Report of the Adjutant General of the State of Illinois,* H.W.Rokker, Printer and Binder, Springfield, IL 1886, Volume 4, page 62-70

57TH ILLINOIS INFANTRY

This is a 34-star regimental flag and was probably manufactured in Cincinnati

*This regimental flag is in poor condition and only a portion of the inscription remains. However, above the stars appears to be the word "*NATIONAL*" and presumably the flag originally was inscribed "*NATIONAL GUARDS*" the name of the Regiment.*

No other flags carried by the 57th Illinois Infantry are known to survive.

58th Illinois Infantry -- "*Lyon Color Guard*"

Soldiers are we from the mountain and valley,
Soldiers are we from the hill and the plain;
Under the flag of our fathers we rally;
Death, for its sake, is but living again.
Then march along, gay and strong,
March to battle with a song!
March, march along![655]

Colonel William F. Lynch, 23-year-old lawyer from Elgin, Kane County, organized the Fifty-eighth Illinois Infantry Regiment at Camp Douglas. The 58th was a multinational regiment with one-third American, one-third of Irish origin and remainder born in Sweden, Norway, Denmark, Holland, Germany, Poland Switzerland, Hungary, Scotland, Italy and France.[656] The volunteers for Companies A and B were from Cook and Kane counties; Company C (*Shabbona Rifles)* from De Kalb County; Company D and F from Cook County; Company E from Cook and De Kalb counties; Company G from Knox County; Company H from Cook and La Salle counties; Company I from Kane County and Company K from Rock Island County. Nine companies were mustered in on December 24 and 25, 1861; Company H was mustered in February 7, 1862. On February 11, 1862, the Regiment departed for Cairo, Illinois where they were supplied with arms and immediately proceeded up the Ohio and Cumberland rivers to Fort Donelson. They were engaged in the assault on the Fort until the rebels surrendered on February 16. After marching to Fort Henry, the Regiment went up the Tennessee River to Crump's Landing, disembarked and went into camp. On March 29 the Fifty-eighth moved by steamer to Pittsburg Landing and went into camp near the river. The battle of Shiloh began on April 6 and the Fifty-eighth was given a position to hold a crossroad and held the position until 4pm when the regiments on the right and left gave way. The Fifty-eighth retreated but was surrounded and a greater portion of the Regiment was taken prisoner. Of the 613 men present for duty on April 6, over 300 were prisoners or missing, over 100 were wounded or sick and only 127 remained for duty.[657,658] The portion of the Fifty-eighth not captured, together with remnants of the 8th, 12th, and 44th Iowa Infantries were organized into the "*Union Brigade*" with the 58th Illinois Infantry forming three of the ten companies of the Brigade. The *Union Brigade* participated in the siege of Corinth April 29 through May 30 and in the pursuit of the retreating rebels. one company from the Fifty-eighth fought at Iuka where they lost 17 men wounded or taken prisoner. In December 1862 the Fifty-eighth was reunited (the remnant of the Regiment from the *Union Brigade* plus those paroled from rebel prisons) at Camp Butler for reorganizing and recruiting. In June 1863 the Regiment was ordered to Cairo, Illinois and garrisoned at Cairo, Mound City, Union City, Columbus, and Paducah. In January 1864 the men re-enlisted as veterans and were ordered to Vicksburg,

---

[655] *March Along* by George Henry Boker

[656] *Chicago Daily Tribune,* February 11, 1862.

[657] O.R. Series I, Volume X/1 (S#10)

[658] The captured prisoners were transferred to various rebel prisons. On May 29, the privates and surgeons of the regiment were released by order of General Beauregard, but General O.M.Mitchell refused to receive them, and they were returned to the rebel prisons. On October 17, 1862, the men were gathered at Libby Prison in Richmond where they were paroled and sent to Annapolis, Maryland. Eddy, TM, DD, *The Patriotism of Illinois,* Volume 2, Clarke & Co., Publishers, Chicago, 1866, page250-1.

Mississippi; then left from Vicksburg in February in the Meridian Campaign. They were the first Regiment to engage the enemy at Queen's Hill and the first to enter Meridian, Mississippi. After return to Vicksburg the Fifty-eighth participated in the Red River Campaign and saw action at Fort DeRussy, where the colors of the Fifty-eighth were the first planted on the works. At Pleasant Hill the Regiment captured over 500 prisoners. In May the Regiment returned to Vicksburg and on June 10 the veterans were furloughed, and the non-veterans sent to Tupelo, Mississippi. On August 6 the veterans rejoined the Regiment at Memphis and in September they began a campaign against Price in Missouri. During October and November, the Regiment marched through Missouri to Kansas, returning to St. Louis in late November. December 1 they were ordered to Nashville, Tennessee and were engaged in battle there. February 1865 the term of the original Regiment expired, and the non-veterans returned home. The veterans and recruits, about 390, were consolidated into four companies known as "Battalion 58th Illinois Infantry." The Battalion joined in operations against Mobile and the investment of Spanish Fort and Fort Blakely, Alabama. The regiment was raised to full strength while at Mobile by the addition of six new companies. The Regiment was ordered to Montgomery Alabama in July 1865 and remained there on garrison duty until mustered out on April 15, 1866.[659,660]

There are four flags of the Fifty-eighth Illinois Infantry in the collection of the State of Illinois Military Museum in Springfield, Illinois—two national flags and two regimental flags. Only a shred remains of one of the regimental flags and is not shown here.

This national flag is of Philadelphia manufacture with evidence of pieces of the flag having been removed, perhaps as mementos, by men of the Regiment.

[659] Eddy, TM, DD, *The Patriotism of Illinois,* Clarke & Co., Publishers, Chicago, 1966, Volume II pages 249-53
[660] *Report of the Adjutant General of the State of Illinois,* Volume 4, pages 101-3.

This national flag has 34 stars and is of New York manufacture. There are no inscriptions on the flag.

The regimental flag contains 34 stars and is of New York manufacture.

A flag was presented to the *Shabbona Rifles* (Company C) at Camp Douglas in Chicago, "the presentation of an eloquent silk flag of the regulation size, the handiwork of Charles Du Bois, to the Shabbona Rifles, Capt. Geo. Kittell, Company C, Col. Lynch's regiment. A very tasteful and appropriate presentation speech was made by Henry S. Monroe, Esq., to which Lieut. Joseph G. Burt responded handsomely. The flag was the gift of the citizens of Shabbona, De Kalb County, in this State and was a splendid token of remembrance and patriotic wishes. The little town of Shabbona, out of two hundred and four voters, has sent eighty-six of her most stalwart and loyal sons to the wars and has not been backward in 'remembering' them in their absence and caring for them in the matter of extra blankets and clothing."[661] The flag is not further described, but is probably not one of the flags shown here.

No other flags known to have been carried by the 58th Illinois Infantry are known to survive.

[661] *Chicago Daily Tribune,* January 16, 1862.

59th Illinois Infantry

"Upholding our banner,
On many a field,
The doom of the traitor
They valiantly sealed;
And worn with the conflict,
Found vigor anew,
Where victory greeted
The Red, White, and Blue."[662]

The Fifty-ninth Illinois Infantry was originally organized at St. Louis September 18, 1861, by Colonel John C. Kelton[663] as the Ninth Regiment Infantry Missouri Volunteers and was intended to be a Zouave unit. However, the companies had been filled with men from Illinois—Company A from Knox County (*Knoxville Rifles)*; Company B from Henderson and Warren counties; Company C from McDonough County; Company D from Madison County; Company E from St. Clair County; Company F from Cumberland County; Company G from Greene County; Company H from Coles and Edgar counties; Company I from Knox, Macoupin and Madison counties and Company K from Cook County (originally known as the *Chicago Sappers and Miners*). In late September the Regiment was ordered to Jefferson City, Missouri and embarked on Fremont's campaign against Springfield, Missouri. In 1862 they campaigned against Price, fighting at Pea Ridge, Arkansas on March 7 and 8 and at Sugar Creek, Missouri on March 10. (During that campaign, on February 12,1862 the Regiment's designation was changed to the Fifty-ninth Illinois Infantry Regiment and command devolved to Colonel P. Sidney Post an Illinois lawyer.) At West Plains, Missouri on March 28, Captain Alfred W. Ellet and fifty men from the Regiment were detached to the Ram Fleet at Cairo, Illinois. They did not return to the Regiment.[664] On May 20 the Fifty-ninth arrived at Cape Girardeau, then left for Hamburg Landing, Tennessee. They

---

662 Lathrop, David, *The History of the Fifty-ninth Regiment Illinois Volunteers,* Hall & Hutchinson, Indianapolis, Ind., 1865, page 243.

663 Colonel John Cunningham Kelton was born in Pennsylvania in 1828 and was a friend and admirer of Thaddeus Stevens. At the age of 19 he was appointed a Cadet at the United States Military Academy at West Point. He graduated in the class of 1851and was appointed Brevet Second Lieutenant in the Infantry. For the next six years he served in the "Frontier Service" at Fort Snelling, Fort Ridgely, Jefferson Barracks and Fort Laramie. In 1857 he was detailed to the Military Academy at West Point as an instructor. In 1861 he was ordered to St. Louis, Missouri and in July, 1861 he was appointed Colonel of the Ninth Missouri Infantry. Extracted from the report of the *Tweny-Fifth Annual Reunion of the Association of the Graduates of the United States Military Academy.* June 12th, 1894

664 Colonel Charles Ellet conceived, organized, and built the Mississippi Ram Fleet or Ellet's Ram Fleet. His brother, Alfred W. Ellet, Captain of Company I of the 59th Illinois Infantry was authorized to act as his second in command and to bring with him reliable men from the 59th. For more information on the Ram Fleet see: Norman E. Clarke, Ed, *Warfare Along the Mississippi, the Letters of Lieutenant Colonel George E. Currie,* Edwards Brothers, Inc., Ann Arbor, Michigan, 1961; Gosnell, H. Allen, *Guns on the Western Waters, The Story of the River Gunboats in the Civil War,* Louisiana State University Press, Baton Rouge, La. 1949, Chapter VIII; Ellet, Alfred W., "Ellet and His Steam-Rams at Memphis, *Battles and Leaders of the Civil War,* Volume I, pages 453-; "Organization of the Mississippi Ram-Fleet" in Part IV of book by .Herr, George W. *Episodes of the Civil War, Nine Campaigns in Nine States,* The Bancroft Company, San Fran Cisco, 1890

participated in the siege of Corinth, Mississippi[665] and after the evacuation of the city; they pursued the retreating enemy as far as Booneville, and then returned to camp near Corinth. In August the Fifty-ninth skirmished with the enemy at Bay Springs, Mississippi, and marched to Murfreesboro, Tennessee in September. On October 8 they fought at Chaplin Hills, near Perryville, Kentucky; here they lost 113 men killed or wounded out of 360 who fought in the battle. The Regiment next skirmished at Lancaster, Kentucky (October 15) and arrived at Nashville in November. In December the Regiment advanced on Murfreesboro, Tennessee and were engaged in the Battle of Stones River December 30, 31 and January 1-3, 1863. They remained at Murfreesboro until June 1863. In the Middle Tennessee Campaign, the Fifty-ninth fought at Liberty Gap June 24-27 then pressed the enemy toward Winchester, Tennessee. In August and September, the Fifty-ninth guarded supply trains during the Battle of Chickamauga then fought in the siege of Chattanooga, Tennessee. They were heavily engaged at the Battles of Lookout Mountain, Mission Ridge, Taylor's Ridge, and Ringgold Gap during November. January 12, 1864, the Fifty-ninth mustered in as a veteran regiment and on February 6 left for Springfield, Illinois on furlough. They returned to Chattanooga in March and in May joined the Atlanta campaign fighting at Tunnel Hill, Rocky Face Ridge, Resaca, Kingston, Dallas, New Hope Church, Allatoona Hills, Pine Top Mountain, Kennesaw Mountain, Marietta, Chattahoochee River, Peach Tree Creek and in the Siege of Atlanta. In a flanking movement south of Atlanta the Regiment fought at Jonesboro on August 31-September 1, and Lovejoy Station September 2-6. They pursued Hood into Alabama in October and fought at the Battle of Franklin in November and Battle of Nashville in December.[666] In the battle of Nashville, at Montgomery Hill, the Fifty-ninth led the assault and the Regiment's colors were the first planted on the enemy's works. In 1865 the Fifty-ninth moved to Huntsville, Alabama and then to New Orleans, Louisiana. In July they moved to Indianola, Texas and were mustered out at New Braunfels, Texas on December 8, 1865.[667,668,669]

The State of Illinois collection of Civil War flags contains two national flags, one regimental flag and one guidon or flank marker from the Fifty-ninth Illinois Infantry. All that remains of one of the national flags is two shreds of a red stripe and the photo is not shown here.

---

665 James W. Archer, First Lieutenant and Adjutant was awarded the Medal of Honor for his actions at Corinth. "Voluntarily took command of another regiment with the consent of one or more of his seniors, who were present, rallied the command, and led it in the assault." Corinth, Mississippi, October 4, 1862. *Civil War Medal of Honor Winners from Illinois,* Published by the Civil War Centennial Commission of Illinois, July 1962.

666 Colonel Philip Sidney Post "Led his brigade in an attack upon a strong position under a terrific fire of grape, canister, and musketry; was struck down by a grapeshot after he reached the enemy's works." At Nashville, Tennessee, December 15-16, 1864. For this he received the Medal of Honor, *Civil War Medal of Honor Winners from Illinois,* Published by the Civil War Centennial Commission of Illinois, July 1962.

667 Ibid. *History of the Fifty-ninth Regiment Illinois Volunteers.*

668 Dyer, Frederick H., *A Compendium of the War of the Rebellion,* Thomas Yoseloff, Publisher, New York, 1959, Part III, page 1073.

669 Herr, Geo. W., Corporal, *Episodes of the Civil War, Nine Campaign In Nine States,* The Bancroft Company, Publishers, 1890

The above national flag has 35 stars in the canton consistent with manufacture late in the war. The inscriptions are all in similar printing suggesting that a single individual did all the scribing. The inscriptions read: PEA RIDGE, PERRYVILLE, STONE RIVER, LIBERTY GAP, LOOKOUT MOUNTAIN, MISSIONARY RIDGE, RINGOLD, 59th REGT. ILLs. VEt. INFTRY, ROCKYFACE, RESACCA, DALLAS, PINETOP, KENNESAW MOUT, RUFFS STATION, PEACHTREE CREEK, ATLANTA, JONESBORO, LOVEJOY, FRANKLIN, NASHVILLE.

A flag was presented to the Regiment as described by Capt. John M. VanOsdel, Co. K of the Fifth-ninth Infantry in his unpublished memoirs--on May 22, 1862, he writes " Drew uniform, hats and shoes. Col. J. G. Kelton sent us colors, as ours was about played out. The name of the Regiment was on the banner and "Pea Ridge"."[670]

Company G of the Fifty-ninth Illinois Infantry received a flag purchased by funds donated by the ladies of Carrollton, Illinois. "Miss Elizabeth Hazle: I hereby acknowledge the receipt of $25, by John Simpson, for the purpose of purchasing a flag for company G, 59th regiment Illinois volunteers. To you and the patriotic ladies of Carrollton, the company return their thanks and best wishes for this token of respect to them and fidelity to our common country…and when the flag which you have furnished the means of procuring is unfurled, to the breeze, with "Pea Ridge" inscribed on its folds..."[671] The descriptions of the two(?) flags are quite similar (both state "Pea Ridge" inscribed) and both presentations are dated about the same time. The flag(s) may be the same flag.

---

[670] A copy of the diary is in the possession of the author.
[671] *Carrollton Gazette*, April 26, 1862

This inscribed guidon measures approximately 30" x 18" and is inscribed: 59th REGT ILLs VET INFTy.

This 35-star regimental flag is characteristic of flags manufactured in Cincinnati.

No other flags carried by this regiment are known to survive.

60th Illinois Infantry

The Sixtieth Illinois Infantry Regiment was mustered into service, under command of Colonel Silas Cox Toler,[672] on February 17, 1862 at Camp DuBois in Anna, Illinois.[673] Companies A and B were from Union County; Company C and I from Jefferson County; Company D from Hamilton and Jefferson counties; Company E from Jefferson and Williamson counties; Company F from Richland and Union counties; Company G from Hamilton County; Company H from Pope County, Company I from Macoupin County and Company K from Johnson County. On February 22 they were ordered to Cairo, Illinois and then in March to action at Island No. 10. In May they moved up the Tennessee River to Hamburg Landing, Tennessee and were engaged in the siege of Corinth, Mississippi from April 29 to May 30. Following the evacuation of Corinth they pursued the retreating enemy to Booneville, Mississippi. In July the Regiment marched to Tuscumbia, Alabama, went into camp, and remained there until the end of August. In September the Sixtieth marched to Nashville, Tennessee and took part in garrisoning that city. On November 5, 1862, Confederate General Morgan at Edgefield, Tennessee attacked the Sixteenth and Sixtieth Illinois Infantry Regiments; the rebels were repulsed. Early in January of 1863 the Sixtieth was sent to Stones River, Tennessee to escort an ammunition train en-route to Nashville. Eight miles from Nashville the train was attacked by a rebel force under General Wheeler, but the rebels were defeated, and the train arrived safely at Nashville. The Regiment moved to Murfreesboro, Tennessee in June and in August marched to Columbia, Athens, Huntsville and Stevenson, Alabama. In September the Sixtieth marched to Bridgeport, Alabama where they remained until October. From October 1 to 17 they were in operations against Confederate General Wheeler in the Sequatchie Valley of Tennessee. On November 25 and 26 the Sixtieth fought in the Battle of Chattanooga and pursued the rebels to Ringgold, Georgia, then marched with General Sherman to Knoxville to raise the siege at that city. They marched back to Chattanooga and went into winter quarters at Rossville Georgia on December 24. February 1864 three fourths of the Regiment re-enlisted, moved toward Dalton, Georgia and fought at Buzzard's Roost where they lost forty-two killed and wounded. On March 6 the veterans left on furlough; when they returned on April 18, they joined the Atlanta Campaign. The Sixtieth subsequently fought at Ringgold, Dalton, Resaca, Rome, Dallas, New Hope Church, Allatoona Hills, Kennesaw Mountain, Nickajack, Peach Tree Creek and the siege of Atlanta. In a flanking movement south of Atlanta they fought at Jonesboro August 31-September 1 and at Lovejoy Station September 2-6. September, October and November the Sixtieth pursued Hood's and Forrest's retreating forces into north Georgia and north Alabama. The Regiment joined Sherman's "March to the Sea" on November 15 and fought at Savannah, Georgia, Averysboro, Bentonville and Goldsboro, North Carolina; they occupied Raleigh. Following Johnston's surrender they marched via Richmond to Washington DC for the Grand Review on May 24. The Sixtieth served as provost guard in Louisville, Kentucky until mustered out July 31, 1865.[674]

There are two national and two regimental flags of the Sixtieth Illinois Infantry in the collection of the State of Illinois Military Museum. Only fragments of one of the Regimental flags remain and a photo is not included of that flag.

---

[672] Colonel Silas Toler was born in Union County, Illinois on July 16, 1831. He was a practicing physician in Jonesboro, Illinois and served as Postmaster from 1856-1858. He died of "chronic hepatitis in March 1863 at the age of 31.

[673] The 54th, 60th, 62nd, and 63rd Illinois Infantry Regiments formed the Kentucky Brigade.

[674] Eddy, T.M., *The Patriotism of Illinois,* Clarke & Co., Chicago, 1865, Volume I, page 409-12, Volume II, page 256.

This national flag displays 35 stars and therefore was manufactured late in the War. The square canton and the star pattern are typical of flags produced in New York. The flag measures 69" on the hoist by 75" on the fly. Inscribed on this flag: ISLAND No. 10, SIEGE OF NASHVILLE, CORINTH, LAVERGNE, CHATTANOOGA, BUZZARD ROOST, ROCKY FACE MOUNTAIN, RESACA, ROME, DALLAS, LOST MOUNTAIN, ACKWORTH, KENESAW MOUNTAIN, PEACH TREE CREEK, NICKAJACK, ATLANTA, JONESBORO, SAVANNAH, AVERYSBORO, BENTONVILLE, GOLDSBORO, RALEIGH.

There are no inscriptions on this flag and the star count is uncertain. The arrangement of the stars, however, is typical of New York manufacture. It measures 69" on the hoist by 72" on the fly.

60TH ILLINOIS INFANTRY

The regimental flag displays 35 stars and is probably of Philadelphia manufacture. It measures 71" on the hoist by 73" on the fly.

No other flags carried by this Regiment are known to survive.

61st Illinois Infantry

"The serried ranks, with flags displayed,
The bugle's thrilling blast,
The charge, the thundr'rous cannonade
The din and shout—were past."[675]

The first three companies of what was to become the Sixty-first Illinois Infantry were organized at Carrolton in Greene County Illinois by Colonel Jacob Fry.[676] These three companies were mustered on February 5, 1862, and at the end of February, the Regiment still being incomplete moved to Benton Barrack, Missouri. There additional companies were added to make up nine full companies and on March 7,1862 the Regiment was given the regimental designation of –"Sixty-first"—Illinois Infantry and mustered into service.[677] The volunteers in companies A (*Woodson Guards)*, B, G and I were from Greene County, Company C (*Jersey County Blues)* and D were from Jersey County, Company E from Cass County, Company F from Scott County and Company H from Lawrence County. (Company K from Greene County did not join the Regiment until March 20, 1864.) On March 30 the Regiment arrived at Pittsburg Landing, Tennessee and fought in the Battle of Shiloh on April 6 and 7 where it lost 80 killed, wounded and missing. From April 29 to May 30, they fought in the siege of Corinth, Mississippi and June 6 the Sixty-first moved to Bethel, Tennessee then on to Bolivar, Tennessee where they remained on duty until May 1863. During that year the Regiment skirmished at Iuka, Mississippi, were in operations against Forest, and saw action at Salem Cemetery near Jackson, Tennessee. On May 31, 1863 the Regiment moved by rail to Memphis then downriver to Chickasaw Bayou, Mississippi northeast of Vicksburg. In the Vicksburg campaign the Sixty-first fought at Satartia and Mechanicsburg, Haines Bluff and finally in the siege of Vicksburg. Following the fall of Vicksburg on July 4, the Regiment moved to Helena, Arkansas and then fought at Bayou Fourche and in the capture of Little Rock, Arkansas. They were stationed at Little Rock until August 1864 and during that time in June 1864 skirmished at Clarendon, Arkansas on the White River. In early 1864 most of the men re-enlisted as Veterans and in March Company K joined the Regiment. In August 1864 the Regiment left on veteran furlough. On return they traveled to St. Louis where companies B, D, and G were detached to Chester, Illinois to guard a crossing on the Mississippi River. On October 14 the detached companies returned to St. Louis and the Regiment was ordered to northeast Missouri where they skirmished with rebel forces. On December 4 the Sixty-first was engaged at Overall's Creek three miles from Murfreesboro, Tennessee; on December 7 they fought at Wilkinson's Pike or the "Cedars" near

---

[675] Stillwell, Leander, *The story of a Common Soldier of the Army Life in the Civil War 1861-1865*, Franklin Hudson Publishing Company, Erie, Kansas, 1920, page 276. Stillwell was born September 10, 1843, grew up on his father's farm and attended the local school in Otterville, Illinois. He was 17 years old when he volunteered for Co. D of the 61st Illinois Infantry. He was mustered out in 1865 and in the fall of 1866, he entered the Albany Law School in Albany New York and graduated in 1867 with a degree in law. He practiced first in Illinois but eventually moved to Erie, Neosho County, Kansas where he became a judge and served in the Kansas legislature. He died at age 90 and was buried in the Ease Hill Cemetery in Erie, Kansas.

[676] Colonel Fry was Swiss born and 30 years old at the outbreak of the civil war.

[677] Ibid., Stillwell, Leander, Page 28.

Murfreesboro where they captured the colors of a Florida[678] or Georgia[679] Regiment. On December 12 the Regiment along with one company of the First Michigan Engineers was sent to escort a train from Murfreesboro to Stevenson, Alabama, and return. On the return trip, on December 15, eight miles from Murfreesboro the train was attacked by an overwhelming rebel force under the command of General Forest. They surrounded the train forcing the surrender of over half of the Regiment as well as Lieutenant Colonel Grass the Regimental commander. This was the last engagement of the Regiment. On February 4, 1865, the non-veterans and recruits rejoined the Regiment and the Regiment moved from Murfreesboro to Franklin, Tennessee. In June recruits from the Eighty-third, Ninety-eighth and One Hundred and Twenty-third Illinois Infantry were transferred to the Sixty-first. On September 8, 1865, the Sixty-first was mustered out at Nashville, Tennessee. [680,681]

The State of Illinois flag collection has four flags of the Sixty-first Illinois Infantry—two are national flags with company and regimental designations and the third national flag is presumably also the national colors of the regiment. There is one regimental flag. It was not until March 7, 1862, that the regiment received its official designation as the 61st Illinois Infantry. According to the regimental history, the Regiment already had company colors prior to that date.[682] These may be two of the flags mentioned as company colors in Leander Stillwell's history.

This flag is inscribed: Co. G, 61st REGt ILLs VOLs and bears a tag stating that it was captured and returned.683 (See below, this may be the flag presented to the Jersey Blues or Jersey County Blues)

---

678 The combined 1st Florida Cavalry and 4th Florida Infantry lost their Confederate Second National flag near Murfreesboro but both the 61st Illinois and 174th Ohio Infantry claimed its capture. The flag is now in the Florida State Museum.

679 Duckett, Drew D., *Glimpses of Glory, The Regimental History of the 61st Illinois Volunteers,* Heritage Books, Bowie Maryland, page 93. The author states that the colors captured were those of the 29th Georgia, his apparent source for this information is the Carrolton (Illinois) Gazette, however, the 29th Georgia is not known to have been in Murfreesboro at that time.

680 Ibid, Stillwell, Leander, *The story of the Common Soldier of the Army Life in the Civil War 1861-1865*

681 *Report of the Adjutant General of the State of Illinois,* HW Rokker, State Printer and Binder, 1886, volume IV, pages 235-237.

682 Ibid, p. 23 "Besides the regimental colors, each company, at that time carried a small flag, which were all fluttering in the breeze….." The small flags mentioned were probably guidons.

683 Companies B and G were presented company flags by the citizens Carrolton and of Greene County at Benton Barracks, Missouri on March 24, 1862. The flags were purchased with each flag costing $40. *Carrolton Gazette,* April 5, 1862.

This flag is inscribed: Co. B, 61st REGt ILLs VOLs *and measures 69" on the hoist by 75" on the fly. The stars and inscription are gold in color.*

This national flag, typical of flags manufactured in Philadelphia, does not have a regimental designation on the center stripe.

The regimental flag is in poor condition and bears no battle inscriptions.

Company A, the "*Woodsman Guards"* were presented a flag by Judge D. M. Woodson on February 11, 1862, at Camp Carrolton. It is referred to, as "stars and stripes" therefore, was presumably a national flag. No other history of this flag is known,

The "*Jersey County Blues*," Company C, were presented a flag— "A subscription having been taken up among our citizens for the purpose, a fine silk flag and three elegant swords, with sashcs and all the necessary accompaniments, have been purchased for the *Jersey County Blues*, now in camp at Carrollton. The flag will be presented to the company, and the swords to Capt. Ihre and Lieuts. Maguire and Hesser by a committee of ladies at 10 o'clock on Christmas day, in the court house yard. The flag cost fifty dollars"[684]

---

[684] *Carrollton Gazette,* December 28, 1861

62nd Illinois Infantry

The Sixty-second Illinois Infantry was organized in southern Illinois at Camp DuBois near Anna in Union County. The Sixty-second along with the 54th, 60th, and 63rd Illinois Infantry Regiments formed the Kentucky Brigade. The men of Company A volunteered from Lawrence County; Company B from Washington County; Company C from Clark County (plus some from Cumberland and Coles County); Company D from White County (plus some from Lawrence and Crawford County); Company E from Fayette County; Company F from Crawford County; Company G from Effingham, Wayne and Clinton counties; Company H from Shelby, Macon and Warren counties; Company I from Schuyler County and Company K from Coles and Edgar counties. Following muster in on April 10, 1862, under the command of Colonel James True,[685]the Regiment was ordered to Cairo, Illinois, then on to Paducah, Kentucky and finally in June to Obion County, Tennessee to guard *the Mobile and Ohio Railroad.* On December 13 the Regiment arrived in Holly Springs, Mississippi and on December 19 left by rail for Jackson, Tennessee. A detachment of 200 men from the Regiment who were sick, convalescent or serving as guards were left behind in Holly Springs. On December 20, twenty-two regiments of Van Dorn's cavalry attacked Holly Springs; the town together with its defenders and their stores surrendered without resistance. One hundred and seventy men, including the Major and three Lieutenants, of the Sixty-second Illinois Infantry were taken prisoners and paroled and the records, papers and equipage of the Regimental detachment destroyed. Regarding the Union defenders at Holly Springs, General Grant issued an order reflecting on the disgraceful surrender, lack of resistance and the fact that the prisoners had taken parole. On December 29 that portion of the Sixty-second Infantry Regiment that had moved to Jackson, left Jackson, and pursued Forrest to Clifton, on the Tennessee River, then returned to Jackson where it remained until June 4, 1863. In May, June and July the Sixty-second was at LaVergne, Tennessee on scouting and guard duty. August 19 the Regiment was ordered to Memphis and by the 28th was on route toward Little Rock, Arkansas. En-route they skirmished with the enemy forces before driving them from Little Rock on September 10. January 9, 1864, the men re-enlisted as veterans and two companies were sent home on veteran furlough. The remainder of the Regiment remained at Little Rock. In April the combined Regiment was ordered to Pine Bluff, Arkansas and remained there until August 1864[686] when eight companies were sent home on Veteran furlough. The Sixty-second remained at Pine Bluff and in April 1865 when the non-veterans of the Regiment were mustered out the remaining veterans and recruits were consolidated into seven companies; they remained on duty at Pine Bluff until July 28, 1865. On that date they moved by river to Fort Gibson.[687] The Regiment returned to Little Rock, Arkansas for muster out on March 6, 1866.[688,689]

The records of the State of Illinois list one national flag and one regimental flag in the holdings of the Illinois flag collection. However, there are one national and two regimental flags.

---

[685] Colonel James Milton True was born in Kentucky in October 1823. He first served as a Captain in 38th Illinois Volunteer Infantry and then as Colonel of the 62nd IIllinois Infantry regiment.

[686] "On yesterday Gen. Forrest sent to Richmond eight guidons and three regimental flags captured in his late expeditions—the flag of the regiment taken at Union City, another at Paducah, and another at Fort Pillow. One of these is the banner of the 62nd Illinois, very large and woven in one solid piece of heavy silk. A golden eagle surmounted the staff." Memphis Daily Appeal (Memphis, Tennessee)Tuesday, May 31, 1864, Pag

[687] Fort Gibson was in the Cherokee Nation Indian Territory near the confluence of the Arkansas and Neosho Rivers. This is just north of present-day Fort Gibson, Oklahoma.

[688] Eddy, T.M., *The Patriotism of Illinois,* Clarke & Co., Chicago, 1866, Volume II, 602-4

[689] *Report of the Adjutant General of the State Of Illinois,* H.W. Rokker, State Publisher and Binder, Springfield, Ill., 1886, Volume IV, page 268-8.

Inscribed on this flag: 62nd Regt. Ills. Vet. Infantry. There are no battle honors on the flag. Since there are 35 gold stars, the flag dates from late in the War. The flag measures 70" on the hoist by 75" on the fly.

This regimental flag contains 34 stars, and it appears that a portion of the flag has been cut off. Based on the shape of the upper scroll, the position and arrangement of the stars and the shape of the lower scroll, this flag is typical of those manufactured by Evans and Hassall of Philadelphia.

62ND ILLINOIS INFANTRY

The pattern of this flag is typical of flags manufactured at the New York depot. The flag shows very little wear and was probably issued late in the War.

No other flags carried by this Regiment are known to survive.

63rd Illinois Infantry

Go look upon the battlefield,
Where shot and shell fly fast
Where freedom's stirring battle cry
Is heard upon the blast;
Go where the lifted sabres flash
And fall on traitor crests;
Where Southern bayonets are dim
With blood from Northern breasts;[690]

The Sixty-third Illinois Infantry was organized at Camp DuBois at Anna, in Union County, Illinois. Along with the 54th, 60th and 62nd Illinois Infantry Regiments it was part of the Kentucky Brigade. The men of Companies A, C, E and I served from Richland County; Company B from Edwards County; Company D from McLean County; Company F from Perry and Franklin counties; Company G from Crawford County; Company H from Macon County; and Company K from Coles County. The Regiment was mustered in for three years' service on April 10, 1862, under the command of Colonel Francis Moro,[691] and ordered to Cairo, Illinois on April 27. The Regiment moved to Henderson Kentucky on July 12 to defend against guerilla attack. In August the Sixty-third moved to Jackson, Tennessee and in November to LaGrange, Tennessee. On November 28, as part of Grant's Mississippi campaign they struck Pemberton's forces causing the rebels to retreat to Grenada, Mississippi. However, the fall of Holly Springs, Mississippi to the enemy forced abandonment of Grant's campaign and the Regiment returned to LaGrange on January 10, 1863 and remained there until May. As part of the Vicksburg campaign, on May 10 the Regiment was ordered to Young's Point, Louisiana for picket duty. Following a march south the Sixty-third crossed the Mississippi River at Warrenton, Mississippi and on May 24 silenced the guns in two of the Confederate batteries in the Vicksburg line of defense. As part of the siege of Vicksburg, the Regiment moved to Milliken's Bend in June and on June 16 participated in the capture and burning of Richmond, Louisiana. Following the fall of Vicksburg on July 4, the Sixty-third moved to Helena, Arkansas in September then on to Memphis and finally Chattanooga -- arriving there November 20. They fought in the battle of Mission Ridge November 23 and 24 and pursued Bragg's forces to Ringgold. Georgia. On December 26 the Regiment went into winter quarters at Huntsville, Alabama. On January 1, 1864, two hundred and seventy-two men of the Regiment re-enlisted in the veteran service and in April left for home on furlough. In May the veterans rejoined the Regiment at Huntsville, Alabama and later in May they were ordered to Triana (Triune), Mississippi on the Tennessee River. On June 22 they moved to Kingston, Georgia to guard the railroad running between Chattanooga and Atlanta. The Sixty-third joined Sherman's "march to the sea" on November 11, 1864, marching from Atlanta to Savannah in time to join in the siege of Savannah. In 1865 the Sixty-third fought in the Carolina campaign and saw action at Salkehatchie Swamp, North and South Edisto Rivers and Columbia, South Carolina and Bentonville, North Carolina. Following Johnston's surrender the Sixty-third marched to Washington, D.C. via Richmond for the Grand Review on May 24. Following the Grand Review, the

[690] Portion of an anti-Copperhead poem published in the *Mattoon Independent Gazetter,* Vol VI, No.32, August 26, 1863.
[691] Colonel Francis Moro was born in New York in September 1824. When he enlisted in the Union Army he was 38 years old and a physician living in Olney Illinois

Regiment was sent to Louisville, Kentucky and was mustered out July 13, 1865.[692,693]

The collection of Civil War flags in Springfield, Illinois has only one flag from the Sixty-third Illinois Infantry—a national flag. There is an inscription on a silver plate attached to the staff of this flag that reads: "Presented by the citizens of Richland County to 63d Illinois Infantry." The oval star pattern is typical of flags manufactured in Philadelphia and the inscriptions are all of similar lettering suggesting that all the inscriptions were added at the same time, either late in the War in time for the Grand Review, or after the War.

The inscriptions on this flag read; SIEGE OF VICKSBURG, MISS., RICHLAND, LA., JUNE 16, 1863: MISSION RIDGE, TENN., NOV 25, 1863; ATLANTA, AUG. 31, 1864; 63rd REGt. ILLs VET. VOL. INFT.; ALLATOONA, OCT 5, 1864; SAVANNAH, DEC. 21, 1864; SALKAHATCHIE, S.C.,FEB. 6,1865; COLUMBIA, FEB. 17,1865; BENTONVILLE,S.C., MARCH 21, 1865. *The shape of the canton and the arrangement of the stars in the canton identify this flag as having been produced by a manufacturer in Philadelphia.*

No other flags carried by this Regiment are known to survive.

---

[692] Eddy, T.M., *The Patriotism of Illinois,* Clarke & Co., Chicago, 1866, Volume II, p 256-8

[693] *Report of the Adjutant General of the State of Illinois*, H.W. Rokker, State Printer and Binder, Springfield, Illinois, 1886, Volume IV, pages 307-8.

64th Illinois Infantry - "*Yates' Sharpshooters*"

There is Yates' Sharpshooters,
Now stationed at Glendale
To do their share of fighting,
They are men that never fail;
The glorious name that they have won
Will last for countless years,
They fought and died like heroes
With the Western Volunteers.[694]

Since there were not the required ten companies to form a regiment, the first battalion of this regiment (Companies A, B, C, D, E and F) was mustered at Camp Butler on December 31, 1861, as "*First Battalion of Yates' Sharpshooters*." [695] The men of Company A were mainly from LaSalle County with some from McDonough County; Company B from Bureau County; Company C from Sangamon and Will counties; Company D from Knox and McDonough counties and Companies E and F from Will County. On January 10,1862 the Battalion was ordered to Quincy, Illinois and here it was armed. February 16 the Battalion moved to Cairo, Illinois and in March via Bird's Point, Missouri moved to New Madrid, Missouri, a rebel fort on the Mississippi River. *Yates' Sharpshooters* were selected to lead the attack on New Madrid and began skirmishing on the evening of March 12. On the 13th they supported the Union siege guns and awoke on March 14 to find that the enemy had evacuated the fort and retreated across the river. The battalion took part in the siege and surrender of Island No.10 and three companies of the battalion assisted the escort of rebel prisoners to Chicago while the other three companies joined Pope's expedition downriver to Fort Pillow on the Mississippi River in Tennessee. Action was not pressed against Fort Pillow and the Battalion moved back up the Mississippi River and up the Tennessee River, disembarking at Hamburg Landing. By then the Battalion had been rejoined by the other three companies and in May they were engaged in the siege of Corinth, with their first battle at Chambers' Creek near Farmington, Mississippi on May 3. On May 30, Corinth was evacuated, the Battalion entered the deserted town and found in addition to camp equipage, two flags belonging to "*McCullough's Avengers*."[696] The Battalion pursued the retreating enemy to Booneville, Mississippi, skirmished at Tuscumbia Creek and returned to camp at Big Springs, six miles from Corinth. In July 1862 General Rosecrans succeeded Pope in command of the Army of the Mississippi and *Yates' Sharpshooters* battalion was chosen as Rosecrans' headquarters and provost guard. In September the Battalion moved to Iuka, Mississippi and engaged the enemy on the evening of September 19. The rebel forces retreated, and the Battalion returned to Corinth. On October 3 enemy forces under Price attacked the Union forces

---

[694] THE WESTERN VOLUNTEERS, this is a verse from a poem written by a soldier when Colonel John Morrill then was in command of the Regiment at the Battle of Corinth and appeared in "*The Hartford Call*" (Hartford, Kansas) November 19, 1880, page 2.

[695] Lieutenant Colonel David E. Williams commanded the regiment from December 31,1861 until his departure on sick leave May 17, 1862. Command of the Regiment devolved to Majot Fred W. Matteson and Matteson commanded the Regiient until his death on April 8,1862. Colonel John Morrill then took command.

[696] Woodruff, George H., *Fifteen Years ago: or the Patriotism of Will County,* Joliet Republican Book and Job Steam Printing House, Joliet, 1876, page 191. Note. These flags probably belonged to Company H of the 10th Mississippi Cavalry—*McCullough Avengers.* The flags are not known to have survived.

at Corinth. The Battalion met the first advance of the enemy and was heavily engaged losing, 73 men killed or wounded out of 233 men who entered the battle, but Union forces took 2000 prisoners and captured Price's supply train. In November the Battalion was ordered to Glendale, Mississippi where it remained for the next year on outpost duty, hunting guerrillas, scouting, and erecting fortifications. On November 4, 1863, *Yates' Sharpshooters* moved to Pulaski, Tennessee and joined Sherman's Army. In January 1864, three quarters of the men re-enlisted and left for home on veteran furlough. During the furlough recruiting replenished the existing companies and added four new companies—Companies G and I from Will County, Company H from Edgar County and Company K from Wabash County. The regiment now consisted of ten full companies and was designated the Sixty-fourth Illinois Infantry Regiment. The Regiment left Ottawa, Illinois and arrived in Decatur, Alabama on March 23, 1864. On May 14 the Sixty-fourth arrived in Chattanooga and entered Sherman's Atlanta Campaign fighting at Resaca, Dallas, New Hope Church, Allatoona Hills and Kennesaw Mountain, where the flags of the Sixty-fourth were the first flags planted on the rebel works. They advanced to fight at the Chattahoochee River and Nance's Creek and on July 22, near Decatur, Georgia the Regiment captured forty prisoners and a battle flag (regiment not identified). During the fighting the flag of the Sixty-fourth was pierced by eighteen musket balls and a bayonet thrust. For the next month the Regiment was engaged in the Siege of Atlanta; then in a flanking movement fought at Jonesboro-August 31-September 1 and Lovejoy Station-September 2-6. After pursuit of Hood into Alabama they joined the "March to the Sea." In December 1864 the Regiment fought in the siege of Savannah and in 1865 marched through the Carolinas with actions at Salkehatchie Swamps, Rivers' Bridge, South Edisto River, North Edisto River, Bentonville, Goldsboro, and Raleigh. They marched in the Grand Review in Washington DC on May 24 and were mustered out July 11, 1865, at Louisville, Kentucky.[697,698]

There are currently two flags identified with the Sixty-fourth Infantry in the collection of the State of Illinois Military Museum. The original inventory[699] also states that one of the flags is inscribed as follows: "Presented by the Governor Richard Yates. New Madrid, Island 10, Siege of Corinth, Chambers' Creek, May 3, 1862; Tuscumbia Creek, Iuka, Sept 19; Corinth, Glendale, Dallas, Nickajack, Snake Creek Gap, Resaca, Big Shanty, Kennesaw Mountain, Siege of Atlanta, July 21, 22, 28; Jonesboro, Savannah." Unfortunately no flag with these inscriptions is now present in the Museum collection.

---

697 Ibid, Woodruff

698 Eddy, T.M., D. D., *The Patriotism of Illinois,* Clarke & Co., Publishers, Chicago, 1866, pages 258-61.

699 *Report of the Adjutant General of the State of Illinois,* HW Rokker, State Printer and Binder, Springfield, Illinois, 1886, Volume 1, page 161,

This 34-star national flag carries no inscriptions. The rectangular canton with seven horizontal rows of stars is typical flags manufactured by Longley & Brother of Cincinnati.

Unfortunately the central portion of this34 star regimental flag is missing. However, the inscription that remains and reads "PRESENTED by the GOVERNOR, RICHARD YATES 1st Battalion, Yates Sharp Shooters. ILL. VOL. It is not inscribed with battle honors as noted in the Adjutant General's records.

No other flags known to have been carried by the 64th Illinois Infantry are known to survive.

65th Illinois Infantry—"*Second Scotch Regiment*"

The Regiment was also known as the *Second Scotch Regiment* or *Cameron's Highlanders* and was organized at Camp Douglas by Colonel Daniel Cameron. Cameron was born in Scotland, an ardent Democrat and a close friend of Stephen Douglas. Many of the Regiment's volunteers were of Scotch descent. The men of Company A (*Bruce Guards*) were from Henderson, Ogle and Stark counties; Company B (Montgo*mery Guards*) from Mercer and Rock Island counties; Company C (*Wallace Guards*) from Marshall County; Company D (*Stanton Rifles*) consisted mainly of men from Michigan; Company E (*Cameron Guards)* from Boone and McDonough counties; Company F (*Fuller Guards*) from Lake County; Company G (*German Rifles*) and H (*Glengarry Guards)* from Cook County; Company I from the State at large and Company K from Cook and Winnebago counties. The Regiment was mustered in for three years' service May 1, 1862, and ordered to Martinsburg, Virginia (now West Virginia). The Regiment remained there until September 1862 and saw action at Darkesville, Virginia on September 7. The Regiment moved to Harper's Ferry on September 12 and was laid siege by Lee's army September 12-15. The Eighth New York and the Twelfth Illinois Cavalry escaped but 12,000 Union troops, including the Sixty-fifth Illinois Infantry surrendered. The Regiment was paroled the following day and sent to Chicago where it remained until April 1863 when it was exchanged. The remainder of 1863 the Regiment served in eastern Kentucky and Tennessee and fought in the defense of Knoxville in November. The Sixty-fifth re-enlisted as veterans and went home of furlough in March 1864. Upon return from furlough the Regiment joined Sherman's Army in the Atlanta campaign. They subsequently fought at Kennesaw and Lost Mountains, Muddy Creek, Noyes Creek, Nickajack Creek, Chattahoochee River and the siege of Atlanta. In the flanking movement around Atlanta the Sixty-fifth fought at Jonesboro August 31-September 1 and at Lovejoy Station September 2-6. Following pursuit of Hood into northern Georgia and Alabama the Regiment moved to Nashville by rail in November then from Nashville to Pulaski, Tennessee. The Sixty-fifth was attacked by Hood and driven back to Columbia, Tennessee where during the battle on November 25 and 26 they lost 3 officers and 50 men killed or wounded. On November 30 they again met Hood at Franklin, Tennessee where they defeated the rebel forces and captured the colors of the Fifteenth Mississippi Infantry. Hood's army fell back to Nashville and December 15 and 16 the Sixty-fifth fought in the battle of Nashville and pursued the retreating rebels to Clifton, Tennessee. On January 15, 1865, the Regiment was sent to Fort Fisher, North Carolina. After engagements at Cape Fear and Fort Anderson they captured Wilmington, North Carolina on February 22. They moved to occupy Goldsboro, North Carolina and Raleigh in March and April. On March 6, 1865, the Companies A-E, except the veterans, were sent to Chicago for muster out. Following the surrender of Johnston's Confederate Army, the remainder of non-veterans were ordered home for muster out. The veterans marched to Goldsboro, North Carolina where they went into camp. May 1, 1865, four new companies of recruits joined the veterans of Sixty-fifth Infantry. On June 4, officers, and men from the Ninety-second, One Hundred Twelfth and One Hundred Seventh Illinois Infantries were assigned to the Sixty-fifth Illinois Infantry. July 13, 1865, the Regiment was mustered out.[700]

There is one national and one regimental flag from the Sixty-fifth Illinois Infantry Regiment in the collection of the State of Illinois Military Museum. Adjutant General Fuller presented a stand of colors to the Second Scotch Regiment in June 1862.[701] Presumably those colors were captured with the Regiment at Harper's Ferry in September. Therefore, the colors shown here must have been presented

---

[700] *1Report of the Adjutant General of the State of Illinois,* H.W. Rokkker, Printer and Binder, Springfield, Ill, 1886, Volume IV, page 307-8.
[701] *Chicago Daily Tribune,* June 9, 1862.

to the Sixty-fifth after that time and are presumably the second set of colors given to the Regiment.

Only portions of a few of the stripes remain from the national flag.

The regimental flag is in poor condition. No inscriptions are seen on this flag.

on.

No other flags carried by the 65th Illinois Infantry are known to survive.

66th Illinois Infantry – *"Birge's Sharpshooters"*

"Our flag still floats on Georgia's shore,
And may it ever wave—
Rebels, if you touch a single star
You will find a speedy grave.
We will fight with noble Sherman,
We will follow night and day—
We will plant our noble banner,
Over many a Rebel gray."[702]

The Sixty-sixth Illinois Infantry was organized at Benton Barracks in St. Louis in September and October of 1861. It was originally designated as "*Western Sharpshooters*" or *"Birge's Sharpshooters*" and the men were to be used as skirmishers. Nine companies comprised the original regiment- Company A from Missouri with added men from Illinois, Wisconsin, Minnesota and Indiana; Company B- from Missouri with added men from Illinois, Iowa, Minnesota; Company C from Bureau and Logan counties in Illinois with added men from Iowa; Company D (*Michigan Sharpshooters)* from Michigan; Company E from Edgar County Illinois; Company F from Missouri with added men from Michigan and Illinois; Companies G and H from Ohio; Company I from Lawrence County Illinois and Company K from Ohio (added in April 1862). The Regiment was mustered into the service as the Fourteenth Missouri Volunteer Infantry on November 23, 1861, with John W. Birge as Colonel. It was Fremont's idea to have a sharpshooting regiment; therefore, each man was supplied with a Dimick American Deer and Target rifle.[703] On December 12, 1861, the Regiment moved by rail from the Barracks to Centralia, Missouri. After minor skirmishing with rebel forces in northeast Missouri, on December 28, five companies of *Birge's Sharpshooters* with five companies from the Third Missouri Cavalry routed a much larger rebel force at Mount Zion Church in Boone County. They then marched twenty miles to Sturgeon, Missouri and remained there until February 4, 1862 when they moved by rail to St. Louis and then by steamer to Fort Henry. They arrived following the fall of Fort Henry but marched on to the siege at Fort Donelson. During the siege they fought by hiding themselves along the entire Union front line and silenced rebel batteries with long range shooting. February 16 the Fort surrendered, and Musician R. F. Larimer of Company E captured the flag of the 18th Tennessee Infantry. In March the Regiment was at Pittsburg Landing, Tennessee and on April 6 and 7 fought in the battle of Shiloh. At the end of April, the Regiment was in the Siege of Corinth; on May 21 they fought at the battle of Phillips Creek and on May 30 entered Corinth. June 23 the Regiment was detailed for guard duty at Grant's headquarters in Corinth. In September the Regiment fought at Iuka, Mississippi, returned to Corinth and on October 3 and 4 fought in the battles at Whitehouse and at Corinth. They pursued the rebels to the Hatchie River, skirmished with them and returned to Rienzi, Mississippi. On November 20, 1862, by order of Secretary of War Stanton the Regimental designation was changed to the Sixty-sixth Regiment

[702] Barker, Lorenzo A. *Military History (Michigan Boys) Company "D" 66th Illinois.* Originally published 1905, New Edition by Blue Acorn Press, Huntington, WV, 1994, page 33. Portion of a poem by Corporal James Burton, Company D, Sixty-sixth Illinois Infantry.

[703] The Dimick was a rifled percussion rifle with an octagonal barrel and was manufactured in St. Louis. The men also carried a bullet-pouch with bearskin covering, a powder horn, bullet mold and patch cutter. They wore the usual Union uniform but with a gray sugar-loaf shaped hat with three squirrel tails attached.

Illinois Infantry. They marched to Tuscumbia Hills, Mississippi and established a stockaded camp (Camp Davies) as an outpost to the garrison at Corinth.[704] While at Camp Davies they skirmished at Tuscumbia Bridge and Hatchie Bridge in December 1862 and at Danville, Ripley Cross Roads, Booneville, Glendale, Jumpertown, Kossuth, Cartersville, Yellow Creek, Seward House and Whiteside's Farm between January and September 1863. In November the Sixty-sixth marched to Pulaski, Tennessee and in December 470 men re-enlisted as veterans and in January 1864 left on veteran furlough. They returned to Pulaski in March and moved by rail to Chattanooga where they joined Sherman's Atlanta campaign in May. Over the next four months they fought at Snake Creek Gap, Resaca, Ley's Ferry, Rome Cross Roads, Dallas, New Hope Church, Allatoona Hills, Kennesaw Mountain, Nickajack Creek, Ruffs Mills, Decatur, and Atlanta. After the non-veterans were mustered out on August 26, 1864, the Regiment fought in the battles at Jonesboro and Lovejoy Station then marched to Rome where they remained until November. For the remainder of the War, they were in the "March to the Sea." In December 1864 they fought at Savannah, Georgia. In 1865 in the Carolina campaign the Regiment fought at Salkehatchie Swamp, the South and North Edisto Rivers and Columbia, South Carolina and Bentonville, North Carolina. Following the surrender of Johnston and his army they marched to Washington D. C. via Richmond and marched in the Grand Review on May 24, 1865. The Sixty-sixth mustered out at Louisville, Kentucky July 7, 1865.[705]

There is a national flag and a regimental flag of the Sixty-sixth Illinois Infantry in the collection of the State of Illinois Military Museum. Neither of the flags bears the Regiment's original designation—14th Missouri Volunteer Infantry. However, the inscriptions on the national flag include battles that were fought when the Regiment had that designation. During the Battle of Atlanta, the colors of the Sixty-sixth were said to have had 65 bullet holes through them.[706] Neither of the flags shows significant battle damage, therefore these flags were probably not those carried in the battle of Atlanta.

The oval pattern of the stars on the national flag is typical of flags manufactured in Philadelphia. The flag bears 35 stars, which places its manufacture late in the War. Note that the inscriptions are all in similar print, which suggests that the inscriptions were all placed on the flag at the same time either late, or after the War. it measures 71" on the hoist by 72" on the fly.

---

[704] It was at this time that two hundred of the men replaced their Dimick rifles with Henry repeating rifles at their own expense--$43.00 each.

[705] Barker, Lorenzo A. *Military History (Michigan Boys) Company "D" 66th Illinois.* Originally published 1905, New Edition by Blue Acorn Press, Huntington, WV, 1994

[706] *Report of the Adjutant General of the State of Illinois,* Volume IV, page 424.

66TH ILLINOIS INFANTRY

Inscriptions on this flag read: Mt ZION, Ft DONELSON, SHILOH, SIEGE of CORINTH, IUKA, CORINTH, SNAKE GAP, LAYS FERRY, ROME X ROADS, DALLAS, KENNESAW, ROUGHS MILLS, ATLANTA 22 & 28 JULY, JONESBORO, LOVEJOY, EDEN X ROADS, SAVANNAH, BENTONSVILLE.

The pattern of this flag with the gold eagle and the rays above the thirteen stars over the eagle is typical of flags manufactured in Chicago by Gilbert Hubbard & Co. It appears that portions of the flag have been cut off, perhaps by the men as souvenirs. Unfortunately, the Regimental designation has been removed. It measures 68" on the hoist by 69" on the fly.

No other flags carried by the 66th Illinois Infantry are known to survive.

67th Illinois Infantry

In May 1862 a call went out to the governors of the northern states for combat ready troops to meet a rumored attack on Washington, D.C. by Confederate forces. The Sixty-seventh Illinois Infantry was one of five regiments raised for three months service to meet this threat. The Sixty-seventh was organized at Camp Douglas in Chicago on June 13, 1862, under command of Colonel Rosell M Hough. The men volunteered from Company A --Winnebago County (*Rockford City Guards);* Company B-- Cook County (*Corcoran Guards)*; Company C --Winnebago County (*Winnebago County Guards)*; Company D --Cook County (*Anderson Rifles)*; Company E-- Cook County (*Ellsworth Zouaves)*; Company F-- Fulton County (*Canton Guards);* Company G-- Tazewell and Peoria counties (*Peoria Guards);* Company H-- Stephenson County (*Washington Light Infantry*); Company I-- Cook County (*Fuller Guards)* and Company K-- Champaign County.[707] The Sixty-seventh spent its three months of service on guard duty at Camp Douglas and was mustered out October 6, 1862.[708]

There are no flags of this Regiment in the collection of the State of Illinois and no record of any flags being presented to this Regiment.

---

[707] *Chicago Daily Tribune,* May 27, June 5 and June 11, 1862

[708] *Report of the Adjutant General of the State of Illinois,* HW Rokker, State Printer and Binder, Springfield, IL, 1886, Volume IV, page 444.

68th Illinois Infantry

In May 1862 a call went out to the governors of the northern states for combat ready troops to meet a rumored attack on Washington, D.C. by Confederate forces. The Sixty-eighth Illinois Infantry was one of five regiments raised for three months service to meet this threat. The Regiment was organized at Camp Butler as a state militia but the officers and men petitioned that the enlistment be in an Illinois Volunteer regiment; therefore, the men were mustered in as the Sixty-eighth Illinois Infantry on June 16, 1862 under the command of Colonel Elias Stewart The volunteers served from Morgan County-Company A, Pike County-Company B, Coles County- Company C, Fayette County-Company D, De Witt County- Company E, McLean County- Companies F and G, Logan County- Company H, Macon County- Company I and Tazewell County-Company K. Following muster in the men were sent to Wheeling, Virginia (now West Virginia) in July and after two days moved to Washington, D.C. The Regiment remained in camp at the "Soldiers Retreat" until August 24 when it was ordered into Alexandria for Provost duty. On September 17 the Sixty-eighth returned to Camp Butler and was mustered out September 27, 1862. The Regiment was never in battle and there is no record of colors being issued to the Regiment. [709]

---

[709] *Report of the Adjutant General of the State of Illinois,* HW Rokker, State Printer and Binder, Springfield, IL, 1886, Volume IV, page 462.

69th Illinois Infantry

In May 1862 a call went out to the governors of the northern states for troops to meet a rumored attack on Washington, D.C. The Sixty-ninth Illinois Infantry was one of five regiments raised for three months service to meet this threat. The Sixty-ninth was organized at Camp Douglas in Chicago on June 14, 1862 under command of Colonel Joseph H. Tucker. The men volunteered from the following counties: Company A-Bureau County, Company B-Kane County, Companies C and E-Cook County, Company D-Henry, Knox and Stark counties, Company F-Rock Island County, Company G-Livingston County (*Dixie Guards)*, Company H-Lee County, Company I-Grundy County, and Company K-Lee and Ogle counties. The Regiment remained on guard duty at Camp Douglas for the three months of its term and was mustered out on September 27, 1862.[710]

The Regiment was presented at least one flag.[711] "PRESENTATION OF COLORS TO THE DIXIE GUARDS--…. Mr. D. McWilliams, of Dwight, Livingston County presented them a magnificent flag, accompanying the presentation with a few well-timed remarks." The *Dixie Guards* were Company G of the Sixty-ninth Illinois Infantry. The whereabouts of this flag today is unknown.

No other flags carried by this regiment are known to survive.

[710] *Report of the Adjutant General of the State of Illinois,* HW Rokker, State Printer and Binder, Springfield, IL, 1886, Volume IV, page 481.
[711] *Chicago Daily Tribune,* June 12, 1862.

70th Illinois Infantry

In May 1862 a call went out to the governors of the northern states for troops to meet a rumored attack on Washington, D.C. The Seventieth Illinois Infantry was one of five regiments raised for three months service to meet this threat. The Seventieth was organized at Camp Butler and was mustered in for three months on July 4, 1862, under the command of Colonel Owen T. Reeves. The men of the Regiment volunteered from Douglas County -Company A, Fayette, Macon, and Sangamon counties- Company B, Montgomery County- Company C, McDonough and Fulton counties- Company D, Jefferson, Sangamon and Macon counties- Company E, Sangamon, St. Clair and Morgan counties- Company F-, Clark and Edgar counties- Company G, McLean, Woodford and Logan counties- Company H, Greene and Pike counties- Company I and Lawrence County- Company K. After serving for three months on guard duty at Camp Butler, the Regiment was mustered out on October 23, 1862.[712]

There is no record of any colors being presented to this regiment and there are no flags of this Regiment held by the State of Illinois.

[712] *Report of the Adjutant General of the State of Illinois,* HW Rokker, State Printer and Binder, Springfield, IL, 1886, Volume IV, page 499.

71st Illinois Infantry

In May 1862 a call went out to the governors of the northern states for troops to meet a rumored attack on Washington, D.C. by Confederate forces. The Seventy-first Illinois Infantry was one of five regiments raised for three months service to meet this threat. The seventy-first was organized at Camp Douglas in Chicago and mustered in for three months service on July 26, 1862, under command of Colonel Othniel Gilbert. The men of the Regiment volunteered from Vermillion County for Company A; Carroll and Stephenson counties for Company B; Champaign County for Companies C, F and H; Fulton, Henry, Livingston and Rock Island counties for Company D, Hancock, Knox and Peoria counties for Company E; Edgar and Menard counties for Company G; Effingham County for Company I; Perry, Franklin, and Hamilton counties for Company K. The Regiment was transported to Cairo, Illinois and left two companies to guard the Big Muddy Bridge on the *Illinois Central Railroad* north of Cairo. The remaining eight companies were detached to guard railroads and bridges: two companies at Mound City, Illinois, three companies at Moscow, Kentucky and three companies at Little Obion bridge in Kentucky. After completion of three months of service the Regiment returned to Chicago and was mustered out on October 29, 1862.[713] There is no record of the Regiment receiving a set of colors.

---

[713] *Report of the Adjutant General of the State of Illinois,* HW Rokker, State Printer and Binder, Springfield, IL, 1886, Volume IV, page 517.

72nd Illinois Infantry – *"First Chicago Board of Trade Regiment"*

A day long to be remembered by those who participated in the events I now write about…Up we started and rushed ahead with a yell and were greeted with a most wondrous volley. Our colors were planted about fifteen feet from the ditch, but we could not go forward, the fire was too severe, men could not live; we laid down and only the wounded fell back…. Our loss in twenty minutes was one hundred and ten killed and wounded. Such was the 22nd of May 1863."[714]

The Chicago Board of Trade and the Young Men's Christian Association raised the Seventy-second Illinois Infantry in Chicago. The volunteers served from the following counties: Company A (*Hancock Guards*) from Cook County; Company B (*Havelock Guards*) from Cook County with some men from the city of Ottawa; Company C (*Scripps Guard*) from Cook County; Company D (*Fuller Guards)*- Cook County; Company E - Cook County; Company F (*Shephard Guards*) - Cook County; Company G (*Brainard Guards*) from Champaign and McDonough counties; Company H from Cook County; Company I from Fulton County and Company K (*Wicker Guards*) from Cook County. The Regiment mustered in for three years of service on August 23, 1862, under the command of Colonel Frederick Augustus Starring[715] and left the following day for Cairo, Illinois. In September they were ordered to Paducah, Kentucky for post duty then ordered to Columbus, Kentucky for guard and picket duty. In January 1863 the Regiment was ordered to Memphis for provost guard duty and in March the Seventy-second marched toward Vicksburg. On May 16 the Regiment arrived at the battlefield of Champion Hills and engaged the enemy. On May 17 they helped rebuild the bridge over the Black River and beginning on May 22 took part in the assault on Vicksburg. Following the fall of Vicksburg, on July 12, the Regiment was transported downriver to Natchez, Mississippi where they were part of the force that took possession of Natchez. They remained as provost guard in Natchez until late October 1863 when they returned to Vicksburg and acted as provost guard at Vicksburg until October 30, 1864. The Regiment was ordered to join Sherman's army at Nashville but arrived too late to join the "March to the Sea," therefore they were ordered to Columbia, Tennessee. Here they were attacked by Hood's army and in the retreat from Columbia, the Seventy-second fought first at Spring Hill and then in the bloody battle at Franklin. In that battle the Regiment lost almost one half of the 325 men who fought. The Union continued to retreat toward Nashville and met and defeated the rebels on December 15 and 16, 1864. The Union forces pursued the Confederate troops as far as Clifton, Tennessee. From Clifton the Seventy-second was transported to Eastport, Tennessee where it remained until mid-February 1865 when it was ordered to New Orleans. In March the Regiment was transported across the Gulf of Mexico to Dauphine Island, Alabama south of Mobile Bay. The next day they crossed to the mainland on the western side of Mobile Bay. On March 27 they attacked Spanish Fort on the eastern side of Mobile Bay and following its surrender moved south to support the capture of Fort Blakely. In April, after Lee's surrender, they marched to Montgomery then in May to Union Springs, Alabama. In July they were ordered to Vicksburg where the veterans were mustered out on August 6, 1865. The remaining 270 non-

---

[714] *War Diary of Brevet Brigadier General Joseph Stockton.* Printed by John T. Stockton, Chicago, 1910, page16. The author was the Lieutenant Colonel of the 72nd Illinois and relates his observations of the ill-fated attack at Vicksburg.

[715] Colonel Starring was born in Buffalo, New York on May 24, 1834. He was educated at Harvard as well as in Paris, Heidelberg and Vienna. He was admitted to the New York bar in 1859 and prior to the civil war he was employed by *the Illinois Central Railroad* and the *Cairo & Fulton Railroad.*

veterans were transferred to the Thirty-third Illinois Infantry.[716,717]

There are no flags of the Seventy-second Illinois Infantry in the collection of the State of Illinois. The flags of the Regiment were deposited at the end of the Civil War at the Chicago Historical Society and were lost in the Chicago Fire of 1871. However, there is record of two sets of colors being presented to the Regiment.

Gilbert Hubbard, & Co., a Chicago flag manufacturer, offered a stand of colors to the first company of the Regiment that filled its ranks. Company E was the first full company, and the flags were presented to Captain W. B. Holbrook.[718]

The second set of colors was presented to the Regiment before departing Chicago in August 1862. The Regiment was drawn up "preparing to receive the flag sent to them by S. T. Suit, Esq., of Kentucky. This flag consists of the ever glorious stars and stripes, emblazoned on the finest silk, and floating from a staff of polished ash. In the center of the flag are inscribed, in golden letters the words, "Presented by S.T. Suit, of Kentucky to the Board of Trade Regiment of Chicago." Accompanying the colors are two beautiful "guidons" of similar material and proportionate size."[719]

Presumably the Regiment received other colors from the Quartermaster of the Army. However, as explained above none of these colors survived the Chicago fire.

No other flags carried by the 72nd Illinois Infantry are known to survive.

---

716 Stockton, Joseph, ibid

717 *Report of the Adjutant General of the State of Illinois,* Volume IV, pages 546-8.

718 *Chicago Daily Tribune,* August 23, 1862.

719 *Chicago Daily Tribune*, August 25, 1862

73rd Illinois Infantry – *"Preacher's Regiment"*[720]

One by one we are passing away,
Love for the blue, and tears for the gray,
Kindness for all, forgiveness is sweet,
Like the hand clasping, wherever we meet
G.O. Pond, Surgeon 73rd Illinois[721]

The Seventy-third Illinois Infantry is unique in that it was raised and organized and commanded by a Methodist minister, Colonel Rev. James Frazier Jaquess, D.D[722].; in addition, the major, six captains and six lieutenants of the Regiment were Methodist ministers. The men of the Seventy-third were recruited from eleven counties in central Illinois, many of the men from Methodist congregations; Company A from Sangamon County; Company B from Tazewell County; Company C from Vermilion County; Company D from Piatt County; Company E from Vermilion County; Company F from Logan County; Company G from -Schuyler County; Company H from Adams and Pike counties; Company I from Sangamon County and Company K from Jackson and Saline counties.[723] The Regiment was mustered in at Camp Butler on August 21, 1862 with Colonel Jaquess in command. They reached Louisville on August 27 but in a few days were detached from their brigade and sent to Cincinnati because of a threat to the city by Bragg's army. The Regiment returned to Louisville in September and began pursuit of Bragg's retreating forces. On October 8, the Seventy-third saw its first action-- at Perryville, Kentucky (Chaplin Hills) then marched to Nashville where it remained until November. December 26 the Regiment advanced towards Murfreesboro, Tennessee and fought in the Battle of Stones River December 30 to January 3, 1863. They remained in camp near Murfreesboro until June then crossed the Cumberland Mountains and the Tennessee River in the Chickamauga Campaign. The Seventy-third fought in the Battle of Chickamauga September 19-20; of the three hundred and eight men who entered the battle, only seventy-three officers and men remained at the end of battle. On September 24th the Regiment embarked on the siege of Chattanooga and in late November fought at Missionary Ridge—charging to the top of the mountain through the entrenched rebel lines. Within days they were marching to relieve Knoxville, Tennessee and were in operations in East Tennessee until February 1864. In May the Seventy-third became part of the Atlanta campaign and fought at Resaca, Adairsville and Kingston, Georgia. In June they battled at Burnt Hickory, Pine and Lost Mountains, Dallas, New Hope Church, Allatoona Hills and Kennesaw Mountain. On July 6 the Regiment reached the Chattahoochee

---

[720] This Regiment was also known as the *Preacher's Regiment,* the *Methodist Preacher's Regiment* the *Persimmon Regiment,* and the *First Methodist Regiment.*

[721] *Minutes of Proceedings of the Tenth Annual Reunion Survivors Seventy Third Regiment Illinois Infantry Volunteers,* held Tuesday afternoon and evening, September 29, 1896-"Old Soldiers Day" at the State Fair, page 8.

[722] Colonel Jaquess was a Mehodist ministerwho known in central Illinois . He had been president of two Mehodist colleges-- Illinois Conference Female Academy (nowMacMurray College in Jacksonville) and Chaddock College in Quincy. He had an examplary military record during the war. Unfortunately after the war he was tried in Kentucky for murder, failed as a carpetbagger, scammed residents in New York to purchase worthless stock. Burnette, Patricia B., James F. Jaquess: Scholar, Soldier and Private agent for President Lincoln. McFarland, 2013.

[723] "More companies and squads wanted to join the 73rd than could be accommodated.; as a result, two more regiments—the 115th and the 117th—were formed." Newlin, William Henry, *A History of the Seventy-third Regiment of Illinois Infantry Volunteers,* Published by the Regimental Reunion Association of the Survivors of the 73rd Illinois Infantry Volunteers, Springfield, 1890, page 20

River and later in the month fought at Peach Tree Creek and in the siege of Atlanta. In August the Seventy-third skirmished at Jonesboro and Lovejoy Station then pursued Hood and Forest into north Georgia and Alabama. In the Nashville campaign of November and December the Seventy-third fought at Columbia and Duck River on November 24-27, Spring Hill November 29 and in the Battle of Franklin on November 30. December 15 and 16 the Seventy-third fought at Nashville, and then joined in the pursuit of Hood to Pulaski, Tennessee. In 1865 the Regiment was in Alabama and Tennessee and was mustered out at Nashville June 12, 1865[724,725]

The collection of Civil War flags belonging to the State of Illinois contains four national flags but no regimental flags of the Seventy-third Illinois Infantry.

This flag is probably "homemade" based on the unusual star pattern and non-standard size of the flag—the flag measures 56" on the hoist by 90" on the fly. Not all the battles on this flag can be read because of deterioration of the paint and silk. Those which can be read are PERRYVILLE, 73d ILLs INFy VOLs, STONE RIVER, CHICKAMAUGA, MISSION RIDGE, DANDRIDGE, ROCKY FACE, ADAIRSVILLE, DALLAS, LOST MT, KENESAW MT, ATLANTA, JONESBORO. *At the twenty-third reunion of the Seventy-third Infantry, the "Survivors" discussed a flag that the Regiment had carried. "Here followed a discussion of the "Old Flag", and among others was Comrade J. O. Joy, who said that the first flag used as a regimental flag was a beautiful silk flag presented by the Loyal Ladies of Loami. He stated that that flag was now in the country near Loami, and the custodian is W. G. Miller of Company I. The white stripes have disappeared, but the red is pretty well intact, and the stars are in good shape. It had been well cared for, he said, and kept well covered except on Memorial Sunday and Decoration Day, when it was unfurled, although it would be shown to anyone who cared to go to see it, and he wished that the comrades would go and see it again. ...He said there were inscriptions on it as to the battles it had been in."*[726] *This may be the flag referred to since the white stripes are missing, the flag is homemade and there are battle inscriptions.*

---

724 Newlin, ibid.

725 *Report of the Adjutant General of the State of Illinois,* HW Rokker, State Printer and Binder, Springfield, Illinois (1886) pp. 569-71

726 *Minutes of the Proceedings of the Twenty-Third Annual Reunion of the Survivors Seventy-Third Regiment Illinois Volunteer Infantry,* Tuesday Afternoon and Evening, October 5th, 1909, State House, Springfield, Ill. Page 5.

Although this flag is in poor condition and the stars cannot be seen the flagstaff does contain a spade-shaped red painted finial. This is typical of flags manufactured by Hugh Wilkens of Louisville. This flag is inscribed with the Regimental ID, plus the battles STONE RIVER AND PERRYVILLE.

This flag carries the Regimental designation only--73rd REGT. ILL'S VOL'S. *The stars are arranged in six horizontal rows with five stars in the top and bottom rows and six stars in the remaining rows. Also, there is a spade-shaped red painted finial on the top of the flagstaff. These two findings are typical of flags manufactured by Hugh Wilkens of Louisville.*

This national flag is designated as having been carried by the Seventy-third Illinois Infantry. There are no inscriptions on the flag. The oval star pattern is typical of flags manufactured in Philadelphia.

At a Regimental reunion of the Seventy-third Infantry – "Comrade A. E. Lewis stated that he had in his possession a relic….This was a star plucked out of the old flag. He had taken it home and pinned it to his discharge. Upon someone asking if it were from Illinois, Comrade Lewis stated it was from the 73rd, which had put down the Rebellion."[727] It is not clear as to which flag the soldier was referring.

At a reunion of the Seventy-third, Comrade Thomas J. Cassady (Company B) summarized the feelings of the men about their cherished flags. "In a room in this capitol building are treasured not only the records of the Regiment, but a tattered and torn banner, faded and yellow with age, with the words inscribed, "73d Regiment, Illinois." This banner, if it could talk, could tell you of how it was saved at Chickamauga from the enemy by the courage and bravery of one of the men, after the bearer had been killed. It could tell how it was carried up the steep rugged sides of Mission Ridge to its crest, where it floated in defiance of a fleeing foe, and how it was carried through the many battles from Chattanooga to Atlanta and through Tennessee in that wonderful run with Gen. Hood for the capture of Nashville."[728]

In 2005, I was fortunate to be contacted by the Regent of the Nancy Ross chapter of the Pittsfield, Illinois NSDAR regarding a civil war flag. The flag had been carried by Company H of the Seventy-third Illinois Infantry from Pike County and in 1930 was in possession of the only two surviving members of the Company –Pvts Edward Penstone and Edwin McAllister. The two elderly soldiers had decided to "present this one beloved flag to the Nancy Ross Chapter of the D.A.R." On August 20, 1930, the men officially turned over their flag to the Chapter.

A flag was presented to Company H in 1862: "On behalf of the Ladies of Griggsville, who fashioned this flag, I take pleasure in now presenting it to Company H. May it go forth with you in battle—every minute of every day may it remind you of the common love and devotion we share for this great nation. Every stitch of this flag is a constant reminder of our loved ones—our fathers, our husbands, our sons, our friends who have placed the love of country above all else. There is a time for all things—

---

727 *Minutes of the Proceedings of the Twenty-Third Annual Reunion of the Survivors Seventy-Third Regiment Illinois Volunteer Infantry,* Tuesday Afternoon and Evening, October 5th, 1909, State House, Springfield, Ill. Pp 4-5.
728 Ibid, page 15.

the Bible tells us that—and your time is at hand."[729]

The flag measures 56" on the hoist by 85 "on the fly. It is made of linsey-woolsey material and is in remarkedly good condition. The canton is 29" on the hoist and 28" on the fly and contains 34 white appliqued stars. On the obverse is inscribed in gold letters, the battles in which the company fought--STONE RIVER, CHICKAMAUGA, ROCKY FACE RIDGE, RESACA, ADAIRSVILLE, NEW HOPE CHURCH, LOST MOUNTAIN, KENASW (sp) MOUNTAIN, PEACH TREE CREEK, ATLANTA, SPRING HILL, FRANKLIN, NASHVILLE, MISSION RIDGE, JONESBORO, LOVEJOY. On the reverse of the flag are the names of the 21 men from Company H who either died in battle, died of disease, or were wounded and died later. The names are in black ink or paint and the names bleed through to the front of the flag.

After the end of the civil war the flag was brought home and kept by the surviving members of Company H until 1930.

"Pittsfield, Illinois

August 20,1930

This is to certify that we, the last surviving volunteers, Edward Penstone-age 89 years and Edwin McAllister-age 90 years, hereby present this our beloved flag to the Nancy Ross Chapter of the Daughters of the D.A,.R and request that Mrs. Mayme Grigsby be made custodian, until, for any reason that, she may like to turn it over to some other member of the chapter.,

Edward Pcnstonc-Company -H

Edwin McAllister-Company -H

The cedar flag staff was made from a rail taken from the Nashville battle field."[730]

The flag has now been conserved and framed and is on display at the Pittsfield All Wars Museum in Pittsfield, Illinois. It remains the property of the Nancy Ross Chapter of the DAR.

No other flags carried by the 73rd Illinois Infantry are known to survive.

---

[729] Portion of a speech delivered August 20, 1862 by Miss Mary Sutphin on the occasion of the presentation of a flag to Company H.

[730] *The Flag with 34 Stars, Company H, 73rd Regiment Illinois Volunteer Infantry*, Berry, Linda, Sweet Prairie Publishing, Dawson, Illinois, 2010,page 10.

74th Illinois Infantry

"*Resolved,* That the report of the committee be approved, and the original flag and banner be forwarded to Springfield and placed in Memorial Hall."[731]

The Seventy-fourth Illinois Infantry was organized at Camp Fuller in Rockford, Winnebago County during the summer of 1862 and mustered into service September 4, 1862 under the command of Colonel John Marsh.[732] Company A volunteered from Knox County; Companies B, C, D, E, F, H and K (*Y.M.C.A. Company)*volunteered from Winnebago County; Company G (*Dutcher's Roughs)* from Ogle County and Company I (*Rock Run Rangers)*from Stephenson County. On September 28 the Regiment was ordered to report to Louisville, Kentucky. It arrived there on September 30 and left the following morning in pursuit of Bragg. The Seventy-fourth was held in reserve at the battle of Perryville Kentucky and from that battle site marched south to Nashville. The Regiment camped at Nashville until late December when it moved to Murfreesboro, Tennessee. From December 30 through January 2, 1863, the Seventy-fourth fought in the battle of Stones River. Following the retreat of the enemy, the Regiment remained in Murfreesboro until June 24, 1863, when they marched south toward Chattanooga. The route took them through Tullahoma and Winchester, Tennessee through the Cumberland Mountains and across the Tennessee River. During the battle of Chickamauga, the Regiment was guarding a supply train at the rear of Bragg's army and did not participate in the battle. They arrived at Peavine Creek near Chattanooga in late September and learned of the Union defeat at Chickamauga. On November 14, the Seventy-fourth received a new flag from the ladies of Rockford and christened the flag at the battle of Mission Ridge November 25. Following the battle, the men marched to relief of Knoxville and remained in East Tennessee until May 1864 when they joined Sherman's Atlanta campaign. In that campaign, the Seventy-fourth saw action at Tunnel Hill, Rocky Faced Ridge, Buzzard's Roost Gap, Dalton, Resaca, Adairsville, Pumpkin Vine Creek, Dallas, New Hope Church, Allatoona, Hills, Kennesaw Mountain,[733] Lost Mountain, Chattahoochee River, Peach Tree Creek, and the Siege of Atlanta. In the flanking movement around Atlanta, the Regiment fought at Jonesboro August 31-September 1 and Lovejoy Station September 2-6, then pursued Hood's and Forest's retreating forces into north Georgia and Alabama. In October the Regiment moved by rail to Pulaski, Tennessee and in November to Nashville. November 29, 1864, the Seventy-fourth fought at Spring Hill and the following day in the battle at Franklin, Tennessee. December 15 and 16 the Regiment fought in the battles around Nashville, which resulted in the defeat of Hood. In 1865 the Regiment was supplied with recruits bringing its numbers back to 343 officers and men. On June 10, 1865, the Regiment was mustered out of service at Nashville and returned

---

[731] At the third reunion of the Regiment in 1885 a committee was appointed to suggest the proper disposition of the regimental flags and banner still held by the Regiment. At the next reunion in 1886 the Regiment marched behind "three stands of their old colors" and voted to send the original flag and banner to Springfield. *Society of the Seventy-fourth Illinois Volunteer Infantry, Proceedings Fourth Reunion,* September 4, 1886, Oregon, Illinois, page 110

[732] Colonel John Marsh was born in Vermont in1807. He taught school in New York, studied law and passed the New York bar exam and served as postmaster in Adams, New York. In 1839 he moved with his family to Rockford Illinois where he continued practicing law. (Information obtained from "Found A Grave" website)

[733] The Seventy-fourth Infantry left Camp Fuller in Rockford, Illinois with 940 men. After the bloody battle at Kennesaw Mountain the Regiment had an effective force of 127 enlisted men and 11 officers.

to Rockford. [734] [735]

The inventory of flags by the Adjutant General as published in the 1886 report does not list any flags belonging to the Seventy-fourth Illinois Infantry. Therefore, these flags now in the collection and shown here must have been deposited after that date. As noted above at the reunion of the Regiment in 1886 the decision was made to send the "original flag and banner" to Springfield. Therefore, the flags now in the collection must be the flags referred to. Unfortunately, both flags are in very poor condition, probably as a result of the weather and battle damage. However, at that same reunion the Regiment marched with "three stands of their old colors"; based on this statement, other sets of colors must have remained with the Regiment after the War and were not deposited in Springfield or "three stands of colors" may refer to three colors.

Only four red stripes remain from this national color.

[734] The flag(s) of the Seventy-fourth may have been the first flags on Mission Ridge. In an address by Robert Simpson, Company I, given at the Fourth Reunion of the Regiment in 1886, he states "Our flag was a new and beautiful one, recently presented to us by the patriotic ladies of Rockford, and here carried for the first time – the brightest and most conspicuous object among all the Federal troops. It was carried by Sergeant Charles Allen, of Company E, who was severely wounded ere he gained many rods from the foot of the Ridge. It was next taken by Alba Miller of Company C, the regimental bugler, who was shortly after shot through the shoulder, and resigned it into the hands of that noble soldier, Samuel Compton, of company D. By him it was carried forward to within a few yards of the crest, when he fell, shot through the body, dying almost instantly. Here the flag was instantly grasped by Corporal F. Hensey, of Company I, and in a moment placed upon the Confederate works on the crest of the Ridge. So far as personal knowledge goes, we cannot say positively that it was the first one there but can say positively that when comrade Compton fell the flag of the Seventy-fourth was conspicuously in advance of any colors in our immediate vicinity." *Society of the Seventy-fourth Illinois Volunteer Infantry Proceedings Fourth Reunion,* September 4, 1886, Oregon, Illinois, page 127.

[735] Report of the Adjutant General of the State of Illinois, HW Rokker, State Printer and Binder, Springfield, IL, 1886, volume IV, Pages 593-8.

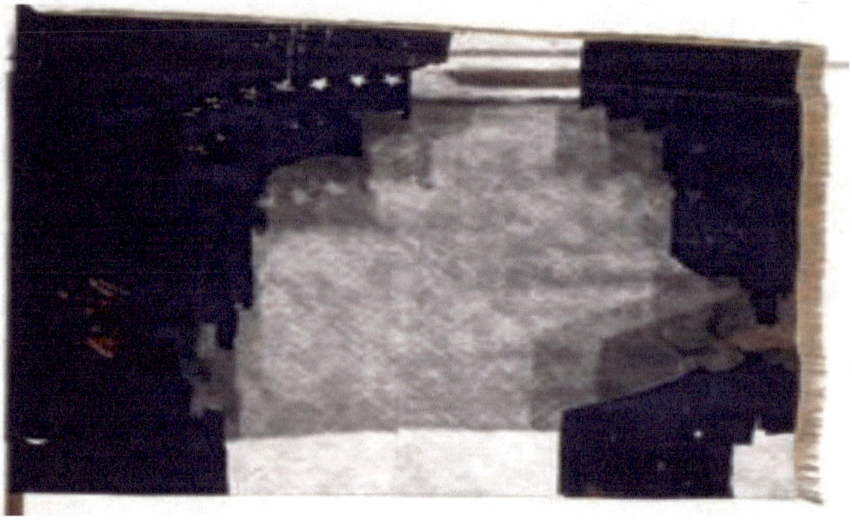

The entire central portion is missing from this regimental color.

A carte de visite (CDV) of a flag of the 74th Regiment Illinois Infantry was sold several years ago. A copy of this photo may be the "third stand of colors" mentioned above. Battle honors on the flag include "CHICKAMAUGA", "MISSION RIDGE", and "SPRING HILL". The photo is identified on the back as being taken by S. M. Fassett of Chicago, Illinois. It is not known the whereabouts of this flag today.

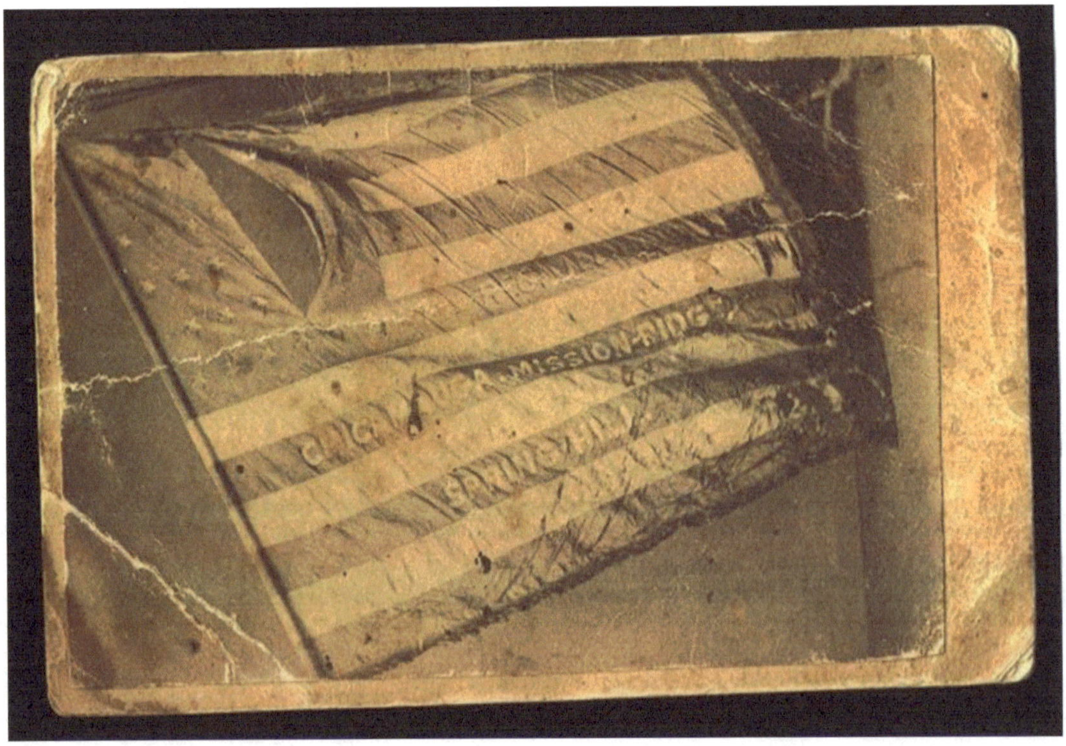

No other flags carried by the 74th Illinois Infantry are known to survive.

75th Illinois Infantry

"HONORED SIR: I have the honor to present you, as the Chief Executive of the State of Illinois, by the hands of David R. Bryson, and John W. Weldon, color Sergeants of this regiment, our Regimental and National Colors, to be placed in the archives of the State. In doing so, allow me to say that these colors were the first ever received by us from the hands of the Government and have been borne the most of the time by these brave Sergeants. They also contain upon their folds the record of the many engagements we have been in...Beneath their starry folds, many of our brave comrades have fallen, and now sleep their last sleep upon the field of their glory. These old banners, though tattered and torn, are sacred mementoes of the past; and in parting with them, we do so with sadness, mingled with pride and pleasure."[736]

The Seventy-fifth Illinois Infantry was organized at Camp Dement at Dixon, Lee County, Illinois.[737] Company A enlisted at Dixon in Lee County; Company B at Lyndon in Whiteside County; Company C from Morrison in Whiteside County; Company D was raised by the Chicago Board of Trade in Whiteside County; Company E volunteered from Lee Center and Sublette in Lee County; Company F came from Amboy in Lee County; Company G volunteered from Franklin Grove in Lee County; Company H was raised from Sterling, Como and Jordan in Whiteside County; Company I (*Reaper Company)*from the townships of Sterling, Erie, Fulton and Ustick in Whiteside County and Company K volunteered from Wyoming in Stark County. The Regiment mustered in on September 2, 1862, under the command of Colonel George Ryan[738] and on September 27 departed for Kentucky to attack Bragg's army. The Seventy-fifth as part of the Thirteenth Brigade, 9th Division of the Army of the Ohio attacked Bragg's troops at Perryville, Kentucky and defeated them on October 8. The Regiment moved south and arrived at Bowling Green, Kentucky on October 31, and Nashville on November 7. They fought at Nolensville and Knob Gap, Tennessee on November 26 and in December the Seventy-fifth marched south to Murfreesboro, Tennessee where on December 30 and 31 and January 1 and 2, 1863 they fought in the battle at Stones River where both Union and Confederate losses were heavy. They remained in Murfreesboro until June 1863 when they marched south toward the enemy positioned north of the Duck River between Shelbyville and Wartrace Tennessee. On June 24 and 25 they met the rebel forces at Liberty Gap and drove them southward skirmishing at Manchester and Winchester, Tennessee. Following their march through the Cumberland Mountains in late August they crossed the Tennessee River at Shell Mound, Battle Creek and Bridgeport and moved into north Georgia. On September 18 the rebel forces attacked the Union army at Chickamauga Creek and after two days of fierce fighting Union forces were forced to retreat to the safety of the approaches of Chattanooga. During the siege of Chattanooga in September, October and November the Seventy-fifth fought in battles at Lookout Mountain and Mission Ridge. Following winter camp at Whiteside, Tennessee on the *Charleston and*

---

[736] Portion of a letter to Governor Richard J. Oglesby from John E. Bennett, Colonel of the Seventy-fifth Illinois, accompanying the Regiment's flags on their retirement to Springfield, Illinois, May 6, 1865. Quoted in-- Dodge, Wm Sumner, *A Waif of the War; or the History of the Seventy-fifth Illinois Infantry,* Church and Goodman, Publisher, Chicago, 1866, page 207.

[737] The first Colonel of the Regiment was George Ryan a physician from Wyoming, Illinois.

[738] Colonel George Ryan resigned 3 months after muster and was replaced by Colonel John E. Bennett. Bennett was born March 18, 1833 in New York. H attended Genesee Wesleyan Sminary in Lima, New York. After he moved to Illinois he was postmaster at Mosrrison, Illinois. . He joined the Union army and became Colonel of the 75th Illinois Infantry after resignation of Colonel Ryan. After the war he was elected Judge in Arkansas, was appointed to the first board of trustees of Arknsas Industrial University and justice of the Arkansas Supreme Court

Memphis railroad, the Seventy-fifth joined Sherman's Atlanta campaign in May 1864. During the next five months the Regiment saw action at Rocky Faced Ridge, Buzzard's Roost Gap, Dalton, Resaca, Adairsville, Pumpkin Vine Creek, Dallas, New Hope Church, Allatoona Hills, Kennesaw Mountain, Pine and Lost Mountain, Ruff's Station, Chattahoochee River, Peach Tree Creek, and Siege of Atlanta. In a flanking movement around Atlanta, they fought in battles at Jonesboro and Lovejoy Station in September. The Regiment pursued Hood into North Georgia and Tennessee and fought at Columbia, Duck River, Spring Hill, Franklin, and Nashville. In December the Seventy-fifth moved to Huntsville, Alabama and remained on duty there until March 1865 then marched to Nashville, Tennessee where they were mustered out June 12, 1865.[739]

There is one national and one regimental flag of the Seventy-fifth Illinois Infantry in the collection of the State of Illinois Military Museum. The national and regimental flags shown here are the flags returned to Springfield by the Seventy-fifth in May 1865.

Battles inscribed on the national flag are: Perryville, Stone River, Chickamauga, Lookout Mountain, Miss----?.(*Originally read* Mission Ridge*). The flag measures 73" on the hoist by 74" on the fly.*

[739] Dodge, Wm Sumner, *A Waif of the War; or the History of the Seventy-fifth Illinois Infantry,* Church and Goodman, Publisher, Chicago, 1866

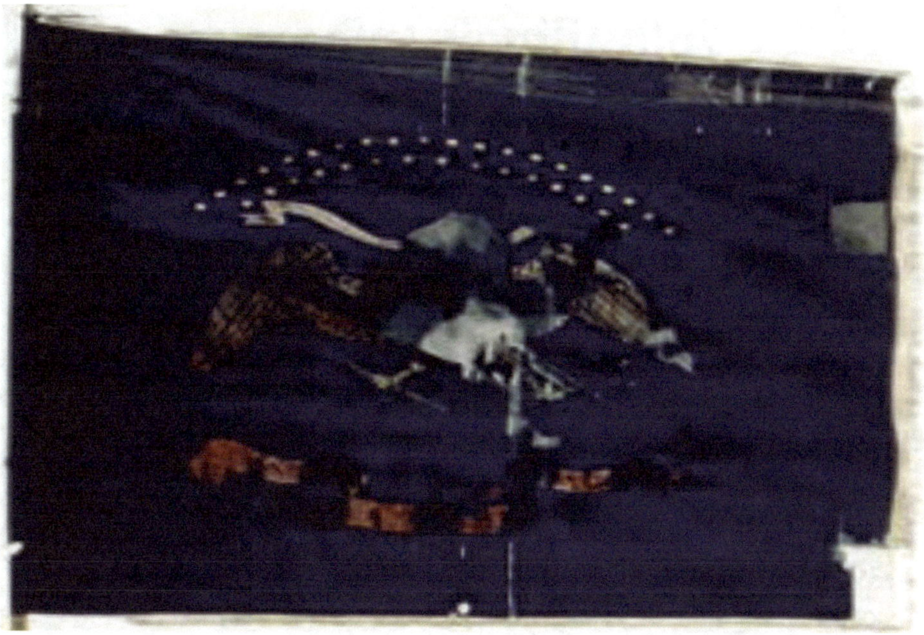

The regimental flag does not bear any inscriptions. It measures 72" on the hoist by 75" on the fly.

No other flags carried by the 75th Illinois Infantry are known to survive.

76th Illinois Infantry

"Veterans of the assault always maintained that the regiment's colors were the first to be planted on the enemy's parapet—but at a terrible cost. Color Sergeant Henry B. Hussey of Company K was killed within twenty feet of the works, and Corporal Charles Goldwood of Company I, who took the colors when Sergeant Hussey fell, was killed on the parapet. When Colonel Busey mounted the parapet, he shot with his revolver a gunner who was about to fire a howitzer loaded with canister."[740]

The Seventy-sixth Illinois Infantry was organized at Kankakee, Illinois and mustered into service on August 22, 1862, under command of Colonel Alonzo Mack.[741] The men of Company A served from Iroquois County; Companies B and G from Champaign County; Company C from Grundy County; Companies D, F, H and I from Kankakee County; Company E from Morgan and Iroquois counties and Company K from Ford and Iroquois counties. On August 26 the Regiment was ordered to Columbus, Kentucky and on October 4 moved by rail to Bolivar, Tennessee. In late November the Seventy-sixth moved on the *Mississippi Central Railroad* to Holly Springs and Waterford, Mississippi; in mid-December they reached Springdale, Mississippi but on December 22 learned that the Union supply depot at Holly Springs had been captured. The Regiment retreated northward, passed through Holly Springs, and went into camp at Moscow, Tennessee on January 11, 1863. In March the Seventy-sixth marched to Memphis where they remained until May 13 when they left by steamer down the Mississippi River for Young's Point, Louisiana. From Young's Point they transported to Chickasaw Bayou on the Yazoo River and marched to Grant's siege lines at Vicksburg. They were in the siege until Vicksburg's surrender on July 4; the following day they marched toward Jackson, Mississippi—skirmished at Big Black River and Champion Hill—then participated in the siege at Jackson. After the rebel forces abandoned Jackson on July 17, the Seventy-sixth marched back to Vicksburg and remained in camp there until August 10, 1863, when the Regiment moved to Natchez, Mississippi. In late November the Regiment returned to Vicksburg and in January 1864 marched to Camp Hebron near the Big Black River. February 3 the Seventy-sixth became part of Sherman's Meridian campaign and returned to Vicksburg in March. In May, as part of an expedition to Yazoo City the Regiment skirmished at Benton, Vaughn's Station and Deasonville then drove the Confederates from Yazoo City and returned to Vicksburg. In July, in a move toward Jackson, Mississippi, the Seventy-sixth was attacked at Jackson Crossroads. In that battle the Regiment lost 102 men killed, wounded and missing. In late July the Regiment was transported to Morganza, Louisiana and in August was moved to Port Hudson. During the remainder of 1864 the Seventy-sixth was in camp in Arkansas at the White River, then at Duvall's Bluff, Arkansas, then at Memphis and finally at Kennersville, Louisiana a few miles above New Orleans. A strike force against Mobile, Alabama was organized in February 1865 and the men were sent by steamer across the Gulf of Mexico to Mobile Point. A third of the Regiment was on board the *George Peabody* when it encountered a storm that severely damaged the steamer. Horses, mules, and wagons were lost but the boat returned to New Orleans with its human cargo. The troops then traveled across Lake Pontchartrain and Lake Borgne and arrived at Fort Barrancas near Pensacola, Florida to rejoin the remainder of the

---

740 This describes the assault of the Seventy-sixth Illinois Infantry on Fort Blakely on April 9, 1865. Behrens, Robert H., *From Salt Fork to Chickamauga, Champaign County Soldiers in the Civil War,* The Urbana Free Library, Urbana, Illinois (1988) p170.

741 Colonel Alonzo Mack was born in Vermont 1822. He was an accomplished physician, lawyer and banker when he become commander of the 76th Illinois Infantry. However, he served only three months and was replaced by Colonel Samuel T. Busey Urbana.

Regiment. From there they marched along the seacoast through pines and swamps to Mobile, Alabama. The Seventy-sixth reached Fort Blakely on April 1 and on April 9 the Regiment assaulted the enemy works and suffered the loss of 98 men out of the 206 engaged. The flags of the 76th were the first flags to fly over Fort Blakely. In June 1865 the Regiment was transported to Galveston, Texas where it was mustered out on July 22, 1865. One hundred and fifty-six recruits whose term had not expired were transferred to the Thirty-seventh Illinois Infantry.[742]

There is one national and one regimental flag of the Seventy-sixth Illinois Infantry in the collection of the State of Illinois Military Museum.

Based on the star pattern of this national flag, it is of Philadelphia manufacture and presumably one of the first Union flags to fly over Fort Blakely. The flag measures 73" on the hoist by 75" on the fly.

[742] Ibid., pages 163-184.

This regimental flag of the Seventy-sixth is one of the first flags to fly over fort Blakely.

No other flags carried by the 76th Illinois Infantry are known to survive.

77th Illinois Infantry

"I am satisfied—satiated with war. It originates in wickedness and it develops wickedness at almost every step. The soldier above all things, needs the truth of God, and the grace of God in his heart, that he may resist temptation, and honor his Christian profession; but when war becomes a necessity to resist encroachments, and drive back the invader, let the Christian church do everything possible to mitigate its evils; for it is true that nowhere is Christian effort more fruitful in blessing than in the army."[743]

The Seventy-seventh Illinois Infantry was organized at Peoria, Illinois and mustered into service on September 3, 1862 under the command of Colonel David Perkins Grier.[744] The men of this Regiment volunteered from: Company A- Knox County; Company B-Putnam County; Company C-Woodford County; Company D-Marshall County; Company E-Peoria County; Company F-Peoria and Woodford counties; Company G-Peoria County; Company H-Woodford, Tazewell and Peoria counties; Company I-Peoria County and Company K-Peoria County. The Regiment was ordered to Cincinnati on October 4 and went into camp across the Ohio River from Cincinnati at Covington, Kentucky. In November the Regiment marched south through Lexington to take possession of Richmond, Kentucky. Two weeks later they were ordered to Louisville and transported downriver to Memphis. December 20 they were transported further down the Mississippi River to the Yazoo River and up the Yazoo about ten miles where they disembarked and soon met the enemy guarding the rear of Vicksburg. After several days of skirmishing, the enemy works could not be carried and on January 2, 1863 the Union troops withdrew to Milliken's Bend upriver on the Louisiana shore. On January 5, 1863 the Seventy-seventh sailed back up the Mississippi River to the White River and up the White River to the cutoff to the Arkansas River to Fort Hindman at Arkansas Post. Here on January 10 and 11 they met and defeated the enemy forces and planted the flag of the Seventy-seventh on the parapet.[745] Following the battle the Regiment returned to Young's Point and remained there until March when they moved to Milliken's Bend. In April the army marched to New Carthage crossed the Mississippi River at Bruinsburg and moved toward Port Gibson. A fight ensued on the march to Port Gibson, but the rebel force was overcome and the town was taken. Further northward movement brought the Regiment to Champion Hills. Here on May 16 the rebel force was routed and the next morning the Seventy-seventh resumed its march toward Vicksburg. They met the enemy at Big Black River and then at the Siege of Vicksburg[746] where on May 22 the

---

743 J.S. McCulloch was the chaplain of the Seventy-seventh Illinois, was taken prisoner at the Battle of Sabine Crossroads and imprisoned at Camp Ford in Texas. The above quote is taken from the conclusion of his diary---*Reminiscences of Life in the Army and As a Prisoner of War.* Privately printed

744 Charles Baliance of Peoria had been offered command of the regiment but he declined the commission and the command devolved to Colonel David Perkins Grier. Grier was born in Danville, Pennsylvania in 1836. When he moved to Peoria he joined the *National Blues* a local Peoria militia. With the outbreak of the war he was commissioned a captian in the 8th Missouri Volunteer Infantry. While serving in the Missouri regiment he fought at Fort Henry, Fort Donelson, Corinth and Shiloh. He returned to Illinois, helped recruit the 77th Illinois Infantry and was named its Colonel. Find A Grave-2023.

745 Bentley, Lieut. W. H., *History of the 77th Illinois Volunteer Infantry, Sept. 2, 1862—July 10, 1865,* Edward Hine, Peoria, Illinois (1883), page116-17. "It was here that we sustained our heaviest loss. It was here that our color bearer, John S. Hornbaker, of Co. C was wounded and left the field. As he fell, Lieut. Philip Jenkins of the same company, seized the colors, and carried them during the remainder of the action, and until the Regiment entered the fort, when Major Hotchkiss took the flag and proudly planted it on the parapet."

746 Winthrop D. Putnam, Corporal of Company A was awarded the Medal of Honor at Vicksburg, "Carried, with others, by hand, a cannon up to and fired it through an embrasure of the enemy's works. Vicksburg, Mississippi, May 22, 1863." *Civil War Medal of Honor Winners from Illinois,* Published by the Civil War Centennial Commission of Illinois, July 1862.

Regiment's colors were lost.[747] After the fall of Vicksburg on July 4, 1863, the Regiment pursued the enemy toward Jackson, Mississippi where they took part in the siege and surrender of Jackson. In August the Regiment was ordered to New Orleans and in December was transported to Pass Cavallo on the Texas Gulf and went into camp at Decrow's Point. They remained there until February 1864, returned to New Orleans, and embarked on Banks' Red River campaign in March. On April 8, 1864, at Sabine crossroads, south of Mansfield, Louisiana, Confederate forces attacked the Union troops. The Seventy-seventh Illinois was sent forward to assist the cavalry and suffered severely in the defeat; their Colonel, Lysander Webb was killed and in additon176 other officers and men were killed, wounded, or taken prisoner, leaving only 125 men in the Regiment. Following retreat down Red River the Regiment went into camp at Baton Rouge; in August the men were transported to Dauphine's Island in Mobile Bay, fought at the capture of Forts Gaines and Morgan guarding Mobile Bay and returned to Morganza above Baton Rouge on the Mississippi River. In October the Regiment moved to New Orleans for provost duty until March 1865,[748] and then was transported to Fort Morgan, Alabama to aid in the siege and capture of Spanish Fort, and Fort Blakely. On July 10, 1865, the Seventy-seventh Illinois Infantry was mustered out at Mobile, Alabama.[749,750]

There is one national and one regimental flag of the Seventy-seventh Illinois Infantry in the State of Illinois collection. There is no record of the return of the colors that the Regiment lost at Vicksburg. At least one flag was presented to the Regiment in Peoria in 1862. Prior to the Regiment leaving Peoria "On the 20th of September the regiment marched to the city for the purpose of receiving a beautiful national flag—a present from the ladies of Peoria. The presentation speech was made by Washington Cockle, Esq., and responded to by Col. Grier, on behalf of the regiment…"[751]

The national flag shown here is inscribed "Co. C 77th Reg. Ill. Vol." This presumably is not the flag mentioned in the presentation above since this flag was the flag of Company C and not the national flag of the Regiment. It is a 34-star flag and measures 66" on the hoist and 72" on the fly.

---

[747] Letter from T.N.Waul, Colonel, commanding Waul's Texas Legion to William T. Rigby dated March 10, 1903. Regimental Files at Vicksburg National Military Park, Vicksburg, Mississippi. The Regimental History also mentions the loss of the colors, "Plant our colors on the ramparts, Colonel Grier shouted; and they were planted amid the shouts of our men—the first and only Union colors planted upon the enemy's ramparts along the whole line….They captured our regimental banner in the sortie, which had floated all day over their fort and had been shot to shreds." Ibid, Bentley, pages 152-3.

[748] In January 1865 the 77th Illinois Infantry and the 130th Illinois Infantry were consolidated as the 77th Illinois Infantry.

[749] Ibid. Bentley

[750] Eddy, T.M., *The Patriotism of Illinois,* Clarke & Co., Chicago (1866) volume II, pages 648-51

[751] Ibid, Bentley, pages 23-24.

This regimental flag carries no inscriptions other than the regimental designation.

Bentley, in his history of the Seventy-seventh details the presentation of a set of colors to the Regiment in September 1863. "It will be remembered that the flags presented to the Regiment before leaving home, were lost in the unsuccessful charge at Vicksburg, May 22d, 1863. The ladies of Peoria, not satisfied with their previous munificence, very generously replaced these flags by others more beautiful, more costly, and under the circumstances, far more acceptable than the first. The flags were presented through Colonel Grier, who had just returned from leave of absence." The officers of the Regiment expressed their thanks for the "two beautiful flags from the "Woman's National League of Peoria" and the Misses of the "Aid Society" of Peoria." [752]

On July 29, 1865, the Regiment returned to Peoria. "At the close of the repast the crowd adjourned to the Court House yard to witness the closing exercises. The old flag that the ladies had given to the Regiment three years ago, that, shot riddled and blood-stained, had been carried through all the battles in which they had been engaged, was now to be given back to its donors as the only one fit to keep it in custody."[753] Based on this information only one flag was returned to the "Ladies of Peoria".

There are two flags in the Missouri Historical Society collections associated with the 77th Illinois Infantry. The first is a regimental flag. It is identified as presented by Misses' Soldiers Aid Society of Peoria, Illinois. The flag measures 77" on the hoist by 77" on the fly, has had conservation treatment and is framed. The second flag is a national flag that measures 61" on the hoist by 73" on the fly. It appears to be homemade.[754] It seems that the two flags presented to the 8th Missouri Volunteer Infantry and/or the 77th Illinois Infantry Regiment by two groups of ladies in Peoria ended up in Missouri Historical Society collection, presumably donated there by the ladies themselves.

No other flags carried by the 77th Illinois Infantry are known to survive.

---

[752] Ibid., Bentley, pages 208-13.

[753] Ibid, Bentley, page 387

[754] Personal communication from Mikall Venso, Military & Firearms Curator,Missouri Historical Society, June 10,2022.

78th Illinois Infantry

I'm glad to meet you comrades,
In our own loyal state,
To clasp your hands as brothers,
What remains of "78."
I saw our old banner
In Chicago, at the fair,
And if this earth was mine, I'd have given it
If each comrade had been there.[755]

The Seventy-eighth Illinois Infantry was organized at Quincy, Illinois and mustered in on September 1, 1862, under the command of William Bennison, a 43-year-old Quincy attorney. The men were recruited from the following counties: Company A from Schuyler County; Company B from Adams County; Company C from McDonough County; Company D from Hancock County; Companies E, F and G from Adams County; Company H from Hancock County; Company I from McDonough and Adams counties and Company K from Adams and Clark counties. The Regiment was ordered to Shepherdsville, Kentucky in October and divided into detachments to guard the *Louisville & Nashville Railroad.* On December 26, 1862, rebel forces in a battle at Muldraugh Hill captured companies B and C.[756] January 1863 the remaining eight companies of the Regiment were transported to Fort Donelson. On their approach to the Fort the enemy retired, and the Regiment returned to Nashville. February thru May the Seventy-eighth was stationed at Franklin, Tennessee and in June marched to Murfreesboro, Tennessee. In actions against Bragg's army the Regiment moved to Shelbyville, Tennessee in July, and further south across the Tennessee River to arrive at Rossville, Georgia in September.

September 17 the Regiment skirmished at Ringgold Georgia and on September 20 fought in the battle of Chickamauga where it lost 156 killed, wounded or missing (40% of the men engaged). Through a blunder by an officer, the pickets covering the Union retreat were not recalled and were captured; by this error the Seventy-eighth lost an additional four officers and fifty-one men from Companies I and F. In late October the Regiment crossed the Tennessee River below Lookout Mountain; the enemy forces met them in the battle of Wauhatchie. The Seventy-eighth was in reserve at the battle of Missionary Ridge but did aid in the pursuit of Bragg's retreating forces. The Regiment was ordered to Knoxville, Tennessee for relief of the siege but en-route received word that the siege was lifted, and the Regiment returned to Chattanooga and went into winter quarters at Rossville, Georgia. The Seventy-eighth joined in the Atlanta campaign in May 1864. In that campaign the Regiment saw action at Rocky Faced Ridge, Buzzard's Roost Gap, Resaca, Rome, Dallas, New Hope Church, Allatoona Hills, Kennesaw Mountain, Chattahoochee River, Peach Tree Creek, and the Siege of Atlanta. In the flanking movement around Atlanta, they fought at Jonesboro-August 31-September 1 and Lovejoy Station-September 2-6. Following actions in north Georgia and Alabama the Regiment joined Sherman's "March to the Sea." In December the Regiment was in the siege of Savannah, Georgia. January 1865 the Seventy-eighth began the campaign through the Carolinas. However, the only significant fighting for the Regiment

[755] Written for a reunion of the 78th by V. B. Faught, Co. D, 78th Illinois

[756] These two companies were sent to St. Louis on parole. They were not exchanged to rejoin the Regiment in Chattanooga until October 1863.

occurred at Averysboro, South Carolina and Bentonville, North Carolina. Following surrender of Johnston's Army the men marched to Washington, D.C via Richmond and marched in the Grand Review May 24, 1865. The Regiment was mustered out June 7, 1865; those men whose terms of service would not have expired by October 1, 1865, were transferred to the Thirty-fourth Illinois Infantry Regiment.[757]

There are two national and one regimental flag of the Seventy-eighth Illinois Infantry Regiment in the collection of the State of Illinois Military Museum. One national flag is not shown here since all that remains is some fringe and a few shreds of red silk.

The following battles are inscribed on this flag: CHICKAMAUGA, CHATANOOGA, RESACA, ROME, KENESAW MOUNTAIN, JONESBORO, ATLANTA, SAVANNAH, AVERYSBORO, BENTONVILLE. *The star arrangement on this flag is characteristic of those manufactured by Gilbert Hubbard of Chicago.*

Evans and Hassall probably manufactured this regimental flag. They supplied flags to both the Philadelphia and Cincinnati depots. The measures 71" on the hoist by 75" on the fly.

No other flags carried by the 78th Illinois Infantry are known to survive.

[757] *Report of the Adjutant General of the State of Illinois,* H. W. Rokker, Printer and Binder, Springfield, Illinois, 1886, Volume V, pages 23-7.

79th Illinois Infantry

The Seventy-ninth Illinois Infantry was organized at Camp Terry, Mattoon in Coles County, Illinois and mustered into service on August 28, 1862, under the command of Colonel Lyman Guinnip[758]. The men volunteered as follows---Company A from Edgar, Vermillion and Randolph counties; Companies B, E, G and part of K from Douglas County; Companies C, D and H from Edgar County; Company F and part of I from Clark County and part of Company K from Coles County. The Regiment was ordered to Louisville, Kentucky on September 13, 1862. October 1 the Seventy-ninth began a march through Kentucky—Frankfort to Perryville to Crab Orchard to Lebanon, Bowling Green and finally arrived at Nashville, Tennessee on November 7. They remained at Nashville until December 26 when they were ordered to Murfreesboro, Tennessee. From December 31 through January 4, 1863, the Regiment was engaged in the battle at Stones River, in that battle the commander of the Regiment, Colonel Sheridan Read, was killed. Other Regimental losses included 24 men killed, 71 wounded and 124 missing. The Regiment remained in camp at Murfreesboro until June 24, 1863, when it moved to Liberty Gap and engaged in fighting there the following day. In that battle, Captain John Patton, Company C, and Captain Hezekiah Martin, Company K, were killed and Captain Robert Lacy, Company I, was wounded; five other men were killed and thirty-nine wounded. The Seventy-ninth occupied middle Tennessee until August 16, then crossed the Cumberland Mountains and Tennessee River and arrived at Chickamauga, Georgia in September. September 19 and 20 the Regiment fought in the Battle of Chickamauga and following the battle retreated with the Union forces to Chattanooga. In the Chattanooga campaign the Seventy-ninth fought at Orchard Knob on November 23-24 and at Mission Ridge on November 25 where they captured two pieces of Confederate artillery. On November 27, 1863, they marched for the relief of Knoxville, Tennessee and were in operations in East Tennessee until April 1864. From May through September 1864 the Seventy-ninth was in the Atlanta campaign. In that campaign they were engaged at Rocky Faced Ridge, Buzzard's Roost Gap, Battle of Resaca, Adairsville, Dallas, New Hope Church, Allatoona Hills, Pine Hill, Lost Mountain, Kennesaw Mountain, Peach Tree Creek, and the siege of Atlanta. In the flanking movement south of Atlanta the Regiment fought at Jonesboro and Lovejoy Station. Following the fall of Atlanta, the Seventy-ninth was ordered to Chattanooga then to Bridgeport, Alabama in October. November 1, 1864 the Regiment marched to Pulaski, Tennessee. In the Nashville Campaign the Regiment was attacked by rebel forces at Spring Hill on November 29 and driven back. The following day at Franklin, Tennessee,[759] out of 210 men of the Seventy-ninth who entered the battle, in four hours of fighting 83 men were killed, wounded and missing. Following the Battle of Nashville December 15 and 16, the Regiment pursued Hood to the Tennessee River. In January 1865 the Regiment was sent to Decatur, Alabama and in March to Bull's Gap east of Knoxville, Tennessee. The Seventy-ninth was mustered out of service on April 22, 1865, at Nashville. [760]

The flag collection of the State of Illinois contains four flags of the Seventy-ninth Illinois Infantry Regiment—three national and one regimental.

---

758 Colonel Guinnip resigned as Colonel of the regiemtnt within two months and was replaced by Colonel Sheridan Read. Colonel Read was killed at the Battle of Murphreysboro on December 31, 1862.and was replaced by Colonel Allen Buckner

759 Lieutenant Presley P. Boyer of Company H captured three battle flags at Franklin on November 30, 1864. *O. R.* Series 1, Volume XLV/1 [Series #93] page 238.

760 *Report of the Adjutant General of the State of Illinois,* H.W. Rokker, State Printer and Binder, Springfield, Ill., 1886, Volume V, page48-9.

79TH ILLINOIS INFANTRY

Based on the shape of the canton and arrangement of stars, his flag is typical of flags manufactured by Gilbert Hubbard & Company of Chicago. Inscribed on this flag are the following, PRESENTED TO THE 79th Ills. Vols. BY THE LADIES OF --?-- LOYAL LEAGUE STONE RIVER, LIBERTY GAP, CHICKAMAUGA, MISSION RIDGE, ROCKY FACE, RESACCA, DALLAS KENASAW, ATLANTA, JONESBORO, LOVEJOY, SPRING HILL, FRANKLIN, NASHVILLE, RUFF'S STATION, PEACHTREE CREEK.

This national flag is smaller than standard size (55" on the hoist by 66" on the fly) and since the stars are arranged in a star-shaped pattern, the flag was not manufactured by one of the Federal flag contractors. Presumably this is a "homemade" flag and was presented to one of the companies of the Regiment. There is a tag affixed to the staff stating that the flag was captured, but no details are given.

The gold painted stars, the gold paint for the unit designation on the center stripe, the square canton and six horizontal rows of stars with 5, 6, 6, 6, 6, 5 stars in each row is typical of flags manufactured by Gilbert Hubbard & Company of Chicago. Inscribed on this flag: 79th REGT. ILL'S VOLS. *There is a tag affixed to the staff of this flag stating that the flag was captured. No information regarding the capture or the return of the flag is known.*

The regimental flag has a shield on the eagle and the overall pattern is like flags manufactured in Cincinnati by John Shilleto & Co.

No other flags carried by the 79th Illinois Infantry are known to survive.

80th Illinois Infantry

"But I doubt if the Union forces ever get another great victory till the leaders in our senate and Congress and the President of these United States shall proclaim liberty to the captives and the opening of the prison doors to them which are bound and obey the command which the Lord gave to Pharaoh saying let my People go that they may serve me. It seems strange to me that we should try to cure anything and still preserve the thing which caused the evil. And it is plain to me that slavery caused this war. Put down slavery and there is no more war. This may not be your view on this subject but certainly it is mine."[761]

The Eightieth Illinois Infantry was organized at Centralia, Marion County, Illinois and mustered in for three years' service August 25, 1862, under the command of Colonel Thomas G. Allen.[762] The men volunteered for this Regiment from the following counties in southern Illinois. Jackson and Randolph counties-company A; Madison County- Company B; Madison and St. Clair counties-Company C; Randolph County-Company D; Jefferson County-Company E; Perry and Randolph counties-Company F; Randolph County-Company G; Jefferson County-Company H; Marion and Washington counties-Company I and Madison and Washington counties-Company K. The Eightieth left Centralia on September 4 for Louisville, Kentucky and on October 1 left Louisville in pursuit of Confederate General Bragg. The Union forces met Bragg at Perryville where the Regiment lost 14 killed and 58 wounded. Continuing the pursuit of Bragg, the Regiment marched to Munfordville, Kentucky to Cave City and back to Munfordville. December 11 the Eightieth marched to Bledsoe Creek, Tennessee and on the 26th began the pursuit of Confederate General Morgan as far as Bear Wallow, Tennessee. January 1863 the Regiment marched to Nashville and on to Murfreesboro, Tennessee arriving there January 10. They remained at Murfreesboro scouting and foraging until April 7 when they returned to Nashville, embarked on steamers, and moved down the Cumberland and up the Tennessee Rivers to Eastport, Mississippi. They marched to Tuscumbia, Alabama, were mounted and left in a raid toward Rome, Georgia, on April 26. The Eightieth was attacked at Day's Gap on April 28, at Sand Mountain on April 30 and at Blount's farm on May 2 and repulsed the enemy each time. However, at Cedar Bluffs, Alabama on May 3 the Union troops were surrounded by Forrest and were forced to surrender.[763] The men were taken to Rome, Georgia, paroled and sent to Atlanta where the officers were sent to Libby Prison and the enlisted men sent via Knoxville, Tennessee to Richmond, Virginia then to Annapolis, Maryland. May 19 they were moved to Camp Chase, Ohio. On June 23, 1863, the Regiment, having been declared exchanged, moved to the Schofield Barracks at St. Louis then to Nashville on June 29. New officers were assigned since the original officers were still in Libby prison. September 8 the Regiment moved to Stevenson, Alabama to guard the railroad and October 16 marched to Battle Creek, Tennessee and on October 23 to Bridgeport, Alabama. During November the Regiment was in Lookout Valley, Tennessee and fought at Mission Ridge on November 24 and 25. November 29 the Eightieth

---

[761] *Dear Lizzie,* privately printed 1978. The letters of James Crawford, Co. G, of the Eightieth Illinois Infantry to his sweetheart Martha Elizabeth "Lizzie" Wilson. The quote is from a letter dated September 17, 1862. The opinion of the writer may be unusual this early in the War, but the writer's father supported the "Underground Railroad" and his sweetheart's family were Southern and owned slaves.

[762] Colonel Thomas Allen was born in Pennsyvania in 1827. He was an early member of the Republican party and a friend of Lincoln. Claybourn, Joshua, Col. Thomas G. Allen of the 80th Illinois Volunteer Infantry. Acaemiaedu updates@academia-mail.com.

[763] In addition to the Eightieth Illinois Infantry, the Third Ohio Infantry and the Fifty-first and Seventy-third Indiana Infantry also surrendered.

started to the relief of Burnside at Knoxville Tennessee but on December 7, Confederate General Longstreet, retreated from his siege at Knoxville and the Eightieth returned to camp in Lookout Valley. December 24 the Regiment moved to Whiteside, Tennessee and remained there until January 27, 1864, then marched to Blue Spring, Tennessee and remained there until May 4. In May 1864 the Eightieth embarked on the Atlanta campaign and participated in actions at Dalton, Tunnel Hill, Buzzard's Roost Gap, Rocky Faced Ridge, Dallas, New Hope Church, Allatoona Hills, Kennesaw Mountain, Pine Hill, Lost Mountain, Ruff's Station, Chattahoochee River, Peach Tree Creek, and the Siege of Atlanta. In the flanking movement south of Atlanta they fought at Jonesboro August 31-September 1 and at Lovejoy Station September 2-6. October 3, the Regiment began the pursuit of Confederate General Hood into Alabama. In November the Regiment fought at Duck River and Spring Hill on November 29 and at Franklin, Tennessee on November 30; December 15 and 16 they fought at the battle of Nashville. The Eightieth moved to Huntsville, Alabama in January 1865 and were mustered out at Nashville June 10, 1865.[764,765]

There is one national and one regimental flag of the Eightieth Illinois Infantry in the collection of the State of Illinois Military Museum. Presumably another set of colors of the Regiment was captured when the Regiment surrendered at Cedar Bluffs, Alabama on May 3, 1863.

The national flag shown here has no inscriptions and is in poor condition. The flag measures 68" on the hoist by 70" on the fly.

---

[764] Eddy, T. M., *The Patriotism of Illinois,* Clarke & Co. Publishers, Chicago, 1866, page 322-4.

[765] *Report of the Adjutant General of the State of Illinois,* H.W. Rokker, State Printer and Binder, Springfield, 1886, Volume V, page 70-1.

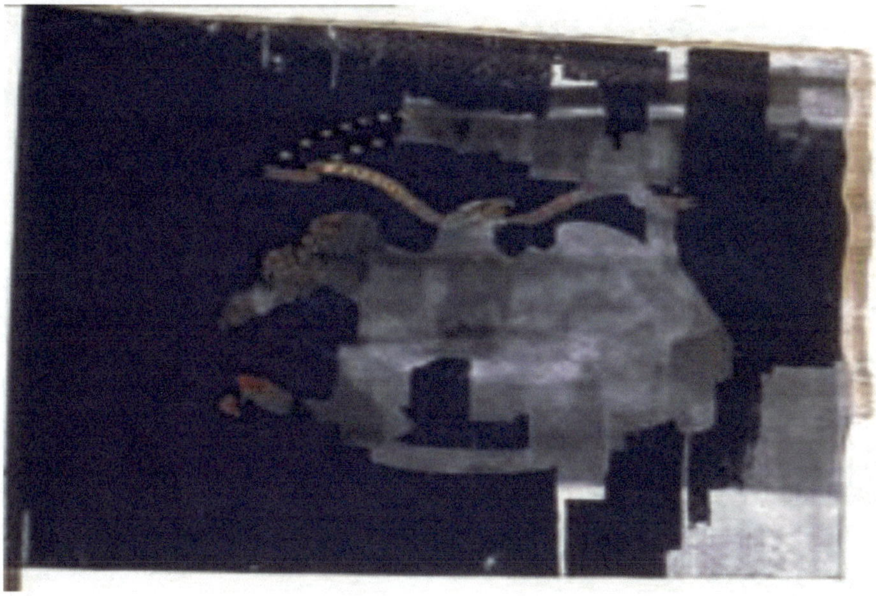

The regimental flag is also in poor condition and the greater part of the eagle is missing. It measures 70" on the hoist by 70" on the fly.

No other flags carried by the 80th Illinois are known to survive.

81st Illinois Infantry

The Eighty-first Illinois Infantry organized at Anna in Union County, Illinois and mustered into service August 26, 1862 under command of Colonel James Jackson Dollins.[766] The men were recruited from counties in southern Illinois as follows, Company A-from Perry County; Company B-from Jackson County; Company C-from Perry County; Company D-from Jackson County; Company E-from Union and Williamson counties; Company F-from Jackson County; Company G-from Williamson County; Company H-from Jefferson County; Company I-from Washington County and Company K-from Madison County. The Regiment was moved by rail to Cairo, Illinois on September 6 and on October 8 sailed downriver to Columbus, Kentucky and moved by rail to Humboldt, Tennessee for garrison duty. The Regiment left by rail on November 12 for LaGrange, Tennessee and remained there until November 29 when they began marching south into Mississippi. They had passed through Holly Springs and had reached Oxford, Mississippi when on December 21, they received word that Confederate General Van Dorn, in their rear, had captured the Union depot at Holly Springs and destroyed the supplies. This resulted in the retreat of the Union forces, including the Eighty-first Illinois, back to Memphis, arriving there January 19, 1863. On February 20 the Regiment broke camp and steamed down the Mississippi River to Lake Providence, Louisiana about 75 miles above Vicksburg, Mississippi. April 17 the Regiment moved further downriver to Milliken's Bend, 20 miles above Vicksburg. The following day a call was made for volunteers to man supply transports to run the batteries at Vicksburg and Grand Gulf. Seven men of Company I were accepted and that night the transports ran the batteries with the sinking of the *Tigress* and *Empire City* and damage to others. April 25 the Regiment marched out of Milliken's Bend, moved cross-county and on April 30 crossed the Mississippi River from Louisiana to Bruinsburg, Mississippi, several miles below Vicksburg. The next day the Eighty-first fought at Thompson's Hill and pursued the retreating rebels through Port Gibson. May 3 the Regiment camped on the Big Black River south of Vicksburg. A week later the men marched toward Raymond and met and defeated the enemy there on May 12 then marched to Clinton, a town on the *Vicksburg and Jackson Railroad.* Destruction of the railroad cut off all communications and supplies between Vicksburg and the remainder of the Confederacy. May 14 the Regiment was engaged in the battle and fall of Jackson, the Mississippi state capitol. Marching toward Vicksburg the Eighty-first fought at Champion Hill on May 16 and at Black River Bridge on May 17. By May 19 they were in the siege lines at Vicksburg; in the assault on Vicksburg, May 22, the Regiment lost 11 men killed and 96 wounded including the death of the Regimental commander, Colonel James J. Dollins. On July 4, 1863, Vicksburg surrendered and the Eighty-first Illinois was one of the Regiments assigned to garrison the city. The Regiment remained at Vicksburg until March 1864 but during that time participated in expeditions to Monroe, Louisiana, Bogue Chitto Creek and Sunny Landing, Arkansas. March 9, 1864, the Regiment departed Vicksburg on the tin-clad steamer *Diana* as part of the Red River Expedition. During that campaign the Eighty-first was engaged in the capture of Fort DeRussy and Alexandria in the advance on Shreveport, Louisiana. After the Union forces, under Banks, were defeated at Mansfield on April 8 the expedition retreated downriver with actions at Pleasant Hill, Cloutierville,[767] Marksville Prairie, Bryce's Plantation, Bayou Boeuf and Yellow Bayou. The Regiment landed back at Vicksburg May 24. On May 27 the

---

[766] Colonel Dollins was born in Benton Illinois in May 1832. He was a Captain of Company C if the 15th Illinois Cavalry Regiment in Septmber 1861; promoted Colonel and became Colonel of the 81st Illinois Infantry on August 26, 1862. He was killed in action on May 22, 1863 at Vicksburg, Mississippi.

[767] On April 23, 1864, the Eighty-first Illinois Infantry captured 250 rebel soldiers in this battle.

Eighty-first was ordered to sail upriver to Memphis, Tennessee where they disembarked and marched southeast into Mississippi. In a battle near Guntown,[768] Mississippi on June 10 the Eighty-first engaged rebel forces under Forrest; although the Regiment captured a rebel flag[769] they were driven from the field and of 371 men in the Regiment, 9 were killed, 18 wounded and 126 taken prisoner.[770] ,[771] In August the Eighty-first moved to Duvall's Bluff, Arkansas and scouted and skirmished throughout the State until September 17 when it joined in the pursuit of Confederate General Price in Missouri. Price having escaped, the Regiment was ordered to Nashville, Tennessee and fought in the defeat of the Confederate army under Hood in the battle of Nashville on December 15 and 16, 1864. The Regiment joined in the pursuit of Hood into Mississippi and remained at camp in Eastport, Mississippi until February 1865. In May they joined the Atlanta campaign and fought at Kennesaw Mountain, Nickajack Creek, Chattahoochee River and in the siege of Atlanta. In the flanking movement south of Atlanta they fought at Jonesboro on August 25-30 and at Lovejoy Station August 31-September 1. After pursuit of Hood into north Georgia and Alabama the Regiment returned to Nashville and moved to New Orleans in February 1865. In March they moved to Mobile, Alabama and campaigned against Spanish Fort, Fort Blakely and Mobile and in April marched to Montgomery, Alabama and remained there until July. They moved to Chicago and were mustered out August 5, 1865.[772]

There are three national flags and one regimental flag of the Eighty-first Illinois Infantry in the collection of the State of Illinois Military Museum. Two of the national flags bear inscriptions.

The pattern of the stars on this national flag is characteristic of flags manufactured in Philadelphia. There are no inscriptions. The flag measures 70" on the hoist by 70" on the fly.

---

[768] This is also known as the Battle of Brice's Crossroads or Tishomingo Creek.

[769] Newsome, Edmund, *Experience in the War of the Great Rebellion,* Printed by E. Newsome, Carbondale, Illinois, 1879, page 87. The rebel color bearer was shot and the flag was recovered by Calvin Brewer of Company B.

[770] *Report of the Adjutant General of the State of Illinois,* H.W.Rokker, State Printer and Binder, Springfield, Illinois, 1886, Volume V, page 98.

[771] Ibid. Newsome. The author of the Regimental History, Captain Edmund Newsome, Company B, was one of those captured and exchanged March 1, 1865, and recounts his experiences as a prisoner in his book.

[772] The history of the Regiment is well covered in Newsome's Regimental History and in the *Report of the Adjutant General of the State of Illinois,* Volume V.

81ST ILLINOIS INFANTRY

This national flag is also of Philadelphia manufacture. It is inscribed: 81st REGT ILLs INF'TY VOLLs

Although identified in the collection of the State of Illinois as a color belonging to the One hundred and Forty-ninth Illinois Infantry, it is probably the color of the 81st with battles inscribed. The One hundred and Forty-ninth Illinois Infantry was mustered in at Camp Butler February 11, 1865, however, the battles inscribed on this flag were fought before that date and are the battles fought by the Eighty-first Illinois Infantry. Inscribed on the flag: CHAMIPION HILL, May 16,1863; BLACK RIVER BRIDGE; VICKSBURG, May 19 & 22; ---?---;CLOUTHERVILLE; OLD RIVER & YELLOW BAYOU; GUNTOWN; NASHVILLE; SPANISH FORT; BLAKELY; MOBILE.

The regimental flag is of typical New York manufacture. It measures 71" on the hoist by 75" on the fly.

No other flags carried by the 81st Illinois Infantry are known to survive.

82nd Illinois Infantry—"*Second Hecker Regiment*"

In December 1862, several weeks after the *Concordia Guards* had been mustered into the Eighty-second Regiment, a scandal concerning "the Jews, as a class" implicated General Ulysses S. Grant. Known as "the Jew Order," General Order No. 11 required Jews "as a class" to leave all areas within the Department of the Tennessee, or else "be arrested and held in confinement until an opportunity occurs of sending them out as prisoners" By the end of the first week in January, however, General-in-Chief Henry W. Halleck had revoked Order No. 11 by order of President Lincoln.[773]

The Eighty-second Illinois Infantry was composed almost exclusively of German volunteers with the exception of Company C (*Concordia Guards)*,[774] a Jewish company and Company I a Scandinavian company (*Asay Guards)*. The men of Companies A (*Butz's Rifles*), B (*Hesing's Sharpshooters),* C, D were from Cook County; Company E from McLean and Peoria counties; Company F from St. Clair County and St. Louis, Missouri; Company G from Cook and St. Clair counties; Company H from Cook and Madison counties; Company I from Cook and La Salle counties and Company K from Cook County. The Regiment was organized at Camp Butler and mustered into service on October 23, 1862, under command of Colonel Friedrich Hecker.[775] On November 3 the Eighty-second was ordered to Washington DC to join the Army of the Potomac. November 19 the Regiment was ordered to Fairfax Court House, Virginia, then in December further south to Stafford Court House on Telegraph Road. They went into camp near Aquia Creek, Virginia on December 19 and remained there until April 27, 1863, except for a three day "Mud March" in January that was abandoned because of impassable roads. On April 27 the Regiment marched northwest and crossed the Rappahannock River at Kelly's Ford, turned south, and crossed the Rapidan River at Germania Ford and formed a line of battle on the Fredericksburg Pike and the Orange Plank Road on April 30. On May 2 the Eighty-second was formed in the second line of battle; that evening the enemy attacked the Division, the first line gave way and but the second line including the Eighty-second held until a new line could be formed it its rear. May 4 the Regiment was again engaged with line of battle on Mineral Spring Road. In the Chancellorsville battle, the Regiment's commander, Colonel Frederick Hecker, and Major Ferdinand Rolshanson were both wounded; in addition, one hundred and fifty-six men were killed or wounded including seven commissioned officers[776]. The Eighty-second returned to camp at Stafford Court House and remained there until June 12 when they embarked on the Gettysburg Campaign.[777] The Eighty-second Illinois, as part of Carl Schurz's XI Corps, arrived shortly after noon on July 1 and deployed north of Gettysburg. They were attacked by Ewell's forces and fell back through Gettysburg and formed a defensive position across the Baltimore Pike with crest of Cemetery Hill. On July 2, while battles raged in the Peach Orchard, the Wheatfield and Devil's Den to their left, Lee attached the Union center and carried a line of rifle pits to the right of the Eighty-second. The morning of July 3, the rebel forces remained in the rifle pits, but

---

[773] Korn, Bertram W.; American Jewry and the Civil War; Atheneum; New York (1970)

[774] Wax, Bernard, "Illinois Jews in the Civil War," *Illinois Civil War Sketches No. 3,* Printed by the State of Illinois, June 1963. This company of Jewish recruits was raised, organized and outfitted by the Concordia Club, a Chicago Jewish organization.

[775] Colonel Hecker was born in Germany and was a lawyer and radical democrat who led a "revolution" in an attempt to introduce democracy in Germany. When he failed he came to the United States, settled on a farm in Summerfield, Illinois. He was a friend of Lincoln and antislavery. There is an excellent paper re Colonel Hecker—"More than Beer and Schnitzel" by Anika Rieper at https://morethanbeerandschnitzel.com/fredrich-heck-german-revolutionary-american-civil-war-hero/

[776] Bigelow, John, Jr., *The Campaign of Chancellorsville,* Yale University Press, New Haven, 1910.

[777] The Eighty-second Illinois Infantry was the only Illinois infantry unit to fight at Gettysburg.

Colonel Salomon led the Eighty-second in a charge on the pits driving the Confederates back and capturing several hundred prisoners. At Gettysburg, out of 312 men engaged, the Regiment lost 131 men killed, wounded, or taken prisoner. Following the pursuit of Lee through Virginia the Regiment, in September, marched to Manassas Junction and on to duty in Lookout Valley, Tennessee in October. They fought at Wauhatchie, Tennessee, at the foot of Lookout Mountain, October 28 and 29 and remained in Lookout Valley until November 22 when they joined Grant's forces and attacked the rebels near Orchard Knob. November 24 and 25 they fought at Missionary Ridge, Tennessee then pursued Bragg's forces to Cleveland, Tennessee. The Regiment was ordered to march to the relief of Burnside at Knoxville, on November 29 but while en route, on December 2, they received word that siege had lifted; therefore, they were ordered back to Chattanooga and went into camp in Lookout Valley on December 17. In 1864 the Eighty-second joined Sherman's campaign against Atlanta and in May fought at Resaca and New Hope Church, Dallas and Allatoona Hills, in June; at Kennesaw Mountain, Pine Hill, Lost Mountain, Muddy Creek and Noyes' Creek, in July and then at the Chattahoochee River, Peach Tree Creek and the siege of Atlanta. Following the fall of Atlanta, the Regiment occupied Atlanta until mid-November when they joined Sherman's "March to the Sea." They skirmished at Milledgeville, and were in the Siege of Savannah, Georgia. In 1865, as the Union Army moved into the Carolinas, the Eighty-second fought at Averysboro and Bentonville, North Carolina. Following the surrender of the Confederacy the Regiment marched through Raleigh en-route to Washington DC to march in the Grand Review on May 24, 1865. The Regiment mustered out in Chicago on June 16, 1865.[778]

There is one national and one regimental flag of the Eighty-second Illinois Infantry in the collection of the State of Illinois.

The national flag bears no inscriptions and measures 69" on the hoist and 7" on the fly.

---

[778] *Report of the Adjutant General of the State of Illinois,* H. W. Rokker, State Printer and Binder, Springfield, Illinois, 1886, Volume V, Pages 120-4.

Only fragments remain of the regimental flag.

The "Concordia Guards," Company C, were presented a flag at Bryan Hall on August 18, 1862. "At the close of his address, Mrs. Mary Leopold came forward, bearing in her hands the flag to be presented to the company, and delivered the following address: *Colonel*, My friends have kindly selected me as the interpreter of our common feelings on the occasion of placing into your hands this banner, to be by you presented on our behalf to this noble band of young men, who are resolved under your leadership to rush in freedom's sacred name to defense of our bleeding country....In behalf of the Israelite ladies of Chicago, Henry Greenbaum, Esq., then presented this flag to the company, accompanying the act with a few appropriate remarks in German...The flag presented to the Concordia Guards is a most beautiful stars and stripes, subscribed for and wrought by the ladies. It has upon its folds in gilt letters the simple legend '82nd Regiment Illinois Volunteers' and is designed to be used as the regimental flag. Its cost was $150."[779] This flag is not in the collection of the State of Illinois and its whereabouts today is unknown.

Another flag was presented to the Regiment, apparently that same evening. "Among the pleasing incidents of the Israelite war meeting on Monday evening was the presentation of a beautiful flag to Col. Hecker by Mrs. Charles Degenhardt, in the name of her two boys who are too young to enter the service; her oldest son, a very promising youth, was drummer in Col. Turchin's Zouave regiment, and was killed last year on the railroad on their passage through Ohio."[780] The flag is not described.

No other flags carried by the 82nd Illinois Infantry Regiment are known to survive.

---

[779] *Chicago Daily Tribune,* August 19, 1862.

[780] *Chicago Daily Tribune,* August 20, 1862.

83rd Illinois Infantry

The Eighty-third Illinois Infantry was organized at Monmouth in Warren County, Illinois. The men filling companies A, B, C, F and H volunteered from Warren County. The men for Company D volunteered from Mercer County and the men for companies E, G, I and K volunteered from Knox County. The Regiment was mustered in for three years' service on August 21, 1862, under the command of local attorney Abner C. Harding.[781] and was transported by rail to Cairo, Illinois on August 25. On September 3 the Regiment moved to Fort Henry and on September 5, five companies moved to Fort Donelson, two companies were sent to Fort Heiman and three were left at Fort Henry. The companies of the Eighty-third subsequently rejoined at Fort Donelson and remained there until September 20, 1863. During that time the Regiment guarded the surrounding area and skirmished daily with guerillas; several full-scale battles were fought including those at Waverly, Tennessee and Kentucky. Also on February 3, 1863, nine companies of the Eighty-third Illinois Infantry along with a battalion of the Fifth Iowa Cavalry and Battery C of the Second Illinois Light Artillery[782] were attacked at Fort Donelson by Forrest and Wheeler but withstood the attack with significant loss to the attacking enemy.[783] In the latter part of the 1863 and 1864 the Regiment was on garrison duty at Fort Donelson and in northwest Tennessee and southwest Kentucky. During that time, they saw action at Budd's Creek, Tennessee on October 28, 1863, and at Pine Bluff, Tennessee River on August 19-20, 1864. The Regiment was on provost duty at Nashville until June 26, 1865, then they were mustered out and moved to Chicago for final discharge.[784]

There are three national and one regimental flag from the Eighty-third Illinois Infantry in the collection of the State of Illinois.

This National color of the 83rd Infantry is misidentified as belonging to the 104th Infantry. The flag is inscribed FORT DONELSON, FEB 3, 1863. *The Eighty-third Illinois was the only Illinois Infantry Regiment at Fort Donelson on that date.*

[781] Colonel Harding was born in Connecticut in 1807 and studied law at Hamilton College in Clinton, New York. He practiced law in New York until 1838 when he moved to Monmouth, Ilinois to continue his practice. He served in the Illinois House in 1848 and after the civil war he was a founder and trustee of Monmouth College. Wikipedia, 2023

[782] Also known as *Flood's Battery* or the *Springfield Artillery*

[783] Eddy, T.M., *The Patriotism of Illinois,* Clarke & Co., Publishers, Chicago, 1865, page 497-99.

[784] *Report of the Adjutant General of the State of Illinois,* H.W. Rokker, State Printer and Binder, Springfield, Illinois, 1886, Volume V, pages 149-50.

There are no inscriptions on this flag and the star pattern is typical of flags manufactured in Philadelphia.

There are no inscriptions on this national flag. The flag is in poor condition, but the star pattern is consistent with Philadelphia manufacture.

The Regimental flag is in very good condition and was probably presented to the regiment late in or after the war. There are no battle honors inscribed; the flag is of New York manufacture.

No other flags carried by the 83rd Illinois Infantry are known to survive.

84th Illinois Infantry

"One day our soldier's trunk came home,
Oh! He will come no more.
For they have left him lying by
That lonely river shore.
And in his trunk they sent the flag
That floated o'er the brave;
I would to God that they had wrapped
It round him in the grave."[785]

The Eighty-fourth Illinois Infantry was mustered in for three years' service at Quincy in Adams County, Illinois on September 1, 1862 under command Colonel Louis H. Waters.[786] Companies A and C were organized at Macomb in McDonough County; Companies B and F were organized at Vermont in Fulton County; Company D was organized at Mount Sterling in Brown County; Company E was organized at Quincy in Adams County; Company G was organized at Oquawka in Henderson County; Company H was organized at Keithsburg in Mercer County, Company I was organized at Clayton in Adams County and Company K was organized at Biggsville in Henderson County but most of the men were from Mercer County. On September 23 the Regiment was ordered to Louisville, Kentucky and on arrival was sent in pursuit of Bragg. On October 8 the Regiment was at the Battle of Perryville, Kentucky but was not engaged. Later in October they marched to Nashville, Tennessee and remained there on duty until December 26. At 7 AM on that morning the Eighty-fourth began a march south on the Murfreesboro Pike through La Vergne to Stones River near Murfreesboro, Tennessee. In the Battle of Stones River, fought December 31 through January 3, 1863, the Regiment lost 155 men[787] out of the 350 who were engaged. Following the battle, the Regiment remained at Murfreesboro until June 24 when they left on the Tullahoma Campaign and campaigned in middle Tennessee until mid- August. August 16, 1863, the Eighty-fourth began marching toward Chattanooga, crossed the Cumberland Mountains and the Tennessee River and arrived at Chickamauga, Georgia in mid- September. The Regiment fought in the Battle of Chickamauga on September 19 and 20 and in two days of fighting lost 100 men. Following the battle, the overwhelmed Union forces retired to a defensive line south of Chattanooga while the Confederate forces held a position from the Tennessee River across Missionary Ridge through the Chattanooga Valley to Lookout Mountain. In October, Grant took command of the forces at Chattanooga and the Union Army took the offense against the surrounding Confederates. The Eighty-fourth took part in these actions—first at Lookout Mountain on November 23 and 24 and then at Missionary Ridge on November 25. The Regiment was next in the pursuit of the rebels to Ringgold Georgia then returned to camp at Whiteside, Tennessee. At the end of January 1864, the Regiment broke

---

785 *Canton Weekly Register,* September 7, 1863. A portion of a poem "They have Sent Back the Flag" by Mary Kirke Dilworth. The *Canton Weekly Register* was published in Fulton County, Illinois. 2nd Lt. Emmor Dilworth and Pvt. Harvey W. Dilworth were from Vermont, Illinois in Fulton County. Both soldiers served in the Eighty-fourth and were presumably related to the author of the poem.

786 Colonel Louis Waters was born in 1827 in Pennsyvania. He had practiced law in Macomb, Illinois since 1848 and had "ridden the circuit" with Lincoln and had served as a Whig in the Illinois legislature in 1854. At the outbreak of the civil war he had recruited and organized the 84th Illinois Infantry and was appointed its colonel. Find A Grave 2023.

787 Simmons, Louis A., *The History of the 84th Reg't Illinois Volunteers,* Hampton Brothers, Macomb, Ill., 1866, page38.

camp and marched east to Cleveland, Tennessee to begin fighting in the battles in north Georgia and Atlanta. In the subsequent campaign the Eighty-fourth fought at Dalton, Tunnel Hill, Buzzard's Roost Gap and Rocky Faced Ridge, Resaca, Kingston, Dallas, New Hope Church and Allatoona Hills, Kennesaw Mountain, Pine Hill, the Chattahoochee River, Peach Tree Creek, and the Siege of Atlanta. In a flanking movement south of Atlanta the Regiment fought at Jonesboro on August 31-September1 and Lovejoy Station September 2-6. Following the pursuit of Hood into north Georgia and Alabama the Eighty-fourth fought at Duck River (Columbia), Tennessee on November 24-27 and in the Battle of Franklin November 30. The rebels attacked the Union forces at Nashville on December 15 and 16 but were routed and retreated to the Tennessee River. The Eighty-fourth after this pursuit moved to Huntsville, Alabama and remained on duty there until March 1865. Following operations in East Tennessee the Regiment marched to Nashville and was mustered out at Nashville on June 8, 1865.[788]

There are two national and one regimental flag from the Eighty-fourth Illinois Infantry in the collection of the State of Illinois.

This national flag is inscribed: 84th REGT. ILL. VOLS. *The 34 stars on the flag are arranged in six horizontal rows of 5, 6, 6, 6, 6, 5 stars per row. It measures 70" on the hoist by 77" on the fly. The finial is spade shaped and painted red. These findings are typical of flags manufactured by Hugh Wilkens of Louisville.*

---

[788] Ibid, Simmons. This regimental history was written by the Quartermaster of the Regiment and is an excellent source of information for the exploits of the Eighty-fourth Illinois Infantry.

Although only a fragment remains of this national flag of the Eighty-fourth Illinois Infantry, the history of this flag is well documented. "The flag of the 84th Illinois Infantry has a remarkable history. There is left of it but a few feet of the shattered staff, and a bit of blood-stained red and white bunting to tell the sad story, --only a strip about two feet long and two inches wide. The regiment though organized and in the field nearly a year before, had seen no active service till the battle of Stones River, and went into that fight nearly eight hundred strong. In four hours, nine balls struck the staff, --their marks are there yet, --and of a color-guard of twenty-two, all were killed but two, and those were wounded. When such a narrow, slender stick of wood as this staff was struck so often, we can but wonder that even a single man could have lived through the shower of lead. Five color-bearers fell during one single hour of that battle, each in turn being replaced by another; scarce did one pair of manly hands loose their death-grip, than others were ready to take their place; and the regiment came out of its terrible baptism of fire and blood with but little more than one-half of its former number. They were supporting the 4th U.S. Artillery, and, God knows, they did their duty well." [789]

---

[789] Reed, Capt. H. B. *The Back-Bone of Illinois in Front and Rear, from 1861 to 1865,* H. W. Rokker, Printer and binder, 1886, page 6.

This Regimental flag has an unusual design with an eagle sitting on a shield surrounded by 13 stars; it measures 66" on the hoist by 72" on the fly. The manufacturer and history of this flag is unknown.

A flag was presented to the Regiment at Cleveland, Tennessee. "On the 21st of February, Capt. Ervin arrived in camp, having been at home for a few days on leave of absence, and brought us a new and very costly and beautiful flag—a present from the citizens of McDonough County. It was gladly, gratefully received; and when the brief address accompanying it, was read that evening by the Adjutant on Dress Parade, it elicited no feeble expression of good feeling; no meager manifestations of joy and pleasure…A unanimous vote of thanks, to the loyal citizens who sent to the field this splendid flag, bearing in gilt letters the names of our hard fought battles, was all the return the Regiment was able at the time to make…."[790] The fate of this flag is unknown. It is not in the collection of the State of Illinois.

No other flags carried by the 84th Illinois Infantry are known to survive.

[790] Ibid. Simmons, page 155-6.

85th Illinois Infantry

"Stand by the flag! Its stripes have streamed in glory,
To foes a fear, to friends a festal robe;
And spread in rhythmic lines, the sacred story
Of Freedom's triumph over the globe."[791]

The Eighty-fifth Illinois Infantry was organized in Peoria, Peoria County, Illinois and mustered into service for three years on August 27, 1862 under the command of Colonel Robert S. Moore.[792] The men of Company A volunteered from Mason County with some from Morgan, Peoria and Tazewell counties; men for Company B enrolled from Mason County with many from Fulton County; Company C and Company D were filled with volunteers from Mason County; Company E was staffed with men from Menard County; Company F recruited from Tazewell and Woodford counties; Company G filled with volunteers from Fulton and Schuyler counties; Company H and Company I had men from Fulton County and Company K had men who volunteered from Mason County. September 7, 1862, the Regiment boarded trains for Louisville, Kentucky and on September 30 moved out in pursuit of Bragg's rebel army. On October 8, the Eighty-fifth saw its first action at the Battle of Perryville, Kentucky where they lost 52 men killed, wounded, and missing. Following the battle, the Regiment pursued Bragg then went into camp at Nashville, Tennessee. They remained on guard and garrison duty at Nashville until March 1863; during that time, they guarded a supply train carrying food and ammunition to the troops at the Battle of Stones River and on January 3, 1863, fought in the Stones River battle. March 9 the Regiment skirmished with the enemy at Franklin, Tennessee and in April relieved the troops stationed at Brentwood, Tennessee. In June they returned to Nashville, in July were sent to Murfreesboro then returned to Nashville where they remained until mid-August. August 20 the Regiment began a march toward Chattanooga passing through Franklin, Columbia, and Pulaski in Tennessee and through Athens and Huntsville, Stevenson, and Bridgeport Alabama, across the Tennessee River and into Chattanooga. On September 15 the Eighty-fifth moved four miles south to Rossville, Georgia; four days later, three miles south of Rossville, the Battle of Chickamauga began and after two days fighting the Union Army was forced to retreat to a defensive line in front of Chattanooga. In October General Grant arrived to take command of the Union Army at Chattanooga and the emphasis changed from one of defense to one of offense. Within four days the Union forces seized Lookout Valley and established a crossing of the Tennessee River to Bridgeport to serve as a supply route. In November Grant began the Ringgold Campaign and the Eighty-fifth fought at Orchard Knob on November 23, at Tunnel Hill November 24-25 and Missionary Ridge on November 25. As soon as the rebel forces were in full retreat, the Eighty-fifth was sent to the relief of Burnside at Knoxville. The Regiment arrived within a

---

791 *Canton Weekly Register,* April 6, 1863.

792 Colonel Moore was born in Green County Kentucky in March 1827. When he was 10 years old his family moved to a farm in Menard County, Illinois. At the breakout of the Mexican-American War he enlisted as a private in the Fourth Illinois regiment. He returned from the war and settled on a farm in Mason County, Illinois and founded the town of Spring Lake. In 1854, he married, and moved to Havana, Illinois and in addition to farming began a business of buying and shipping grain. At the outbreak of the civil war, he recruited a company and entered the service as Captain of Company E of the 27th Illinois Infantry. He was wounded at the Siege of Corinth and while home on leave was authorized by Governor Yates to raise an infantry regiment. He raised the Eighty fifth Illinois Infantry Regiment and was appointed Colonel of the regiment. Aten, Henry J., *History of the Eighty-fifth Regiment Illinois Volunteer Infantry,* Compiled and Published under the auspices of the Regimental Association, Hiawatha, Kansas, 1901.page332

few miles of Knoxville on December 6 and learned that Longstreet had retreated, therefore, the Eighty-fifth returned to Chattanooga and went into camp at McAfee's Church near Chickamauga until May 1864. While at camp they were in skirmishes at Tunnel Hill and Buzzard's Roost Gap. In May 1864 the Eighty-fifth joined Sherman's Atlanta campaign. The Regiment saw action at Resaca, Dallas, New Hope Church, and Allatoona Hills in May and at Pine Hill, Lost Mountain and Kennesaw Mountain in June and at Ruff's Station, Chattahoochee River, Peach Tree Creek[793]and the Siege of Atlanta in July. In the flanking movement south of Atlanta they fought at Jonesboro August 25-30 and at Lovejoy Station September 2-6. Following the fall of Atlanta, the Eighty-fifth pursued Hood and Forest into north Georgia and Alabama, then joined the "March to the Sea" in November. Following the Siege and fall of Savannah, Georgia in December the Regiment was in the Campaign of the Carolinas and fought at Averysboro, the Battle of Bentonville and at Bennett's House and the surrender of Johnston's army. They marched to Washington D. C. via Richmond and marched in the Grand Review on May 24. The Eighty-fifth Illinois Volunteer Infantry was mustered out at Washington D. C. on June 5, 1865.[794]

The State of Illinois holds one national and two regimental flags of the Eighty-fifth Illinois Infantry in its collection.

Since this national flag has 35 stars and since it bears the battle honors of the Carolina campaign it was presumably manufactured late in the War or after the War, possibly to be carried in the Grand Review. The following inscriptions appear on this flag: PERRYVILLE, STONE RIVER, CHICKAMAUGA, POLK CREEK, BUZZARD ROOST, RESACA, ROME, KENESAW, PEACH TREE CREEK, ATLANTA, JONESBORO, SAVANNAH, GOLDSBORO, RALEIGH. *The flag measures 69" on the hoist by 76" on the fly.*

[793] On July 19, 1864 the colors of the 85th Illinois were captured along with Color Corporal Edward Scattergood of Company F and Pvt. William Hohamer of Company E.. The colors were probably captured by either a Confederate Arkansas or Mississippi Regiment. Personal communication Robert D. Jenkins, Sr. Dalton, Ga.

[794] Aten, Henry J., *History of the Eighty-fifth Regiment Illinois Volunteer Infantry,* Compiled and Published under the auspices of the Regimental Association, Hiawatha, Kansas, 1901.

This is one of the few flags in the Illinois collection manufactured by the Wilkins Company of Louisville, Kentucky.

Longley & Brothers of Cincinnati probably manufactured this regimental flag of the Eighty-fifth Illinois Infantry. The pattern of the stars over the eagle, the shield on the eagle and the lowered center section of the lower scroll are all typical of this manufacturer. It measures 71" on the hoist by 74" on the fly.

No other flags carried by the 85th Illinois Infantry are known to survive.

86th Illinois Infantry

"The flag of your country was never disgraced while under your protection and never went down but once and that at Kennesaw Mountain, Georgia, but Sergt. A. J. Deihl went down with it severely wounded, when Private Wm. A. McGinley of company H rescued it as it lay close to the rebel works. He flaunted it in the face of the victorious rebels all battered and torn by the enemies balls, but it floated over our dead until the enemy were driven from their works and our victory complete."[795]

The Eighty-sixth Illinois Infantry was organized in August 1862 at Camp Lyon in Peoria, Peoria County, Illinois and mustered into service for three years on August 27, 1862 under the command of Colonel David D Irons.[796] The men of Company A volunteered from Woodford County; Companies B and E from Marshall County; Companies C, D, H, I and K from Peoria County; Company F from Knox County and Company G from Tazewell County. September 7 the Regiment boarded a train for Louisville, arrived at Jeffersonville, Indiana on September 9 and crossed the Ohio River to Louisville, Kentucky on September 14. The Eighty-sixth remained at Louisville until October 1 when they marched in pursuit of Bragg. The pursuing Union forces met Bragg on October 8 at the Battle of Perryville, Kentucky. The Eighty-sixth took only a minor role in the Battle. After the battle Bragg was again pursued through Danville and Lancaster en-route to Crab Orchard, Kentucky. Pursuit was broken off and the Regiment was ordered to Nashville, Tennessee where they arrived November 7 and went into winter quarters. June 30, 1863, the Eighty-sixth left Nashville and marched to Murfreesboro, Tennessee, remained several days and returned to Nashville.[797] August 20, 1863, the Regiment marched south from Nashville to repair the *Nashville and Decatur Railroad.* At West Harpeth, Tennessee they were ordered to Columbia, Tennessee[798] then on through Lynnville, Pulaski Tennessee arriving at Huntsville, Alabama on September 8. On September 10 the Regiment left Huntsville, marched east through Stevenson and Bridgeport, Alabama, crossed Lookout Mountain and joined their brigade at Rossville, Georgia, six miles south of Chattanooga. September 19-20 the Eighty-sixth fought in the Battle of Chickamauga then retreated with the Union army to a defensive line south of Chattanooga. Four days later they were ordered north of the Tennessee River where they remained on guard duty until late November. Following a skirmish with Bragg at Sheppard's Point, the regiment accompanied Sherman in his march to Knoxville to relieve Burnside. However, while in route, the rebels retreated from their siege at Knoxville and the Eighty-sixth returned to camp above Chattanooga on December 18. On December 26 the Regiment

---

[795] From an address by Col. Allen Fahnestock as reported in *Fourth Reunion of the Eighty-sixth Regiment, Illinois Volunteer Infantry,* Peoria, Illinois, August 27 and 28, 1890. Page 19. Sergt. Deihl was wounded three times during the War but survived and was mustered out with the Regiment. The Private William McGinley referred to was actually Private William Maginly.

[796] Colonel David Irons was born in 1816 in New York state. The Irons family moved to Michigan and then on to Peoria, Illinois. For the next several years David Irons served as Deputy Sheriff for Peoria County and in 1861 is listed as Secretary of the City Insurance Company of Peoria. During the summer of 1862 David Irons was selected to command a new regiment of Illinis Infantry—the Eighty-sixth. Men were recruited from aross central Illinois and gathered at the Peoria County fairgrounds, then called Camp Lyon for training and muster.Colonel Irons died of "fever" at Nashville on August 11, 1863. Information is from Find A Grave wesite, 2023.

[797] The commander of the Regiment, Colonel David D. Irons, died at Nashville on August 11, 1863.

[798] "It was at Columbia Colonel Magee procured the famous whistle that ever afterwards remained with the regiment. By the mandates of this little instrument, in the hands of its successive commanders, the actions of the Eighty-sixth were controlled…Other regiments have prided themselves in their eagles and pets, and the Eighty-sixth too, had long since concluded she 'paid too dear for the whistle,' not to cherish it in lasting remembrance." Kinnear, J.R., *History of the Eighty-sixth Illinois Volunteer Infantry, During its Term of Service,* Tribune Company's Book and Job Printing Office, Chicago, 1866, page 20.

moved south across the Tennessee River and went into camp at McAfee's Church near Chickamauga Station. In February 1864 the Regiment skirmished at Buzzard's Roost and in March moved to Lee's and Gordon's Mills eight miles from McAfee's Church camp site. In May 1864 the Eighty-sixth joined Sherman's Atlanta campaign and marched twelve miles to Ringgold, Georgia; on May 10 they fought at Buzzard's Roost, then moved through Snake Creek Gap to fight at Resaca on May 14 and 15. Further fighting by the Eighty-sixth in the Atlanta campaign occurred at Rome, Dallas, New Hope Church, Allatoona Hills, Pine Hill, Lost Mountain, Kennesaw Mountain,[799] Ruff's Station, Chattahoochee River, Peach Tree Creek and the Siege of Atlanta. In the flanking movement south of Atlanta they fought at Jonesboro and Lovejoy Station in August and September. Following actions against the retreating rebels in north Georgia and Alabama, the Eighty-sixth joined the "March to the Sea" in November. The Regiment saw further action at Savannah, Georgia in December and in the Campaign in the Carolinas in early 1865 when they fought at Averysboro, Taylor's Hole Creek, Bentonville and at the Bennett's House where Johnston's army surrendered. They marched to Washington D. C. via Richmond and marched in the Grand Review on May 24, 1865, and were mustered out in Washington D.C. on June 6, 1865.[800]

There is one national and one regimental flag of the Eighty-sixth Illinois Infantry in the collection of the State of Illinois Military Museum.

Inscribed on the national flag: "86th Regt. Ill's Vol Infantry." *The stars on this flag are arranged in 7 horizontal rows (5,5,5,5,5,5,4 stars per row), which is typical of a flag manufactured by Longley & Brother of Cincinnati. However, the finial on the flagstaff has a spade shape and is painted red--this is typical of flags manufactured by Hugh Wilkens of Louisville. I suspect that the manufacturer was Longley & Brother, and the finial has been replaced.*

[799] Alonzo P. Webber, Principal Musician of the Regiment, volunteered as an advanced sharpshooter in the Battle of Kennesaw Mountain. On June 27, 1864, he found shelter behind a tree twenty-seven yards from the rebel line and remained there from nine o'clock in the morning until six o'clock in the evening. For his actions he received the Medal of Honor. Beyer, W. F., Keydel, O. F., *Deeds of Valor,* Longmeadow Press, Stamford, St., 1993, page 372.
[800] Ibid Kinnear.

This fragment is all that remains of the regimental flag. This is presumably in part due to the battle damage described in the battle at Kennesaw Mountain.

No other flags carried by the 86th Illinois Infantry Regiment are known to survive.

87th Illinois Infantry

"Capt. Anderson with 75 men went down to Friar's Point 15 miles below here and stade (sp) 4 days & nights to try to get Capt Stanly or a band of Gurillers (sp) that was in that portion of the state of Miss but the Citizens saved us the truble (sp) by capturing him and his men and shooting them."[801]

In August 1862 the Eighty-seventh Illinois Infantry was organized at Camp Logan in Shawneetown, Gallatin County, Illinois.[802] The men of Companies A and E volunteered from Hamilton County; Company B from Edwards and White County; Company C, F, G, I and K from White County; Company D from Wayne County and Company H from Edwards County. The Regiment was mustered for three years' service on October 3, 1862, but with an effective date of muster of August 2, 1862. The Regiment was commanded by Colonel John E. Whiting.[803] For the first five months the Regiment remained at Camp Logan but raided rebel camps across the Ohio River into Kentucky. On January 31, 1863, the Eighty-seventh boarded transports for Memphis. During the trip measles broke out resulting in the loss of 250 men dead and disabled from the disease. The Regiment remained at camp near Memphis for three months and during that time, along with the Sixty-third Illinois Infantry, raided a Confederate partisan cavalry at Hernando, Mississippi. On May 9, 1863 the Eighty-seventh boarded the *Crescent City* for the trip to Young's Point, Louisiana, and the Vicksburg Campaign. After several days of picket duty, they were ordered to cross the Mississippi River south of Vicksburg at Warrenton, Mississippi and to join the Union forces at Vicksburg. The Regiment was in heavy fighting during the assault on rebel lines on May 22 and 23, regrouped at Warrenton in early June and returned to the siege lines at Vicksburg on June 23. They remained in the trenches until the fall of the city on July 4 when they marched to Jackson, Mississippi to fight in the siege there July 10-17. In August the Regiment was ordered to Natchez, Mississippi and on August 3 raided a cotton warehouse at Kingston, Mississippi. Later in August the Eighty-seventh moved further down river to New Orleans and in September crossed the River to Algiers, Louisiana and moved with the Army of the Gulf into Western Louisiana along the Atchafalaya and Teche Bayous. In November the Regiment was mounted[804] and continued scouting duties near Franklin and New Iberia, Louisiana. In March 1864 the Eighty-seventh led the cavalry on Banks Red River campaign. During the campaign, at Wilson's plantation, near Pleasant Hill, on April 7 the Union cavalry met and defeated General Green's Confederate cavalry. In this action the Eighty-seventh lost 30 men killed and wounded. The Eighty-seventh saw action the following day at the Battle of Sabine Cross Roads (Mansfield); April 9 they again fought at Pleasant Hill, then skirmished at Cloutiersville, Monett's Ferry at Cane River and at Alexandria. They fought at Marksville on May 15 and went into camp at Morganza Bend, Louisiana on May 21 until Fall of 1864. While at Morganza Bend the Regiment was busy scouting and skirmishing on the bayous between the Mississippi River and the Atchafalaya River to the west; part of the time was also spent on the steamer *Baltic* as part of the Marine Brigade. In August 1864, Captain Thomas Sheridan, Company K, with a detachment of 50 men was captured near Williamsport, Louisiana. On September 4, 1864, the Regiment departed on the *Ohio Belle* for White River Island, Arkansas[805] and remained there until January 1865, when the Regiment

[801] Springhouse Magazine Online. A Civil War Letter submitted by Doris Cottingham Nelson; the letter is from McAllister. S. Hooker a soldier of the 87th Illinis Infantry to his wife Elizabeth

[802] Camp Logan was in Southern Illinois just across the border from Kentucky on the Ohio River.

[803] Colonel Whiting was born in 1821 in Indiana and became a practicing lawyer in Carmi, Illinois

[804] The Eighty-seventh Illinois Infantry remained a mounted unit until the end of the War.

[805] Three companies were sent detached for duty at St. Charles, Arkansas.

moved to Helena, Arkansas. The Eighty-seventh was mustered out at Helena on June 16, 1865.[806]

There are three national flags and one regimental flag of the Eighty-seventh Illinois Infantry in the collection of the State of Illinois.

The inscriptions on this national flag are difficult to read but the inscriptions probably read as follows: VICKSBURG, JACKSON, WILSON HILL, MANSFIELD, PLEASANT HILL, CANE RIVER, 87th REGT. ILLs, --(UNREADABLE)--, BAYOU LETSWORTH

There are no inscriptions of this national flag of Philadelphia manufacture.

[806] *Report of the Adjutant General of the State of Illinois,* H. W. Rokker, State Printer and Binder, Springfield, Ill., 1886, Volume V, pages 239-40.

This national flag is torn and bears no inscriptions.

Based on the position of the stars, the eagle and scroll under the eagle, this flag is characteristic of those manufactured for the New York depot. It measures 72" on the hoist by 71" on the fly.

No other flag carried by the 87th Illinois Infantry Regiment are known to survive.

88th Illinois Infantry - *"Second Chicago Board of Trade Regiment"*

"Thus was the contest raging and every moment brought us nearer and nearer to the crest. Hoarse cries mingled with the rattle of musketry; eyes stared out of smoke and powder-begrimed faces were wild with the light of battle. The proud old flag had nearly reached the rebel lines; many of its brave and noble bearers had been struck down, but ere it could fall to the ground it was raised boldly aloft and moved steadily on....There on the crest waved the war-worn banners, and in rapid succession and with scarcely an interval, the regimental banners of my brigade, nine in number were planted....Hers (Eighty-eighth Illinois Infantry) was the flag that first floated over Mission Ridge, borne by the gallant Sgt. John Cheevers."[807]

The Eighty-eighth Illinois Infantry was organized in Chicago by Colonel Francis Sherman[808] and mustered in for three years on September 4, 1862. The men of Company A (*Wicker Guards*) were from 14 counties in Illinois and from Indiana; Company B (*Arthur Guards)* from La Salle County with some men from Cook and Champaign counties; Company C (*Nelson Guards)* primarily from Iroquois County with some men from La Salle and Mc Lean counties; Company D (*Kimbark Guards)* from Bureau County; Company E (*Holden Guards)* from Cook County with some men from DeKalb and Grundy counties; Company F (*Chickering's Company)* from Hancock, Kankakee and Woodford counties; Company G (*Hubbard Guards)* from Cook County; Company H (*Crosby Guards)* from Cook, LaSalle, Macoupin and Marion counties; Company I from Cook County and Company K (*Barnard's Company)* from Cook and Woodford counties. September 4, 1862, the Regiment moved by train first to Cincinnati, then across the Ohio River to Covington, Kentucky. When it was evident that Cincinnati was not threatened the Eighty-eighth moved down the Ohio River to Louisville, Kentucky. October 1 the Regiment left in pursuit of Bragg and fought the rebel forces at Perryville, Kentucky on October 8. The Union troops continued pursuit of the rebel forces to Crab Orchard, Lebanon and Bowling Green, Kentucky then went into camp south of Nashville. December 26 the Regiment was ordered south to Murfreesboro and fought in the Battle of Stones River December 30-January 2, 1863; the Eighty-eighth was heavily engaged and according to Colonel Sherman's diary the Regiment lost 150 of the 420 men who went into action.[809] Following the battle the Regiment remained in camp at Murfreesboro until June 1863. During June and July, the Eighty-eighth was engaged in the campaign to drive Bragg from Middle Tennessee. They were then ordered east across the Cumberland Mountains and Tennessee River in the Chickamauga Campaign. On September 19 and 20 the Regiment fought in the Battle of Chickamauga and with the defeat of the Union forces, fell back to a defensive line south of Chattanooga. The Union troops remained at Chattanooga through October and part of November but took the offense in late November and fought at Orchard Knob on November 23 and 24 and at Mission ridge on November 25.

---

[807] Aldrich, C. Knight, editor, *Quest for a Star, the Civil War Letters and Diaries of Colonel Francis T. Sherman of the 88th Illinois,* The University of Tennessee Press, Knoxville, 1999, page 87-8. From a letter written by Sherman to his brother-in-law "Bill" in January 1864.

[808] Colonel Sherman was born in December 1825 in Newtown Connecticut. He came to Chicago with his family in 1834; in the Spring of 1849 he went to California in the gold rush but returned to Chicago in 1850 none the richer. Prior to the civil war he was engaged in manufacturing and railroad building. In 1861 he entered the Union army as Major of the 12th Illinois Cavalry. In that role, in March 1862, he commanded a detail that arrested the Confederate spy—Belle Boyd. In August, 1862 he was commissioned Colonel of the 88th Illinois Volunteer Infantry.

[809] Ibid, Aldrich, page20.

The day following the battle the Regiment marched to the relief of Knoxville, Tennessee and then campaigned in east Tennessee until May 1864 when the Regiment joined the Atlanta campaign. Subsequently the Eighty-eighth saw action at Buzzard's Roost Gap, Rocky Faced Ridge, the battle of Resaca, battles at Dallas, New Hope Church and Allatoona Hills, Pine Mountain, Lost Mountain, Kennesaw Mountain, Ruff's Station, Chattahoochee River, Buckhead, Peach Tree Creek and the Siege of Atlanta. In a flanking movement south of Atlanta the Regiment fought at Jonesboro on August 31-September 1 and Lovejoy Station on September 2-6. They joined in the pursuit of Hood into north Georgia and Alabama in October and fought at Duck River, Spring Hill, and the Battle of Franklin[810] in November. December 15-16 the Eighty-eighth fought at the Battle of Nashville[811] and pursued the fleeing Hood until the end of December. The Regiment went into winter camp at Huntsville, Alabama in January 1865, operated in East Tennessee in March and April and moved to Nashville for muster out on June 9, 1865. [812]

There are two national flags and one regimental flag of the Eighty-eighth Illinois Infantry Regiment in the holdings of the State of Illinois.

*This heavily worn national flag of the Regiment is inscribed "*88th REGT ILL, PERRYVILLE.*" Because of its obvious damage this may have been the flag planted on Mission Ridge.*

---

[810] Referring to "trophies captured by this division during the engagements of Spring Hill, Franklin, and Nashville: Corpl. J. K. Merrifield, Company C, Eighty-eighth Illinois Volunteers, captured and brought from the field two rebel flags, one a brigade and the other a regimental flag, without designation. Corpl. Peter M. Woolf, Company A, Eighty-eighth Illinois Volunteers, captured a rebel battle-flag inscribed 'First and Fourth Missouri Infantry.'" O. R. Series 1, Volume XL/1 [Series #93] page 237. James K. Merrifield claimed capture of the flag of the 1st Missouri and General Featherston's headquarters flag. The flag of the 1st Missouri "I sent to Chicago and during the great fire there it was destroyed." Featherston's flag I "have it now in a frame on the wall in my house in St. Louis." *Confederate Veteran*, Vol. 24, No. 12, Dec 1916, pp 551-2. For his actions Merrifield received the Medal of Honor. *Civil War Medal of Honor Winners from Illinois,* Civil War Centennial Commission July 1962, page 8.

[811] The Eighty-eighth captured five rebel flags at Nashville. "The five flags captured by the Eighty-eighth Illinois and the one captured by Corporal Case were the only ones that were brought from the field by the captors. Those captured by the Eighty-eighth Illinois were displayed before General Thomas when he visited the brigade on its arrival at Nashville. They were ordered to be returned and will be forwarded to department headquarters as soon as they arrive." O. R. Series 1, Volume XL/1 [Series #93} page 238.

[812] *Report of the Adjutant General of the State of Illinois,* H. W. Rokker, State Printer and Binder, Springfield, 1886, Volume V, pages 260-1.

88TH ILLINOIS INFANTRY

*This 35-star national flag of Eighty-eighty Illinois Infantry is in very good condition. The inscriptions on the flag read "*2nd BOARD OF TRADE REGT. 88TH ILL. VOL. INFANTRY. CHAPLIN HILLS, STONE RIVER, CHICKAMAUGA, MISSION RIDGE, ATLANTA, FRANKLIN, NASHVILLE.*" All the printing of the inscriptions is similar suggesting that all the inscriptions were applied at the same time. Since the flag has 35 stars and the last battle inscribed was fought in December 1864, this flag must have been acquired by the Regiment late in the War or after the War. The fact that the stars are gold and arranged in six horizontal rows and the Regimental inscription is gold are typical findings of flags manufactured by Gilbert Hubbard & Company, 205 & 207 South Water Street in Chicago. The flag measures 72" on the hoist by 74" on the fly.*

This regimental flag of the Eighty-eighth is a 35-star flag in good condition. The star pattern above the eagle, the shield on the chest of the eagle and the down turned center portion of the lower scroll are all characteristic of flags manufactured by John S. Shilleto & Company, 101 West 4th Street, Cincinnati. It measures 71" on the hoist by 75" on the fly.

There are references to other flags having been presented to this Regiment.

"Our company's patron, Mr. Murry Nelson, of Chicago, had formally, and with some ceremony and a speech, presented us with a costly silken banner, but it was after some time given over to storage and safety. (A gentleman with a better memory than I appear to have had informs me that our flag was with us in the field and used there as a battle flag. Mr. Nelson holds the flag now as a valued memento of the war.)"[813] Since the flag mentioned was presented in 1862 it would have been a 34-star flag and therefore not the "battle flag" shown above. The whereabouts of this flag today is unknown.

Capt. Alexander C. McClurg, Company H, in his memoirs mentions the flags of the Regiment[814]. "The morning of that day was bright and fine and, headed by its drum corps, the regiment marched from camp at about nine o'clock, with drums beating and flags flying, for, contrary to regulations, through the patriotic ardor of the citizens, and particularly of the ladies, every company in the regiment had its flag and its color-bearer. We marched down Michigan Avenue to the Illinois Central station. We were sure we made a fine show and were fully conscious of the soldierly appearance we presented to the admiring thousands gathered to see 'another Chicago regiment off for the front.'" Based on this report each company had a flag, probably a national flag. What happened to these flags after the War is unknown.

Details are available regarding flag presentations to two of the Companies. "FLAG PRESENTAION TO THE KIMBARK GUARDS-A large number of the lady friends of the *Kimbark Guards*, Capt. Chandler, of the Second Board of Trade regiment, repaired to the camp a day or two since, and gave to the company an elegant supper. The occasion was made doubly interesting to the company and its friends by the presentation to them by their patron Geo. M. Kimbark, Esq., of a beautiful silk banner, full regulation size bearing upon its folds, in letters of gold, the name of the company, and very appropriate inscription, 'God is with us.' Capt. Chandler responded in very appropriate terms."[815]

"THE CROSBY GUARDS' JUBILEE—The Crosby Guards, on Thursday afternoon, were made the recipients of an elegant and costly silk flag by their patron, W. H. Crosby, esq. The remarks attending the presentation were made by E. W. Russell, esq., to which Capt. McClurg, responded most touchingly and eloquently."[816]

---

[813] Francis, Charles Lewis, *Narrative of a Private Soldier in the Volunteer Army of the United States, During a Portion of the Period Covered by the Great War of the Rebellion of 1861,* William Jenkins and Company, Brooklyn, N. Y. 1879, page 46. Charles Francis was private in Company B or the Eighty-eighth Illinois Infantry.

[814] "American Volunteer Soldier" (Extract from the memoirs of General Alexander C. McClurg) published in *Reminiscences of Chicago During the Civil War,* The Citadel Press, New York, 1967, page 130.

[815] *Chicago Daily Tribune,* August 30, 1862. The *Kimbark Guards* were Company D of the Regiment.

[816] *Chicago Daily Tribune,* August 29, 1863. The *Crosby Guards* were Company H of the Regiment.

89th Illinois Infantry—"*Railroad Regiment*"

"THE HEADLIGHT."—The Railroad Regiment, the 89th Illinois Infantry, have commenced the publication, at Shelby, Ky., of a spicy little sheet bearing the above title. Capt. H. M. Blake is 'Official Director,' and Sergeant Major John M. Farquhar, 'Chief of Corps' Of the five compositors four belong to the Railroad Regiment.[817]

Robert Forsyth of the *Illinois Central Railroad* and W. D. Manchester of the *Lake Shore and Michigan Southern Road* organized the "Railroad Regiment" in Chicago during the summer of 1862. The Regiment was composed mainly of men working for the railroads in Illinois; in addition to the two railroads mentioned above the superintendents of the following railroads were also active in the recruitment and organization, *Chicago, Burlington and Quincy*, *Pittsburg, and Fort Wayne, Michigan Central, Chicago and North-Western, Chicago and Rock Island, Chicago and Milwaukee, Chicago Alton and St. Louis* and the *Galena and Chicago Union.* The men of Company A (*Scammon Light Infantry)* volunteered from 12 counties in northern Illinois; Company B (*Du Quoin Company)* from Franklin County; Company C (*Milwaukee Railroad Company)* from Cook County; Company D (*Forsyth Guards)* mainly from Jackson and Perry counties in southern Illinois and Iroquois County in upper central Illinois (from cities on the *Illinois Central RR*); Company E (*Chicago, Burlington & Quincy Railroad Company)* from Kane and Knox counties; Company F from Rock Island County; Company G (*Captain Whiting's Company)* from Knox County; Company H (*Kendall Guards)* from Kendall County; Company I (*Amboy Guards)* from Lee County and Company K(*Galena Railroad Guards)* from Cook County. Thirty-two-year-old accountant, Colonel John Christopher commanded the Regiment. On August 27, 1862, the Eighty-ninth was mustered into service and on September 4 was ordered to Louisville, Kentucky. On October 1, the Regiment under General Buell marched in pursuit of the rebel forces under Bragg. During the pursuit, on October 7, the Eighty-ninth met the rebel forces at Lawrenceburg, Kentucky and the following day fought at Perryville, Kentucky. November found the Regiment in Nashville, and on December 26 the Regiment, as part of Johnson's division marched toward Murfreesboro, Tennessee. In the subsequent Battle of Stones River from December 31 to January 2, 1863, the Eighty-ninth lost 149 men killed, wounded and missing.[818] Following the battle the Regiment remained on duty at Murfreesboro until June. In the Middle Tennessee or Tullahoma Campaign, they fought at Liberty Gap, Tennessee June 22-27 and remained in middle Tennessee until mid-August. From mid-August until late September the Eighty-ninth marched across the Cumberland Mountains, crossed the Tennessee River, and arrived at Chickamauga, Georgia on September 17. In the following Battle of Chickamauga, Georgia on September 19-20, the Eighty-ninth was heavily engaged and lost 109 men killed, wounded, and missing including in officers killed, Lieutenant-Colonel Duncan J. Hall and Captains Rice, Spink, and Whiting. The Union Army fell back to a defensive position south of Chattanooga following their defeat at Chickamauga but in November went on the offense and the Eighty-ninth fought at Orchard Knob on November 23-24 and at Mission Ridge on November 25. On November 28 the Regiment marched with the Union forces for the relief of Knoxville, Tennessee and from late

---

[817] *Chicago Daily Tribune,* October 29,1862

[818] John M. Farquhar, Scotland born Sergeant Major of the Eighty-ninth received the Medal of Honor for his actions at the battle on December 31, 1862. "When a break occurred on the extreme right wing of the army of the Cumberland, this soldier rallied fugitives from other commands and deployed his own regiment, thereby checking Confederate advance until a new line was established." *Civil War Medal of Honor Winners from Illinois, published* by the Civil War Centennial Commission, July 1962, page 11.

December 1863 until April 1864 campaigned in East Tennessee against Longstreet's forces. In May 1864 the Eighty-ninth joined Sherman's Atlanta campaign. In that campaign the Regiment saw action at Rocky Faced Ridge, Buzzard's Roost Gap, Dalton, Resaca, Pickett's Mill, Adairsville, Dallas, New Hope Church, Allatoona Hills, Pine Hill, Lost Mountain, Kennesaw Mountain, Ruff's Station, Peach Tree Creek, and the Siege of Atlanta. In a flanking movement south of Atlanta the Eighty-ninth fought at Jonesboro on August 31-September 1 and Lovejoy Station on September 2-6. The Regiment marched into Atlanta September 8, and then went in pursuit of Hood into north Alabama and Georgia September 22-November 3. In the Nashville Campaign of November-December 1864 the Regiment fought at Columbia (Duck River) November 24-27, the Battle of Franklin on November 30 and in the Battle of Nashville December 15-16. Following the pursuit of Hood to the Tennessee River the Eighty-ninth was ordered to Huntsville, Alabama and then to Nashville where the Regiment was mustered out on June 10, 1865. [819]

There is only one flag, a national flag, of the Eighty-ninth Illinois Infantry in the holdings of the State of Illinois. However, both a national and regimental flag were presented to Eighty-ninth—described as follows. "THE RAILRAOD FESTIVAL. The Fair and Festival for the Railroad regiment passed off very finely at Bryan Hall last evening…... At the left of these hung a splendid banner bearing the American eagle. Over his head were the Stars of the Union as it shall be, and beneath his talons the time-honored motto of our nation. At the opposite end of the wall the national flag hung; on its centre stripe the number and name of the regiment were beautifully executed in gold. Its stars encompassed the railroad man's most expressive phrase, 'Clear the Track.' Resting against the wall back of all these was the finest banner and most expressive design of all, corresponding in size with the first. Its center displays a magnificent engine, 'The Union.' The engineer is at his post, throttle bar in hand, and all steam on. At the opposite end of the device is a brakeman, at his post. These beautiful banners are the handiwork of the well-known artist, Chas. L. Dubois, and are exceedingly creditable to him"[820]

Two days later this follow up appeared. "RAILROAD REGIMENT—This fine regiment, Col. John Christopher, was yesterday afternoon presented by a committee of ladies' with an elegant stand of colors. The flags were purchased by the ladies of Chicago from the proceeds of a festival given by them at Bryan Hall a few evenings since, and are in every respect worthy the fair donors."[821] Based on the above it appears that the "stand of colors" presented by the ladies of Chicago consisted of the regimental flag and national flag described at the Fair in Bryan Hall. The regimental flag is apparently lost but the national flag is probably the flag shown here. On this flag "CLEAR THE TRACK" is written on the lower edge of blue canton, just below the stars and the regimental designation is printed in gold on the center red stripe. This fits the description of the national flag described in the *Chicago Daily Tribune* of 1862.

---

[819] There is not a good history of this Regiment. Sources used were *Report of the Adjutant General of the State of Illinois,* H. W. Rokker, State Printer and Binder, 1886, Volume V, pages 288-290; Eddy, T. M., *The Patriotism of Illinois*, Clarke & Co. Chicago, 1866, Volume II, pars 335-341; George, Charles, B., *Forty Years on the Rail,* R. R. Donnelley & Sons, Chicago, 1887, pages 99-128.
[820] *Chicago Daily Tribune*, September 2, 1862.
[821] *Chicago Daily Tribune,* September 4, 1862.

89TH ILLINOIS INFANTRY

The National flag is inscribed in the blue canton-"CLEAR THE TRACK" *The remainder of the inscriptions read-*89th ILLINOIS INFANTRY, STONE RIVER, LIBERTY GAP, ORCHARD KNOB, CHICKAMAUGA, MISSION RIDGE. *The flag measures 70" on the hoist by 74" on the fly.*

No other flags carried by the 89th Illinois Infantry are known to survive.

90th Illinois Infantry- "*Irish Legion*"

The *Irish Legion* on at least two occasions had a disguised woman in its regiment. Frances Elizabeth Quinn was discovered the first time by Colonel O'Meara and the second time when she was wounded at the battle of Stones River and captured by the Confederates. [822]

The "*Irish Legion*" was recruited in the summer of 1862 at Camp Douglas in Chicago with most recruits of either Irish or Irish-American descent. Unique in the history of this Regiment is the fact that the Ninetieth Illinois Infantry was largely recruited and organized by an Irish Catholic priest—Very Rev. Dunne who at the time was Catholic Vicar General of the Diocese of Chicago. Apparently, his efforts were not only patriotic in nature but were influenced by his desire to disprove the charge that the Irish Catholic element was disloyal. Governor Yates offered Dunne a commission as Colonel of the regiment, but the honor was declined as being inconsistent with Dunne's holy office. The men of the Regiment were recruited as follows: Company A (*Mulligan Guards*) from Rockford in Winnebago County; Company B (*Ryan Guard)* from Galena in Jo Daviess County; Company C (*Lockport Guards)* from Lockport in Will County; Company D from Joliet in Will County; Company E, F, G and H from Chicago in Cook County; Company I from Belvidere in Boone County and Freeport in Stephenson County and Company K from Ottawa in La Salle County. The Regiment was mustered in on September 7, 1862, under the command of Colonel Timothy O'Meara[823] and remained in Chicago on guard duty at Camp Douglas until November 27, 1862.[824] On that date the Regiment was ordered to LaGrange, Tennessee and went into camp there on December 2. Three days later the Ninetieth was ordered south to Cold Water, Mississippi to guard Grant's supply line on the *Mississippi and Tennessee Railroad.* On December 20, rebel forces under Van Dorn having already captured Holly Springs, Mississippi, attacked the Ninetieth at Cold Water.[825] However, the Regiment was prepared, and Van Dorn was repulsed. Nevertheless because of the loss of the supply base at Holly Springs, Grant ordered the Ninetieth back to camp at La Grange, Tennessee. In early 1863 the Regiment was ordered west to Lafayette, Tennessee to guard the *Memphis and Charleston Railroad.* In June they were ordered to move via Memphis down the Mississippi River to Vicksburg, Mississippi; there they were engaged until the fall of Vicksburg on July 4. From July 4 until July 17 the Regiment took part in fighting around Jackson, Mississippi and then went into camp on the Big Black River until September 27. They were then ordered upriver by boat to Memphis and then marched along the route of the Mem*phis & Charleston Railroad* toward Chattanooga. On October 11, 1863, at Germantown, Tennessee the Regiment boarded platform cars to hurry forward to Collierville, Tennessee to relieve General Sherman who was surrounded by Confederate cavalry. On arrival they found that the enemy had retreated, and the Regiment resumed their march toward Chattanooga. On November 24 the Ninetieth crossed the Tennessee River and the following morning were heavily engaged in the assault on Mission Ridge. In the fighting that day, Colonel O'Meara, and Lieutenant James Conway (Company A) were killed; total casualties for the Regiment

---

[822] Cordell, Melinds (2016)"Hurrah for God's Country" *Courageous Women of the Civil War.*Chicago Review Press

[823] Colonel Timothy O'Meara was born in County Tipperary, Ireland in August 1836. He immigrated with his family to New York and enlisted in the United States army as a member of the New York Regiment of Mounted Rifles. In 1861 he enlisted in the 42nd New York Inantry; in August 1862 he traveled to Chicago to inquire re the Irish Regiment being organized there. Due to his experience he was given the command of the *Iruh Legion.* https://chickamaugablog.wordpress.com.

[824] During the three months that the Regiment was on guard duty in Chicago over three hundred of the men deserted.

[825] A detachment from the Second Illinois Cavalry and four companies from the One Hundred and First Illinois Infantry had escaped from Holly Springs before its surrender and joined the Ninetieth in defending against Van Dorn's forces attaching Cold Water.

were approximately one hundred men killed, wounded or missing. The following day the Ninetieth joined in the pursuit of Bragg's forces as far as Graysville, Georgia and then marched to relief of Burnside's army under siege at Knoxville, Tennessee. However, Longstreet's siege of Knoxville was lifted while the Ninetieth was en-route therefore the Regiment was ordered to return to Chattanooga. In late December 1863 the Regiment was ordered to Scottsboro, Alabama and remained there until May 1864 when ordered to Resaca, Georgia to join Sherman's Atlanta campaign. In that campaign the Ninetieth fought at Resaca, Dallas, New Hope Church, Allatoona Hills, Kennesaw Mountain, Nickajack Creek, Chattahoochee River, the Siege of Atlanta and Ezra Chapel. In the flanking movement south of Atlanta the Regiment fought at Jonesboro August 31-September 1 and at Lovejoy Station on September 2-6. Following the fall of Atlanta, the Regiment joined in the pursuit of Hood into North Georgia and Alabama and skirmished with the enemy at Rome, Georgia and Gadsden, Alabama. Mid November the Ninetieth joined in Sherman's "March to the Sea" and saw action in Georgia on the Ogeechee River on December 7, Bryant's Court House on December 8, and Fort McAllister in Savannah December 10-21[826]. In the 1865 campaign in the Carolinas the Ninetieth fought at Salkehatchie Swamps, South and North Edisto Rivers and Columbia in South Carolina and then in the Battle of Bentonville, North Carolina on March 20-21. They were at Bennett's House for the surrender of Johnston's rebel forces on April 26 and then marched via Richmond, Virginia for the Grand Review in Washington D. C. on May 24, 1865. June 7 the Ninetieth departed Washington D. C. for Chicago where it was mustered out June 10, 1865.[827]

There is one national and one regimental flag from the Ninetieth Illinois Infantry Regiment in the holdings of the State of Illinois.

The national flag is not inscribed. The double elliptical arrangement of stars is characteristic of flags manufactured for the Philadelphia Depot. The flag measures 70" on the host by 73" on the fly.

[826] The colors of the Ninetieth were the first colors planted on the Fort following its surrender.
[827] *Report of the Adjutant General of the State of Illinois,* H. W. Rokker, Printer and Binder, Springfield, 1886, pages 311-13.

After the Battle of Cold Water, Mississippi General Grant honored the Ninetieth for its stand against the rebel forces. “For their gallant stand on this occasion, the 90th, then only twenty days in the field, received the thanks of Gen Grant in general order No. 4, and also a splendid national flag." [828] Although not specifically identified the national flag shown here may be that presentation flag.

Company K was presented a company flag on their departure from home. "Earlville sends a fine company of soldiers to the wars, to go in the Irish Brigade of Chicago. On Saturday, last, before their departure for Chicago, a splendid flag was presented them in behalf of the ladies by Miss Fidelia Norton."[829] There is no further description or history of this flag and its whereabouts today is unknown.

The regimental flag of the Ninetieth Illinois Infantry Regiment is unique, and its origin is known. The flag was presented to the Regiment early in the War and the battles were inscribed during and after the War. "While at Joliet, Co. D was presented with a large and very costly silk flag by the ladies of Joliet, Mrs. E.C.Fellows, making the presentation speech. It was an exceedingly beautiful gift, and an appropriate one also, notwithstanding the fact that several of the Irishmen in the company were Dutchmen. These fought as well ‘under the green’ as if ‘to the manor born.’"[830]

The four corners on the obverse of the flag are inscribed: SAVANNAH, Ga., RESACA, Ga., JACKSON, Miss., FORT McALLISTER, Ga. *Surrounding the central corps symbol-"*40 ROUNDS*" - the following battles are inscribed:* Siege of Atlanta, Ga July 19th 22nd 28th Aug 3rd 17th, 1864; Missionary Ridge, Tenn Nov 25th, 1863; Siege of Vicksburg, Miss.; Siege of Knoxville, Tenn; Coldwater, Miss; Kennesaw Mt, Ga; Dallas, Ga.; Jonesboro, Ga; Lovejoy Station, Ga; Chattahoochee River, Ga; New Hope Church, Ga; Ogeechee River, Ga; Gadsden, Al; Salkehatchie River, N.C.; Bentonville, N.C.; Rome, Ga.

---

828 Woodruff, George H., *Fifteen Years Ago or the Patriotism of Will County,* Joliet Republican Book and Job Steam Printing house, Joliet, 1876, page 373.
829 *Aurora Beacon,* June 20, 1861,
830 Ibid, Woodruff, page 368

90TH ILLINOIS INFANTRY

On the reverse of the flag the four corners are inscribed: IRISH LEGION, 15th ARMY CORPS, MUSTERED OUT June 1865, ARMY OF THE TENNESSEE. *In the center is the Regimental ID:* 90th ILL VOLUNTEER INFT.

No other flags carried by the *Irish Legion* are known to survive.

91st Illinois Infantry

The Ninety-first Illinois Infantry Regiment was organized at Camp Butler in Springfield, Illinois in the summer of 1862 by 35-year-old, Colonel Henry Day, a merchant from Morris, Illinois, and mustered into service on September 7, 1862. The men volunteered from the following counties, Company A from Litchfield in Montgomery County; Company B from Terre Haute and Raritan in Henderson County; Company C from Oquawka in Henderson County; Company D from Morris in Grundy County, Company E from Lisbon in Kendall County, Company F from Lawrenceville in Lawrence County, Company G from Carrollton and White Hall in Greene County and Manchester in Scott County and Company H, I and K from Greene County. The Regiment left Camp Butler on October1 and arrived at Shepherdsville, Kentucky on October 7 to scout for Morgan in Kentucky and to guard the line of the *Louisville and Nashville* Railroad. On December 26, Morgan captured three companies of the Regiment on detached service at Nolin Station and Bacon Creek Station on the rail line south of Shepherdsville and on December 27 captured the remainder of the Regiment at Shepherdsville. The men and officers were paroled and returned to St. Louis or to home. On February 28, 1863, two thirds of the Regiment reunited at Benton Barracks in St. Louis and remained at Benton Barracks until June 5, 1863 when the Regiment was exchanged and reequipped. On July 8 the Ninety-first boarded the steamboat *Nebraska* and in company with the Twenty-ninth Illinois Infantry sailed downriver to Vicksburg. The Regiment was heavily engaged at Vicksburg; however, most of the Regiment's losses were due to disease rather than enemy fire. Following the fall of Vicksburg in July, the Regiment was ordered downriver to Port Hudson, Louisiana and then in August ordered to New Orleans. On September 5, 1863, the Ninety-first was transported upriver to Morganza Bend and then marched west toward the Atchafalaya River. They met the enemy near the River on September 7 and 8, took prisoners and retired back to the Mississippi River at Morganza Bend. In October the Regiment returned to New Orleans and on October 23 sailed for Point Isabel, Texas, arrived there November 3 and moved inland to Brownsville, Texas on the Rio Grande River just north of the Mexican border. Brownsville was occupied to interfere with trade between the Confederate States and Mexico. The Regiment remained at Brownsville until July 1864 when they marched to Brazos Santiago on the Texas coast. While on garrison duty there, on September 11, the Ninety-first fought rebel forces at Bagdad, Texas. In late December the Ninety-first sailed back to New Orleans where they remained until February 1865. On February 24 the Regiment boarded the *Katie Dale* and were transported to Mobile Point, Alabama. The Ninety-first took an active role in the capture of Spanish Fort, Fort Blakely, and the city of Mobile. Following the surrender of Mobile the Ninety-first led the pursuit of the retreating rebels to Eight Mile Creek near Whistler's Station, Alabama. There the rebels were routed; the Regiment then marched to Nana Hubba Bluffs on the Tombigbee River, went into quarters, and began construction of Fort Granger. When news of surrender of the Confederate Army reached the Regiment on May 9, the Regiment returned to Mobile where they were mustered out on July 12, 1865.[831]

The collection of the State of Illinois has five flags identified as having been carried by the Ninety-first Illinois Infantry—two national flags and three regimental flags.

---

[831] *Report of the Adjutant General of the State of Illinois,* H. W. Rokker, State Printer and Binder, Springfield, 1886, Volume V, pages 334-6; Eddy, T. M, *The Patriotism of Illinois,* Clarke & Co., Chicago, 1866, Volume II, pages 343-4.

The inscription on this 35-star national flag reads: 91st REGT.ILL's INF'TY. VOLS. *Based on the square canton and 5 horizontal rows of stars, the flag was manufactured for the New York depot.*

Little remains of this national flag.

This 35-star regimental flag is not inscribed with battle honors. The design of the flag is typical of flags manufactured for the New York depot. The flag measures 72" on the hoist by 76" on the fly.

Only shreds remain of the above two regimental flags.

At least two of the Companies raised in Greene County were presented company colors.

"FLAG PRESENTATION TO CAPT. CULVER'S COMPANY—*Messrs. Price & Clay:* On Saturday last, we had quite an interesting ceremony in the presentation of a beautiful flag to Capt. Culver's company, by the ladies of Whitehall and vicinity. The flag was made by Pearson & Co. of St. Louis, at a cost of $50, and is a well-wrought ensign of heavy silk and appropriately ornamented. The presentation was made by Miss Lora Worcester, of Whitehall, in behalf of the ladies."[832]

In the same issue of the newspaper appears the following: "Capt. Lakin's company had a flag presentation, supper, and ball at Wilmington, on Tuesday. Miss Minerva Patterson presented the flag, and responsive speeches were made by Capt. Lakin and Major Smith of the 91st." There is further description of the flag presented to Company H in the history of the Regiment.[833] "The flag was made of beautiful silk, and was presented to Company H by the citizens of Patterson, the presentation being made by a young lady by the name of Minerva Patterson, daughter of Judge L. J. Patterson, and afterwards the wife of our Lieutenant, John Jones....The beautiful silk flag that is here spoken of can be

[832] *Carrollton Gazette*, August 30, 1862. The article goes on to describe this flag as a national flag. Captain Culver was Captain of Company I and Captain Lakin the Captain of Company H, both Greene County companies.

[833] Beverly, James M., *A History of the Ninety-First Regiment Illinois volunteer Infantry 1862-1865,* Pearce Printing col, White Hall, Illinois, (1913), page, 35-6.

found at the home of our Lieutenant, John Jones, in the city of Roodhouse, Greene county, Illinois. It shows age, is tattered and torn, and is not so bright as it was when presented to Company H fifty years ago. When Lieutenant Jones was compelled to quit the arms at New Orleans on account of broken health, the Company voted that he should retain the flag." Presumably none of the flags shown above are the flags presented to Company H.

No other flags carried by the 91st Illinois Infantry are known to survive.

92nd Illinois Infantry - "*Abolition Regiment*"[834]

The 92nd Illinois Infantry Regiment, along with the 98th Illinois Infantry, 123rd Illinois Infantry, 17th Indiana Infantry, 72nd Indiana Infantry and 18th Independent Battery Indiana Light Artillery formed *Wilder's Lightning Brigade*. The Infantry regiments were all mounted and armed with 7 shot Spencer lever-action repeating rifles.

The Ninety-second Illinois Infantry Regiment was organized at Rockford, Winnebago County, Illinois in the summer of 1862 commanded by Colonel Smith D. Atkins.[835] The men of the Regiment volunteered as follows: Company A, F and G from Stephenson County; Company B, D, E (*Applington Avengers*), H and K from Ogle County and Company C and I from Carroll County. The Regiment mustered in for three years' service on September 4, 1862, at Camp Fuller in Rockford and departed for Cincinnati on October 10. The Ninety-second was stationed at Mr. Sterling, Kentucky with orders to guard Western Kentucky from the rebel guerrillas. November 16 the Regiment was ordered to Nicholasville, Kentucky and on November 26 to Danville, Kentucky. In late January 1863 the Ninety-second marched to Louisville and boarded steamers for a trip to the mouth of the Cumberland River at Smithland, Kentucky. Hearing firing to the south, the Regiment marched to Fort Donelson and arrived in time to see the withdrawal of rebel forces that had attacked the Union held Fort. The Regiment was ordered to Nashville, Tennessee and camped south of the city on February 6; in March they moved to Franklin, Tennessee and immediately began pursuit of the rebel forces under Van Dorn to south of Columbia, Tennessee. They returned to garrison Brentwood, Tennessee and in April moved to Franklin, Tennessee. Van Dorn's cavalry attacked the Union forces at Franklin on April 10, but the rebels were repulsed. On June 1 the Regiment moved to Triune, Tennessee, where Wheeler's cavalry attacked them on June 11; on July 1 the Ninety-second occupied Shelbyville, Tennessee. While at Shelbyville the Regiment was detached from the 1st Brigade and attached to Wilder's Mounted Infantry Brigade. Wilder's Brigade moved over the Cumberland Mountains and into the valley of the Tennessee River north of Chattanooga in August, driving the rebel cavalry before it. The Ninety-second was then temporarily detached from Wilder's Brigade, recrossed the Mountains and rejoined General Thomas at Trenton, Alabama; on September 9 the Ninety-second led the advance on Chattanooga drove the rebels from Lookout Mountain, entered Chattanooga and became the first Regiment to raise its flag in the captured city.[836] The Regiment pursued the retreating rebels and met Forrest' cavalry at Ringgold Georgia on September 11. The Ninety-second served as scouts prior to the Battle of Chickamauga and was engaged in the Battle on September 20 when they covered the retreating Union forces of McCook to Chattanooga. The Regiment guarded the Tennessee River north of Chattanooga until October 27 when it was ordered to Huntsville, Alabama for winter quarters. In April 1864 the Regiment was ordered to Ringgold, Georgia to participate in the Atlanta campaign. In that campaign the Ninety-second saw

---

834 *Ninety-Second Illinois Volunteers,* Journal Steam Publishing House and Bookbindery, Freeport, Illinois, 1875, page 360. The Ninety-second was called the *Abolition Regiment* because of the willingness of the men in the Regiment to refuse return of slaves to their owners. At the Second reunion of the Regiment, the President of the Reunion Association and former Captain of Company C, Robert M. A. Hawk, explained the name as follows— "Our stay in Kentucky was made up of a mixture of pleasant and unpleasant events. We were called *the Abolition Regiment*—using the more chaste language of Kentucky loyalists, we were "*nigger thieves.*'"

835 Smith D. Atkins was an attorney in Freeport, Illinois and had been elected State's Attorney of the 14th Judicial Circuit. He had served as a Major in the Eleventh Illinois Infantry but had resigned due to illness.

836 Eddy, Thomas Mears, *The Patriotism of Illinois,* Clarke & Co. Chicago, 1866, Volume II, page 372; Ibid. *Ninety-Second Illinois Volunteers,* page 101.

action at Tunnel Hill, Resaca, Dallas, New Hope Church, Allatoona Hills, Kennesaw Mountain, the Chattahoochee River and in the Siege of Atlanta. As part of Kilpatrick's Cavalry Raid around Atlanta, the Regiment fought at Jonesboro and Lovejoy Station. In October, in the pursuit of Hood's retreating forces, the Regiment struck the enemy at Powder River and Powder Springs north of Atlanta. In mid-November, as part of Kilpatrick's Division the Ninety-second participated in the "March to the Sea" and in the Campaign in the Carolinas. In Georgia the Regiment fought at Macon, Buckhead Creek, Waynesboro, Fort McAllister, and Savannah; in the Carolinas they fought at Barnwell, Averysboro, Bentonville, Raleigh and Bennett's House. Following the surrender of Johnston, the Ninety-second remained in North Carolina until muster out on June 21 at Concord, N.C.

Despite references to the Regimental colors in the histories of the Ninety-second Illinois Infantry,[837],[838] there are no flags in the State of Illinois collection that are identified as having been carried by the Regiment and no flags carried by the 92nd Illinois Infantry are known to survive elsewhere.

---

[837] Ibid, *Ninety-Second Illinois Volunteers,* page 33. "On the morning of the tenth, in full strength, with blankets rolled and knapsacks packed, the Ninety-Second, with music and with colors flying marched down through the streets of Rockford, and embarked on a special train for Chicago, reaching there at 3 P. M.."

[838] Ibid. *Ninety-Second Illinois Volunteers,* page 364-5. "You have seen our beautiful banner carried forward to victory. You gaze today upon its bright stars and ample folds, and there you read in characters of living light and beauty, those sublime words:--*Liberty and Union—Now and Forever—One and Inseparable!'"*

93rd Illinois Infantry

"Old Banner, farewell! To our friends now return
Who gave your bright folds to our care;
Return to our friends; though ragged and torn,
No marks of dishonor you bear.
Tell our friends, should they fill up our ranks again,
And give us a banner once more
That its folds shall float over Georgia's plain,
To the far off Atlantic's shore."[839]

The companies of the Ninety-third Illinois Infantry were organized in Chicago during August 1862 and mustered in October 13, 1862 under the command of Colonel Holden Putnam.[840] The companies were recruited as follows: Company A at Camden Mills in Rock Island County; Company B (*Bureau County Rifles*) at Dover in Bureau County; Company C (*Wyanet Union Guards*) at Wyanet in Bureau County; Company D at Freeport in Stephenson County; Company E (*Tiskilwa Tigers)* at Tiskilwa in Bureau County; Company F (*Tornado Company)* at Albany in Whiteside County; Company G at Cedarville in Stephenson County; Company H (*Bureau County* Tigers) at Neponset in Bureau County; Company I (*Princeton Light Infantry*) and Company K (*Princeton Guards*) at Princeton in Bureau County. On September 2 all the companies except Company A assembled in Princeton at Camp Bureau; on September 17, the nine companies were ordered to Camp Douglas in Chicago where they were joined by Company A from Rock Island. On October 9, the regiment was ordered to Memphis, Tennessee. After a month in camp at Memphis, on November 26, the Ninety-third embarked on Grant's campaign into Northern Mississippi. Despite heavy marching, the Regiment saw no action during this campaign; due to the surrender of the Union forces at Holly Springs, the Campaign was terminated, and the Regiment returned to Memphis, arriving there December 20. January and February 1863 were spent in camp east of Memphis. On March 2 the Regiment boarded the steamer *Henry Von Phul* and moved down the Mississippi River to Grand Lake, Arkansas and two days later moved back up the river to a point opposite Yazoo Pass. The Regiment boarded the steamer *Jesse K. Bell* on March 22, passed through Moon Lake, and traveled the Yazoo Pass to Cold Water River and down the Tallahatchie River to survey the enemy's fortifications at Greenwood, Mississippi. After General McPherson decided the fortifications could not be reached because of water and swamps, the Regiment retuned to Helena, Arkansas. On April 13 the Regiment sailed down river to Milliken's Bend, Louisiana, camped there until April 25 and then marched to Bruinsburg, Louisiana and crossed the Mississippi on the steam ram *Lafayette.* The Ninety-third marched toward Jackson, Mississippi and on May 14 met the rebel forces at the Battle of Jackson. On May 16 the Regiment while en-route to Vicksburg met the enemy at the Battle of Champion's Hill; the Regiment entered the battle with less than 500 men and lost 164 men killed, wounded and missing. The Ninety-third was next involved in the disastrous assault on Fort Fisher at Vicksburg on May 22; the Confederates repulsed the Union assault that then led to the Siege of Vicksburg. Following the fall of Vicksburg on July 4 the men of the Regiment remained in the area until

[839] Trimble, Harvey, *History of the Ninety-third Regiment Illinois Volunteer Infantry from Organization to Muster Out,* The Blakely Printing Co., Chicago, 1898, page86-87. Portion of a poem written by Lieutenant Colonel Nicholas Buswell and included in a letter accompanying the battle torn flag returned to Bureau County on December 24, 1863.

[840] Colonel Putnam was born in Middlexes Vermont in 1821. He came to Illinois in the 1850s. He was a harness maker and banker before the civil war.

September and then were transported upriver to Memphis where they arrived on October 1, 1863. On October 4 the Regiment departed Memphis on the Chattanooga Campaign. They passed through Glendale, Burnsville, and Iuka Mississippi, Bear Creek and Dixon's Station, Alabama, crossed the Tennessee River and the Cumberland Mountains and on November 19 marched within range of Confederate batteries on Lookout Mountain, Tennessee. Wednesday November 25, 1863, the Ninety-Third Illinois was heavily involved in the Battle of Mission Ridge. In the charge up Mission Ridge, Colonel Putnam was killed carrying the colors and the colors were torn to shreds;[841] of two hundred and ninety-three men who went into the battle ninety-six were killed, wounded or missing. Following the battle, after a brief pursuit of General Bragg's forces the Regiment marched to Stevenson, then Huntsville, Alabama. In early 1864 the Ninety-third remained in Alabama and Tennessee. On February 23 the Regiment marched to Ringgold, Georgia and then on the 24th and 25th fought at Tunnel Hill and near Dalton, Georgia. From April through September the Regiment campaigned in Alabama, Tennessee and Georgia but saw no actual fighting. However, on October 4, 1864, the Ninety-third, as part of Sherman's Atlanta Campaign, was camped at Allatoona, Georgia; Atlanta had fallen, and the retreating Confederates attacked the Union forces at Allatoona. The Union forces were outnumbered and surrounded but refused surrender and in 13 hours of fighting saved the supplies at Allatoona and forced Confederate retreat; the Ninety-third lost 89 men or one third of the Regiment. In mid-November the Ninety-third joined Sherman's "March to the Sea." They fought in the siege of Savannah December 10-21,1864 and then in 1865 moved into the Carolinas where they fought at Salkehatchie Swamps, South and North Edisto Rivers, Columbia, Bentonville and at Bennett's House and the surrender of Johnston and his army. They marched via Richmond, Virginia to Washington D. C. and marched in the Grand Review on May 24, 1865. The Ninety-third Illinois Infantry was mustered out on June 23, 1865, at Louisville, Kentucky.[842]

There is one national and one regimental flag of the Ninety-third Illinois Infantry Regiment in the collection of the State of Illinois.

Inscribed on this national flag: JACKSON, May 14, 1863; CHAMPION HILL, May 16, 1863; VICKSBURG, May 19 and 22, July 4, 1863; 93rd ILLINOIS INFANTRY; MISSION RIDGE Nov 25, 1863; DALTON, Feb 25, 1864. *The rectangular canton with seven horizontal rows of stars is characteristic of flags manufactured in Cincinnati. It measures 72" on the hoist by 76" on the fly.*

[841] "I was within a few feet of your husband when he fell. He was mounted at the time, and had just asked the Adjutant for the colors, the color bearer being badly wounded, and the colors torn into more than twenty pieces; the Adjutant had only placed them in his hand when he fell off his horse at my feet with the fatal wound in his head." Letter from Capt. Taggart to Mrs. Putnam as reported in *Freeport Weekly Bulletin*, December 10, 1863.
[842] Ibid, Trimble.

The regimental flag of the Ninety-third Illinois Infantry shows significant damage. It appears that the edge of the flag has been carefully cut off-presumably by the men of the 93rd. There are no inscriptions on the flag other than the regimental designation in the lower scroll. The position and number of the stars above the upper scroll, the shape of the upper scroll and the shape of the lower scroll is characteristic of flags manufactured by the Evans and Hassall Company of Philadelphia. The flag measures 69" on the hoist by 74" on the fly.

A stand of colors was presented to the Ninety-third before it left for War. "We hear that the citizens of Princeton, Bureau County, will make an excursion to this city to visit the soldiers at Camp Douglas, and to present the 93d Illinois regiment with a beautiful stand of colors."[843] This stand of colors was returned to Bureau County and is not the stand of colors shown above. "Six times the colors of this regiment went out of sight. Once, the life of the brave Colonel Putnam went out when the flag went down; twice, the spirits of brave sergeants took their flight when the banner fell; and three times, wounded and bleeding heroes relinquished it to other hands. The staff that supported its shining folds was splintered and shivered and shot in twain. The banner itself was riddled and tattered and torn into shreds. Not a twentieth part of it remained upon the broken staff. Carried away by shot and shell, it(s) fragments were scattered on the mountain side among the dead and bleeding heroes who followed it there."[844] On December 24, 1863, the remnants of the flag were sent home to Bureau County, Illinois. The letter accompanying the banner home, states that War Committee of Bureau County had presented the flag to the Regiment.[845,846] A flag remnant is on a staff in the Bureau County Historical Society and Museum.

A flag was presented to Company F during a picnic given in honor of the Company in August 1862. "The ladies of Albany, Fulton and Garden Plain, gave a picnic supper to the Company of Volunteers, on Friday afternoon last, in Albany…After supper a flag was presented to the company by

[843] *Chicago Daily Tribune,* October 10, 1862.

[844] Ibid, Trimble, pages 77-8.

[845] Ibid, Trimble, page 84

[846] After the GAR in Bureau County closed, the flag was given to the Bureau County Historical Society. The flag is now in their possession and is hanging/rolled on a staff in a display case with a gold eagle finial. The flag has not been unwrapped and there are no photos of the flag itself. A tag attached to the flag reads: "Battle scarred remnants of the regimental flag of the 93rd Regiment, Illinois volunteer Infantry. Returned to Bureau County December 24, 1863 by W. C. Buswell Lieutenant Colonel commanding Regiment. Colonel Holden Putnam was killed while carrying these colors to rally the troops at the Battle of Tunnel Hill November 25, 1863."

the ladies of Albany, through the Rev. Mr. Lackey in a brief and appropriate speech, which was received with cheers."[847] The flag is not described, and its further history is unknown.

No other flags carried by the 93rd Illinois Infantry are known to survive.

[847] *The Fulton Weekly Courier,* August 22, 1862

94th Illinois Infantry- "*McLean County Regiment*"

"There is one spot on the battlefield where bayonets gleam brighter, where swords dart quicker, where guns and cannon flash faster, where soldiers fight not with the desperation of demons but with the undaunted courage and bravery of heroes. Why is it? *They are guarding their colors.* The honor of a nation is in their hands."[848]

The men who served in the "*McLean County Regiment*" were all from McLean County except for a few men in Company I from Tazewell County. The Regiment was organized in Bloomington, McLean County, and mustered in on August 20, 1862, under command of Colonel William W. Orme.[849] August 25 the Ninety-fourth left for Benton Barracks in St. Louis and was assigned to the "Army of the Frontier." Three weeks later the Regiment moved by rail to Rolla, Missouri and then to Springfield, Missouri; at Springfield the men spent six weeks in company and battalion drill. Word reached the Union forces at Springfield on December 1 that Union General Blunt was surrounded by Confederate irregulars at Cane Hill in northwest Arkansas. The forces at Springfield, after a forced march of 120 miles, attacked the Confederate forces at Prairie Grove, Arkansas, and though the Union forces were outnumbered 30,000 to 4,000 drove the enemy from the field. Two weeks later the Ninety-fourth as part of an expedition to Van Buren on the Arkansas River captured and burned two rebel steamers. The Regiment remained in camp near Rolla, Missouri until June 1863 and then moved to St. Louis and down the Mississippi River to Young's Point in the Vicksburg Campaign. The Ninety-fourth was in the lines below the city and was present when Vicksburg surrendered on July 4. Following the surrender, the Regiment was ordered to Yazoo City, Mississippi and encamped there from July 12-21 and then moved to Port Hudson, Louisiana. In September the Ninety-fourth was ordered to Morganza, Louisiana and saw action at Morgan's Ferry on the Atchafalaya River on September 7. October 25 the Regiment sailed to Brownsville, Texas on the Rio Grande River where they remained for the next nine months on guard detail. In July 1864 the Regiment was withdrawn from Texas and joined the Union forces in the siege and capture of Fort Morgan on Mobile Bay, Alabama. Following the fall of the Fort the Ninety-fourth remained at Navy Cove near Fort Morgan until March 1865. In the campaign against Mobile and its defenses the Regiment fought in the siege and capture of Spanish Fort and Fort Blakely during March and April and escorted prisoners from the captured Forts to Ship Island, Mississippi. They went into camp south of Mobile until June 18 and then moved to Galveston, Texas for garrison duty until mustered out on July 17, 1865.[850]

Although the original inventory of the flags held by the Adjutant General of the State of Illinois listed both a national and regimental flag of the Ninety-fourth Illinois Infantry in their holdings, at this time only a national flag remains. The Regiment was presented a stand of colors at Rolla, Missouri, "Judge Funk arrived in Camp Heron a short time after we pitched our tents and was welcomed in old fashioned order. The next morning before our departure, he presented the regiment, in behalf of McLean

---

[848] *Daily Pantograph,* October 8, 1862. A series of letters appeared in this Bloomington, Illinois newspaper from a member of the Regiment who signed the letters "Boswell." This quote is from a letter dated October 1 from Springfield, Missouri.

[849] Colonel Orme was born in Washington D.C. in February 1832. He moved to Bloomington,Illinois and practiced law in the firm of Swett & Orme. Through his partner Leonard Swett, Orme became acquaited with Abraham Lincoln and David Davis. Orme formed and commanded the 94th Illinois Infantry. While fighting in Mississippi, Orme contracted tuberculosis and died of the disease in 1866. Wikipedia, 2023.

[850] *Report of the Adjutant General of the State of Illinois,* H. W. Rokker, State Printer and Binder, Springfield, 1886, Volume V, pages 409-11.

county, a stand of colors gorgeously beautiful…Each flag bore the inscription, in gold letters, 'From McLean county to 94th regiment Illinois volunteers.'"[851] According to the records of the Adjutant General of Illinois the stand of colors cost $500.[852]

As seen in the photo, the National colors are in tatters and no inscriptions are seen on the portions of the flag that remain.

No other flags carried by the 94th Illinois Infantry are known to survive.

[851] *Daily Pantagraph,* September 29, 1862.
[852] Ibid, *Report of the Adjutant General of the State of Illinois, page 409*

95th Illinois Infantry

Nineteen-year-old 5'3" Albert D. J. Cashire[853] volunteered for Company G of this Regiment in August 1862. After the war Cashier worked as a laborer and voted in elections (elections were not open to women voters). As the result of an auto accident in 1911 Cashier was examined and found to be a woman. She had been born in Ireland on December 25, 1843, with the given name of Jennie Irene Hodgers. She died October 10,1915 and is buried in Sunny Slope Cemetery in Saunemin, Illinois.[854]

The Ninety-fifth Illinois Infantry was organized at Camp Fuller in Rockford, Winnebago County, Illinois during the summer of 1862. Companies A (*Avery Guards)*, C, D, E, F, H, and I volunteered from McHenry County while Companies B, G and K volunteered from Boone County. The Regiment mustered in on September 4, 1862, under the command of Colonel Lawrence S. Church[855] and departed November 4 for Jackson, Tennessee. They remained in camp at Jackson until the end of November when ordered to Grand Junction, Tennessee to join the Army of the Tennessee and march south in Grant's Northern Mississippi campaign. However, when Confederate General Van Dorn captured the Union stores depot at Holly Springs, Mississippi, the campaign was canceled, and the Union forces retreated toward Memphis. The Ninety-fifth reached Moscow, Tennessee (east if Memphis) on December 30; continuing the march west, the Regiment arrived at Collierville on January 2, 1863, and at Memphis on January 13. A week later the Regiment boarded the *Maria Denning* and along with 14 other steamers moved downriver to Milliken's Bend, 15 miles above Vicksburg. On February 2, the Regiment moved back upriver to Lake Providence, Louisiana to assist in digging a canal one half mile long from the Mississippi River to Lake Providence. Lake Providence was at a much lower level than the flood-stage Mississippi River; the plan was that by raising the water level of the Lake (by diverting water from the Mississippi River through the canal) the bayous entering the Lake would widen and deepen permitting Union gunboats and transports to by-pass the Vicksburg defenses. This in turn would allow landings below Vicksburg and open a second military front. However, in mid-February when the canal was completed, and the levee opened neither the canal nor the bayous were deep enough nor wide enough for the steamboats to pass and the project was abandoned. May 1 the Ninety-fifth moved to Milliken's Bend and then through Richmond and Hard Times, Louisiana to the Mississippi River below Vicksburg opposite Grand Gulf, Mississippi. The Regiment crossed the river on steamers, marched north and arrived a Champion Hill on May 16 just as the battle was raging. Although the Regiment did not fight at Champion Hill, they did pursue the retreating enemy to the Big Black River. On May 19 when a charge was ordered on the enemy's works before Vicksburg, the Ninety-fifth advanced to within 100 yards of the rebel lines but was forced to fall back that same evening. A second charge was ordered on May 22, again the Ninety-fifth held an advanced position, but again they were forced to withdraw; of the three hundred and sixty-seven officers and men of the Regiment who had gone into the charge, one third were killed or wounded. After six weeks of siege, on July 4, Vicksburg surrendered, and the Ninety-fifth was one of the first Regiments to enter the city. From mid-July until October the Regiment

---

[853] The AG records spell the name Cashire but her tombstone in the Sunny Slope Cemetery in Saunemin, Illinois reads Albert D. J. Cashier

[854] There is a photo of Cashier on the Sunny Slope Cemetery tombstone.

[855] Colonel Lawrence Church was born in New York in 1820. At an early age he moved to McHenry County Illinois where he had two married sisters. After studying law in Illinois he passed the necessary exam and began the practice of law in Woodstock Illinois. He was elected a representative to the Illinois legislature in 1856. At the outbreak of the civil war he he was instrumental in recruiting men for the an Illinois regiment and in 1862 he was elected Coonel of the 95th Illinois Infantry. Hill, Elizabeth, "1885 Bio",Colonel Lawrence Smith Church, Ancestry.com

occupied Natchez then returned to Vicksburg for garrison and construction duty until February 1864.[856] During February the Ninety-fifth was part of an expedition to Jackson, Mississippi to remove the enemy from the state capitol. In March the Regiment, in the Red River Expedition, boarded the steamship *John Raines* and moved up the Red River. Fort DeRussy was captured on March 14; however, General Banks' Union forces were defeated at Pleasant Hill and in the retreat down the Red River, the Ninety-fifth participated in battles at Cloutiersville, Natchitoches, Alexandria, Boyce's Plantation, Bayou Boeuf, Morganza, Mansura and Yellow Bayou, Louisiana. On May 21 the Regiment reached the mouth of the Red River and on May 22 sailed back to Vicksburg and then on to Memphis.[857] There the Regiment was placed under the command of Brigadier General Samuel Sturgis and ordered on an expedition to Guntown, Mississippi against rebel general Forrest. In the battle at Guntown the Union forces were mismanaged and soundly defeated; Colonel Humphrey, commander of the Ninety-fifth was mortally wounded, and the Regiment returned to Memphis shattered by the defeat. Due to its loss of men and equipment at Guntown the Regiment was relieved of all duties for several weeks to allow for recruitment of new members and replacement of lost arms and equipment. In early August 1864 the Regiment embarked on the steamer *White Cloud* and moved down the Mississippi River from Memphis to St. Charles, Arkansas. On September 1, they moved up the White River in Arkansas and skirmished with rebel forces at Augusta, returned down river to Duvall's Bluff and then on September 6, moved by rail to Brownsville and pursued Price through Arkansas and Missouri for the next two months. The Regiment moved to Nashville in late November and fought in the Battle of Nashville December 15 and 16, 1864. Following the battle, the Regiment joined the pursuit of Hood to the Tennessee River and then moved to Eastport, Mississippi for duty there until February 1865. On February 6, 1865, the Ninety-fifth boarded the steamer *Adam Jacobs* for the trip to New Orleans; they were on duty there until March 12 when the Regiment joined the campaign against Mobile, Alabama. Following the fall of Spanish Fort, Fort Blakely, and the city of Mobile, the Ninety-fifth was on duty at Montgomery, Alabama until July and then was ordered to Vicksburg, Mississippi. In August the Regiment was ordered to Springfield, Illinois for muster out on August 17, 1865.[858]

The current collection of flags held by the State of Illinois Military Museum contains one national and two regimental flags of the Ninety-fifth Illinois Infantry. However, in the Adjutant General Report of 1886[859] a national flag is listed with the following inscriptions, "CHAMPION HILL, May 16, 1863; BLACK RIVER BRIDGE; VICKSBURG, May 19 and 22, JULY 4, 1863; RED RIVER EXPEDITION; CLOUTIERVILLE, OLD RIVER AND YELLOW BAYOU; GUNTOWN; NASHVILLE; SPANISH FORT; BLAKELEY; MOBILE." That national flag is no longer in the collection and its whereabouts today is unknown.

---

[856] While at Vicksburg, a recruiting party from the Regiment was sent North to recruit volunteers to fill the depleted ranks of the Regiment.

[857] Wood, Wales, W., *A History of the Ninety-fifth Regiment Illinois Infantry Volunteers,* Tribune Company's Book and Job Printing office, Chicago, 1865, pages 131-33. The "Georgia Detachment" of the Ninety-fifth Illinois Infantry consisted of the convalescents who were left behind at Vicksburg at the time of the Red River Expedition together with new recruits sent from Illinois in May 1864. This detachment of 100-200 men accompanied the 17th Army Corps in the Atlanta campaign and fought at Kenesaw Mountain, Nickajack Creek, Howell's Ferry, Leggett's Bald Hill, the siege and battle of Atlanta, Jonesboro and Lovejoy Station. They rejoined their Regiment in November 1864 at Nashville, Tennessee.

[858] Ibid, Wood, Wales W.

[859] *Report of the Adjutant General of the State of Illinois*, H. W. Rokker, State Printer and Binder, Springfield, Illinois, 1886, Volume I, page 162.

The national flag shown here is not inscribed. The double oval arrangement of the stars on the flag is characteristic of flags manufactured in Philadelphia.

On this Regimental flag, the star pattern above the upper scroll, the shape of the upper scroll, the shape of the eagle and the shape of the lower scroll are characteristics of flags manufactured in Philadelphia.

The pattern of the lower scroll and the position of the stars above the upper scroll on this Regimental flag is consistent with New York manufacture.

A silk flag was presented to Company A (*Avery Guards*) of the Ninety-fifth. "FLAG AND SWORD PRESENTATION AT MARENGO, ILL. We take pleasure in noticing a very pleasant little affair which occurred in the patriotic town of Marengo on Monday last. In the presence of a large concourse of people from the town and surrounding country, R. M. Patrick, a citizen of the place, after a few stirring remarks, presented as his individual donation to the A*very Guards*, a beautiful silk flag; after which, the Rev. D. E. Halteman, in behalf of the citizens and members of the company, presented Captain Avery with an elegant sword and sash, and each Lieutenant with a beautiful sash."[860]

The flag is not described further and its whereabouts today is unknown.

The Abraham Lincoln Presidential Library and Museum in Springfield, Illinois has two fragments from a flag of the 95th. The first is a gold painted star with a diameter of 2 ¾ inches on a fragment of frayed blue silk. An attached note reads "A star from the 1st Flag of the 95th Regt Ills Vol Inft Badly shattered during the charges of May 19 & 22 and Siege of Vicksburg 1883" The second fragment is a blue silk fragment 3 ½ inches long. An accompanying note reads" From Blue flag 95th Regt Ills Vol Inft Nov 1863."

No other flags carried by the 95th Illinois Infantry are known to survive.

[860] *Chicago Daily Tribune,* September 4, 1862. Marengo is in Mc Henry County.

96th Illinois Infantry

"Wave the old flag, exultant, o'er the land,
Spread out its folds of beauty toward the sea;
Bid softest winds its blood-bought charms expand;
Hail it with shouts—the banner of the free.
Count all the stars, the stripes, both white and red,
Where'er on sea or land the flag is seen.
They tell how God our growing States has led—
Stars forty-two, and stripes, the old thirteen."[861]

The Ninety-sixth Illinois Infantry was recruited during the summer of 1862 and organized at Camp Fuller in Rockford, Winnebago County, Illinois. The men for Companies A and E (*Apple River Company*)[862] F and H (*Burnett's Company)*, I and K (*Champion's Company)* were recruited from Jo Daviess County and the men for Companies B, C, D and G from Lake County. The Regiment was mustered in for three years' service on September 6, 1862, under the command of Colonel Thomas E. Champion.[863] Wednesday morning, October 8 the Regiment marched to the depot of the *Chicago & North-Western Railway* and departed for Cincinnati. Companies D, I and K were detached from the Regiment[864] and sent to guard forts in the area and on October 19 Companies A, E, F, G and H were ordered to escort a commissary train into northern Kentucky and then proceed to Lexington, Kentucky. In November the Regiment moved to Harrodsburg, Kentucky to guard rebel prisoners and in December attempted to intercept Confederate troops commanded John H. Morgan at Lebanon Junction but the rebels escaped, and the Ninety-sixth returned to Danville, Kentucky. January 1863 the Ninety-sixth was ordered to join the Army of the Cumberland and arrived at Fort Donelson on the Cumberland River on February 3. From Fort Donelson the Regiment moved to Nashville and in March to Franklin, Tennessee. The Ninety-sixth skirmished with enemy forces under Van Dorn during March and April and in June marched to Triune, Tennessee where they skirmished with Wheeler on June 11. Also, while at Triune, the Army of the Cumberland was reorganized and the Ninety-sixth became part of the 1st Brigade, 1st Division of the Reserve Corps, under the command of General Gordon Granger.[865] The Regiment saw little action during the Tullahoma Campaign of June and July but in September embarked on the Chattanooga Campaign. The men passed through Anderson and Stevenson, Tennessee, through Bridgeport, Alabama and crossed the Tennessee River. The columns turned east, marched through Whiteside, Georgia, crossed the nose of Lookout Mountain and moved into Lookout Valley. September

---

[861] *Reception of the Ninety-Sixth Regiment Illinois Infantry Volunteers, at the Residence of General John C. Smith, December 16, 1889,* Knight & Leonard Co., Printers, Chicago, 1890, page 21. Read as part of an address by General Smith.

[862] Companies E and H each had ten men who volunteered from Wisconsin.

[863] Colonel Champion was born in Wayne County New York in August 1825. When he was 12 years old he moved with his family to Michigan. In his leisure time he studied medicine, graduated from the Unversity of Michigan in 1847 and began practicing medicine that same year. He and his wife moved to Warren, Illinois in 1851 and although he continued practicing medicine he developed an interest in law, and after study of law he passed the exam to becme an attorney in Illinois. In 1862 he helped raise Company K of the 96th Illinois Infantry and was elected Colonel of the regiment. CivilWarTalk.com "Thomas Emmet Champion"

[864] The companies of the Regiment were reunited November 4, 1862, at Lexington, Kentucky.

[865] General Order 177 issued when the Reserve Corps was formed stated that the flags carried by the Reserve Corps should have red, white, and blue diagonal backgrounds. These flags are discussed in detail below.

17, the Ninety-sixth marched south from Rossville, Georgia and September 19-22 was heavily engaged in the Battle of Chickamauga—the Regiment's first battle. The Ninety-sixth, as part of Granger's Reserve Corps saved the right of the Union lines but paid dearly—losing fifty per cent in killed and wounded or taken prisoner.[866] Following the Battle the Ninety-sixth marched through Chattanooga crossed the Tennessee River and went into camp on Moccasin Point. In the Chattanooga-Ringgold campaign of late November the Regiment fought at Lookout Mountain on November 23-24 and at Missionary Ridge on November 25. They encamped at Nickajack Cove until January 26, 1864, when the Regiment was ordered to cover the working party repairing the *East Tennessee RR*. In late February the Ninety-sixth was ordered to join Union forces operating against Dalton, Georgia, in that action the Regiment fought at Buzzard's Roost Gap and Rocky Faced Ridge. In April 1864 the Ninety-sixth joined Sherman's Atlanta campaign and fought at Resaca, Kingston, New Hope Church, Dallas, Pine Mountain, Kennesaw Mountain, Smyrna Camp Ground, Peach Tree Creek, Atlanta, Jonesboro and Lovejoy Station. In October the Regiment marched in pursuit of Hood into north Georgia and Alabama. In the Nashville campaign the Regiment was heavily engaged November 30 at the Battle of Franklin. Falling back to Nashville, the Regiment fought in the Battle of Nashville December 15 and 16 and is credited with being the first to plant their colors on the enemy's earthworks.[867] The Ninety-sixth joined in the pursuit of Hood to the Tennessee River then moved to Huntsville, Alabama. After operations in East Tennessee in March and April 1865 the Regiment moved to Nashville where it was mustered out on June 10, 1865.[868] The enlisted men of the Ninety-sixth Regiment Illinois Infantry whose term had not expired were transferred to the Twenty-first Regiment Illinois Infantry.

In the collection of the State of Illinois there are five flags identified as flags of the Ninety-sixth Illinois Infantry—two inscribed national flags, two regimental flags and one guidon or flank marker.

Charles Partridge on the frontispiece of his Regimental history has a photo of four tattered flags belonging to the Ninety-sixth Illinois Infantry. He explains that the Regiment from its organization carried two of the colors until the close of 1863 and that both colors were torn to shreds at Chickamauga. The other two colors are those colors carried in the Atlanta campaign and at the battles of Franklin and Nashville.[869] He further relates that one color from each set is in the possession of General J. C. Smith[870] and one color from each set is in the Court House at Waukegan, Lake County, Illinois.[871],[872]Finally Partridge states that the flags in Springfield are those flags carried by the Regiment during the closing

---

[866] Partridge, Charles A., Ed., *History of the Ninety-Sixth Regiment Illinois Volunteer Infantry,* Brown, Pettibone & Co., Printers, Chicago (1887), page 211-2. Ten color guards from the Ninety-sixth bore flags in the Battle of Chickamauga. Only one, Corporal J. W. Swanbrough, returned after the battle, the other nine were killed or wounded.

[867] Ibid, Partridge, page 445

[868] *Report of the Adjutant General of the State of Illinois,* H.W. Rokker, State Printer and Binder, Springfield, Illinois, (1886), Volume V, pages 466-71.

[869] Ibid, Partridge, page 283. The presentation of the set of colors carried in the Atlanta campaign is mentioned— "January 1, 1864, a pair of new flags from Lake and Jo Daviess Counties were presented to us by a committee consisting of Judge J. D. Platt, L. P. Woodworth and Edgar Scace, of Jo Daviess, who received in return our tattered battle flags, which were separated and taken one to each county. The new flags were very beautiful, with their gilded lettering shining on their unsullied silken folds; but the old flags, though torn and soiled, had a beauty which the new did not possess, for they had not passed through Chickamauga, and had they not led us over Lookout."

[870] *Chicago Legal News*, December 21, 1889. In its report on the reunion of the Ninety-sixth the newspaper confirms that General Smith did have the colors. "The spacious home was decorated throughout with American flags, military emblems and trophies. In the library were the colors of the regiment, bearing abundant evidence of being riddled with shot and shell."

[871] Ibid, Partridge, Frontispiece and page 938.

[872] The Bess Bower Dunn Museum (formerly Lake County Discovery Museum) has the remains of a 96th Illinois Infantry flag. The silk is completely shattered and the remaining fragments very fragile. There are no records or provenance found to tell how or when the flag came into the collection.

months of the War. In that regard a member of the Regiment, in his letter to the editor of the *Weekly Northwestern Gazette* described the presentation of new colors to the Regiment on December 31, 1863, while the Regiment was camped at Nickajack Cove, Georgia.[873]

The two national flags shown below are both 35-star flags, in keeping with their manufacture late in the War. Since they are both in relatively good condition these colors were probably carried by the Regiment when they returned to Chicago and appeared at the Chicago Northwest Sanitary Fair. Presumably the flags were inscribed at that time.

This national flag has a double oval start pattern characteristic of flags manufactured by William Scheible for the Philadelphia depot. The inscriptions on this flag read, SPRING HILL; TRIUNE; LOOKOUT MOUNTAIN; ROCKEY FACE RIDGE; 96th REG ILLs INFANTRY; DAYTON; RESACA; KINGSTON; NEW HOPE CHURCH; PINE MOUNTEIN (sp?); KENESAW MOUNTAIN; PEACH TREE CREEK; ATLANTA; JONESBORO; LOVEJOY STATION; FRANKLIN; NASHVILLE.

The square canton with five horizontal rows of stars is typical of flags manufactured for the New York depot. Inscriptions on this flag read, CHICAMAUGA (sp?); LOOKOUT MOUNTAIN; BUZZARD ROOST; 96th Regt. Illinois Infty.; The ATLANTA CAMPAIGN; FRANKLIN; NASHVILLE.

---

873 *Weekly Northwestern Gazette,* January 19, 1864

This regimental flag of the Ninety-sixth Illinois Infantry was photographed with a guidon (on the left side of flag) from Battery H of the Second Illinois Light Artillery. The guidon is unique since it is the diagonal red, white and blue pattern required under General Order 177 for flags of the Granger Reserve Corps. Both the Ninety-sixth Illinois Infantry and Battery H of the Second Illinois Light Artillery were members of this Corps. This Regimental flag has no inscriptions, and the pattern is consistent with Cincinnati manufacture.

This34-star Regimental flag is typical of flags manufactured for the New York depot. The flag is not inscribed with battle honors.

This guidon or flank marker is identified as having carried by the Ninety-sixth Illinois Infantry. It measures 27"by 40" and is of the type manufactured after 1862.

According to the history of the Regiment, at least three companies received a flag before leaving home. Company E, the *Apple River Company*, "the good ladies gave the soldiers a grand picnic dinner, and presented them with an elegant flag."[874] Company H received a "beautiful flag,--a present from the patriotic ladies of Warren."[875] And, "Before leaving Warren, the citizens presented beautiful flags to the two Companies, known as *Champion's Company* (Company K) and *Burnett's Company* (Company H)."[876] Based on this information at least three Companies—E, H and K—received company flags. There is no record as to what happened to these flags.

No other flags carried by the 96th Illinois Infantry are known to survive.

---

874 Ibid, Partridge, page 776

875 Ibid, Partridge, page 844

876 Ibid, Partridge, page 889.

97th Illinois Infantry

"The time for assault was set for 5:30 in the afternoon. Precisely at that hour Lieutenant-Colonel Victor Vifquain, commanding the Ninety-seventh Illinois Infantry, gave the command: 'Forward, Ninety-seventh!' and with irresistible dash they advanced and sprang with him over the parapet, with loud cheers charging the line as skirmishers upon the enemy. Colonel Vifquain made for the rebel battle flag on the works, with his color company at his heels, and being the first man upon the works immediately cut the halyard with his sword and pulled it down, while his color bearer planted the Union flag beside the rebel staff. For a moment Vifquain was hidden by the folds of the victorious banner mingling with those of the rebel standard and when he had freed himself from their folds he found his brave color-bearer lying by his side pierced with a bullet."[877]

The Ninety-seventh Illinois Infantry was organized at Camp Butler near Springfield in Sangamon County, Illinois. The men of Company A volunteered from Macoupin County; Company B from Cumberland County; Company C from Calhoun and Fulton County; Company D from Cumberland and Jasper County; Company E from Jasper County; Company F from Fayette County; Company G from Madison County; Company H from Fayette and Jersey County; Company I from Fayette and Madison County and Company K from Jersey County. The Regiment was mustered in for three years' service on September 16, 1862, under the command of Colonel Friend Smith Rutherford.[878] On October 1 it was ordered to Covington Kentucky to march south for relief of Union forces at Cumberland Gap. En-route it was learned that the forces at Cumberland Gap had been relieved, and the Ninety-seventh was ordered to Louisville, Kentucky and then down to Memphis, Tennessee. In late December 1862 the Regiment was moved by steamer down the Mississippi River and landed on the left bank of the Yazoo River where they immediately moved against the enemy at Chickasaw Bluffs. On December 31, 1862, and January 1, 1863, the Ninety-seventh formed the right of the Union line at Chickasaw Bluffs; when the enemy retreated the Ninety-seventh boarded steamers bound for Young's Point, Louisiana and then moved to Milliken's Bend, Louisiana. During January 3-10, 1863, the Regiment moved by water up the Arkansas River against Arkansas Post and on January 10 and 11 were in the assault and capture of Fort Hindman (Arkansas Post).[879] On January 14 the men boarded steamers and sailed back to Young's Point, arriving on January 22. By March 6, the camp at the Point had become rampant with disease and the men were moved by boat to Milliken's Bend. On April 15 the Regiment moved to Hard Times Landing on the Mississippi River opposite Grand Gulf and on the night of April 30 crossed the Mississippi River at Bruinsburg and set out for Vicksburg, Mississippi. May 1 the Ninety-seventh met the enemy at Port

---

[877] Beyer, W.F., Keydel, O. F., eds, *Deeds of Valor, How America's Civil War Heroes Won the Congressional Medal of Honor,* Longmeadow Press, Stamford, CT (1993), page534. Description of the actions before Fort Blakely, Alabama on April 9, 1865, for which Vifquain was awarded the Medal of Honor. The State of Illinois does not have a Confederate flag in their collection identified as the flag captured by the Ninety-seventh Illinois Infantry at Fort Blakely. The color sergeant was Edwin D. Lowe, "Edwin D. Lowe, color sergeant 97th Illinois Volunteer Infantry, was killed at Blakely, Ala., while planting the colors of the 97th on the breastworks and was the first man on the works. He was shot in the stomach and died next morning. Lowe Post, No. 295, G.A. R., Jerseyville, Ill., was named after him." *The Back-Bone of Illinois in Front and Rear, From 1861-1865,* H.W. Rokker, Printer and Binder, Springfield, Ill., (1886), page39.

[878] Colonel Rutherford was born in New York in September 1820. He studied law in Troy, New York; he moved west and by 1857 he was practicing law in Alton, Illinois. In June 1862 he joined the Union Army. as captain commissary subssistaence. He resigned 3 months later, helped organize the 97th Illinois Infantry Regiment and was commissioned as colonel.Wikipedia, The Free Encyclopedia, "Friend Smith Rutherford"

[879] The Ninety-seventh Illinois rushed into the Fort and planted its colors on the ramparts.

Gibson but the following morning the rebel outposts were abandoned, and the Regiment pursued and subsequently engaged the enemy at Champion's Hill on May 16, Big Black River on May 17 and in the siege of Vicksburg May 18 [880] until July 4. Following the fall of Vicksburg on July 4 the Ninety-seventh marched with the Thirteenth Corps to Jackson, Mississippi and fought in the siege and capture of Jackson from July 10 to 17; then returned to camp at Vicksburg. August 27 the Regiment was ordered to New Orleans, Louisiana where they camped until October. During October the Ninety-seventh marched in the Western Louisiana campaign but on November 1 was ordered back to New Orleans for provost duty and remained there until May 1864.[881] From May to September the men were on picket duty at Morganza, Louisiana; this included a skirmish at nearby Atchafalaya Bayou in September. The Ninety-seventh returned to New Orleans in October and was transported by boat first to Dauphin Island in Mobile Bay and then landed at Pascagoula, Mississippi to begin the march toward Mobile, Alabama. The Regiment met rebel outposts twenty miles from Mobile, fell back and finally returned to Pascagoula where they remained until February 1, 1865. They boarded steamers and traveled to Barrancas, Florida.[882] March 20 the Ninety-seventh as part of the Thirteenth Corps began the march northward through the swamps of Florida and Alabama planning to attack Mobile from the north. On April 2 the Union forces struck the rebel outposts of Fort Blakely on the Blakely River north of Mobile. On April 9 a general assault was ordered, led by the Ninety-seventh; within twenty minutes the Fort was captured along with five thousand enemy prisoners and thirty-five pieces of artillery. The loss to the Ninety-seventh was eighty killed and wounded.[883],[884] By April 12 the Union forces occupied the city of Mobile, and on April 28 the Ninety-seventh was transported up the Alabama River to Selma, Alabama. Early in the morning of May 1 the Regiment was ordered to proceed to Cahawba, Alabama downriver to intercept a train at Marion Junction said to be carrying Jefferson Davis. Although Marion was captured, and the depot burned there was no sign of Jefferson Davis. The Regiment returned to Mobile and then was sent to Galveston, Texas where the Ninety-seventh was mustered out on July 29, 1865.[885]

The colors of the Ninety-seventh Regiment of Illinois Infantry held by the State of Illinois Military Museum, consist of one national and one regimental flag. There is no documentation regarding the history of these flags, but they are probably the flags carried by the Ninety-seventh at the Battle of Fort Blakely. Both flags show significant damage. The national flag is inscribed with battle honors.

---

[880] During the assault of May 22, 150 volunteers were needed to attack the fort in front of the Union Fifteenth Army Corps. Since the possibility of returning alive seemed poor the attack was called "forlorn hope." Three men volunteered from the Ninety-seventh Illinois Infantry—Carlos N. Colby, Sgt. David Dickey, and Capt. Richard H. Wood. They received the Medal of Honor— "Was a member of a "Volunteer Storming Party' which made a most gallant assault upon the enemy's works. Vicksburg, Mississippi, May 22, 1863." Beyer, W. F. and Keydel, O. F. *Deeds of Valor,* reprint by Longmeadow Press, Stamford, CT., 1993, pages 190-7.

[881] While en route back to New Orleans, November 3, 1863, the Regiment was in a railroad accident. Eighteen men were killed and sixty-seven injured. When the Regiment arrived in New Orleans for duty only four hundred men were available for duty. However, officers were sent North for recruiting and the Regiment was refilled to almost full strength.

[882] Fort Barrancas was located on Tartar Point directly south of Pensacola, Florida.

[883] The commander of the Ninety-seventh Illinois Infantry, Lt.-Col. Victor Vifquain, received the Medal of Honor for his leadership during that charge. See description above.

[884] O.R. Series 1, Volume XLIX/1 [S#103] page 212-214. According to the report of Lt.-Col. Vifquain, the Ninety-seventh captured "1 battle-flag, 1 headquarters flag (French's Division), and another battle-flag (Missouri brigade, General Cockrell's)."

[885] *Report of the Adjutant General of the State of Illinois,* H. W. Rokker, State Printer and Binder, Springfield, Illinois, (1886), Volume V, page 494-8.

Although a portion of the canton and several of the stars are missing, it appears that the canton was originally square and contained five rows of stars. This is consistent with flags of New York manufacture. The flag is inscribed, CHICKASAW BLUFF, December 29, 1862: ARKANSAS POST, January 11, 1863: PORT GIBSON, May 1, 1863: CHAMPION HILL, May 16, 1863: BLACK RIVER BRIDGE, May 17, 1863: VICKSBURG, May 18, 19, 20, 22, July 4, 1863: JACKSON, July 10, and 15 1863.

The regimental flag is heavily damaged, but the overall pattern of the flag is characteristic of flags manufactured for the New York depot.

No other flags carried by the 97th Illinois Infantry Regiment are known to survive.

98th Illinois Infantry[886]

The Ninety-eighth Illinois Infantry was organized at Centralia, Marion County, Illinois and mustered in for three years of service on September 3,1862 under the command of Colonel John J Funkhouser, a 28-year-old merchant from Effingham, Illinois. The men volunteered from the following counties: Company A from Clay County; Companies B, G and H from Richland County; Company C from Effingham County; Company D and E from Crawford County; Company F from Clay and White County; Company I from Jasper County and Company K from Effingham County. On September 8 the Regiment was ordered to Louisville, Kentucky. While passing through Bridgeport, Illinois the train derailed resulting in the death of 8 and injuring of 75 men from the Regiment. By September 30 the Ninety-eighth had reached Elizabethtown, Kentucky, moved to Frankfort in October and then to Bowling Green via Bardstown, Mumfordville and Glasgow, Kentucky. In late November the Regiment marched south to Gallatin and Castalian Springs, Tennessee and in December to Bledsoe Creek, Tennessee. In late December the Ninety-eighth marched northward in pursuit of Morgan, arriving at Glasgow, Kentucky on December 31. In 1863 the Regiment continued scouting and guarding forage trains in southern Kentucky and northern Tennessee. In March 1863 the Regiment was mounted and in early April was sent on a scouting mission to Rome, Lebanon and Snow's Hill, Tennessee and returned to Murfreesboro. Later in April the Ninety-eighth was sent to McMinnville, Tennessee where they destroyed a cotton mill and captured a railroad train. After capturing many horses and mules the Regiment returned to Murfreesboro. On June 4 the Ninety-eighth moved out on the Liberty Road and attacked the First Kentucky and Eleventh Texas Cavalry and on June 10 attacked the enemy at Liberty, Tennessee driving them back to Snow's Hill. During the Middle Tennessee Campaign of June and July 1863 the Regiment fought at Hoover's Gap on June 24-26 and Decherd on June 29. While in the vicinity of Wartrace and Decherd during July and August the Regiment captured over 1,000 horses and mules. In August the Ninety-eighth crossed the Cumberland Mountains and in September forded the Tennessee River advancing toward Ringgold and Tunnel Hill, Georgia, skirmishing with Forrest en-route. September 13 and 14 the Regiment skirmished at Stephenson's Gap and on September 18-21 fought in the battle of Chickamauga, Georgia.[887] Following the battle, in October, along with the First and Second Brigades of Cavalry the Regiment pursued Wheeler north through Sequatchie, Tennessee to the Cumberland Mountains where they attacked the enemy and drove them through McMinnville to Shelbyville, Tennessee; the enemy was routed during engagements at Shelbyville and Farmington. The pursuit of Wheeler ended when he crossed the Tennessee River and the Ninety-eighth went into camp at Maysville, Tennessee until mid-November. The Ninety-eighth was ordered to Chattanooga on November 21, upon arrival the Regiment numbered only 200 mounted effective men. From Chattanooga the Regiment was active in East Tennessee and on December 1 they led the advance of Sherman's Army and drove the enemy through Loudon, Knoxville and Maysville, Tennessee finally going into camp at Calhoun, Tennessee on the Hiawassee River northeast of Chattanooga. In February 1864 Colonel Funkhouser who had been wounded at Chickamauga, rejoined the Regiment in time to command the Ninety-eighth during their campaign in north Georgia. During February the Regiment saw action at Tunnel Hill, Buzzard's Roost Gap, and Rocky Faced Ridge; in March the men were placed on picket duty at Ringgold, Georgia. From May through July 1864, as part of Sherman's Atlanta campaign, the

---

[886] The 92nd, 98th, and 123rd Illinois Mounted Infantry Regiments, the 17th and 72nd Indiana Mounted Infantry Regiments and 18th Indiana Battery formed *Wilder's Lightning Brigade* (1st Brigade, 4thDivision, 14th Army Corps).

[887] In the battle of Chickamauga, the commander of the Ninety-eighth, Colonel John Funkhouser was severely wounded.

Ninety-eighth fought at Resaca, Rome, Dallas, Pumpkin Vine Creek, New Hope Church, Allatoona Hills, Big Shanty, Noonday Creek, Lattimer's Mills, Assault on Kennesaw Mountain, Nickajack Creek, Chattahoochee River, Siege of Atlanta—July 22-August 15, Kilpatrick's Raid south of Atlanta and the Chattahoochee River Bridge. Following the fall of Atlanta, the Regiment joined in the pursuit of Hood into North Georgia and Alabama. November 1 the Regiment turned their horses over to Kilpatrick and during November and December the Ninety-eighth was in Louisville, Kentucky for refitting; they moved to Nashville in early 1865.

On January 12 the Regiment marched via Columbia and Mount Pleasant to Gravelly Springs, Alabama and remained in camp until March. In the Spring of 1865, the Ninety-eighth skirmished with enemy forces in Alabama and Georgia and returned to Nashville for muster out June 27, 1865. The recruits were transferred to the Sixty-first Illinois Infantry.[888]

The flag collection of the Adjutant General of the State of Illinois holds one national flag, one regimental flag and one guidon or flank marker.

This national flag is not of Quartermaster issue. The thirty-four stars are arranged in an unusual pattern; undoubtedly this flag was homemade and presented to the Regiment. The flag is inscribed: HOOVER'S GAP; CHICKAMAUGA; FARMINGTON; KENESAW MOUNTAIN; SIEGE OF ATLANTA; ROME; SALEM; 98th ILLs VOLs MTD INFTY. *The inscriptions are all similar color and lettering; therefore the inscriptions were presumably all added at the same time either late in the War or even after the War.*

[888] *Report of the Adjutant General of the State of Illinois,* H. W. Rokker, State Printer and Binder, Springfield, IL (1886), Volume V, page 521-4.

The Regimental flag of the Ninety-eighth is not inscribed with battle honors. The V-shaped upper ribbon with stars extending to the edge of the ribbon and a lower three-piece scroll with a lowered center section as seen on this flag are characteristics of flags produced by Evans and Hassall for the Philadelphia depot.

The guidon or flank marker of this Regiment measures approximately 18" x 24".

The Ninety-eighth assisted in the capture of Macon, Georgia and one of the brass cannons, captured at Macon was given to the Regiment. The cannon was subsequently presented to the State of Illinois by the Regiment and was placed on exhibit in Memorial Hall.

No other flags carried by the 98th Illinois Infantry are known to survive.

99th Illinois Infantry

"The 99th once owned a flag and it was bright and gay,
But it is now in Springfield and going to decay;
We all remember when it was that it was given to us,
In August, 1862, by the ladies of St. Louis,
It was given to us to keep as a sacred trust,
And we were never to trail it in dishonor's dust."[889]

The Ninety-ninth Illinois Infantry Regiment was organized in Pike County, Illinois during August 1862 and mustered in at Florence, Pike County for three years' service on August 23, 1862, under the command of Colonel George W. K. Bailey, a 38-year-old farmer from Pittsfield, Illinois. The men in the Regiment all volunteered from Pike County, Company A from Pittsfield Township, Company B from Perry Township, Company C from Pittsfield and Pleasant Hill Townships, Company D from Barry Township, Company E from Hardin Township, Company F from Fairmount, Griggsville and Perry Townships, Company G from Derry, Pittsfield, Martinsburg and Pleasant Vale Townships, Company H from Barry and Kinderhook Townships, Company I from Detroit and Montezuma Townships and Company K from New Salem Township. August 24, 1862, the Regiment arrived at St. Louis, Missouri and was ordered to the Benton Barracks to receive equipment. The Regiment remained on duty in Missouri until March 3, 1863. During that time a detachment of Ninety-ninth skirmished at Beaver Creek and Hartville. On March 15 the Regiment was ordered to Milliken's Bend, Louisiana north of Vicksburg on the Mississippi River. April 30 the Ninety-ninth crossed the Mississippi River below Vicksburg at Bruinsburg and met the rebel forces the next day at Magnolia Hills near Port Gibson, Mississippi. The Regiment then marched north and was engaged at Champion's Hill on May 16 and at Big Black River Bridge on May 17. The Ninety-ninth was heavily engaged on the historic assaults on Vicksburg of May 19 and 22. In the assault of May 22 the Regiment lost 103 men killed and wounded of the 300 men who entered the battle; Sergeant Thomas J. Higgins, Company D, was awarded the Medal of Honor for his actions in that assault.[890] Following the surrender of Vicksburg on July 4, 1863 the Ninety-ninth joined Sherman's pursuit and siege of Johnson's retreating rebel forces at Jackson, Mississippi then returned to Vicksburg where they remained until August. In late August the Regiment was ordered to Brashear City, Louisiana to guard the *New Orleans, Opelousas, and Great Western Railroad* on the Atchafalaya River. During the subsequent campaign in Western Louisiana the Regiment was in several skirmishes including one at Grand Coteau and then returned to New Orleans. November 16 the Ninety-ninth was transported to Mustang Island, Texas, marched up to Matagorda Island and joined in the attack and surrender of Fort Esperanza on November 29. The Regiment remained in Texas until June 1864 when it was ordered to Algiers, Louisiana to perform garrison duty on the Mississippi River. September 1864 the Ninety-ninth was ordered to St. Charles, Arkansas and in October and November was at Duvall's Bluff. In November the Regiment moved to Little Rock and then Memphis where the Regiment was consolidated into a Battalion of five companies (Companies A, B, C, D and E of the Ninety-ninth Consolidated Infantry Regiment). The Battalion guarded the *Memphis and Charleston*

[889] Portion of a poem written by Solomon T. Johnston, Company E, and read at the reunion of the Regiment in 1903.
[890] Canadian born Higgins was awarded the Medal-- "When his regiment fell back in the assault, repulsed, this soldier continued to advance and planted the flag on the parapet, where he was captured by the enemy." *Civil War Medal of Honor Winners,* Published by the Civil War Centennial Commission of Illinois, July 1962, page 11.

Railroad at Germantown, Tennessee during November and December. January 1865 the Battalion embarked for New Orleans and on February 1 was transported to the Dauphine Islands, Alabama at the mouth of Mobile Bay. In the Mobile Campaign the Battalion fought at Spanish Fort, Fort Blakely and the Battle of Mobile. In June 1865 the Battalion accompanied the Division up the Red River to receive the surrender of Kirby Smith. The Ninety-ninth Consolidated Infantry Regiment was mustered out at Baton Rouge on July 19, 1865.[891]

There are three national and one regimental flag of the Ninety-ninth Illinois Volunteer Infantry in the holdings of the State of Illinois. The poem by Solomon Johnston quoted above, goes on to describe the history of the flag presented by the ladies of St. Louis.

"That Flag was duly honored by both the rank and file,
And guided us over many a hundred weary mile;
When we crossed the river at Bruinsburg that Flag it led the way,
Where we found the enemy in force before the break of day;
General Grant was with us, all honor to his name,
And this was our first battle in the Vicksburg campaign.
On that unfortunate twenty-second day of May,
The 99th got into the engagement quite early in the day,
The air was filled very soon with Rebel shot and shell,
When William Sitton, our noble color bearer fell;
Thom Higgins grasped the colors and took them up before,
And left the gallant Sitton lying bleeding in his gore."

Based on information at Vicksburg National Military Park, Thomas Higgins and the flag of the Ninety-ninth Illinois were captured by the Second Texas Infantry.

According to the Illinois Adjutant General's Record, "In 1873 the colors were sent from Richmond, Va., to Philadelphia, Pa., thence to Springfield, Ill., where they now are. Stains of blood can yet be seen on them, the patriot blood of the lamented Sitton."[892]

Of the three national flags shown below, the first flag is the only flag in the Illinois holdings that appears to be home made and thus may be the flag presented by the "ladies of St. Louis" to the Ninety-ninth Illinois Infantry.

However, a second homemade flag was presented to the Regiment. On August 10, 1862, in Perry, Pike County, Illinois a flag, made by the ladies of Perry, was presented to the Captain of Company B, Benjamin L. Matthews of Perry. Miss Mary Sutphin presented the flag:

"On behalf of the ladies of Perry, I take pleasure in presenting you with this flag. It is our farewell gift, and we mean it shall keep you in mind of our deep interest and sympathy on your behalf. We know that the honor of this flag will be safe in your keeping. Old Pike sends none but brave men to the battle field. Take this flag, the sacred colors of which have been arranged and its stars of glory set for you. Remember it is the emblem of your country which traitorous hands are raised to destroy."[893] The flag is

---

[891] *Report of the Adjutant General of the State of Illinois,* H. W. Rokker, State Printer and Binder, Springfield, Ill. (1886), volume V, Pages 543-5.
[892] Ibid, page 544
[893] Personal communication from Kathy Zimmerman

not further described but suggests that the flag was a national flag.

This national flag measures 51" x 110", has 34 stars in the canton and has a fold in the center of the flag. It is not of typical Quartermaster design or issue and is presumably homemade. Based on Solomon Johnston's information this may be the flag that Sergeant Thomas Higgins planted on the Confederate works at Vicksburg.

This national flag, with a square canton is inscribed on the fifth red stripe, "99th REG. ILL' VOL." *The square canton is typical of New York manufacture.*

This 35-star flag with a square canton is typical of New York manufacture. It is not inscribed.

The Regimental flag of the Ninety-ninth is in excellent condition and is typical of New York manufacture.

No other flags carried by the 99th Illinois Infantry are known to survive.

100th Illinois Infantry—"*Will County Regiment*"

"Loudly the bugles are merrily blowing,
Exulting, the guns beat a gay reveille—
Today must the mountain yield to our prowess,
The traitor flag bow to the flag of the free.
And proudly the banners are floating and soaring;
Proudly may they be, though all tattered and torn,
And they flutter and struggle as if to be free—
And kiss in their rapture, the breeze of the morn."[894]

The Will County Regiment was organized at Camp Irwin in Joliet, Will County, Illinois during August 1862 and was mustered in on August 30 for three years' service under the command of Colonel Frederick A. Bartleson[895]. On September 2, 1862, the men broke camp, boarded cars of the *Chicago, Alton & St. Louis Railroad* bound for St. Louis but en-route orders were changed to send them to Louisville, Kentucky. While in camp at Louisville, the Regiment received their arms, equipment and sixty rounds of ammunition. The One Hundredth had its first skirmish at Bardstown, Kentucky on October 4; when the Regiment charged, the rebels broke and ran and abandoned Bardstown. The Regiment joined in the pursuit of Bragg as far as Wild Cat, Kentucky then marched south and arrived in Nashville, Tennessee on November 26. By the end of December, due to disease, the Regiment was down to 600 men designated as fit for duty. On the evening of December 26, the One Hundredth broke camp and joined the forces moving south toward LaVergne and Murfreesboro, Tennessee. The following day, the Regiment, in their first experience under fire, charged the enemy at LaVergne, driving them from the town. Further south, at Murfreesboro from December 31, 1862, through January 3, 1863 the One Hundredth fought in the Battle of Stones River. In that action the Regiment lost over 100 men killed, wounded or missing. The One Hundredth remained at Murfreesboro, foraging, and performing guard duty until June 24 when as part of the Army of the Cumberland, the Regiment marched southeast toward Chattanooga. The Army reached Chattanooga on September 9 then moved further south and camped on the banks of Chickamauga Creek. Preliminary skirmishes led to the Battle of Chickamauga beginning on Friday, September 18, 1863. Over the next two days the Regiment lost 165 men of the 315 who went into the battle, every color guard but one was killed, and Colonel Bartleson and Surgeon Henry Woodruff and twelve others were taken prisoner.[896] The Regiment, with Major Charles Hammond in command fell back toward Chattanooga and joined the defensive line south of the city and remained there until late November. On November 23 and 24 the One Hundredth supported the storming party that secured Orchard Knob; on the morning of November 25 the Regiment, after learning that Hooker had captured Lookout Mountain, was ordered to advance on Mission

[894] Portion of a poem written by Colonel Frederick Bartleson of the One Hundredth Illinois Infantry while he was a prisoner at Libby prison. Colonel Bartleson, a Joliet attorney, had lost his left arm at Shiloh when he was serving as Captain of Company B of the 20th Illinois Infantry. He was taken prisoner at Chickamauga while leading the One Hundredth Illinois Infantry and was subsequently killed at the Battle of Kennesaw Mountain. Woodruff, George H., *Fifteen Years Ago: or the Patriotism of Will County,* Joliet Republican Book and Job Steam Printing House, Joliet, Ill., (1876) page 310.

[895] "The Kennasaw Line: In Memory of Frederick Bartleson" Emerging Civil War (https://emergingcivilwar.com)

[896] Apparently, a color of the 100th was destroyed. "Captain W. A. Munger of the Hundredth Illinois Regiment has sent to his father in this place an interesting relic of the battle of Chickamauga. It consists of one star of the Battle Flag of that Regiment which was shot away by the enemy fire." (Iowa) *State Daily Register,* October 9, 1863. p. 3

Ridge. The Regiment charged the Ridge directly in front of Orchard Knob and did not stop until they had reached the top of the Ridge, captured guns and continued the pursuit of the enemy as far as Chickamauga Creek. Although the Regiment suffered many wounded, only one man was killed in the assault. Immediately following the battle, the One Hundredth joined in the march toward Knoxville, Tennessee for the relief of Burnside. They arrived at Knoxville, on December 7 to learn that Longstreet had retreated into West Virginia. December 12 the Regiment marched to Louisville and operated in East Tennessee until February 1864.[897] During February, March and April the One Hundredth was stationed at Athens, Tennessee to guard the *Knoxville and Chattanooga Railroad.* In May the Regiment entered the Atlanta campaign. During May the Regiment saw action at Rocky Faced Ridge, Buzzard's Roost Gap, Dalton, Resaca, Adairsville, Dallas, New Hope Church and Allatoona Hills.[898] June and July the Regiment was in actions at Marietta, Kennesaw Mountain, Pine Mountain, and Lost Mountain; in action at Kennesaw Mountain on June 23, an enemy sharpshooter killed Colonel Bartleson, the commander of the Regiment. The One Hundredth was in heavy fighting during July at Smyrna Campground, the Chattahoochee River, Buckhead, Peach Tree Creek, and the Siege of Atlanta. In a flanking movement south of Atlanta the Regiment fought at Jonesboro and Lovejoy Station in August and September, then joined in the pursuit of Hood into North Georgia and Alabama in October. The One Hundredth saw its next action at Spring Hill on November 29 when they drove off charging rebel cavalry; the following day, November 30, at Franklin, Tennessee the Regiment repulsed five charges of the rebel forces. Following the fighting at Franklin the Union forces fell back to Nashville and, in the final battle of the War in the West, defeated Hood at the Battle of Nashville December 15 and 16. On December 17 the One Hundredth joined in the pursuit of Hood to the Tennessee River then moved to Huntsville, Alabama where they remained in camp until March 1865. Following operations in East Tennessee in March and April the Regiment was ordered to Nashville where it was mustered out on June 12, 1865.[899]

There is one national and one regimental flag of the One Hundredth Illinois Volunteer Infantry in the collection of the State of Illinois Military Museum.

This 35-star national flag has a star pattern in the rectangular canton typical of flags manufactured by William F. Scheible for the Philadelphia depot. The flag is inscribed: LAVERGNE; STONE RIVER;

[897] During December Surgeon Henry Woodruff, who had been captured at Chickamauga, rejoined the Regiment.
[898] Colonel Bartleson rejoined the Regiment at the end of May.
[899] Woodruff, George H., *Fifteen Years Ago: or the Patriotism of Will County,* Joliet Republican Book and Job Steam Printing House, Joliet, Ill., (1876)

100TH ILLINOIS INFANTRY

CHICKAMAUGA; MISSION RIDGE; ROCKY FACE; RESACA; DALLAS; LOST MOUNTAIN; KENESAW; JONESBORO; PEACH TREE CREEK; ATLANTA; LOVEJOY; SPRING HILL; FRANKLIN; NASHVILLE; 100th ILLs VOLs.

The regimental color of the One Hundredth Illinois Volunteer Infantry is not inscribed with battle honors.

George Woodruff in his history of the One Hundredth mentions two flag presentations to the Regiment. "While at this camp the regiment was presented with a banner from the ladies of Joliet. It was accompanied by an eloquent letter from Judge Parks, to which the colonel sent a suitable reply."[900]

"April 27th, Lt. Col. Waterman who had been in command of the 100th since Bartleson was taken prisoner, having received from the government a new stand of colors, sent home the old flag, which the regiment had carried since leaving Joliet, now all tattered and torn, so that it could no longer be unfurled, accompanied by a most eloquent letter giving briefly the scenes through which it had passed."[901]

At least one other flag was damaged or perhaps "parted out" to the soldiers in 1863. As reported in the *State Daily Register* (afterwards the *Des Moines Daily Register*)[902] "A Relic—Captain W. A. Munger of the Hundredth Illinois Regiment has sent to his father in this place an interesting relic of the battle of Chickamauga. It consists of one star of the Battle Flag of that Regiment which was shot away by the enemy fire. The Regiment was opposed to a portion of Longstreet's line and lost 150 men out of 314."

No other flags carried by the 100th Illinois Infantry are known to survive.

---

[900] Ibid, Woodruff, page 227-28, The camp referred to was Camp Yates located on the farm of John C. Breckenridge four miles from Louisville. This presentation would have taken place in September 1862.

[901] Ibid, Woodruff, page 69-70. The new stand of colors would have been received in April 1864 and apparently at that time the flag that had been presented by the ladies of Joliet was returned to Joliet. Presumably the flags shown here are the flags received by the Regiment in April 1864.

[902] *State Daily Register*, October 8, 1863, p3, col.1. William A. Munger enlisted from Joliet, Illinois and was Captain of Company G.

101st Illinois Infantry—"*Morgan County Regiment*"

"Ring the good ol' bugle, boys, we'll sing another song,
Sing it with the spirit that will start the world along,
Sing it as we used to sing it 50,000 strong
While we were marching through Georgia."

"Hurrah, hurrah, we bring the jubilee!
Hurrah, hurrah, the flag that makes you free!
So we sang the chorus from Atlanta to the sea
While we were marching through Georgia!"[903]

The One Hundred and First Illinois Infantry was organized in Jacksonville, Morgan County Illinois at Camp Duncan during August 1862. All the men who served in the One Hundred and First volunteered from Morgan County except for 20 men in Company G who volunteered from Sangamon County. The regiment was mustered in for three years' service on September 2, 1862, under the command of Jacksonville lawyer, Colonel Charles H Fox, but remained at Camp Duncan for drilling and equipping until October 7 when they departed for Cairo, Illinois. They remained in Cairo guarding prisoners until November 26 when the men boarded the steamer *Universe*, steamed down the Mississippi River, debarked at Columbus, Kentucky, and boarded cars of the *Mobile & Ohio Railroad* for the trip to Davis' Mills, Mississippi[904] to join the Army of the Tennessee. On November 28 the Regiment joined the march north to Lumpkin's Mills and then occupied Holly Springs, Mississippi. Company A was sent with rebel prisoners to Cairo on December 13. The Confederate forces captured Holly Springs on December 20. The field officers and Companies B, C, E, F and I were captured while Companies D, G, H and K, who were stationed outside of Holly Springs guarding the *Tennessee & Ohio Railroad*, retreated to Cold Water, Mississippi. Company A was not captured since it was in Cairo. The captured men were paroled and sent to Benton Barracks, Missouri where they remained until exchanged in June 1863. Companies D, G, H and K were formed into a Battalion and attached to Fourteenth Illinois Infantry until March 1863. Company A rejoined the Battalion of the other four Companies at Memphis in February 1863. Early in March 1863 the Battalion was ordered to Vicksburg; on arrival the Battalion was broken up as follows—Company A assigned to the ram *General Bragg,* Company D to the tinclads *Cricket* and *Rattler,* Company G to the ram *Switzerland,* Company H to the *Lafayette* an ironclad ram side-wheeler and Company K assigned as provost guard at General Grant's headquarters. In June 1863, Companies B, C, E and F of the One Hundred and First were exchanged and were sent to New Madrid, Missouri and then in July to Columbus, Kentucky. All the Companies of the Regiment were rejoined at Union Tennessee on August 25, 1863. On September 24, 1863, the Regiment was ordered to Louisville and on September 27 ordered on to Bridgeport, Alabama. In further movement toward Chattanooga, thc One Hundred and First departed Bridgeport on October 27, crossed the Tennessee River, entered Lookout Valley and on October 28 and 29 fought the rebels at Wauhatchie Station on the *Nashville & Chattanooga Railroad.* The Regiment camped in Lookout Valley until late November when they

[903] The 101st Illinois Infantry was one of the regiments that marched with General Sherman "to the sea". Marching Through Georgia, composed by Henry Clay Work, published by Root & Cady, Chicago.
[904] Davis' Mills lay just north of Meridian, Mississippi.

marched to Chattanooga and participated in the battle of Chattanooga. Immediately after the battle the Regiment was ordered to the relief of the Union forces under Burnside at Knoxville; however, the rebels abandoned the siege at Knoxville before the Union relief arrived and the Regiment returned to Lookout Valley in late December. On January 1, 1864, the One Hundred and First was ordered to nearby Kelly's Ferry on the Tennessee River to relieve the Sixteenth Illinois Infantry. In February the Regiment was ordered to Bridgeport, Alabama and remained there until early May when they joined Sherman's Atlanta campaign. Over the next five months the One Hundred and First Illinois Infantry fought at Resaca, New Hope Church, Pumpkin Vine Creek, Dallas, Allatoona Hills, Kennesaw Mountain, Pine Hill, Lost Mountain, Muddy Creek, Noyes Creek, Kolb's Farm, Ruff's Station, Chattahoochee River, Peach Tree Creek and the siege of Atlanta. The Regiment was one of the first to enter Atlanta after its fall and remained in Atlanta until mid-November. On November 15, 1864, the Regiment joined Sherman's "March to the Sea." During the remainder of the War the Regiment saw action at Savannah, Georgia, Averysboro and Bentonville, North Carolina and at the surrender of Johnston's army at Bennett's House, Virginia. The One Hundred and First Illinois Infantry marched to Washington DC for the Grand Review on May 24, 1865, and then went into camp at Bladensburg, Maryland until muster out on June 7, 1865.[905,906,907]

There are remnants of one national and one regimental flag of the One Hundred and First Illinois Infantry in the collection of the State of Illinois. No associated history is available for either flag.

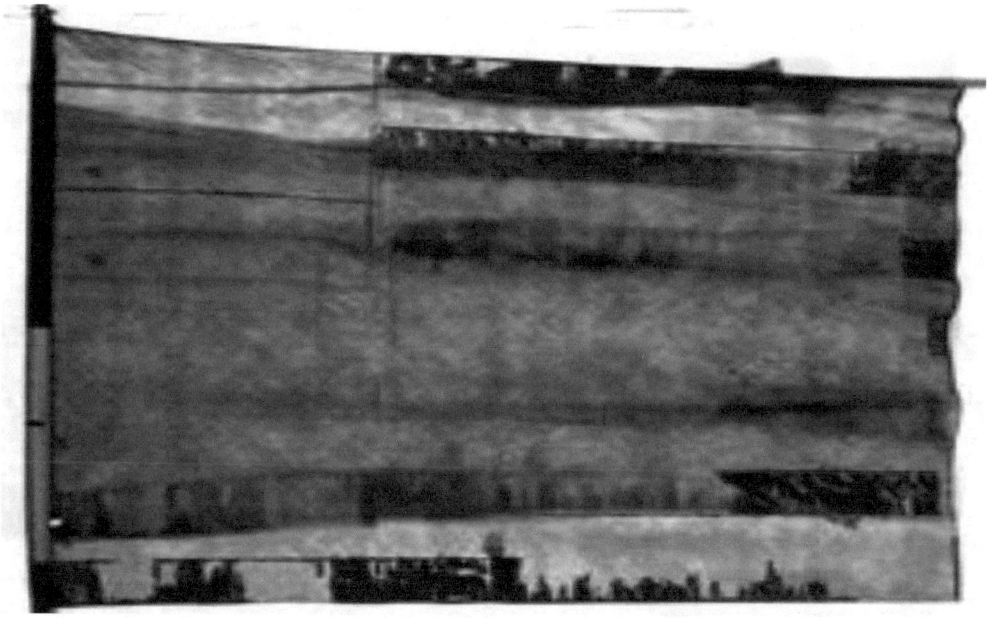

Little remains of this national flag.

---

[905] *Report of the Adjutant General of the State of Illinois,* HW Rokker, State Printer and Binder, Springfield, Ill, (1886), Volume V, pages 596-98.
[906] Potter, Rev. John, *Reminiscences of the Civil War in the United States,* The Globe Presses, Oskaloosa, Iowa, (1897)
[907] *Jacksonville Daily Journal,* Sunday, May 30, 1909.

Only shreds remain of the regimental flag.

No other flags carried by the 101st Illinois Infantry are known to survive.

102nd Illinois Infantry

"Near the commencement of the charge, our flag bearer, Sergt. R. L. Carver was severely wounded. The banner bearer Corpl. P. F. Dillon was shot through the breast an instant later. When he fell, Adjutant J. H. Snyder gallantly caught up the banner and carried it into the redoubt. Subsequently it was planted by him on the parapet of the earthwork. During the action fifty balls cut the banner, and two shattered the staff."[908]

The One Hundred and Second Illinois Infantry was organized at Knoxville in Knox County, Illinois during August of 1862 and mustered in for three years' service on September 1, 1862 under the command of Colonel William McMurtry.[909] The men of the Regiment volunteered from the following counties; Companies A, D, F, H and I from Knox County; Company B from Knox and Warren Counties and Companies C, E, G and K from Mercer County. September 22 the Regiment moved to Peoria, Illinois and on October 1 boarded cars of the *Peoria, Burlington and Logansport Railroad* for Louisville, Kentucky. From Louisville, the Regiment marched south to Gallatin, Tennessee via Frankfort and Bowling Green, Kentucky. During 1863, the Regiment saw action at Woodbury, Tennessee on April 27. In June the Regiment was ordered to La Vergne, Tennessee to guard the *Nashville & Chattanooga Railroad;* Companies E, G and K guarded the railroad bridge at Stewart's Creek, Company H was stationed at Overall's Creek, Company B at Smyrna and the remaining five companies were at Regimental headquarters near the Stewart's Creek bridge. In August five companies of the One Hundred and Second were mounted and furnished with Spencer repeating rifles and with either a Colt or Remington revolver. February 25, 1864, the Regiment broke camp at La Vergne and marched via Stevenson, Alabama to Wauhatchie Valley, Tennessee—adjacent to Lookout Mountain. While there the five mounted companies gave up their horses but retained their Spencer rifles. In May 1864 the Regiment joined Sherman's Atlanta campaign. After demonstrations at Dalton and Rocky Faced Ridge, the Regiment fought in the Battle of Resaca on May 14 and 15 and at New Hope Church and Allatoona Hills May 24-June 5. During June and July, the One Hundred and Second fought at Pine Hill, Lost Mountain, Muddy Creek, Noyes Creek, Kennesaw Mountain, Ruff's Station, Chattahoochee River, Peach Tree Creek and in the Siege of Atlanta. Following the fall of Atlanta, the Regiment remained in Atlanta until November 15—then joined Sherman's "March to the Sea". In the Georgia campaign the Regiment occupied Milledgeville, fought at Ogeechee River in late November and joined in the siege of capture of Savannah in December. The year 1865 saw the One Hundred and Second Illinois Infantry in the Campaign of the Carolinas. In January the Regiment occupied Hardeeville, South Carolina and in February Lawtonville, South Carolina. After occupying Fayetteville, North Carolina in March the Regiment was engaged in the battle at Averysboro on March 16 and in the Battle of Bentonville, North Carolina on March 19-21. In April the Regiment occupied Raleigh until after the surrender of the Confederate forces. On May 24 the One Hundred and Second Illinois Infantry marched in the Grand Review in Washington DC and was mustered out of service on June 6, 1865.[910,911]

---

[908] Fleharty, S. F., *Our Regiment. A History of the 102nd Illinois Infantry Volunteers,* Brewster & Hanscom, Printers, Chicago, Illinois (1865) page59-60. Description of the action at the Battle of Resaca, Georgia on May 15, 1864.

[909] Colonel McMurtry was 61 years old when he became Colonel of the 102nd Illinois Infantry. He was a friend of Stephen Douglas, was the 11th Lieutenant Governor of Illinois, and had been a member of the Illinois House of Representatives and member of the Illiinois Senate. Due to poor health he served only 6 weeks in the 102nd Infantry and was replaced by Colonel Franklin Smith.

[910] Ibid., Fleharty.

[911] *Report of the Adjutant General of the State of Illinois,* H. W. Rokker, State Printer and Binder, Springfield, IL (1886) Volume 5, pages 620-21.

102ND ILLINOIS INFANTRY

There are two national and two regimental flags of the One Hundred and Second Illinois Volunteer Infantry in the collection of the State of Illinois.

This national flag was probably produced by Hugh Wilkins of Louisville, Kentucky. It is inscribed "102nd REGT. ILL. VOLS.

This national flag is typical of those produced by Evans & Hassall and by Horstmann Brothers & Company for the Philadelphia depot. Since it is a 35-star flag it was produced late in the War.

The pattern of this regimental flag with 13 stars above the eagle is characteristic of flags manufactured by Hugh Wilkins of Louisville, Kentucky. The scroll beneath the eagle is inscribed: 102nd REGT. ILL. VOLS.

This regimental flag is in poor condition with loss almost one half of the silk..

There was a presentation of at least one flag to one of the companies in the Regiment. "FLAG PRESENTATION—The ladies of Rock Island and Mercer counties residing in Coal Valley, Preemption, Berlin, Richland Grove and vicinity, have presented captain Shedd's company, with a beautiful flag….The flag was received on the part of the company, by Mr. Snyder, who paid a high compliment to the ladies of Rock Island and Mercer counties for the taste they had displayed in getting up the flag, and for their patriotic devotion to the cause of the union."[912] Frank Shedd of Richland Grove was the Captain of Company C. No further details are known about this flag.

No other flags carried by the 102nd Illinois Infantry are known to survive.

[912] *Argus* (Rock Island) August 25, 1862. Signed "Old Flag." Franklin Shedd of Richland Grove was Captain of Company C.

103rd Illinois Infantry – "*Babcock's Infantry*"

"Farewell to home and kindred till the traitor cause is dead—
Till Floyd has dropped his plunder and Jeff has lost his head—
When this victory is ours we'll return with shouts of joy
To our lovely wives and sweethearts in old Fulton, Illinois.
Then shout out your loudest huzzas,
Three cheers for old Fulton's brave boys!
For our flag shall ever be
The Banner of the free,
While upheld by the brave Fulton boys."[913]

In response to the August 1862 call by President Lincoln for 300,000 additional troops, the One Hundred and Third Illinois Infantry was recruited and organized in Fulton County, Illinois. The men volunteered from the following areas within the County: Company A-Lewiston and Liverpool townships; Company B-Ellisville and Young Hickory townships; Company C (*Fulton Guards)*-Farmington, Buckheart, Canton and Banner townships; Company D (*Halleck Guards)*-Fairview township; Company E-Banner, Putnam and Orion townships; Company F-throughout the County, Company G-Canton and Lewistown townships; Company H-Lewistown township; Company I-Bernadotte and Pleasant townships and Company K-Canton township. On September 7 the Regiment arrived at camp in Peoria and was mustered in for three years of service on October 2, 186 under the command of Colonel Willard Dickerman.[914] On October 30, the One Hundred and Third boarded cars of the *Illinois Central Railroad* for the trip to Bolivar, Tennessee via Cairo, Illinois. By early November, the Regiment was encamped at La Grange performing escort and scouting duties. In December, the Regiment moved south through Holly Springs, Mississippi then on to Waterford, Mississippi for garrison duty and to fortify and guard the bridges of the *Mississippi Central Railroad.* In January 1863 the Regiment moved north by rail to Jackson, Tennessee, then threatened by Forrest; In March they returned to La Grange. On April 17, as part of the diversion regarding the true intent of "Grierson's Raid," the garrison at La Grange (including the One Hundred and Third) under command of General W. Sooy Smith, moved south to Holly Springs and then on to Waterford, Mississippi and back to La Grange. On June 1 the Regiment was placed on guard at various points on the *Mississippi Central Railroad* and on June 5 were ordered to Memphis, placed on the steamer *Henry von Phul,* and moved downriver to Snyder's Bluff on the Yazoo River. From the end of June until the fall of Vicksburg on July 4, the Regiment guarded the rear approaches to Vicksburg at Oak Ridge, Mississippi; following the fall of Vicksburg, the One Hundred and Third marched to the Black River and then on to the siege of the retreating Confederates at Jackson. When the rebel army retreated the Regiment returned to camp near Vicksburg. September 25 the Union troops were ordered to Chattanooga, Tennessee to the relief of the Army of the Cumberland. On October 1, the Regiment boarded the Marine Boat *Diana*, reached Memphis on October 11 and marched east through LaGrange, Tennessee, through Corinth, Iuka and Eastport, Mississippi; they arrived at Florence, Alabama on October 30. Turning north the men reached

[913] *Fulton County Ledger*, October 28, 1862. This is a portion of a song—words by Corp. S. Spellman, set to music by Sergt. A. E. Wagstaff and bugler R. R. Marshall of Company C and sung by the 103rd Regiment.
[914] Colonel Amos Babcock had been elected commander of the regiment, however after the regiment had been recruited, Babcock resigned his commission and the regiment was mustered in under the comman of Colonel Willard A. Dickerman.

Fayetteville, Tennessee on November 7, Winchester on November 11, crossed the Cumberland Mountains and arrived at Bridgeport, Alabama on the west bank of the Tennessee River on November 15. At Bridgeport, Companies C and G were mounted and detached from the Regiment. The remainder of the Regiment left Bridgeport on November 17, crossed the Tennessee River, and camped at Sand Mountain, Georgia. On November 24 and 25 the Regiment was part of a Union force that attacked Confederate positions north of Missionary Ridge-Tunnel Hill; despite a desperate charge the enemy forces could not be dislodged and the One Hundred and Third retreated to the base of the Hill. Following the battle, the enemy withdrew; after a brief pursuit the Regiment was ordered to Knoxville, Tennessee for the relief of Union troops under General Burnside. By the time the Union troops reached Maryville, 16 miles south of Knoxville, the Confederates under Longstreet had quit the siege at Knoxville and retreated to Virginia. The One Hundred and Third returned to Chattanooga, passed through, and spent Christmas 1863 at Stevenson, Alabama.[915] In February 1864 the Regiment was ordered to Cleveland, Tennessee, and from there moved south to Dalton, Georgia and skirmished at Tunnel Hill, Buzzard's Roost Gap and Rocky Faced Ridge. In May the One Hundred and Third became part of the Atlanta campaign and in the next five months fought at Resaca, Dallas[916], New Hope Church, Allatoona Hills, Kennesaw Mountain, Nickajack Creek, Ruff's Mills, Chattahoochee River and at Atlanta. After the fall of Atlanta, the Regiment pursued the retreating rebel forces under General Hood as far as Gaylsville, Alabama and returned to the Smyrna Camp Ground, Georgia in early November. In mid-November 1864 the One Hundred and Third joined in Sherman's "March to the Sea." The Regiment fought at Griswoldville on November 22, at the Ogeechee River December 7 and in the Siege of Savannah in December. From January to April 1865 in the Campaign of the Carolinas the Regiment fought at Salkehatchie Swamps, South and North Edisto Rivers, Congaree Creek, Columbia, and Bentonville. Following the surrender of Johnston and Lee's forces the Regiment was ordered to Washington D.C. and marched in the Grand Review on May 24, 1865. The Regiment mustered out on June 14, 1865.

In the history of the regiment, the author quotes from the diary of Captain Charles Wills of Company C, "Aug. 10 (1864), Our "color" that has floated over the 103rd for nearly two years has become much worn and torn. One shell and bullets innumerable have passed through it. It is entitled to be inscribed with the following battle: Vicksburg, Black River, Jackson, Miss., Mission Ridge, Dalton, Resaca, Dallas, New Hope, Big Shanty, Kennesaw Mt., Atlanta. It has been under the enemies' fire 72 days on this campaign. Nearly 300 of the Regt. have fallen under it."[917] However, the whereabouts of this flag today is unknown. The national flag shown here was presented at the time of the Grand Review. "The original flags of the regiment were presented to them at Holly Springs, Miss[918]., and are now battered beyond recognition. One of them has received a bullet through the staff and bears other equally certain proofs of having been where bullets were not strangers, nor unfrequent visitants. At Washington the other day, the regiment received a fine new banner, upon whose folds are inscribed the following glorious record: Tallahatchie River, Siege of Vicksburg, Black River, Jackson, Mississippi, Mission Ridge, Knoxville, Dalton, Resaca, Dallas, New Hope Church, Noonday Creek, Kennesaw Mountain, Atlanta, July 22nd, Ezra's Chapel, Lovejoy Station, Jonesboro, Gadsden, Ala., Griswoldville, Savannah, Congaree Creek, Columbia, Bentonville and Raleigh, N.C."[919]

---

[915] The mounted companies, C and G, rejoined the Regiment at this time.

[916] The Regimental commander, Colonel Willard Arms Dickerman, was wounded on May 28, 1864, near Dallas and died May 30.

[917] *Reminiscences of the Civil War from Diaries of Members of 103d Illinois Volunteer Infantry,* Compiled by a Committee of the Regiment, J. F. Leaming & Co., Chicago, 1904, page 112.

[918] The One Hundred and Third Infantry Regiment was first in Holly Springs, Mississippi in December 1862.

[919] Ibid. *Reminiscences of the Civil War.* Page 236.

Presumably this is the National flag that the Regiment received at Washington DC before the Grand Review. The flag is inscribed as follows: 103 REGT. ILL. VOL. INFANTRY: TALLAHATCHIE RIVER; SIEGE VICKSBURG; BLACK RIVER; JACKSON; MISSIONARY RIDGE; KNOXVILLE, DALTON; RESACA; DALLAS, NEW HOPE CHURCH; NOON DAY CREEK, KENESAW MOUNTAIN, ATLANTA, JULLY 22; EZRA'S CHAPEL; JONESBORO; LOVEJOY; GADSEN, ALA; GRISWOLDVILLE, CONGAREE CREEK, S. C.; COLUMBIA; BENTONVILLE; RALEIGH

The Regimental flag of the One Hundred and Third is in very good condition and was also probably received by the Regiment prior to the Grand Review. The design of the flag is typical of those produced by Evans and Hassall.

A flag was presented to Company C and later probably also to Companies G and K in September 1862. "A beautiful silk flag was presented Capt. S. A. Stockdale's company in the morning. Miss M. E. Anderson on behalf of the ladies presented the flag making a few beautiful and appropriate remarks.

Capt. Stockdale received the flag and returned the thanks of the company promising that it should never be disgraced. P. C. Stearns, Esq., also notified the other Canton companies that flags were being prepared for them and as soon as ready, would be presented to them at Peoria. The time was too short to get them ready before the time to leave here. We learn that Miss Cordelia Steel, Miss Lida Piper, Miss Helen Lamond and Miss Lucy Shinn have taken the most active part in raising the money and preparing the flags."[920] No further history is available regarding these flags.

No other flags carried by the 103rd Illinois Infantry are known to survive.

[920] *Fulton County Ledger,* September 9, 1862.

104th Illinois Infantry - "*La Salle County Regiment*"

"Then shout after shout for bold Sherman
Went up from each valley and glen,
And the bugles re-echoed the music
That fell from the lips of the men;
For we knew that the stars on our banner
More bright in their splendor would be,
And that blessings from Northland would greet us
When Sherman marched down to the sea."[921]

The One Hundred and Fourth Illinois Infantry Regiment was organized at Ottawa, La Salle County, Illinois during August of 1862. The men in all the companies volunteered from La Salle County with the exception of a few men in Company I who volunteered from nearby Evans Township in Marshall County. The Regiment was mustered in at the "Fair Grounds" in Ottawa on August 27, 1862, under the command of Colonel Absalom B. Moore.[922] Although the Regiment was presented a flag by the patriotic ladies of Ottawa on September 1, 1862[923] it wasn't until the men reached Camp Holt, near Louisville, Kentucky in mid-September that they received uniforms and arms. In mid-October, the Regiment left Camp Holt, in pursuit of Confederate General Bragg; after passing through Frankfort, Bowling Green and Glasgow Kentucky, they reached Hartsville, Tennessee on the Cumberland River on December 1. Early on the morning of December 7, Confederate forces commanded by General Morgan attacked the Union forces including the One Hundred and Fourth. The inexperienced Union troops were soon outflanked, surrounded, and captured. The Regiment was marched to Murfreesboro, Tennessee and paroled but Lieutenant-Colonel Douglas Hapeman and Major John Widmer of the Regiment were sent to Libby Prison in Richmond where they remained until paroled in April 1863. The Regiment was sent from Nashville to Camp Douglas in Chicago to guard rebel prisoners. On April 12, 1863, the Regiment was exchanged and departed for Tennessee to join the Army of the Cumberland. Major Widmer, who had been released from Libby Prison, rejoined the Regiment at Brentwood, Tennessee on April 18 and Lieutenant-Colonel Hapeman rejoined the Regiment on April 21. On June 7, the One Hundred and Fourth joined the Army of Cumberland at Murfreesboro, Tennessee. The morning of June 24 the Army of the Cumberland embarked on the Tullahoma Campaign. In that Campaign, the One Hundred and Fourth skirmished at Hoover's Gap, at Manchester on the Elk River[924] and then went into camp at Decherd, Tennessee. In mid-August the Union forces advanced toward the Cumberland Mountains in the Chickamauga Campaign. They reached Stevenson, Alabama on August

[921] Calkins, William Wirt, *The History of the One Hundred and Fourth Regiment of Illinois Volunteer Infantry,* Donohue & Henneberry, Chicago (1895) page 277. Portion of a song composed by Lieutenant Byers of the Fifth Iowa Cavalry while in prison and quoted by Calkins.

[922] Colonel Moore was 34 years old when he mustered into the 104th Illinois Infantry. He had been born in New Jersey and came to LaSalle county, Illinois 1856,

[923] Ibid, Calkins page 13

[924] On July 2, 1863, the bridge over the Elk River was guarded by a small rebel stockade. Volunteers from Company D were selected to capture the stockade. Sergt. George Marsh led John Shapland, Richard Gage, Oscar Slagle, Lemuel Holland, George Houghton and Reuben Smalley in a successful capture of the stockade. For this action the men received the Medal of Honor. *Civil War Medal of Honor Winners from Illinois,* published by the Civil War Centennial Commission of Illinois, July 1962.

30, crossed Sand Mountain on September 3, and Lookout Valley September 6. Due to the Union advance, the Confederate forces under Bragg evacuated Chattanooga and marched south. The Union forces followed and on September 10 the One Hundred and Fourth arrived at the Davis' house west of Dug Gap in Pigeon Mountain, Georgia. Skirmishing began on September 11 with fighting at Davis' Cross Roads near Dug Gap and led to the Battle of Chickamauga on September 18 and 19, 1863. In that battle the Regiment was heavily engaged and when the Union forces were forced to retreat the One Hundred and Fourth retired to Rossville, Georgia. For the next two months the Union Army held a defensive position on the outskirts of Chattanooga, and it was during this time that General Grant took command of the Northern forces. On November 23 the Union forces advanced to Orchard Knob and to the foot of Lookout Mountain; on the 24th the One Hundred and Fourth took part in the assault and capture of Lookout Mountain and on the 25th the Regiment's flag is said to have been the first planted on Missionary Ridge.[925] Bragg retreated and the One Hundred and Fourth joined in the pursuit of the rebels to Grayville and Ringgold, Georgia and then returned to Chattanooga. The Regiment remained at Chattanooga until February 1864 when it was ordered to Nashville; in March the Regiment returned to Chattanooga and in May 1864 joined Sherman's Atlanta Campaign. During May the Regiment fought at Rocky Faced Ridge, Buzzard's Roost Gap, Resaca, and Pumpkin Vine Creek and in June at Dallas, New Hope Church, Allatoona Hills, Pickett's Mills, Pine Mountain, Lost Mountain, and Kennesaw Mountain. To stem the retreat of the Confederate forces, General Hood replaced Confederate General Johnston. The Army of the Cumberland, containing the One Hundred and Fourth Illinois advanced across the Chattahoochee River and bivouacked on Peach Tree Creek. On July 20, 1865, the Regiment crossed Peach Tree Creek at Howell's Mills and formed in line of battle on the left of the Brigade. The men began building works and about 4 pm were attacked on their right line and their right flank. Half of the men in right five companies were killed or wounded but the Regiment reformed and aided by the Tenth and Twenty-first Wisconsin drove the rebels back and captured the rebel colors which had been planted on the Union works.[926] The decimated ranks of the One Hundred and Fourth fought in the siege of Atlanta and at Jonesboro until Hood ordered the evacuation of the city on September 1; they then entered the city and remained there until October 3. Following the pursuit of Hood's retreating forces into north Georgia and Alabama, the Regiment joined Sherman's "March to the Sea" in mid-November. After the siege of Savannah December 10-21, 1864, the Regiment moved north into the Campaign of the Carolinas and in 1865 saw action in the Salkehatchie Swamps, South Edisto River, North Edisto River, Congaree Creek, the Battle of Bentonville and occupied Goldsboro and Raleigh. Following Johnston's surrender the Regiment marched to Washington D. C. via Richmond and marched in the Grand Review on May 24. The Regiment moved to a camp across the river from Washington and was mustered out on June 6, 1865.[927]

There is one national and one regimental flag from the One Hundred and Fourth Illinois Infantry Regiment in the collection of the State of Illinois. Both flags bear a tag stating that the flag had been captured.

---

[925] Corporal Lemuel Holland carried the flag and the correspondent of the *Nashville Press* reported that the flag of the One Hundred and Fourth Illinois Infantry was the first to be planted on Missionary Ridge. *Ibid*, Calkins, page181.

[926] Lieutenant-Colonel Douglas Hapeman, in command of the One Hundred and Fourth, was awarded the Medal of Honor – "With conspicuous coolness and bravery rallied his men under a severe attack, re-formed the broken ranks, and repulsed the attack at Peach Tree Creek, Georgia, July 20, 1864." *Civil War Medal of Honor Winners from Illinois,* Published by the Civil War Centennial Commission of Illinois, July 1962.

[927] Ibid., Calkins.

The square canton and star arrangement on this national flag is typical of those flags manufactured by Hugh Wilkins of Louisville, Kentucky. The inscription on the flag reads: 104th REGT.ILL'S VOL'S. *"On September 1st, 1862, the One Hundred and Fourth, pursuant to orders, marched to the court house in Ottawa, and were presented with a beautiful silken flag by a number of Ottawa's patriotic ladies."*[928] *Whether the flag mentioned in the history of the regiment is the flag displayed above is unknown. In Regiment's history, Calkins details the capture of the Regiment's flagß at the Battle of Hartsville, Tennessee and the Knoxville Register reports that the Ninth Kentucky captured the flag of the One Hundred and Fourth Illinois Infantry.*[929]

The regimental flag is characteristic of flags manufactured in Louisville by Hugh Wilkins.

No other flags carried by the 104th Illinois are known to survive.

---

[928] Ibid, Calkins, page 13

[929] *Knoxville Register,* December 14, 1862

105th Illinois Infantry – "*Dixon Regiment*"

Hunger and thirst, hunger and thirst
Give me my pipe, let 'em do their worst.
Cold and wet, cold and wet
Give me my pipe, I soon can forget
Sickness and pain, sickness and pain;
Give me my pipe, and I won't complain.
Wounds and death, wounds and death;
I'll draw on my pipe with my dying death.[930]

The One Hundred and Fifth Illinois Infantry Regiment was mustered into service at Dixon, Lee County, Illinois on September 2, 1862, under command of Colonel Daniel Dustin.[931] The men of Companies A, C, E, G and H volunteered from DeKalb County, the men of Companies, B, D, F (*Bryan Blues)*, and Company I from DuPage County and the men of Company C from both DeKalb and DuPage counties. On September 8 the Regiment moved to Camp Douglas in Chicago and in October to Frankfort, Kentucky. While at Frankfort the Regiment skirmished with Morgan's forces and then in November moved southwest to Bowling Green and Scottsville, Kentucky. In late November the One Hundred and Fifth moved to Gallatin, Tennessee and in early December to nearby South Tunnel, Tennessee. In February 1863 the Regiment returned to Gallatin where they remained until June when they marched south to La Vergne, Tennessee. After a brief stay in Murfreesboro, Tennessee the Regiment returned to Gallatin and then in August to Nashville, Tennessee where they were assigned to Fort Negley for guard duty. In February 1864 they marched toward Chattanooga. They arrived at Wauhatchie, Tennessee, southwest of Chattanooga in March. In May, as part of the Atlanta Campaign the Regiment was ordered into north Georgia and en-route to Resaca passed through Gordon's Mills, Leet's farm, Taylor's Ridge, and Snake Creek Gap. May 14 and 15 the One Hundred and Fifth fought at Resaca and then pursued the retreating enemy south through Calhoun and Cassville, crossed the Etowah River and advanced on and fought at Dallas, New Hope Church and Allatoona Hills[932] near Pumpkin Vine Creek. During June and July, the Regiment saw action at Pine Hill, Lost Mountain, Golgotha Church, Muddy Creek, Noyes Creek, Kolb's Farm, Kennesaw, Ruff's Station, and Smyrna Campground. On July 17 the One Hundred and Fifth crossed the Chattahoochee River and moved forward to Peach Tree Creek. In the Battle of Peach Tree Creek on July 19-20 rebel forces were repulsed, and the flag of the Twelfth Louisiana Infantry captured.[933] From July 22 until August 25 the Regiment

[930] Strong, Robert Hale, *A Yankee Private's Civil War,* Henry Regnery Company, Chicago (1981). Written on front piece.

[931] 42 year old Daniel Dustin was a native of Vermont. He had previously been a Captain and Major of the 8th Illinois Cavalry. He next he took command of the 104th Illinois Infantry; he was a practicaing physician in Sycamore, DeKalb County Illinois before the civil war.

[932] Surgeon Horace Potter of Chicago was killed in battle on June 2, 1864

[933] George Cram and Melvin Smith crawled forward and secured the rebel colors after the color bearer had been shot. According to Lieut. Col. E. F. Dutton of the One Hundred and Fifth "Among the trophies captured by my command was one set of colors, said by rebel wounded to belong to the Twelfth Louisiana Infantry." *O.R.* Series 1-Volume XXXVIII/2 [S#73] page 364. This Louisiana flag is one of the Confederate flags returned at Memorial Hall in New Orleans on April 13, 1905. It is a second pattern Confederate national flag measuring 88 by 50 inches. Painted in black on the white field: "12th La. flag, captured July 12, 1864, at the battle of Peach Tree Creek, by the 105th Regt. Illinois Infty. Vols., 20th Corps, Army of the Cumberland." *The Returned Battle Flags,* (Edited by Richard Rollins) reprinted by Rank and File Publications, Redondo Beach, California, 1955. However, the veterans of the 12th Louisiana stated that their colors had never been captured. The captured flag was more likely the flag of the 55th or 57th Alabama regiment of Scott's Brigade. (Personal

was in the siege of Atlanta and then fell back to north of the Chattahoochee River on picket and guard duty. After Atlanta surrendered on September 2, the Regiment remained near Atlanta until November 15 when it joined in Sherman's "March to the Sea" and the siege of Savannah in December. In 1865 the One Hundred and Fifth, in the Campaign of the Carolinas was engaged at Hardeesville, Lawtonville, and Salkehatchie, South Carolina and at Averysboro, Bentonville and Goldsboro, North Carolina. Finally, following the surrender of Johnston's army the One Hundred and Fifth marched to Washington D.C. via Richmond, Virginia and marched in the Grand Review on May 24, 1865. The Regiment remained at Washington until mustered out on June7, 1865.[934,935]

The collection of the State of Illinois Military Museum contains one national and three regimental flags from the One Hundred and Fifth Regiment Illinois Infantry.

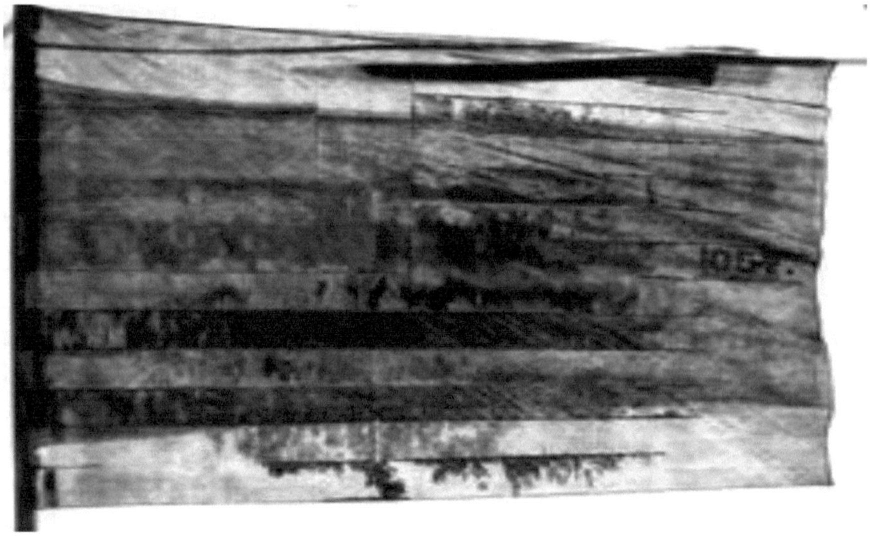

Little remains of the National flag. "105th" can be seen on one stripe at the far right of the flag. "The regiment was presented with a beautiful stand of colors, by Hons. T. B. Bryan and H. C. Childs, of DuPage, upon whose folds were inscribed, in golden letters, 'Strike together:'--words destined to become actualized in the conduct of the men before the enemy."[936] Too much of the National colors shown here is lost to tell whether this is one of the flags presented.

There is a National Flag of the One Hundred and Fifth Illinois Infantry in the DuPage County Historical Museum in Wheaton, Illinois. According to the accession records at the Museum the National flag was presented to Capt. T. E. Rogers, Co B by the ladies of Downers Grove on August 26, 1862. The center stripe of the flag is inscribed "ELLSWORTH" in gold block letters with black shading. The flag is made of silk with field machine stitching. There is a 2 1/2" gold fringe. the original flag probably measured 54" x 70" but is now in tatters. No photo is available.

(Personal communication Hugh Simmons).

934 Boies, Henry, *History of DeKalb County, Illinois,* O. P. Bassett, Printer, Chicago, (1868), pp113-202

935 *Report of the Adjutant General of the State of Illinois*, H. W. Rokker, State Printer and Binder, Springfield, IL, (1886), Volume V, pp 692-4.

936 Ibid, Boies, page 116

This 34-star Regimental flag is in very good condition and shows little sign of wear. It is characteristic of flags manufactured in Philadelphia.

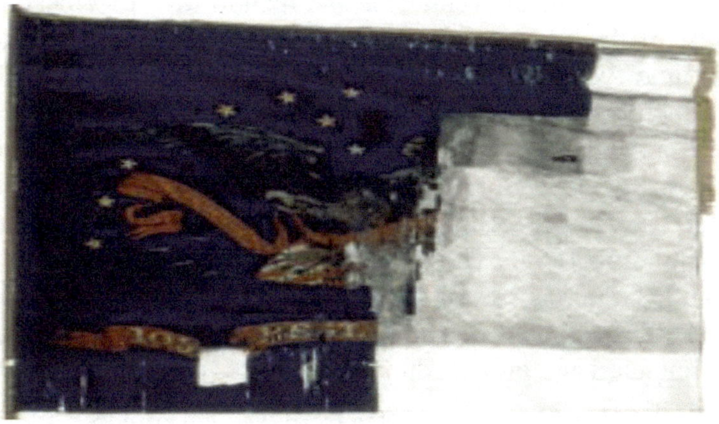

It appears that this Regimental flag has had portions cut off. It is characteristic of flags manufactured by Hugh Wilkins of Louisville, Kentucky.

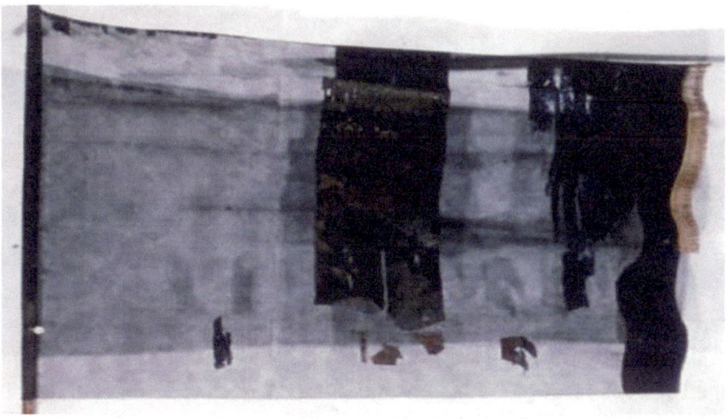

Only shreds of this Regimental flag remain. Since it shows the most wear of all the Regimental flags it was probably issued early in the War. Perhaps this is one of the flags in the stand of colors described above and presented by Du Page County.

No other flags carried by the 104th Illinois Infantry are known to survive.

106th Illinois Infantry

The One Hundred and Sixth Illinois Infantry was organized following Lincoln's call for troops in July 1862. The men of Company A volunteered from Sangamon County and the men of Company K from Menard County; the men of Companies B through I all volunteered from Logan County. The Regiment was mustered in at Camp Lincoln on September 18, 1862, under command of Colonel Robert B Latham.[937] Following six weeks of drill, the Regiment was ordered to Columbus, Kentucky on the Mississippi River and then on to Jackson, Tennessee. At Jackson the Regiment was broken into detachments and Companies H, I and K were detailed to guard the *Mobile & Ohio Railroad* at the Forked Deer River bridge 30 miles north of Jackson; Companies C and G were sent to Carroll Station eight miles north of Jackson; the remaining five companies were scattered as guard along the railroad and acted as provost guard at Jackson. On December 17, the Regiment was attacked by General Forrest and on December 20, Companies C and G surrendered.[938] Companies H, I and K were also attacked by Forrest but repulsed the attack from behind fortified breast works.[939] In March 1863 the remaining men of the Regiment were ordered to Bolivar, Tennessee and then in May to Vicksburg, Mississippi. From June 9 until the surrender of the city on July 4, the One Hundred and Sixth fought in the Siege of Vicksburg. During that time the Regiment was ordered up the Yazoo River forty miles to repel an advancing rebel force. On July 29 the Regiment was ordered to Helena, Arkansas, on August 13 to Clarendon, Arkansas and on August 22 to Duvall's Bluff. From September 1 through 10 the Regiment joined in Steele's expedition against Little Rock, Arkansas. Following the capture of Little Rock, they were on duty in Little Rock, Duvall's Bluff, Hot Springs, Lewisburg, St. Charles, Dardanelle and Brownsville, Arkansas until July 1865. The One Hundred and Sixth Regiment Illinois Infantry was mustered out at Pine Bluff, Arkansas on July 12, 1865. Except for the Siege of Vicksburg and the engagement at Little Rock, the Regiment was not in any of the major battles of the War and spent most of the time foraging and on guard and provost duty.[940,941]

There is one National and one Regimental flag from the One Hundred and Sixth Regiment Illinois Infantry in the collection of the State of Illinois. The National flag is in poor condition and most of the canton and most of the stripes are missing. However, the flag is inscribed with two battle honors, reflecting the only two engagements in which the Regiment saw significant action.

---

[937] Colonel Latham was born in 1818 in Union County Kentucky. He was an early settler in Illinois, one of the founders of the town of Lincoln Illinois, a friend of Abraham Lincoln , David Davis and Stephen Douglas and one of the founders of the Republican party.

[938] The prisoners were paroled by Forrest and sent to Benton Barracks in St. Louis; they were exchanged in the summer of 1863.

[939] In the battle, German born Sergeant Henry Fox of Company H crossed the Forked Deer River bridge to secure reinforcements from Jackson. "When his command was surrounded by a greatly superior force, voluntarily left the shelter of the breastworks, crossed an open railway trestle under a concentrated fire from the enemy, made his way out and secured reinforcements for the relief of his command." *Civil War Medal of Honor Winners from Illinois,* Published by the Civil War Centennial Commission, July 1962, page 11.

[940] *Report of the Adjutant General of the State of Illinois*, H. W. Rokker, State Printer and Binder, Springfield, IL, (1886), Volume VI, pp 24-5.

[941] Dyer, Frederick H., *A Compendium of the War of the Rebellion,* Volume III, Sagamore Press, In., Thomas Yoseloff, Publisher, New York, 1959, page 1092.

Inscribed on the flag are: 106 ILL. VOL.; VICKSBURG; LITTLE ROCK.

This Regimental flag is typical of those manufactured by Horstmann Bros. of Philadelphia. It is in very good condition but is not inscribed with battle honors.

No other flags carried by the 106th Illinois Infantry are known to survive.

107th Illinois Infantry

"Hurrah for the Red, White and Blue!
Hurrah for the Red, White and Blue!
Our Country's at stake, and 'tis time to awake
And show to the world what we'll do.
Awake to your Country's call!
'Tis time for the Boys to be true."[942]

The One Hundred and Seventh Regiment Illinois Infantry was mustered in at Camp Butler, near Springfield, Illinois on September 4, 1862 under the command of Colonel Thomas Snell.[943] The men of Company A volunteered from Waynesville and Wapella townships of De Witt County; Company B from Clinton in De Witt County; Company C from Monticello and Centerville in Piatt County; Company D from Clinton in De Witt County; Company E from Monticello Township in Piatt County; Company F from De Witt Township in De Witt County; Company G from Marion in Williamson County; Company H from Bement and Monticello Townships in Piatt County; Company I from Santa Anna Township in De Witt County and Company K from Piatt and Sangamon Counties. September 30 the Regiment left Camp Butler for Jeffersonville, Indiana where it remained in camp for drill instruction until mid-October; on October 12 the Regiment crossed the Ohio River to Louisville, Kentucky. October 18 the Regiment was ordered to Elizabethtown, Kentucky to meet the advance of General John Morgan who was moving into that area. Following a skirmish with Morgan's men the One Hundred and Seventh moved south to Mumfordsville, Kentucky, remained there until March 1863 and then moved further south to Glasgow, Kentucky. On June 18, 1863, the One Hundred and Seventh, as part of the Twenty-third Corps, marched for Tompkinsville, Kentucky on the Cumberland River to meet General Morgan's anticipated invasion of Kentucky. However, in July, Morgan crossed the Ohio River and raided through southern Indiana and Ohio. Union troops, including the One Hundred and Seventh defeated Morgan's forces near Buffington Island on the Ohio River and captured Morgan at Salineville, Ohio one week later. The Regiment returned to Lebanon, Kentucky in late August; they then marched in Burnside's campaign in East Tennessee. This included a march and countermarch from Loudon to Greenville, Tennessee, sixty miles east of Knoxville followed by a march to Sweetwater, Tennessee where they learned that the Battle of Chickamauga had been fought earlier that month. In the Knoxville Campaign the Regiment saw action at Huff's Ferry on the Holston River, Campbell's Station on the *Tennessee and Georgia Railroad,* and the siege of Knoxville. The One Hundred and Seventh was with the Union forces that pursued Longstreet after his retreat from Knoxville and engaged the rebel forces at Dandridge, Tennessee, 20 miles east of Knoxville on December 21. The Regiment returned to Knoxville where they remained until April 1864 when they joined Sherman's Army in the Atlanta Campaign. Between May and September, the One Hundred and Seventh was engaged at Dalton, Rocky

[942] *The Clinton Public,* June 4, 1863.
[943] Colonel Thomas Snell was born in Cincinnati, Ohio in 1818. He moved with his family to Illinois, settling in Clinton in 1845. In 1852 the *Illinois Cetral Railroad* was beginning to be built and Snell took an active part—building 800 miles of the *Illinois Central.* He afterwards built 150 miles of the *Chicago & Northwestern RR,* , 40 miles of the *Ohio & Dayton RR,* 80 miles of the *Bloomington, Lafayette & Mississippi RR,* and 101 miles of the *Lafayettte Road* -- Champaign to Havana. With the outbreak of the civil war he raised the 107th Illinois Infantry and was appointed colonel of the regiment. Find A Grave, Colonel Thomas Snell, Sr.

Faced Ridge, Resaca, Dallas, New Hope Church, Allatoona Hills, Kennesaw Mountain, Lost Mountain, Muddy Creek, Noyes' Creek, Kolb's Farm, Nickajack Creek, Chattahoochee River, the Siege of Atlanta and Lovejoy Station. Following the fall of Atlanta, the Regiment pursued Hood into north Georgia and Alabama in October and early November. On November 18, 1864, the Regiment marched to Colombia, Tennessee and met the advance of Hood's army on November 22. The two armies skirmished until November 28 finally resulting in the Battle of Franklin on November 30. In that Battle the commander of the Regiment, Colonel Francis Lowry, was killed; the Regiment captured two stands of enemy's colors, and had its own colors captured but Pvt. Bailey Walker of Company G recovered these.[944] The Regiment arrived at Nashville on December 1 and went into camp near Fort Negley on picket duty; December 15 the Regiment broke camp and charged the enemy's lines. The rebel forces were routed and the following morning the One Hundred and Seventh pursued the fleeing enemy toward Franklin. The Regiment arrived at Columbia, Tennessee on December 20 and in January 1865 departed for Clinton, Tennessee. January 26 the Regiment was ordered to Washington D.C. and arrived there on February 2; February 9 the Regiment moved to Alexandria, Virginia and on February 11 sailed by streamer to Fort Fisher, North Carolina on the Cape Fear River Inlet. On February 18 the Regiment moved on Fort Anderson, drove in the skirmishers, and held a siege position the following day. On February 20 the Regiment attacked the Fort at daylight but found it evacuated. The Regiment captured the Fort's colors and pursued the retreating enemy to Wilmington. After a two-week occupation of the city the Regiment was ordered to Goldsboro and then to Raleigh, North Carolina. Following the surrender of Johnston's Army at Bennett's House on April 26 the One Hundred and Seventh remained on duty at Salisbury, North Carolina where it was mustered out on June 21, 1865.[945,946]

The collection of the State of Illinois contains no National flags and one Regimental flag from the One Hundred and Seventh Illinois Infantry.

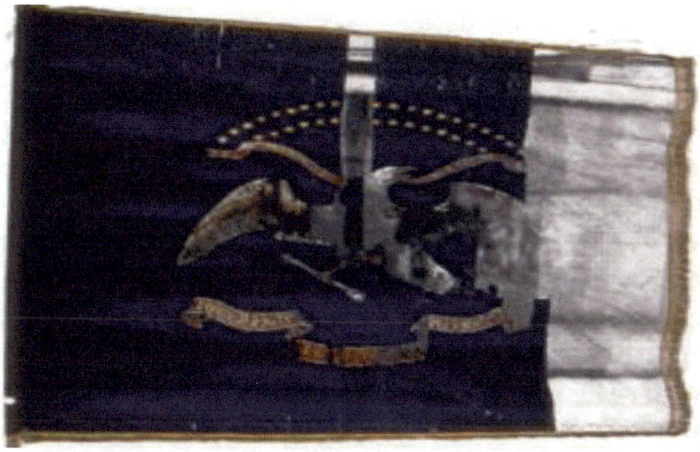

This Regimental flag is inscribed on the lower scroll: 107 REGT. ILLS VOLS. No battle honors are inscribed on the flag.

No other flags carried by the 107th Illinois Infantry are known to survive.

---

[944] *O.R. Series 1—Volume SLV/1 [S#93] pp378-83.* Although the records confirm that two stands of the enemy's colors were captured, the Confederate units are not identified.

[945] *Report of the Adjutant General of the State of Illinois,* H. W. Rokker, State Printer and Binder, Springfield, IL (1886), pp 47-9

[946] Dyer, Frederick H., *A Compendium of the War of the Rebellion,* Thomas Yoseloff, Publisher, New York, (1959), Part III, p 1092

108th Illinois Infantry

Colonel John Warner[947] organized the One Hundred and Eighth Regiment Illinois Infantry in Peoria, Illinois in the summer of 1862 and mustered with the regiment into service at Peoria on August 28, 1862. The men of Company A volunteered from the eastern townships of Tazewell County; Company B came from Pekin in Tazewell County; Company C enlisted from Peoria County; Company D and E volunteered from Woodford County; Company F came from Bath and Lynchburg townships in western Mason County; Company G came Peoria County; Company H enrolled from Logan, Mason and Tazewell counties; Company I volunteered from Peoria County and Company K volunteered from Peoria, Tazewell and Woodford counties. On October 6 the Regiment left camp in Peoria and joined the Army of Kentucky at Covington, Kentucky. The One Hundred and Eighth left Covington on October 17 and marched south through Lexington to Nicholasville, Kentucky and remained there until mid- November; the Regiment next marched to Louisville, Kentucky and boarded transports for Memphis, Tennessee where they arrived on November 26. While in camp at Memphis on picket and guard duty, the Regiment was assigned to the Army of the Tennessee. On December 20, 1862, the men of the One Hundred and Eighth boarded the *City of Alton,* proceeded down the Mississippi River to the Yazoo River, steamed up to Johnson's Landing where they disembarked and formed in line of battle against the enemy forces situated on the bluffs, northeast of Vicksburg. On the morning of January 2, 1863, the Regiment withdrew, re-embarked[948] on the *City of Alton*, proceeded down the Yazoo River to the Mississippi River, up the Mississippi River to the White River and up the White River through a cut-off to the Arkansas River to Arkansas Post or Fort Hindman.[949] In the battle on January 11, the Regiment for the first time was under heavy fire, but assisted in the capture of the Fort. Following the capture, the Regiment proceeded back downriver to Young's Point, Louisiana opposite Vicksburg. For the next two months the men were on guard duty and participated in the digging of the "canal" to bypass Vicksburg; one hundred and thirty-four men died of disease during that time. March 10 the Regiment boarded the steamer *Spread Eagle* and proceeded north to Milliken's Bend, Louisiana where they went into camp; on April 15 the One Hundred and Eighth marched south through Louisiana to Hard Times Landing on the west bank of the Mississippi River. After crossing the Mississippi on the ironclad *Lafayette,* the Regiment hurried to the battlefield at Port Gibson and participated in the victory of the Union forces. Following the battle at Champion's Hills on May 16, the Regiment was detached to guard and transport 4,500 prisoners to Memphis, Tennessee. The Regiment returned to camp at Young's Point where it remained until after the fall of Vicksburg. In late July the One Hundred and Eighth proceeded to Memphis and in August moved by rail to La Grange, Tennessee. In October the Regiment marched to Pocahontas, Tennessee and then in November moved by rail to Corinth, Mississippi. January 1864 the One Hundred and Eighth moved by rail to Memphis where it remained until June. On June 1 a Union force departed Memphis in search of Confederate General Forrest. The opposing forces met at Brice's Crossroads in northeastern Mississippi where, on June 10, the outnumbered Confederate forces soundly defeated the Union troops.[950] The Union forces retreated to Memphis and the One Hundred and Eighth remained there until February 1865. In March 1865 the Regiment was ordered to New Orleans, Louisiana where, on March

---

[947] Colonel Warmer was born in Ohio in 1826. Prior to the Civil War he was involved in mustiple business in Peoria; after the war he was an eight time mayor of Peoria.

[948] This Union force was called the Army of the Mississippi and commanded by Major General John A. McClernand of Illinois.

[949] An earthen bastion built by the Confederates on the Arkansas River.

[950] Captain Henry Sommers of Company G was captured on June 13, 1964, and murdered by a guerrilla force led by Dick Davis. Davis was subsequently captured and hung for the murder of Sommers.

12, it embarked on the ocean steamer *Guiding Star* and sailed to Fort Gaines on Dauphine Island at the entrance to Mobile Bay, Alabama. March 21 the Regiment moved across the mouth of the Bay to Fish River then upriver and disembarked at Danby's Mills. The Regiment marched north and on March 27 drove the enemy within his works at Spanish Fort and took part in the thirteen-day siege of the Fort. The Fort was captured on April 8 and on April 9 the One Hundred and Eighth participated in the assault and capture of Fort Blakely. April 13 the Regiment was ordered to Montgomery, Alabama for provost duty, on July 18 was ordered to Vicksburg and was mustered out at Vicksburg on August 5, 1865.[951]

In the collection of the State of Illinois, there are two National flags of the One Hundred and Eighth Regiment of Illinois Infantry but no Regimental flags.

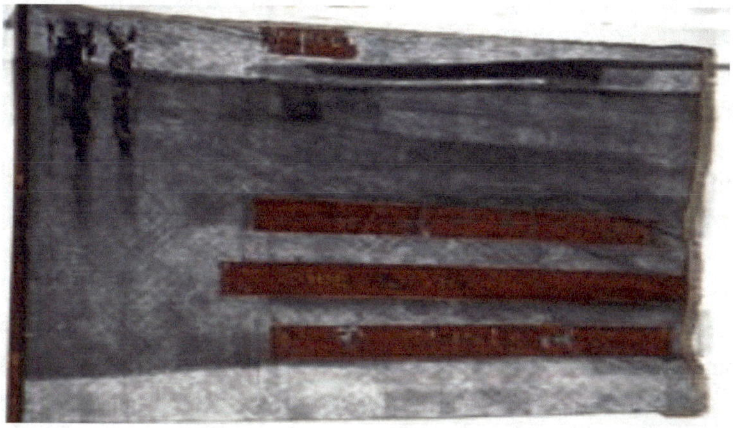

Little remains of this National flag. It is inscribed "SIEGE OF VICKSBURG."

This 35-star National flag is typical of flags manufactured for the New York depot. It is inscribed, "108 REGT ILLS VOLS."

No other flags carried by the 108th Illinois Infantry are known to survive.

[951] *Report of the Adjutant General of the State of Illinois*, H. W. Rokker, State Printer and Binder, Springfield, Ill., (1886), Volume VI, pp 71-4.

109th Illinois Infantry - *"Phillips Invincibles"*[952]

"It having been alleged that the 109th Regiment Illinois Infantry Volunteer, has shown indications of disloyalty, and many members of the regiment having voluntarily hunted up citizens in the neighborhood of their camp to surrender and obtain paroles from, is hereby placed in arrest. The Regiment will be disarmed by the commander of the brigade to which the regiment is temporarily attached, and the arms and ammunition of the regiment turned over to the Ordinance Officer, Lieut. Carter, to be disposed of as may hereafter be ordered."[953]

The One Hundred and Ninth Regiment Illinois Infantry was mustered in on September 11, 1862, at Jonesboro, Illinois under the command of Colonel Alexander Nimmo.[954] The men of Company A volunteered from the Jonesboro area of Union County; Company B from Alexander, Jackson and Union counties; Company C from the area of Jonesboro and South Pass in Union County; Company D from Johnson and Union Counties; Company E, F, G, H and I from Union County and Company K from Pulaski and Union Counties. At the time of muster the Regiment was furnished inferior weapons of a variety of caliber and models ranging from 1812 to 1853. On October 20 the Regiment was ordered to Columbus, Kentucky by way of Cairo, Illinois. At Columbus the One Hundred and Ninth was placed on guard duty for three weeks, then ordered to Bolivar, Tennessee. Two weeks later the Regiment was ordered to Moscow, Tennessee and then on toward Holly Springs, Mississippi. However, because of the poor condition of their weapons the men of the One Hundred and Ninth were detached and left at Lumpkin's Mill.[955] In January 1863 the Regiment moved to Holly Springs and then on to Memphis; in late March the One Hundred and Ninth was ordered to Lake Providence, Louisiana. By the time the Regiment reached Lake Providence on April 1, 1863, the ranks were greatly reduced in numbers by death and desertion. Because of this, the remaining men of the One Hundred and Ninth Regiment Illinois Infantry were consolidated with the Eleventh Regiment of Illinois Infantry.[956]

While the above information is taken from the "official" history of the Regiment as detailed by the Illinois Adjutant General, there is another "history" of the Regiment, which may be closer to the truth. According to T. M. Eddy[957] the Regiment, which was organized at Anna in far southern Illinois, was composed of members of the Knights of the Golden Circle.[958] Special Order No. 6, quoted below, supports Eddy's contention.

---

[952] Chicago Daily Tribune; September 4, 1862: "The regiment has been named "Phillips Invincibles" in honor of David L. Phillips, esq., the energetic and efficient United States marshal for this district, whom we are proud to claim as hailing from this county and in whom our regiment recognize the ardent and devoted friend of the soldier, the able and popular office and the high toned honorable gentleman."

[953] "The Arrest of the 109th Illinois" in the *Daily Illinois State Journa.*22 January 1863.

[954] Colonel Alexander Nimmo was born in Union County, Illinois in 1822. He served in the 2nd Illinois Infantry in the Mexican-American war and served as the constable, sheriff, and county clerk of Union County.

[955] "We still have the same miserable arms which were furnished us at Jonesboro—there is uniformity in neither the manufacture nor caliber-ranging from the model of 1812 to that of 1853, and all of an inferior stock. *Jonesboro Gazette,* Jonesboro, Illinois, December 20, 1862, and quoted by Ken Baumann in *Arming the Suckers, 1861-1865,* Morningside House, Inc, Dayton, Ohio (1996) page 200.

[956] *Report of the Adjutant General of the State of Illinois*, H. W. Rokker, State Printer and Binder, Springfield, Ill., (1886), Volume VI, page 93

[957] Eddy, T. M., DD, *The Patriotism of Illinois,* Clarke & Co., Publishers, Chicago (1866), Volume II, pp 385-6.

[958] This organization was dedicated to the expansion of slaveholding in the Southern United States and Mexico. Although there is no evidence that the Knights existed in Southern Illinois, Republicans used the terms Knights of the Golden Circle and Copperheads interchangeably and charged that the Democrats supported these organizations.

SPECIAL ORDERS No. 6.[959]

LAKE PROVIDENCE, *LA., April* 10, 1863.

The officers of the One hundred and ninth Regiment Illinois Volunteers, except those of Company K, having been reported as utterly incompetent to perform the duties of their respective commissions, and evincing no disposition to improve themselves, are hereby discharged from the service of the United States.

This is the regiment which was within a few miles of Holly Springs when attacked by the rebels and failed to march to the support of their comrades, but drew in their pickets and stood ready to surrender. From nine companies 347 men deserted, principally at Memphis, and but 1 from Company K.

To render the men efficient, it is necessary to transfer them to a disciplined regiment, and they are accordingly transferred to the Eleventh Regiment Illinois Volunteers, Company K to make the tenth company.

The officers thus discharged are Col. A. J. Nimmo, Maj. T. M. Perrine, First Lieut. C. B. Dishon, regimental quartermaster; Capts. J. C. Hunsaker, Hugh Andrews, Samuel M.P. McClure, S. O. Lewis; First Lieuts. James P. McLane, Morgan Stokes, Jacob A. Milikin, Charles Barringer, B. F. Hartline, Josiah Toler, Abram L. Misenheimer, R. B. Bartleson; Second Lieuts. M. A. Goodman, Squire Crabtree, T. T. Robinson, Henry Gassaway, Charles Klutts, Andrew Colvin.

Chaplain P. H. Kroh will, as the regiment has been consolidated, be mustered out of the service. Surg. John L. Dewey, Asst. Surgs. John W. Henley and George H. Dewey will be assigned by Major-General Grant to other Illinois regiments to fill vacancies.

By order of the Secretary of War:

L. THOMAS,

Adjutant-General.

Based on Special Order No. 6, the history as given by Eddy appears to be the more accurate history of the One Hundred and Ninth Regiment Illinois Infantry.

There are no flags of the One Hundred and Ninth Regiment of Illinois Infantry in the collection of the State of Illinois. It is not known if any flags were ever issued to this Regiment.

---

959 O.R.—Series I—Volume SVII/2 [S#25] pp590-1.

110th Illinois Infantry

On September 11, 1862 the One Hundred and Tenth Regiment Illinois Infantry mustered in for three years' service at Jonesboro, Union County, Illinois under the command of Colonel Thomas Casey.[960] The men of the Regiment volunteered from the following counties: Company A from Franklin County, Company B from Mount Vernon Township in Jefferson County, Company C from Marion in Williamson County, Company D from Washington County, Company E from Saline County, Company F from Franklin County, Company G from Jefferson and Wayne County, Company H from Perry County, Company I from Franklin County and Company K from Franklin and Hamilton County. On September 23 the Regiment was ordered to Louisville, Kentucky and was assigned to the Army of Ohio. Four days later the Regiment joined in the pursuit of Confederate General Bragg to the Salt River, forty miles south of Louisville. A line of battle was formed on October 8, but no order was given to advance. During the night Bragg retreated and the following morning the enemy lines were deserted. Over the next several weeks the Regiment skirmished with Bragg's forces at Danville, Crab Orchard, Mount Vernon, and Camp Wild Cat in Kentucky and moved to Nashville, Tennessee in November. In late December, the One Hundred and Tenth departed Nashville and marched south toward Murfreesboro, Tennessee. On December 30 the Regiment was in line of battle and from December 31 –January 3, 1863, was engaged in the Battle of Stones River. The Regiment lost 8 men killed and fifty wounded in the fighting; the Regiment remained in camp at Readyville, 20 miles east of Murfreesboro until June 1863. During that time the Regiment skirmished with rebels at Woodbury, Kentucky on April 3.

On May 7, 1863, due to loss of men in battle, sickness and discharges, the Regiment was consolidated into a battalion of four companies. The battalion joined in the occupation of Middle Tennessee in the summer of 1863, and then in August was ordered east in the Chattanooga-Chickamauga campaign. The battalion fought in the Battle of Chickamauga, Georgia on September 19-20 and the Siege of Chattanooga during September, October, and November. Subsequently, they fought at Orchard Knob on November 23, Tunnel Hill on November 24-25, and Mission Ridge on November 25. Following the pursuit of the enemy as far as Graysville, Tennessee they marched to the relief of Knoxville, Tennessee in December. The battalion was in camp at North Chickamauga until May 1864 when they joined Sherman's Atlanta campaign. In that campaign the battalion was engaged at Tunnel Hill, Resaca, the siege of Atlanta, Utoy Creek and the battles at Jonesboro and Lovejoy Station. Following the fall of Atlanta, the consolidated One Hundred and Tenth Illinois Infantry pursued the retreating rebel forces into north Georgia and Alabama before joining in Sherman's "March to the Sea". Savannah fell to the Union forces in mid-December, the battalion then marched north in the Campaign of the Carolinas in early 1865 and fought at Averysboro, Bentonville and Raleigh, North Carolina and witnessed Johnston's surrender at Bennett's House on April 26, 1865.

The Consolidated One Hundred and Tenth marched in the Grand Review in Washington D. C.

[960] Colonel Casey was born in 1832 in Jefferson County Illinois.He attened Mount Vernon Seminary and McKendree College and after studying law with Hugh Montgomery he was admitted to the Ilinois bar before the Illinois Supreme Court. He practiced law in Mount Vernon and was elected state's attorney. In 1862 Casey raised a full regiment—110th Illinois Infantry—and was elected Colonel.

on May 24 and was mustered out June 5, 1865.[961],[962] [963]

There are no flags carried by the One Hundred and Tenth Regiment Illinois Infantry in the holdings of the State of Illinois and no flags carried by this regiment are known to survive.

---

961 *Report of the Adjutant General of the State of Illinois*, H. W. Rokker, State Printer and Binder, Springfield, Ill., (1886), Volume VI, pp110-12.

962 Eddy, T. M., DD, *The Patriotism of Illinois,* Clarke & Co., Publishers, Chicago (1866), Volume II, pp 396-8.

963 Dyer, Frederick H., *A Compendium of the War of the Rebellion,* Thomas Yoseloff, Publisher, New York, (1959), Part III, p 1092-3.

111th Illinois Infantry

"About 4 P. M., the bugle sounded the charge and in less than 20 minutes the fort was ours. We claim for our regimental flag the honor of being the first inside the fort. The commanding officer of the fort surrendered to Colonel Martin, and Captain Castle, of Company H, now has in his possession the garrison flag of the fort." [964]

On September 18, 1862, at Camp Marshall, Salem Township in Marion County, Illinois the One Hundred and Eleventh Regiment Illinois Infantry was mustered in for three years' service under the command of Colonel James Stewart Martin[965]. The men of the Regiment volunteered from the following counties: Company A—Marion County, Company B—Washington County, Company C—Clay and Wayne County, Company D—Clay and Wayne County, Company E—Marion County, Company F—Patoka Township of Marion County, Company G—Clinton, Marion and Washington County, Company H—Marion County, Company I—Clinton County and Company K—Marion and Wayne County. The Regiment departed Camp Marshall on October 31 and moved via the *Illinois Central Railroad* to Cairo, Illinois and then via river transports to Columbus, Kentucky. The Regiment was assigned garrison duties at Columbus and remained there until March 12, 1863, when ordered to Fort Heiman[966] for garrison duty and for raids into the surrounding countryside. At the end of May the Regiment was ordered to Paducah, Kentucky for garrison duty and remained there until ordered to Gravelly Springs in northeastern Alabama in October. In November the One Hundred and Eleventh marched to Pulaski, Tennessee where they went into winter quarters until February 1864. On February 25 the Regiment was ordered to Decatur Junction, Alabama, on the north side of the Tennessee River directly opposite Decatur, Alabama; during the night of March 7, the Regiment entered the River above Decatur, floated downriver and captured the city. In mid-March the One Hundred and Eleventh reported to Huntsville, Alabama and was ordered to Larkinsville, Alabama where it remained on garrison duty until May. On May 1, 1864, the Regiment broke camp and joined Sherman's Atlanta campaign. Over the next five months the One Hundred and Eleventh fought at Snake Creek Gap, Resaca,[967] Dallas, New Hope Church, Allatoona Hills, Kennesaw Mountain, Nickajack Creek, Chattahoochee River, The Battle and Siege of Atlanta, Ezra Chapel, Jonesboro, and Lovejoy Station. Following the fall of Atlanta, the Regiment joined in the pursuit of Hood into north Georgia and Alabama and in mid- November joined Sherman's "March to the Sea." Following the Siege and capture of Fort McAllister and Savannah in December 1864 the Regiment campaigned north into the Carolinas in 1865. During the Carolina campaign the Regiment saw action at Salkehatchie Swamps, South Edisto River, North Edisto River, Congaree Creek, and Bentonville.

---

[964] . *Report of the Adjutant General of the State of Illinois,* H. W. Rokker, State Printer and Binder, Springfield, Ill (1886) Volume VI, page 147. This describes the December 13, 1864, capture of Fort McAllister located on the Great Ogeechee River guarding the southern approach to Savannah, Georgia. Confederate Major George Wayne Anderson commanded the Fort with a garrison of 230 men. The garrison flag was a Second National Flag and is now displayed at Fort McAllister and labeled "Captured by Captain G.E.Castle 111th ILLs. Vols. At the Assault on Fort McAllister"

[965] Colonel Martin was born in 1826 Scott County, Virginia. He attended Emory and Henry College in Emory, Virginia and moved to Salem, Illinois in1846. Colonel Martin served in Company C of the 1st Regiment of Illinois Volunteers in the Mexican-American war; he studied law and was admitted to the Illinois bar in 1861. During the civil war he served as Colonel of the 111th Illinois Infantry. Excerpted from Wikipedia, The Free Encyclopedia.

[966] Fort Heiman was opposite Fort Henry on the Tennessee River.

[967] Marcellus J. Newman, Company B, was awarded the Medal of Honor for his actions at the Battle of Resaca, Georgia, May 14, 1864. "Voluntarily returned, in the face of a severe fire from the enemy and rescued a wounded comrade who had been left behind as the regiment fell back." *Civil War Medal of Honor Winners from Illinois,* Published by the Civil War Centennial Commission of Illinois, Springfield, July 1962, page 4.

Following the surrender of Johnston at Bennett's House the Regiment marched through Richmond to Washington D.C and marched in the Grand Review on May 24, 1865. The One Hundred and Eleventh Regiment Illinois Infantry was mustered out on June 7, 1865, at Alexandria, Virginia.[968,969]

The collection of the State of Illinois holds one National and two Regimental colors of the One Hundred and Eleventh Regiment Illinois Infantry. Possibly the first two flags shown below are those flags, which were the first to fly over Fort McAllister.

The National flag has a flag pattern in the canton typical of flags manufactured for the Philadelphia depot. There are no inscriptions on the flag.

This Regimental flag is severely damaged with loss of more than 50% of the original material. It appears to be of New York manufacture.

---

968 Ibid, *Report of the Adjutant General of the State of Illinois* pages 144-48

969 Dyer, Frederick H., *A Compendium of the War of the Rebellion,* Thomas Yoseloff, Publisher, New York, (1959), Part III, p 1093.

111TH ILLINOIS INFANTRY

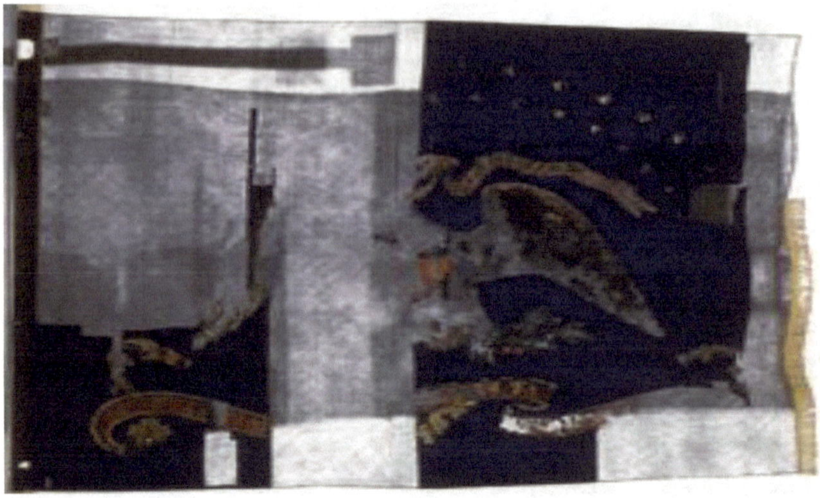

This Regimental flag bears a tag stating that it was captured; no details are given. In the history of the Regiment, the only time that the Regiment was overrun, and a significant number of men captured was on July 22nd, 1864, during actions before Atlanta. Approximately 80 men of the Regiment were captured and sent to Andersonville. However, no records have been found mentioning the loss of a flag to the enemy. The flag is typical of those manufactured for the New York depot.

No other flags carried by the 111th Illinois Infantry Regiment are known to survive.

112th Illinois Infantry

"Upon the National flag and Colors of the regiment were inscribed: "Kentucky," "Monticello," "East Tennessee," "Campbell's Station," "Knoxville," "Bean's Station," "Dandridge," "Atlanta," "Resaca," "Kennesaw," "Utoy Creek," "Nashville," "Columbia," "Franklin," "Wilmington," "Fort Anderson;" and these inscriptions marked the places of their fall."[970]

The One Hundred and Twelfth Regiment Illinois Infantry was mustered into service at Peoria, Illinois on September 20, 1862 under command of Colonel Thomas J. Henderson.[971] The men for the Regiment volunteered from the following counties: Company A—Annawan Township in Henry County; Company B—Osceola Township in Stark County; Company C—Henry Country; Company D—Henry County; Company E—Stark County; Company F—Stark County; Company G—Galva Township in Henry County; Company H—Henry County; Company I—Henry County and Company K—Henry County. On October 8 the Regiment was ordered to join the Army of Ohio at Cincinnati. In late October the Regiment served as guard on a supply train and marched south to Big Eagle, Kentucky and then on to Lexington, Kentucky arriving there on October 24. They remained at Lexington on guard and provost duty[972] until March 1863 when ordered to Danville, Kentucky. Shortly thereafter, the Union command believed that an overwhelming rebel force was invading Kentucky and ordered the Union forces to fall back north to the Kentucky River. The "invading force" proved to be only rebel forces gathering horses, mules and cattle and no fighting took place. In April the One Hundred and Twelfth returned to Danville where on April 24, Companies B, C, D, E, G and K became mounted infantry. Two days later the Regiment was ordered to Somerset, Kentucky and while enroute additional horses were obtained to mount Companies H and part of I. In May the Union forces moved further south and the One Hundred and Twelfth aided in driving a rebel force from a fortified position near Monticello, Kentucky. This was the first time that the Regiment was under fire. The Regiment returned to Somerset, Kentucky[973] and remained there until July.[974] During July, a detachment of one hundred and twenty-nine men under command of Captain Augustus Dunn of Company D scouted the area of Columbia, Albany and Burkesville, Kentucky and across the border into northern Tennessee capturing scattered rebel forces of Morgan's command. July 26, 180 men of the One Hundred and Twelfth were detailed with other Union troops to move from Danville to Richmond, Kentucky to intercept rebel forces moving north from London, Kentucky. At Richmond on July 28 the Union forces were attacked and surrounded but fought their way out and retreated to Lexington with little loss of men. On July 29 word reached Lexington that the rebel forces were at Winchester; Union forces, including the One Hundred and Twelfth advanced to

---

[970] Thompson, Bradford F., *History of the 112th Regiment of Illinois Volunteer Infantry, in the Great War of the Rebellion, 1862-1865,* Printed at the Stark County News Office, Toulon, ILL, (1885) page 330-1.

[971] Colonel Thomas Henderson was born in 1824 in Haywood County Tennessee. At age 11 he moved with his parents to Stark County, Illinois. He studied law and was admitted to the Illiinois bar in 1852 and opened a practice in Toulon, Illinos. Colonel Henderson served in the State House of Representatives in 1855 and 1856 and the State Senate in 1857-1860.At the outbteak of the civil war he was commissioned Colonel of the 112th Illinois Infantry.United States Congress. "Thomas J. Henderson", *Biographical Directory of the United States Congress*

[972] February 23, 1863, a mounted force of twenty-five men of the Regiment, while on patrol, was captured by Morgan's rebel cavalry and sent to Parole Camp at Benton Barracks in St. Louis where they remained until exchanged the following September.

[973] At this time the remainder of the Regiment was mounted.

[974] On June 10, two hundred men from the One Hundred and Twelfth along with four hundred men from other regiments rode on a 16-day raid into East Tennessee capturing prisoners and horses and destroying bridges, railroads, and supplies. From the Regiment five men were drowned and nine men were captured.

Winchester and succeeded in driving the rebel force of 1500 back across the Cumberland River, capturing 500 prisoners and recovering horses and other property taken by the rebel raiders. In late August the Regiment crossed the Crab Orchard Mountains and arrived in Kingston in East Tennessee on the Tennessee River on September 1. During the remainder of 1863 and early 1864 the Regiment was in various actions in East Tennessee including those at Cleveland, Athens, Calhoun, Charleston, Leiper's Ferry on the Holston River, Campbell's Station, the Siege of Knoxville, Bean's Station, Blain's Crossroads and the area near Dandridge.[975] In late January 1864 the Regiment moved to Marysville, Tennessee and while en-route skirmished at Flat Creek and Kelly's Ford; on February 4 the Regiment was ordered to Knoxville and dismounted. From Knoxville the Regiment marched north to Mount Sterling, Kentucky and remained there until April 6. In May the One Hundred and Twelfth arrived at Tunnel Hill, Georgia and from then on fought in Sherman's Atlanta campaign. Following demonstrations against Dalton the Regiment fought in the battle of Resaca losing forty-six men either killed or wounded. Moving south the men also fought at Dallas, New Hope Church, Allatoona Hills, Lost Mountain, Muddy Creek, Noyes' Creek, Cheyney's Farm, Kennesaw Mountain, Chattahoochee River, the Siege of Atlanta, Utoy Creek, Jonesboro and Lovejoy Station. Following the fall of Atlanta, the Regiment remained in camp in Decatur, Georgia until October 4, when they marched in pursuit of Hood into north Georgia and Alabama. However, when it was reported that Hood had moved into Tennessee the One Hundred and Twelfth broke camp at Rome on October 30 marched to Dalton, Georgia and then moved by rail to Nashville, Tennessee. From Nashville the Regiment marched to Thompson Station and then on to Pulaski, Tennessee arriving there in mid-November. When Hood crossed the Tennessee River and attempted to flank the Union forces, the Union troops, including the One Hundred and Twelfth retreated to Columbia and crossed the Duck River to Spring Hill skirmishing en-route. At Franklin, on November 30, the Union army, with their backs to the Harpeth River, was attacked by Hood's forces resulting in heavy losses on both sides. During the night the Union forces drew back across the Harpeth River toward Nashville. December 14-15 the Regiment participated in the Union victory at Nashville and then joined in the pursuit of Hood's forces to the Tennessee River. In January 1865 the One Hundred and Twelfth, as part of the Twenty-third Army Corps, moved by steamboat and rail to Washington DC and then by steamship *Atlantic* to Fort Fisher guarding the Cape Fear River inlet at Wilmington, North Carolina. The Fort had already fallen and the One Hundred and Twelfth moved up the cape and assisted in the capture of Fort Anderson and then Wilmington, North Carolina. In the campaign in the Carolinas the Regiment marched on Goldsboro, North Carolina, and rejoined Sherman's Army. Following Johnston's surrender the Regiment remained on duty at Goldsboro until muster out on June 20, 1865.[976]

There is one National and one Regimental flag of the One Hundred and Fourth Regiment Illinois Infantry in the collection of the State of Illinois Military Museum.

[975] These actions were often fought by only a detachment of the Regiment. However, in the East Tennessee campaign losses were heavy and more than 50 men were captured by the enemy—many of the captured subsequently died at Andersonville. At Knoxville, the One Hundred and Twelfth had two hundred and eleven men on the field and at the end of the fighting over one third were killed, wounded or missing.

[976] Ibid, Thompson.

This 35-star National flag has seven horizontal rows of stars in a rectangular canton typical of flags manufactured for the Cincinnati depot. The flag is inscribed 112th REGT ILL. VOL. INFANTRY.

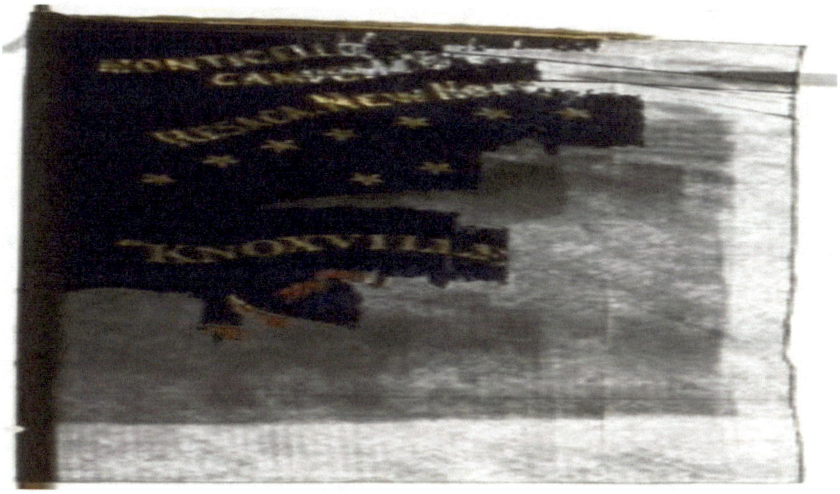

This is a remnant of a Regimental flag and is probably the flag mentioned in the Regimental history as being inscribed with the battle honors. The complete list of inscriptions is given in the introduction to this Regiment's history above.

George McKinney Dunkle, Company D, in his diary reports that flags were presented to three of the companies in the Regiment. "Saturday, August 30. Flags presented to the companys. To Dunn's by Miss Soule, To Shroufe's by Miss Davenport. To Bond's by Miss Ayres."[977] Augustus A. Dunn was Captain of Company D; Emery S. Bond was originally Captain of Company C but was subsequently elected Lieutenant Colonel of the Regiment; George W. Shroufe was Captain of Company H. All three companies were enrolled at Cambridge in Henry County. There is no further description of the flags, and the location of these flags today is unknown.

---

[977] Dunkle, George McKinney, *Diary of George McKinney Dunkle: August 25, 1862 to November 29, 1864: 112th Regiment Illinois Volunteers Company D,* Transcribed by Mary Whitson Greenwald. M.T. Publishing Co., Evansville, Ind. (1957) page 1.

Dunkle also describes a unique flag presented to Company G. "October 1, 1862. Witnessed the presentation of a beautiful flag to Company G, Galva Capt. Albro. The flag was of blue silk. A beautiful large eagle in the centre. Over it immediate a lightish color. Sunrise in appearance. Still above this a half circle of the thirty-four stars. Gilt. Below the eagle is a scroll or stripe of beautiful red. On it in gilt letters 112 reg Ills Vols. Presentation address by Lawyer Bennett. A short address in return by Capt. Albro followed by Col. Henderson Then by a Mr Payne and Lawyer Hinmon."[978] The flag mentioned probably no longer exists. The description is typical of regimental flags manufactured by Gilbert Hubbard & Company of Chicago.

No other flags carried by the 112th Illinois Infantry are known to survive.

---

[978] Ibid, Dunkle, page 4.

113th Illinois Infantry --"*Third Board of Trade*"

"When I again stood up, I could see a fragment of our line only, to my left, with which I recognized our colonel and regimental colors. I started towards our flag but had gone only a few steps when one of the enemy's shells exploded in front of me, and when the smoke had lifted a little, I saw that our regimental flag and the colonel had gone down."[979]

The One Hundred and Thirteenth Regiment Illinois Infantry was organized at Camp Hancock in Chicago during August of 1862 and mustered in on October 1, 1862 under the command of Colonel George B Hoge.[980] The men volunteered from the following counties: Company A-Cook County, Company B-Iroquois and Kankakee County, Company C-Cook County, Company D-Iroquois County, Company E-Cook County, Company F-Iroquois County, Company G (*Governor's Guards)*-Cook County, Company H-Kankakee County, Company I-Iroquois County and Company K-Kankakee County. The Regiment spent the month of October guarding rebel prisoners at Camp Douglas in Chicago and on November 5 was ordered to Memphis to join the Army of the Tennessee. As part of Grant's Central Mississippi campaign against invading rebel forces, the Regiment marched south from Memphis on November 24 into northern Mississippi. They arrived at Wyatt, Mississippi on December 2, crossed the Tallahatchie River and arrived at College Station, Mississippi on December 5. However, the rebel forces had retreated, and Grant ordered Sherman to return to Memphis and then proceed downriver to Vicksburg. Sherman ordered the division containing the One Hundred and Thirteenth back to Memphis; the Regiment arrived there December 12. One week later the One Hundred and Thirteenth boarded the steamer *Ed. Walsh* and sailed downriver as a part of Sherman's Yazoo Expedition. The Union forces reached Milliken's Bend on December 25 and were ordered up the Yazoo River to Johnson's Plantation.[981] December 26-29 the Regiment joined in the fighting at Chickasaw Bayou and Chickasaw Bluff. In early January 1863, the Union forces boarded transports and sailed down the Mississippi River and up the White River to the "cut off" to the Arkansas River and then disembarked just below Fort Hindman or "Arkansas Post." The Fort was captured on January 11 and Companies C, D, F, I and K were ordered to guard prisoners being transported North.[982] The remaining five companies[983] were ordered to Young's Point, Louisiana where they remained on provost duty until March. On March 3, 1863, the five remaining companies of the One Hundred and Thirteenth moved by transport to Vicksburg where they were put to work on Butler's canal—the canal being dug in an attempt to divert the Mississippi River from its course past Vicksburg. However, on March 19, word was received that rebel forces on Deer Creek north of Vicksburg had surrounded Admiral Porter and six gunboats. A Union force, commanded by General Sherman, and including the One Hundred and Thirteenth Illinois Battalion moved by transport and by forced march to the assistance of Porter.[984] The enemy was driven into the

---

[979] From a description of the assault on Vicksburg, May 19, 1863, in which the commander of the Regiment, Colonel George Hoge was wounded. Kellogg, Capt. John Jackson, *The Vicksburg Campaign and Reminiscences,* Printed by the Evening Journal, Washington, Iowa, 1913, page 29.

[980] 28 year old Colonel Hoge was commissioned Captain of Company F of 25th Missouri Volunteer Infantry on July 25, 1861 and continued until October 25, 1862 when he was commissioned Colonel of the 113th Illinois Volunteer Infantry.

[981] Sherman, General W. T., *Personal Memoirs of Gen'l W. T. Sherman,* Charles L. Webster & Co., New York, (1892), Volume I, pp. 301-31.

[982] These five companies of the Regiment remained in the North performing guard duty and recruiting until they rejoined the Regiment at Memphis in October 1864.

[983] The remaining five companies fought as a Battalion.

[984] Ibid, Sherman, page 332-41.

woods, relieving the gunboats. In April the Battalion marched to the rear of Vicksburg and skirmished at Haines' and Drumgould's Bluffs on April 29-May 2.[985] The Battalion next moved to Jackson, Mississippi via Grand Gulf, fought at Champion's Hill on May16 and then fought in the assaults on Vicksburg on May 19 and 22.[986] Following the surrender of Vicksburg on July 4, the Battalion marched to Jackson, Mississippi and fought in the Siege of Jackson from July 10 - 17. From Jackson the One Hundred and Thirteenth was ordered to Memphis and then on to Corinth, Mississippi where they performed provost duty until January 1864. From January 1864 through June 1865 the One Hundred and Thirteenth was based in Memphis primarily on guard duty. On June 10, 1864, the Battalion, as part of Sturgis' expedition into Mississippi, was routed by rebel forces under Forrest at the Battle of Brice's Cross Roads (or Tishomingo Creek) near Guntown, Mississippi. During the following days Sturgis' forces skirmished with Forrest at Ripley, Davis's Mill, Okolona, and Ashwood, Mississippi. On August 21, 1864, the One Hundred and Thirteenth joined in repulsing Forrest's attack on Memphis; in October 1864 the five Companies that had gone north to guard prisoners and to recruit rejoined the Regiment at Memphis bringing the Regiment back to full strength. In early October 1864 the Regiment was ordered to Eastport, Mississippi on the Tennessee River to cut communications at Eastport by tearing up the tracks of the *Memphis & Charleston Railroad* and destroying the bridge over the Tennessee River. The men were transported down the Mississippi River and Tennessee River by wooden gunboats but when they landed three miles below Eastport a masked battery shelled them. The gunboats hastily withdrew and many of the men were left "to shift for themselves."[987] The Regiment returned to Memphis and was mustered out at Memphis on June 20, 1865.

There is one National and One Regimental flag of the One Hundred and Thirteenth Regiment Illinois Infantry in the collection of the State of Illinois Military Museum. Both are in excellent condition; therefore, they were probably issued later in the War.

---

[985] William W. Burritt, Private of Company C, received the Medal of Honor for "Voluntarily acted as fireman on a steam tug which ran the blockade and passed the batteries under a heavy fire." Vicksburg, Mississippi, April 27, 1863. *Civil War Medal of Honor Winners From Illinois, Civil* War Centennial Commission of Illinois, July 1962, page 5-6.

[986] During the assault of May 22, 150 volunteers were needed to attack the fort in front of the Union Fifteenth Army Corps. Since the possibility of returning alive seemed poor the attack was called "forlorn hope." Four men volunteered from the One Hundred and Thirteenth Illinois Infantry—Pvt. Newton T. Gould, Sergt. James Henry, Sergt. Elisha Johns and Sergt. Jacob C. Miller. For their bravery, they received the Medal of Honor— "Was a member of a "Volunteer Storming Party which made a most gallant assault upon the enemy's works. Vicksburg, Mississippi, May 22, 1863." *Civil War Medal of Honor Winners from Illinois,* Published by the Civil War Centennial Commission of Illinois, July 1962

[987] Sergt. John S. Darrough, Company F was escaping across the Tennessee River in a small boat when he noticed one of the men of his Regiment lying exhausted on the river bank. He returned to rescue his comrade who he discovered was Captain A. W. Becket of Company B. Darrough received the Medal of Honor— "Saved the life of a Captain, Eastport, Mississippi, October 10, 1864. Beyer, W. F., and Keydel, O. F. eds, *Deeds of Valor, How America's Civil War Heroes Won the Congressional Medal of Honor,* Longmeadow Press, Stamford, CT. (1993) reprint of 1903 edition, pages 437-9.

The 35-star national flag with a square canton is characteristic of flags manufactured for the New York depot. Inscribed on this flag: 113TH REGT. ILL. VOL.

Based on the position of the stars, the eagle and the embellishments on the lower scroll, this flag was manufactured for the New York depot. Inscriptions on the flag read: CHICKASAW BAYOU; ASSAULT THE 19th and 22nd; ARKANSAS POST; SIEGE OF VICKSBURG; BLACK BAYOU; GUNTOWN; MEMPHNIS; EASTPORT.

"PRESENTATION OF COLORS—A splendid black horse was presented to Col. Hoge, and a stand of colors to his regiment, the 113th, by the Board of Trade on Tuesday evening, with appropriate ceremonies."[988] The colors are not described, and their further history is unknown.

No other flags carried by the 113th Illinois Infantry are known to survive.

---

988 *The Tribune,* Thursday, October 16, 1862.

114th Illinois Infantry

"The bullet riddled flag of the 114th Illinois as it came from the fight at Guntown, Miss., and the retreat which followed, was received by Gov. Yates, on Saturday last. It bears upon its folds abundant evidence of the fierceness of the struggle through which it was so bravely borne by the gallant regiment to which it belonged. Though torn by defeat, it has never been trailed in dishonor..."[989]

The One Hundred and Fourteenth Regiment Illinois Infantry was organized during the summer of 1862 and mustered into service at Camp Butler on September 8, 1862, under the command of Colonel James W. Judy. [990] The men volunteered from the following counties: Company A-Cass County; Company B and C-Sangamon County; Company D—Cass County; Company E—Sangamon County; Company F—Menard County with a few men from Sangamon County; Companies G, H and I—Sangamon County and Company K—Menard County. On November 8 the Regiment was ordered to Memphis for picket duty; on November 26 as part of Grant's Central Mississippi campaign against invading rebel forces, the Regiment marched south into northern Mississippi. They crossed the Tallahatchie River and arrived at College Station, Mississippi on December 4 and remained there until December 23. When word was received that Confederate General Forrest was threatening western Tennessee, the Regiment was ordered to Jackson, Tennessee and arrived there on January 8, 1863. Following a month of picket duty, the Regiment returned to Memphis to guard the *Memphis & Charleston Railroad.* The One Hundred and Fourteenth left Memphis on March 17 and was transported downriver to Young's Point, Louisiana and then ordered into camp at Duckport, Louisiana where they remained until May 2. On that date the Regiment left for the rear of Vicksburg via Grand Gulf, Louisiana and fought at Jackson, Mississippi on May 14. The Regiment was next involved in the ill-fated assaults on Vicksburg on May 19 and 22 and then from May 26-June 4 was on the Mechanicsburg campaign northwest of Vicksburg.[991] After the surrender of Vicksburg on July 4, the Regiment moved to Jackson, Mississippi and fought at Birdsong Ferry on the Big Black River and then in the Siege of Jackson. When rebel General Johnson evacuated Jackson, the One Hundred and Fourteenth joined in the pursuit of Johnson as far as Brandon, Mississippi and then returned to Vicksburg and remained there until September 3 when ordered to Oak Ridge, Mississippi. While at Oak Ridge[992] the Regiment performed picket duty and skirmished with guerrillas. During that time—October 14-20—they were sent on an expedition to Canton and the Bogue Chitto Creek north of Jackson, Mississippi. In late November the Regiment was transported to Memphis for provost duty. While at Memphis, in February 1864, the One Hundred and Fourteenth accompanied a scouting force to the Tallahatchie River; the Union forces moved into northwestern Mississippi, crossed the Coldwater River, and engaged the rebel forces at Wyatt, Mississippi on the Tallahatchie River, thus allowing General W. S. Smith's Union Cavalry forces to cross the River at New Albany. April 30-May 9 the Regiment was part of Sturgis' Expedition to Ripley, Mississippi; but after two weeks of hard marching, they returned to Memphis. On June 1 a second expedition under Sturgis was ordered to destroy the *Ohio & Mobile Railroad* between Corinth and

---

[989] Correspondence from "Headquarters 114th Reg. Ill. Vol., Memphis, Tenn. June 18, 1864" as quoted in Satterlee, John L., *The Journal of the 114th, 1861 to 1865,* Phillips Brothers, Inc., Springfield, Ill., (1979) page 268.

[990] Colonel James Judy was born in Clark County, Kentucky in 1822. He moved with his parents to Montgomery Couny Illinois when he was 14 and began farming in Mendard County in 1851. He became an expert in Shorthorn cattle breeding and selling, and later was president of the Illinois State Board of Agriculture and helped establish the Illinois State Fair. *History of Menard & Mason Couties, Illinois, 1879*, Chicago: O. J. Baskin & Co. Historical Publishers, Chicago

[991] Bearss, Edwin Cole, *The Vicksburg Campaign,* Morningside Press, Dayton, Ohio (1986), Volume III, pages 995-1006.

[992] Oak Ridge was 29 miles northeast of Vicksburg on the Benton Road.

Tupelo. The Union force was intercepted and repulsed by rebel forces under Forrest at Guntown, Mississippi[993] on June 10 and the One Hundred and Fourteenth served as rear guard during the first night's retreat. Further skirmishing occurred between the opposing forces at Ripley and Davis Mills. Following another two weeks' of picket duty at Memphis the Regiment left on an expedition to Tupelo, Mississippi under command of General A. J. Smith.[994] In that expedition the Regiment was commended for its actions in fighting at Harrisburg, Mississippi during July 13-15.[995] Again the Regiment returned to Memphis. In August the Regiment joined a second expedition, commanded by General Smith, to Oxford, Mississippi; this included skirmishes at the Tallahatchie River and at Abbeville. In September the One Hundred and Fourteenth moved to Duvall's Bluff, Arkansas and then pursued rebel General Price's forces through Arkansas and Missouri. Late in November 1864 the Regiment moved to Nashville, fought in the Battle of Nashville on December 15-16, and then pursued Hood's retreating forces to Pulaski, Tennessee. January 1865 found the One Hundred and Fourteenth at Eastport, Mississippi where they remained until February when ordered to New Orleans and then on to Dauphin Island, Alabama at the entrance to Mobile Bay. Subsequently the Regiment campaigned against Battery Huger, Battery Tracy, Spanish Fort, and Fort Blakely, occupied Mobile, Alabama and then in April marched to Montgomery, Alabama. In July the Regiment was ordered to Vicksburg where the One Hundred and Fourteenth Regiment Illinois Infantry was mustered out on August 3, 1865.[996]

There are two national and one regimental flag carried by the One Hundred and Fourteenth Illinois Infantry in the collection of the State of Illinois.

This national flag is well worn and is probably the flag returned to Governor Yates of Illinois in June 1864. The flag had been issued to the Regiment September 18, 1862, and had been carried through battles at Jackson Mississippi, Vicksburg, Messengers Ford, Brandon Mississippi, and Guntown Mississippi.[997] Based on the remaining canton and star pattern, the flag is of Philadelphia manufacture.

---

993 Known as the Battle at Brice's Crossroads. In that action the One Hundred and Fourteenth lost 205 men killed, wounded, or missing out of 397 men who entered the action. Among those killed was Asst. Surgeon Alvin. S. French.

994 By this time only 120 men remained to comprise the One Hundred and Fourteenth.

995 In this action the Regiment lost another 40 men.

996 *Report of the Adjutant General of the State of Illinois,* H. W. Rokker, State Printer and Binder, Springfield, Ill (1886) Volume VI, pages 222-3.

997 Ibid, Satterlee, page 268

This National flag of the Regiment is a 34 -tar flag also of Philadelphia manufacture. It bears no inscriptions.

The Regimental flag is of New York manufacture. It shows significant signs of wear. The date of issuance of this flag is unknown.

No other flags carried by the One Hundred and Fourteenth Illinois Infantry are known to survive.

115th Illinois Infantry - "*Second Methodist Regiment*"[998]

"The gallant Steedman, seizing the colors of a regiment, led his men to the attack. With loud cheers they rushed upon the enemy, and, after a terrific conflict lasting but twenty minutes, drove them from their ground and occupied the ridge and gorge. The slaughter of both friend and foe was frightful…This attack was made by our troops, very few of whom had ever been in an action before, against a division of old soldiers, who largely outnumbered them; yet with resolution and energy they drove the enemy from his strong position, occupied it themselves, and afterward held the ground they had gained with such terrible losses. The victory was dearly won, but to this army it was a priceless one."[999]

The One Hundred and Fifteenth Regiment Illinois Volunteer Infantry was organized at Camp Butler in the summer of 1862 and mustered on September 13, 1862 under the command of Colonel Jesse H. Moore.[1000] The men were recruited and organized as follows: Company A at Taylorville with men from Christian County; Company B at Shelbyville with men from Shelby County; Company C from Friendsville and Bridgeport in Wabash and Lawrence Counties; Company D from Bainbridge, Frederick and Woodstock townships of Schuyler County; Company E mainly from Macon County with some men from Christian County; Company F from Macon and Shelby Counties; Company G from Shelby County; Company H from Tazewell County; Company I from Morgan County and Company K from Menard and Sangamon County. On October 4 the Regiment boarded cars of the *Chicago & Alton Railroad* for the trip to Covington, Kentucky. Following further instruction, on October 18, the Regiment, now part of the Army of Kentucky, was ordered south to Lexington and on October 28 went into camp at Sandersville, three miles from Lexington. November 13 the Regiment moved further south to camp at Richmond, Kentucky and during the next two months scouted the area between Richmond and the Cumberland Mountains. In late December the One Hundred and Fifteenth was ordered to Danville, Kentucky and December 26-31 joined in the pursuit of Rebel General Morgan to Lebanon Junction, Kentucky. Due to the shift of Confederate threat to middle Tennessee, the Army of Kentucky was ordered to Louisville in late January 1863. The One Hundred and Fifteenth reached Louisville on January 31, boarded the steamers *Leonora* and *New York* and sailed up the Cumberland River past Fort Donelson and landed at Nashville, Tennessee on February 7. The Regiment remained at Nashville until March 5 when word was received of Union General Coburn's defeat near Franklin, Tennessee. The Regiment was immediately moved to Franklin for the pursuit of rebel General Van Dorn's forces, skirmished with the rebels at Spring Hill and

---

[998] Royse, Isaac Henry Clay, *History of the 115th Regiment Illinois Volunteer Infantry,* published by the author, Windsor & Kenfield Publishing Co., Chicago (1900) page14, "In religion the Methodist Church predominated to such an extent that we were sometimes called the "" Second Methodist Regiment"" the 73rd being known as the first."

[999] O.R.—Series 1—Volume XXX/1 [Series 50] page 855. From the report of Maj. Gen. Gordon Granger, Commander of the Reserve Corps at the Battle of Chickamauga. Brig. Gen. James Steedman, commanded a Division of three Brigades in the Reserve Corps. During the battle on September 20, both Granger and Steedman left their reserve positions in the rear to rush troops and ammunition to support General George Thomas on Snodgrass Hill. Steedman personally rallied and led inexperienced troops in a counterattack. The flag he carried was that of the One Hundred and Fifteenth Regiment Illinois Volunteer Infantry.

[1000] Colonel Jesse Moore was born near Lebanon, Illinois in 1817 and attended McKendree College in Lebanon in 1842. Moore taught school from 1842 to 1848 and then studied for the ministery and became an ordained Mrehodist minister in 1849. He was appointed Colonel of the 115th Illinois Volunteer Infantry in 1862. Wikipedia, The Free Encyclopedia, 2023.

returned to Franklin.[1001] The Union forces at Franklin were attacked by Van Dorn's Cavalry on April 10, but they were driven off. On June 2 the Regiment moved to Triune, Tennessee.[1002] Early in the morning of June 23, 1863, the Regiment was ordered south on the Shelbyville Pike, passed through Guy's Gap on June 29 and went into camp at Wartrace, Tennessee on July 3. In August the Regiment moved to camp on the Elk River and then on to Tullahoma. From Tullahoma, on August 23 the men moved north to Fosterville and went into camp there until September. The One Hundred and Fifteenth left Fosterville September 5 and marched south through the Cumberland Mountains to Stevenson, Alabama and then on to Bridgeport, Alabama on the Tennessee River. They then marched south of Chattanooga and arrived in Rossville, Georgia on September 14. September 19-20 the Regiment fought in the Battle of Chickamauga, playing a significant part in the actions of September 20 on Snodgrass Hill.[1003] Union forces retreated from Chickamauga into Chattanooga and the city was laid siege by the Confederates. On November 1, the One Hundred and Fifteenth left Chattanooga to protect Union supply lines at Shell Mound, Tennessee. January 26, 1864, the Regiment left Shell Mound and in February skirmished at Dalton, Tunnel Hill, Buzzard's Roost Gap and Rocky Faced Ridge in Georgia. During March and April, the Regiment was in camp at Cleveland, Tennessee. In May the Regiment joined Sherman's Atlanta campaign and during the next five months fought at Tunnel Hill, Rocky Faced Ridge, Buzzard's Roost Gap, Resaca, Kingston, Cassville, Dallas, New Hope Church, Allatoona Hills, Pine Mountain, Lost Mountain, Kennesaw Mountain, Ruff's Station, Chattahoochee River, Peach Tree Creek, Siege of Atlanta, Jonesboro, and Lovejoy Station. Following the fall of Atlanta, the Regiment joined in the pursuit of Hood into north Georgia and Alabama. October 13 Hood attacked a blockhouse held by Company D guarding Buzzard's Roost Gap. The forty-one men of the Company held the blockhouse for 12 hours against 30,000 enemy with 70 pieces or artillery, thus delaying Hood's passage through the gap and allowing Sherman's forces to gain on the rebel forces.[1004] In the Nashville campaign the One Hundred and Fifteenth fought in the Battle of Franklin on November 30 and the Battle of Nashville December 15-16. After the pursuit of Hood to the Tennessee River the Regiment was ordered to Huntsville, Alabama and remained there until March 1865. Following operations in East Tennessee in April and May the Regiment marched to Nashville and was mustered out June 11, 1865.[1005]

The State of Illinois military flag collection holds two flags from the One Hundred and Fifteenth Regiment Illinois Volunteer Infantry—one National and one Regimental flag. According to the Regimental History, "On the 17th (*July 1863*, ed), the 115th received its regulation regimental colors and battle flags; prior to that time a flag belonging to Company I had been used."[1006] The men of Company I were from Jacksonville and Meredosia (Morgan County) but there is no record of a flag presentation to that Company.

[1001] At this time the One Hundred and Fifteenth became part of General Gordon Granger's Reserve Corps of the Army of the Cumberland.

[1002] In May, after only eight months' service, 94 enlisted men had died, 90 had been discharged due to physical disability and 28 had deserted. In addition, 2 officers had died, 1 was dismissed and 16 had resigned. Thus, a total of 231 men or 23% of the Regiment had been lost. Royse, Isaac Henry Clay, *History of the 115th Regiment Illinois Volunteer Infantry,* Published by the author, Windsor & Kenfield Publishing Co., Chicago (1900) page 64.

[1003] Lt. Col. Kinman of the Regiment was killed and in total the Regiment lost 183 men killed wounded or missing out of 426 effectives.

[1004] Captain Samuel Hymer of Company D was brevetted as major for his actions on October 13 and on March 28, 1896 was awarded the Medal of Honor for "With only forty-one men under his command, defended and held a blockhouse against the attack of Hood's Division for nearly ten hours, thus checking the advance of the enemy and insuring the safely of the balance of the regiment, as well as that of the 8th Kentucky Infantry, then stationed at Ringgold, Georgia, October 13, 1864." *Civil War Medal of Honor Winners from Illinois,* Published by the Civil War Centennial Commission of Illinois, July 1962.

[1005] Ibid., Royse. Royse's history of the Regiment is well-written and accurate portrayal of the Regiment's exploits.

[1006] Ibid. Royse, page 80.

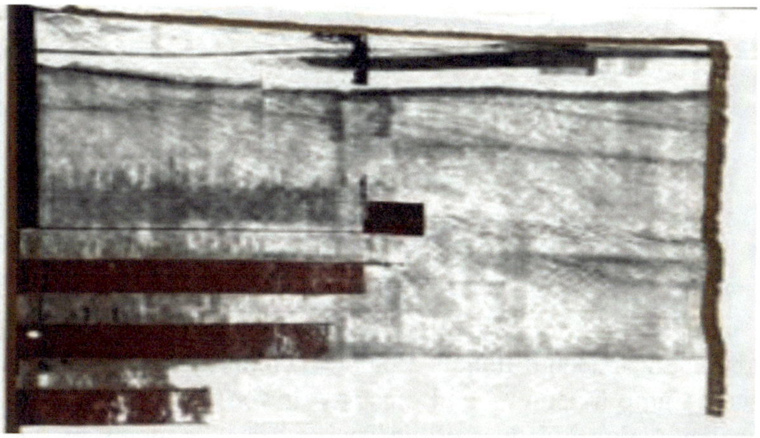

Little remains of the National flag of the Regiment. This is possibly one of the flags received by the Regiment in July 1863.

The Regimental flag is a 35-star flag; therefore, it was manufactured later in the War and could not be one of the flags received by the Regiment in 1863. Based on the position of the stars above the eagle and the lower scroll pattern this flag was manufactured for the New York depot.

The Regimental History relates the presentation of two other flags to the Regiment. "On the 20th the ladies of Macon presented the company (*Company E,* ed.) a beautiful flag, Miss Sarah Coltra making the presentation speech, to which Captain Lane responded on behalf of the company."[1007]

"On September 3rd the ladies of Delavan presented the company (*Company H,* ed.) a fine silk flag and a splendid supper, which were accompanied with speeches and much general merriment."

There is no description of either flag and the further history of these two flags is unknown.

No other flags carried by the 115th Illinois Infantry are known to survive.

[1007] Ibid. Royse, page 18.

116th Illinois Infantry - "*Macon County's Own*"

"Flag of the free hearts hope and pride!
By angel hands to valor given
Thy stars have lit the welkin dome,
And all thy hues were born in heaven."[1008]

The One Hundred and Sixteenth Illinois Volunteer Infantry Regiment was organized in Decatur, Macon County, Illinois at Camp Macon[1009] and mustered in for three years of service on September 30, 1862 under the command of Colonel Nathan Willis Tupper.[1010] The men for Companies A, B, C, D, E, I and K volunteered from Macon County while the men for Company F volunteered from Cheney's Grove Township in McLean County, the men for Company G volunteered primarily from Macon County with some from Christian County and the men for Company H volunteered from Christian County and Shelby County. The regiment departed on the *Illinois Central Railroad* for Memphis on November 8. At Cairo, Illinois the Regiment transferred to the steamer *Tigress* and arrived at Memphis on November 11. From Memphis the Regiment joined Grant's Central Mississippi Campaign and left on the march to the Tallahatchie River on November 26. There was no contact with the enemy and the Regiment returned to Memphis December 13 in time to join in the operations being planned against Vicksburg. On December 20 the Regiment boarded the steamer *Planet* as part of Sherman's Yazoo River Expedition and traveled down the Mississippi River and then 15 miles up the Yazoo River to engage in the Regiment's first military action at Chickasaw Bayou on December 26-28 and Chickasaw Bluff on December 29. January 1, 1863, the Union forces steamed back down the Yazoo River to the Mississippi River and then up the Arkansas River to assault the rebel forces holding Arkansas Post (Fort Hindman). The Fort was surrounded and captured in the battle of January 10-11. Following the battle the Regiment sailed back downriver and disembarked at Young's Point, Louisiana where the men were assigned the task of continuing to dig a canal that would allow Federal gunboats to bypass the loop of the Mississippi River on which Vicksburg was situated.[1011] In March Porter's gunboats were bottled up on Deer Creek, north of Vicksburg and the One Hundred and Sixteenth along with the Sixth and Eighth Missouri moved via Muddy, Steele's and Black Bayous for the relief of the gunboats. Following demonstrations against Haines' and Drumgould's Bluffs north of Vicksburg April 29-May 2, the One Hundred and Sixteenth marched south through Louisiana and crossed the Mississippi River below Vicksburg. The Regiment fought at Jackson, Mississippi on May 14, Champion's Hill on May 16, and Big Black River Bridge on May 17. The evening of May 18, while charging the enemy' s works at Vicksburg, Corporal Sam Baty, flag bearer of Company A was killed, and the Regiment's colors destroyed by a bursting shell. In the siege of May 22, eight men from the One

---

[1008] *Decatur Magnet*, April 20, 1865. Included in a report of a flag being made for presentation to the Macon Regiment by the ladies of Decatur.

[1009] Now known as Fairview Park

[1010] Colonel Tupper was born in St. Lawrence County, New York in 1830. "When Mr. Tuppers brother Ansel Tupper was killed in the battle of Shiloh, Nathan Tupper was then practicing law in Decatur and was a part of the late Judge W. E. Nelsons firm and although he was crippled in his arm as soon as he received word of his brothers' death he organized a regiment, which became the one Hundred and Sixteenth Illinois and went to the front. He was elected colonel of the regiment and served until 1864. Information furnished by Janet Donner, Illinois Civil War Detail Report in "Find A grave " for Col Nathan Willis Tupper.

[1011] The Regiment lost 58 men of disease while at Young's Point. *Report of the Adjutant General of the State of Illinois,* H. W. Rokker, State Printer and Binder, Springfield, Ill, (1886) Volume VI, pages 248-271.

Hundred and Sixteenth volunteered to lead an assault against a fort on the enemy lines.[1012] The assault was unsuccessful, and all eight men survived. After the fall of Vicksburg on July 4, the Regiment joined in the pursuit of rebel General Johnston's forces, fighting them at Jackson, Mississippi and pursuing them to the Pearl River. The Regiment returned to camp near the Black River and remained there until they embarked for Memphis in October. From Memphis the XV Corps marched to Chattanooga, arriving there on November 21. During the night of November 23, the One Hundred and Sixteenth Illinois and the Sixth Missouri floated down the Tennessee River to the mouth of Chickamauga Creek and secured the position until the entire XV Corps could cross the river. November 24-25 the Regiment fought at Tunnel Hill, Georgia and the following day at Missionary Ridge, Tennessee. Immediately following these actions, the Regiment joined in the march to Knoxville to relieve Burnside. In December the One Hundred and Sixteenth was ordered to camp in Larkinsville, Alabama and remained there until May 1864. In May the Regiment joined Sherman's Atlanta campaign and during the next five months fought at Resaca, Dallas, New Hope Church, Allatoona Hills, Marietta, Kennesaw Mountain, Nickajack Creek, Chattahoochee River, siege of Atlanta, Ezra Chapel, Jonesboro, and Lovejoy Station. After pursing Hood into north Georgia and Alabama the Regiment joined the "March to the Sea" and skirmished at Clinton, Ball's Ferry, Statesboro, Savannah and Fort McAllister, Georgia during November and December 1864. In the 1865 Carolina campaign the men fought at Salkehatchie Swamps, South and North Edisto River, Columbia and the Battle of Bentonville. Following Johnston's surrender the Regiment moved north and marched in the Grand Review in Washington, D. C on May 24 and were mustered out on June 7, 1865.[1013]

The collection of the State of Illinois holds one National and one Regimental flag of the One Hundred and Sixteenth Illinois Volunteer Infantry Regiment.

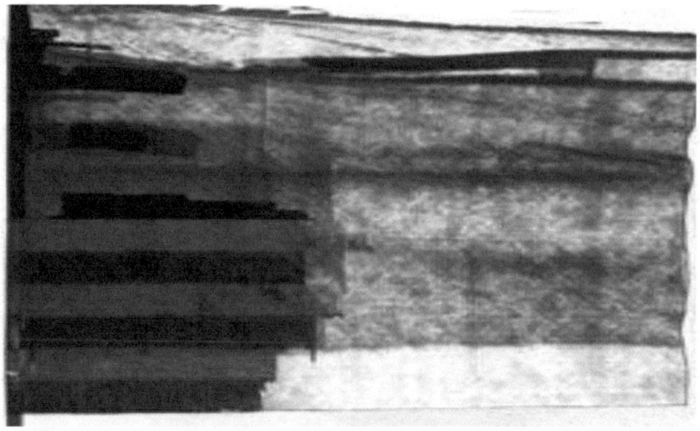

"The original flag of the 116th was destroyed by a shell at Vicksburg. Remnants of it, stained with the blood of Samuel Baty, are in the State of Illinois Military Museum at Springfield. Baty was killed while carrying the flag when the shell exploded. When the women of Decatur learned of the fate of the

[1012] The mission was called "forlorn hope" and only volunteers were accepted since there seemed little probability that the men would survive. Their task was to build a bridge over a ditch protecting the front of the fort and plant scaling ladders against the walls. From the Regiment, Martin K. Davis, Nicholas Geschwind, Andrew Johnson, Charles W. Rundle, Benjamin W. Schenck, Benona Sprague, Thomas Ward and Andrew J. Widick volunteered to be part of the storming party. For "Gallantry in the charge of the "Volunteer Storming Party" Vicksburg, Mississippi, May 22, 1863," the men received the Medal of Honor. Beyer, W. F., Keydel, O. F., *Deeds of Valor, How America's Civil War Heroes Won the Congressional Medal of Honor, Longmeadow* Press, Stamford, CT, (1993) (Reprint of 1903 edition), pp 190-7.

[1013] Dyer, Frederick H., *A Compendium of the War of the Rebellion,* Thomas Yoseloff, New York, (1959), Volume III, p1095-6.

flag, they immediately made a new one for the regiment, meeting in Powers Hall for the work. Many put stitches in that flag. It went through the rest of the war and the remains of it, torn with bullet holes, was long a cherished possession of the regiment."[1014] The National flag shown here is presumably the flag carried by Color-bearer Samuel Baty.

The "replacement" for this flag is probably lost. However, the story of the flag and its subsequent history is preserved. "George W. Lyons[1015] of 539 West Green Street, Decatur has in his possession a flag which went through all the vicissitudes of war and victory, as the Regimental insignia of the One Hundred and Sixteenth Regiment, Illinois Volunteers, after the original regimental flag furnished by the state, was 'put out of commission by shot and shell, though never surrendered.' ...Mr. Lyons said the flag was made in the Powers building by a number of Decatur women. About $400 was raised for the making of the flag. It was first presented to the One Hundred and Sixteenth Illinois regiment in Decatur. However, when the regiment went away, they left the flag in Decatur and were carrying a state flag.[1016] This state flag saw service through many hard fights and was considerably shot and torn. It was put out of service at Camp Sherman, where the new flag came to the regiment. In October the regiment left the camp and were carrying the flag that was made in Decatur. This flag then followed them through all of their battles. At the final mustering out the flag was given to Dr. Ira N. Barnes[1017] to keep until the first annual reunion of the regiment. At this reunion it was voted to George W. Lyons, and he has kept it since then, and still has it in his possession...Later, I learned almost by accident that the flag I had been in pursuit of was a bundle of rags and tatters, reverently and lovingly cared for by George W. Lyons."[1018] The whereabouts of this flag today in unknown.

Although only 50% of the Regimental flag remains, enough is seen to identify it as a flag typical of those produced for the New York depot. This Regimental color bears no inscriptions.

[1014] Banton, OT, Editor, *History of Macon County 1976,* Printed by the Macon County Historical Society, Decatur, Ill. (1976), page 182-3.

[1015] George W. Lyons had been a musician in Company A of the Regiment

[1016] This probably denotes a flag supplied by the state since Illinois did not have a "state flag" at that time.

[1017] Dr. Ira N. Barnes of Decatur was the Regimental Surgeon.

[1018] Johns, Jane Martin, *Personal Recollections of Early Decatur Abraham Lincoln, Richard J. Oglesby and The Civil War,* Published by the Decatur Chapter Daughters of the American Revolution, (1912), pages 198-200.

Two other flags are known to have been presented to Companies of the Regiment—one from the women of Mount Zion and one from the women of Harristown.

In August 1862 the ladies of Mt. Zion collected money to buy a flag to present to the men enlisting in their area. The women thought that they did not have enough time to make a flag themselves. Miss Sarah Price presented the flag to the Company (later designated as Company C) on August 13 and Nathan M. Baker accepted the flag on behalf of the Company. The flag was carried by the Company throughout the War but usually packed in a strong box with the Company's books and papers; however, at the end of the War the flag was carried by the Company in the Grand Review. After the War the flag was kept in a Hall in Mt. Zion but was brought out at Regimental reunions. In 1874, Nathan Baker had married Sarah Price and at a reunion of the Regiment in 1880 the flag was given to the Bakers. The flag in question was a national flag with 34 stars and measured "six feet two inches long and five feet ten inches wide and bears the following inscription: 'Presented to Co. C, 116th Ills. Vol. Inf. By the Ladies of Mt. Zion and Vicinity.' In letters of gold on the white stripes are the names of the battles in which this Company bore an honorable part: 'Chickasaw Bayou, Arkansas Post, Champion Hills, Vicksburg, Jackson, Mission Ridge, Resaca, Dallas, Kennesaw Mtn., Atlanta, Jonesboro, Ft. McAllister and Bentonville.'"[1019] This flag is probably the flag that is in the Texas Civil War Museum in Fort Worth, Texas.

In 1862 Misses Martha Eyman and Mary McGuire of Harristown solicited funds to purchase a flag for the men of Company E. Since silk was not available in Decatur, the material was ordered from New York. Both men and women worked on sewing the flag and it was presented to Captain Lewis Eyman of Company E at Camp Macon before the Company was mustered into service. At the company's first engagement—the Battle of Arkansas Post—Captain Eyman was killed, and the flag entrusted to Captain John S. Windsor and then Captain Richard M. Hamilton. The flag was brought home by Captain Hamilton and in 1912 was in the possession of his daughter Miss Katherine Hamilton, principal of Pugh School in Decatur.[1020] The whereabouts of this flag is unknown.

No other flags carried by the 116th Illinois Volunteer Infantry are known to survive.

[1019] Ibid, Johns, page 201-3. The information is from Nathan Baker who had the flag in his possession in 1912.
[1020] Ibid, Johns, page200-1

117th Illinois Infantry - "*McKendree Regiment*"[1021]

"As we were passing through town, a pretty little girl was standing at the gate of her yard. Hieronymus[1022] stepped from the ranks and let the banner fall over her. She tried to catch hold of it & said 'I will tear a piece of that to make my doll a dress.' It showed she was thinking of nice & fine clothes as the flag was a rich blue silk."[1023]

The One Hundred and Seventeenth Regiment Illinois Infantry was organized at Camp Butler, near Springfield, Illinois in September of 1862. The men of the Regiment volunteered from the following counties, Company A from Mount Hope and Funk's Grove Townships of McLean County; Company B from Montgomery County; Company C from Lebanon Township of St. Clair County; Company D from Madison County; Company E from the area of Aviston and Trenton in Clinton County; Company F and G from Madison County; Company H, I and K from St. Clair County. The Regiment was mustered in at Camp Butler on September 19, 1862, under command of Colonel Risdon Moore.[1024] Following the issue of equipment and training, the regiment departed for Alton, Illinois on November 10. From Alton the Regiment floated downriver, passed Cairo, Illinois and arrived at Memphis, Tennessee on November 17. The One Hundred and Seventeenth went into camp near Fort Pickering just south of Memphis on the Mississippi River. In late November the Regiment was ordered inside the Fort to man the heavy guns mounted in the breastworks guarding the Mississippi River. In mid-February 1863 a five-company detachment from the Regiment scouted across the River to Hopefield, Arkansas looking for the rebel guerilla Captain Barton but returned empty handed. In May the Regiment received its first set of colors and on May 14 the Regimental color guard was selected. On July 5 and again on July 18 a detachment of the Regiment went downriver in search of rebel forces but did not contact the enemy. For the next six months the One Hundred and Seventeenth remained at Fort Pickering on guard duty with detachments occasionally scouting into Arkansas. Christmas Day companies B, E, F, G and K were ordered to Moscow, Tennessee via the *Memphis & Charleston RR* and on December 27 marched toward Holly Springs, Mississippi in search of rebel General Forrest. Failing to find the elusive enemy the men returned to Fort Pickering on December 31. January 26, 1864, the Regiment departed Fort Pickering aboard the steamer *Sioux City* for the trip to Vicksburg, Mississippi. One week after arriving at Vicksburg, the One Hundred and Seventeenth departed Vicksburg on the Meridian Campaign[1025] under command of General Sherman. The Union troops marched east through Jackson and arrived at Meridian on February 14 but by then the Confederates had abandoned the city. Sherman ordered destruction of much of the city and the two railroads that passed through city; on February 20 Sherman's forces began the march back to Vicksburg. Enroute, at Canton, Mississippi the Regiment tore up a mile

---

[1021] This Regiment was known as the McKendree Regiment since a nucleus of the Regiment came from the faculty and students of McKendree College in Lebanon, Illinois.

[1022] Benjamin Hieronymus, 1st Lt. Company A from McLean County

[1023] "From the letters of Sid Risdon Moore" Sidney Z. Robinson, 2nd Lt, Company D. The letters are in the Illinois State Historical Library at Springfield, Illinois and quoted in the book *117th Illinois Infantry Volunteers 1862-1865* by Edwin G. Gerling privately printed by the Author, 1992, page 64.

[1024] Colonel Moore was born in 1827 near Cahokia, Illinois, He atteneded McKendree Collge 1845-1850 and graduated with a Master of Art's Degree. Following graduation he tutored in Latin and Greek and was Professor of Mathematics. He joined the 117th Illinois Infantry and was elected Colonel of the Regiment. McKendree University web site, McKendree's Civil War History.

[1025] The Meridian Campaign was designed to destroy the town of Meridian, Mississippi where the *Mobile & Ohio Railroad* crossed the *Southern Railroad of Mississippi.*

of track of the *Mississippi Central Railroad* and arrived back at Vicksburg March 4. Five days later the One Hundred and Seventeenth, as part of the ill-fated Red River Campaign, boarded the *Thomas E. Tutt,* sailed down the Mississippi River and up the Red River. The Regiment debarked at Simmesport, Louisiana on March 13 and the following day marched north to capture Fort DeRussy. They reboarded the *Thomas E. Tutt* and sailed further up the Red River to Alexandria in time to see the Confederates abandon the town. When the Union troops marched further upriver, they were trapped and defeated near Mansfield at the Battle of Sabine Crossroads. The Union forces retreated downriver, being harassed by the Confederates the entire way; the One Hundred and Seventeenth reached Alexandria on April 26 and fought off a rebel attack on April 28. May 22 the Regiment reached the Mississippi River, boarded the *Thomas E. Tutt* and returned to Vicksburg. On June 3 the Regiment boarded the sidewheel packet *Leviathan* for the trip to Memphis, arriving June 9. June 23 the Regiment moved by rail to Moscow, Tennessee and then marched south into Mississippi through Pontotoc towards Tupelo. The Regiment was involved in fighting at Tupelo on July 14 and assisted in destroying track of the *Mobile & Ohio Railroad.* They marched back to Tennessee and arrived in Memphis July 24. August 6 the Regiment was ordered to Oxford, Mississippi, and en-route skirmished at Hurricane Creek; however, after arriving at Oxford it was learned that General Forrest had attacked Memphis, and the Regiment was ordered to return to Memphis. The Union forces were attacked by the enemy at the Tallahatchie River near Abbeville, Mississippi on August 23, but arrived without further incident at Memphis on August 30. In September the Regiment was transported upriver to St. Louis and camped at Jefferson Barracks. During October and November, the One Hundred and Seventeenth pursued Price through Missouri, then returned to St. Louis. November 25 the Regiment sailed downriver on the steamer *Mollie McPike* then up the Ohio and Cumberland Rivers to Nashville. The Regiment immediately built breastworks and on December 15 and 16[1026] was heavily involved in the Battle of Nashville. The rebels were routed and the One Hundred and Seventeenth joined in the pursuit of Hood to the Tennessee River. The Regiment was ordered to Eastport, Mississippi in January 1865 and to New Orleans in February. On March 7 the Regiment boarded the steamer *Empire City* for the trip to Dauphin Island at the entrance to Mobile Bay. They moved up the Bay, participated in the siege of Spanish Fort and in the assault and capture of Fort Blakely. The Regiment occupied Mobile on April 12 and then marched to Montgomery, Alabama where they remained on duty until July 16 when ordered to Selma, Alabama. From Selma the Regiment moved by train to Meridian and then on to Jackson and finally Vicksburg, Mississippi. On July 24, 1865, the Regiment boarded the steamer *Pauline Carroll,* arrived at St. Louis on July 29 and boarded trains for Camp Butler. On August 5, 1865, the Regiment was mustered out at the State Capital Building in Springfield, Illinois.[1027]

There are two national and one regimental flag in the collection of the State of Illinois Military Museum. The Quartermaster of the Regiment, Henry Fike of Mascoutah Illinois, secured a set of colors from the Quartermaster of the US Army while the Regiment was in Memphis. Henry Fike wrote,

May 7, 1863, "I went up in the city today and I got a splendid National Color and Regimental Color."

May 9, 1863, "Today I went up and took our new National Color and Regimental Color to a shop to have the number and name of our regiment inscribed."

[1026] Private Wilbur F. Moore, Company C, on December 16, 1864, captured a Confederate battery flag and the Confederate color bearer of the flag during the battle of Nashville. For his actions he was awarded the Medal of Honor. There is an Army of Tennessee Battle flag (catalog WD 215) in the American Civil War Museum that is ID'd as having been captured by Pvt. Moore at the Battle of Nashville. The flag is possibly that of Dent's Alabama Battery.

[1027] Ibid, Gerling.

May 15, 1863, "I am going up in the city after dinner to get our new colors."[1028]

The colors supplied by the Army Quartermaster Department did not come with the regimental designation on the flags. These had to be inscribed by the individual regiment. Henry Fike apparently secured the colors from the Quartermaster Department in Memphis and then had the regimental designation applied. Presumably this was the first set of colors for the Regiment since it was not until this time that a color guard was selected.

This is a 35-star national flag and is typical of those flags supplied to the Quartermaster Department from the New York depot. However, this is not one of the flags secured by Henry Fike since flags manufactured in May 1863 only had 34 stars.[1029] *The inscriptions on this flag read:* 117th REGT. ILLs INFY. VOLs; FORT DERUSSY, MARCH 14, 1864; PLEASANT HILL, LA., APRIL 6, 1864; TUPELO, MISS., JULY 14, 1864; NASHVILLE, DECEMBER 15 & 16, 1864; FORT BLAKELY & MOBILE, ALA., APRIL 9, 1865.

The national flag shown here is a 34-star flag, also of New York manufacture. The inscription has been partially lost but can be read as: 117th REGT. IL___. *Based on the number of stars, this flag would have been manufactured during the years 1861-1863 and may well be the flag secured in Memphis by Quartermaster Fike.*

[1028] From the letters of Henry Fike in the Spencer Research Library, Kansas Collection, University of Kansas Library, Lawrence, Kansas

[1029] The 35th state, West Virginia, did not join the Union until June 20, 1863.

The regimental flag is a 35-star flag of New York manufacture. It is in very good condition and was probably obtained at the same time as the 35-star national flag.

The One Hundred and Seventeenth left Camp Butler with at least two Company flags—the flag of Company K and flag of Company I. A Company flag may have been used as the Regiment's color until the set of colors was secured in Memphis.

"This company met on Thursday afternoon, the 14th inst, in the School Commons, and was sworn in, and then proceeded to the election of its officers...As soon as the election was over, the company was marched and drawn up in front of the Schoolhouse, where the ladies of town presented them with a superb silk flag...The presentation was made by Miss Lizzie Spies, and the address delivered by Mrs. Phebe H. White, and responded to by Lieut. Fike."[1030] This flag for Company K was probably a national flag and not one of the flags shown above.. Its whereabouts today is unknown.

"The ladies of O'Fallon, Ridge Prairie and Shiloh presented a silk flag to the company. (Company I, ed.) The company left O'Fallon and marched 114 miles to Camp Butler arriving there on August 25, 1862".[1031] The flag is not described, and no further history of this flag is known.

No other flags carried by the 117th Regiment Illinois Infantry are known to survive.

[1030] *The Weekly Advocate* (Belleville), August 22, 1862. 1st. Lieutenant Henry C. Fike was from Mascoutah, Illinois, and a member of Company K.

[1031] O'Fallon Historical Society: Civil War' O'FallonCentennial Souvenir Program, Page1.

118th Illinois Infantry[1032]

The One Hundred and Eighteenth Regiment Illinois Infantry was organized in the summer of 1862. The men volunteered from the following counties, Company A from Fountain Green and Hancock Townships of Hancock County; Company B from Carthage and Prairie Townships of Hancock County; Company C from the town of Hamilton and surrounding Wythe and Montebello Townships in Hancock County; Company D from Adams County; Company E from Warsaw Township in Hancock County; Company F from Richfield Township in Adams County; Company G from Hancock and Henderson Counties; Company H from Hancock County; Company I from. Gallatin County and Company K from Mendon Township in Adams County. Major John G. Fonda[1033] of Hancock County was in command of Camp Butler in October 1862 and was elected Colonel of the One Hundred and Eighteenth Illinois Infantry. The Regiment was mustered into service at Camp Butler on November 7, 1862, and remained at the Camp training and guarding rebel prisoners until December 1 when ordered to Memphis, Tennessee. The Regiment arrived at Memphis December 11 and was assigned to the 13th Army Corps. On December 20 the Regiment moved downriver on the steamer *Northerner* as part of Sherman's Yazoo River Expedition. The steamer reached Milliken's Bend on the Mississippi River on December 25 and then proceeded up the Yazoo River to capture Vicksburg from the rear. The One Hundred and Eighteenth participated in the fighting at Chickasaw Bluffs, Mississippi from December 26-29 but when the Confederate lines could not be broken the Union forces, under heavy fire, retreated down the Yazoo River. January 1863, under command of General McClernand, the Regiment was ordered up the Arkansas River to Arkansas Post, Arkansas (Fort Hindman). In two days of fighting, January 10-11, the Fort was captured along with 6,000 prisoners. The Regiment returned to Young's Point, Louisiana and from January 23 until March 9 was involved in digging "Grant's Canal" that was being dug to bypass Vicksburg by water. With the failure of the canal the Regiment was ordered to Milliken's Bend, Louisiana and went into camp. On April 2, as part of General Osterhaus' Division, the One Hundred and Eighteenth departed south through Louisiana on an expedition against Vicksburg. After capture of Richmond, Louisiana the Regiment moved further south through New Carthage and then along the west side of the Mississippi River to Bruinsburg where the Regiment crossed the Mississippi River on the morning of April 30. The following day at the battle of Thompson's Hill[1034] the One Hundred and Eighteenth along with other regiments charged the enemy lines several times leading to rout of the enemy forces. Leading Osterhaus' Division the Regiment marched northeast and went into camp west of Raymond, Mississippi on May 13. On May 15 the Regiment was at Bolton on the *Vicksburg & Jackson Railroad* when word was received that Confederate forces were advancing from the west. On May 16 the Union and Confederate forces met at Champion's Hill, where after heavy fighting the Confederate forces were routed and retreated towards Vicksburg. The One Hundred and Eighteenth, in pursuit of the retreating rebels fought the next day at Black River Bridge.[1035] During the siege of Vicksburg, the Regiment was in the assaults on Vicksburg's lines on May 19 and May

---

[1032] This Regiment was also known as *Smith's Guerrillas*

[1033] Colonel Fonda was born on the Mohawk river in New York, March 8, 1822. He came to Hancock county Illinois with his parents in 1835 and began practicing surveying in 1840. In 1847 Fonda served in a company of Illinois Mounted Volunteers in the Mexican-American war. In 1861 he volunteered for the Uniion Armys serving first in the 2nd Illinois Cavalry, next in the 12th Illinois Cavalry and then in October 1862 he was made Colonel of the 118th Regiment of Illinois Infantry. Posted on Find A Grave, "Gen John Giles Fonda" information extracted from History of Hancock County, Illinois, together with an outine history of the State, and digest of State laws., Thomas Gregg, Chicago, C.C. Chapman, 1880

[1034] Also known as the Battle of Port Gibson.

[1035] Also known as the First Battle of Big Black River.

22 but on May 24 was ordered east to guard Black River Bridge from the troops of General Joseph Johnston.[1036] Following the fall of Vicksburg on July 4, the One Hundred and Eighteenth joined in pursuit of the retreating rebel army toward Jackson, Mississippi and fought in the Siege of Jackson July 10-17. The Regiment went onto camp at Vicksburg on July 25 and remained there until ordered to Carrollton, Louisiana.[1037] September 5, 1863, the Regiment crossed the Mississippi River and on September 16 arrived at Brashear City on Berwick Bay, Louisiana. In early October the Regiment was mounted and on October 12 began moving north to rejoin the 13th Army Corps at New Iberia. At Vermillionville, Louisiana on October 15 the Union forces skirmished with the rebel forces and later that day reached Carrion Crow Bayou[1038]. The remainder of October was spent scouting and skirmishing around Opelousas and Washington Louisiana. During November and December, the Regiment continued reconnaissance in western Louisiana around Vermillionville, New Iberia and Donaldsonville—arriving at Port Hudson, above Baton Rouge on the Mississippi River in early January 1864. The One Hundred and Eighteenth remained at Port Hudson until July. During that time the Regiment was involved in daily scouting and skirmishing and captured a significant number of prisoners. July 3 the Regiment was ordered to Baton Rouge and in an expedition to Clinton, Louisiana fought at Comite River on August 25. In September the Regiment marched to Donaldsonville, Louisiana and scouted and skirmished in the Amite River area. In November detachments from the Regiment were involved in raids on Liberty, Brookhaven and Summit, Mississippi and as far east as Augusta, Alabama. In December the Regiment was ordered to Baton Rouge for outpost duty until May 1865 and then provost duty until October 1865. The Regiment was mustered out at Baton Rouge on October 1, 1865.[1039, 1040, 1041]

The collection of the State of Illinois holds one national and one regimental flag of the One Hundred and Eighteenth Regiment Illinois Volunteer Infantry.

The 35-star national flag is of New York manufacture. Inscribed on the flag are, GRAND COTEAU; 118TH REGT. ILL. VOL. INF.; CHICKASAW BLUFF; BLACK BRIDGE; ARKANSAS POST; VICKSBURG; PORT GIBSON; JACKSON, CHAMPION HILLS.

1036 June 10 a Battalion of the Regiment was mounted and participated in a raid on Brookhaven, Mississippi in early July. The Battalion was dismounted in late July.

1037 Carrollton is now part of the residential area of New Orleans on the Mississippi River.

1038 Also known as Bayou Burbent. The fighting here was called the Battle of Grand Coteau.

1039 *Report of the Adjutant General of the State of Illinois,* H.W.Rokker, State Printer and Binder, Springfield, IL, (1886), Volume VI pages 315-8.

1040 Eddy, T. M., *The Patriotism of Illinois,* Clarke & Co., Chicago, IL (1866), Volume II, pages 655-62.

1041 Dyer, Frederick H., *A Compendium of the War of the Rebellion,* Thomas Yoseloff, New York, (1959), Volume III, 1096-7.

The regimental flag is also of New York manufacture and has 35 stars. There are no battle honors inscribed on this flag.

No other flags carried by the 118th Illinois Infantry are known to survive.

119th Illinois Infantry

The One Hundred and Nineteenth Illinois Infantry was organized at Quincy in Adams County during September 1862 and mustered in for three years October 10, 1862 under command of Colonel Thomas Jefferson Kinney.[1042] The men in the Regiment volunteered from the following areas, Company A from Ellington, Mendon and Quincy townships of Adams County; Company B from Browning and Rushville townships of Schuyler County; Company C from Rushville Township of Schuyler County; Company D from Mount Sterling, Cooperstown and Versailles townships of Brown County; Company E from Mound Station[1043]in Brown County; Company F from Camden and Huntsville townships of Schuyler County; Company G from Lima Township in Adams County; Company H from Bushnell, Macomb and Mound townships of McDonough County; Company I primarily from Clayton Township in Adams County with men also from Augusta Township in Hancock County and Company K from Pitman[1044] in Adams County and Augusta Township of Hancock County. On November 2, 1862, the Regiment departed camp for Jackson, Tennessee. On arrival the Regiment was ordered to various points on the *Mobile & Ohio Railroad* to guard against attack by the forces of Confederate General Forrest. On December 21 Forrest captured Company G at Rutherford Station and Company K at Dyer's Station. The remaining companies fell back to Union City, Tennessee and remained there until February 1863. In March the Regiment was ordered to Buntyn Station near Memphis to guard rail approaches and in May the One Hundred and Nineteenth moved into Memphis for guard and provost duty. In late January 1864 the Regiment moved down the Mississippi River to Vicksburg, Mississippi; from Vicksburg the Regiment marched with General Sherman's forces in the Meridian Mississippi Campaign. At Meridian on February 14-15 the rebel forces were routed, and the lines of the *Mobile & Ohio Railroad* destroyed. The Union troops returned to Vicksburg and remained there until March 4. On that date the Regiment moved down the Mississippi River to the Red River and disembarked at Simsport on the Atchafalaya River. With gunboat support the Union forces captured Fort DeRussy on March 14 and occupied Alexandria, Louisiana on March 16. The Union forces continued upriver toward Shreveport but were intercepted and defeated at Sabine Crossroads on April 8.[1045] The following day, the One Hundred and Nineteenth fought at Pleasant Hill,[1046] Louisiana and even though the Union forces had a tactical victory, the forces were demoralized and withdrew back down the Red River. In the retreat downriver the Regiment was engaged at Natchitoches, Cloutiersville, Alexandria, Bayou Lamourie and Yellow Bayou. The Expedition reached the Mississippi River on May 21 and boarded transports for the trip back to Vicksburg. From Vicksburg the Regiment moved further north and disembarked at Lake Chicot, Arkansas where on June 6 they routed forces of Confederate Brigadier General John Marmaduke. June 24 the One Hundred and Nineteenth moved by rail to Lagrange, Tennessee and on July 5 marched into Mississippi in Smith's Tupelo Expedition in pursuit of Forrest. Smith's forces met the Confederates at

---

[1042] Colonel Thomas Kinney was born in Madison County New York in 1828. He came to Illinois and was elected circuit clerk of Schyler County, Illinois. At Lincoln's call for volunteerrs, Kinney organized the 119th Illinois Infantry and became its colonel

[1043] Mound Station is now Timewell in Lee Township, Brown County.

[1044] Now known as La Prairie in Northeast Township of Adams County

[1045] The Battle of Sabine Crossroads is also known as the Battle of Mansfield. The One Hundred and Nineteenth Illinois Infantry was serving as rear guard and did not fight at Sabine Crossroads.

[1046] England born; Sgt. John H. Cook of Company A was awarded the Medal of Honor for his bravery at Pleasant Hill on April 9, 1864. His citation reads; "During an attack by the enemy, voluntarily left the brigade quartermaster, with whom he had been detailed as a clerk, rejoined his command, and, acting as first lieutenant, led the line farther toward the charging enemy."

Tupelo, Mississippi July 14-15[1047] and a piecemeal attack on the part of the Confederates allowed the Union forces to prevail. The Union forces did not follow up on the victory but instead returned to Memphis. On August 10 the Regiment was part of the raid to the Tallahatchie River, Mississippi by way of Holly Springs. No contact was made with Forrest's forces and the Regiment returned to Memphis. Because of threatening movements of Confederate forces under Price in Missouri, the One Hundred and Nineteenth was sent to St. Louis in September and on October 2 marched in pursuit of Price. However, after marching 700 miles through Missouri no contact was made with the enemy and the Regiment returned to St. Louis in late November 1864. In early December the Regiment boarded steamboats for the trip to Nashville. At Nashville, defenses were built in anticipation of attack by Hood and on December 15 Hood attacked. In two days of fighting the enemy was routed and the One Hundred and Nineteenth captured a battery of brass guns. Following the pursuit of Hood through Franklin and Columbia the Regiment spent Christmas and New Year's at Clifton, Tennessee on the Tennessee River. In January 1865 the Regiment was transported to Eastport, Mississippi and in February transported to New Orleans. In the final stages of the War the One Hundred and Nineteenth accompanied Union forces sent to Alabama. On March 12 the Regiment landed at Dauphin Island in Mobile Bay and participated in the siege and capture of Spanish Fort and Fort Blakely[1048,1049,1050] from March 26-April 8. The Regiment occupied Mobile on April 12 and then marched to Montgomery, Alabama. They returned to Mobile and were mustered out at Mobile on August 26, 1865.[1051,1052,1053]

In the collection of the State of Illinois Military Museum there are two national and three regimental flags that were carried by the One Hundred and Nineteenth Illinois Infantry.

[1047] Also known as the Battle of Harrisburg.

[1048] Captain Samuel McConnell Company H received the Medal of Honor for his actions during the battle of Fort Blakely on April 9, 1865. His citation reads: "While leading his company in an assault, Capt. McConnell braved an intense fire that mowed down his unit. Upon reaching the breastworks he found that he had only one member of his company with him, Pvt. Wagner. He was so close to an enemy gun that the blast knocked him down a ditch. Getting up, he entered the gun pit, the gun crew fleeing before him. About 30 paces away he saw a Confederate flag bearer and guard which he captured with the last shot in his pistol." Beyers, W. F., and Keydel, O. F., *Deeds of Valor,* Longmeadow Press, Stamford, Ct., (1993-reprint of 1903 edition), page 535-6

[1049] Pvt. John Whitmore Company F received the Medal of Honor for his deed at Fort Blakely on April 9, 1865. His citation simply states: "Capture of flag." *Civil War Medal of Honor Winners from Illinois,* Published by the Civil War Centennial Commission of Illinois, July 1962.

[1050] Sgt. George F. Rebbmann Company B received the Medal of Honor for bravery at Fort Blakely on April 9, 1865. His citation reads: "Capture of flag." Ibid, *Civil War Medal of Honor Winners from Illinois.* The flag captured was that of the 2nd Alabama Infantry.

[1051] *Report of the Adjutant General of the State of Illinois,* H.W. Rokker, State Printer and Binder, Springfield, IL, (1886), Volume VI, pages 339-43.

[1052] Dyer, Frederick H., *A Compendium of the War of the Rebellion,* Thomas Yoseloff, New York, (1959), Volume III, 1097.

[1053] Eddy, T. M., *The Patriotism of Illinois,* Clarke & Co., Chicago, IL (1866), Volume II, pages 480-1

This 35-star national flag does not have any inscriptions. The square canton and stars arranged in five horizontal rows is typical of flags manufactured for the New York depot.

This 35-star national flag is similar to the flag above and also lacks inscriptions.

Due to damage the number of stars on this regimental flag is uncertain. However, the flag is typical of regimental flags manufactured for the New York depot.

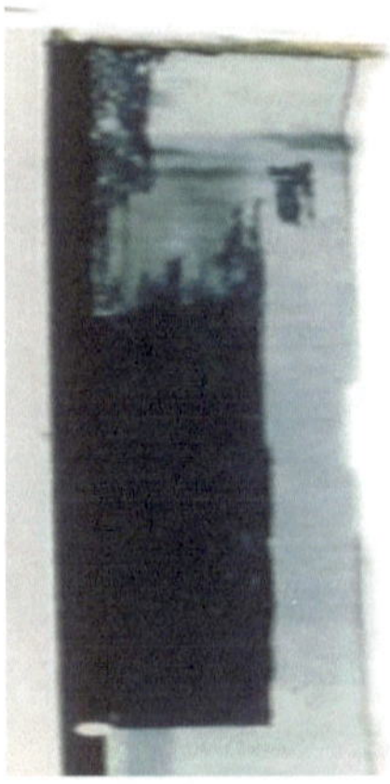

Only a few shreds remain of this regimental flag.

This regimental flag is also of New York manufacture but due to damage the number of stars is indeterminate.

"A Relic of The War. On Tuesday last John L. Sweeney received a relic by mail that will be of interest to the veteran soldiers who were members of Co. B 119th Illinois Infantry. It was the flag made and presented by you ladies of Rushville to that company just previous to its departure for the seat of war in 1862, an incident that will be remembered by many of our citizens. The flag was sent to Mr. Sweeney by Geo. Owens, who was a member of the company, now a resident of Clayton. He gave no information in regard to how he happened to have it in his possession but simply says he had neglected forwarding it sooner. On the field worked in white thread, are the initials of the ladies who donated the flag to Co. B. Of course, it is badly used but it is a relic worth keeping." [1054] George Owens was 1st Lieutenant in Company B and apparently kept the flag after the end of the war. The flag is not described and there is no further history of this flag.

No other flags carried by the 119th Illinois Infantry are known to survive.

[1054] A copy of an article originally published in the Rushville Times. *Nauvoo Rustler, December 20, 1892 , Page 1, Nauvoo, Illinois.*

120th Illinois Infantry

The One Hundred and Twentieth Regiment Illinois Infantry was organized in southern Illinois during the summer of 1862. LTC John G. Hardy from Vienna in Johnson County recruited companies A, B, C, E, G, I and K for the Regiment and went into camp at Vienna on August 13. Colonel George W. McKeaig, lawyer from Shawneetown in Gallatin County recruited companies D, F and H and went into camp at Shawneetown on August 15. In September both camps were ordered to Camp Butler and all the companies were consolidated as the One Hundred and Twentieth Regiment Illinois Infantry. The men in the Regiment volunteered from the following counties, Company A from Pope County; Company B from Johnson and Massac County; Company C from Johnson County; Company D from Gallatin County; Company E from Pope County; Company F from Saline County; Company G from Johnson and Williamson Counties; Company H from Gallatin and White Counties and Company I and K from Johnson County. The Regiment was mustered in for three years' service on October 28, 1862, under command of Colonel McKeaig[1055] and immediately was ordered to guard the *Great Western Railroad* bridge over the Sangamon River at "Jimtown."[1056] November 9 the Regiment moved by rail to Alton, Illinois and then boarded the steamship *Stephen Decatur* for the trip downriver to Memphis. In late November the Regiment was assigned garrison duty at Fort Pickering adjacent to Memphis.[1057] In mid-January 1863 the One Hundred and Twentieth was part of an expedition to Arkansas that routed the enemy forces at Marion and then returned to garrison duty at Memphis. In May the Regiment was ordered downriver to Vicksburg and landed at Young's Point Louisiana upriver from Vicksburg. A week later the Regiment marched south to Hard Times Landing, crossed the Mississippi River at Grand Gulf and marched north to Warrenton, Mississippi. On June 7 the Regiment was ordered upriver to Milliken's Bend to rescue a detachment of Union soldiers who were being attacked by Price. On June 13 the Regiment was sent further upriver to Lake Providence, Louisiana to protect Union interests from attacks by Price's army. The Regiment remained at Lake Providence until July 28 when ordered back to Memphis. The following day the Regiment rode the *Memphis and Charleston Railroad* to Lagrange, Tennessee, arriving there on August 2. But when General Forrest raided Memphis the One Hundred and Twentieth was immediately recalled to the defense of the city. The Regiment was again sent to Lagrange in October and from there ordered on a ten-day scout into Mississippi. Rebel forces were found at Cold Water and Davis' Mills but were routed by the Union cavalry before infantry forces arrived. The One Hundred and Twentieth returned to Memphis but left again on October 30, ordered to guard a wagon train to Corinth, Mississippi. The Regiment arrived without incident at Corinth on November 2 and returned by rail to Memphis. Five days later the Regiment was ordered back to Corinth for guard duties and remained there until the town was evacuated by the Union forces on January 25, 1864. The One Hundred and Twentieth was again placed on provost duty at Memphis until May 30 when ordered to join Sturgis' Expedition to Guntown, Mississippi. Sherman had ordered Sturgis to find and destroy the

---

[1055] Colonel George McKeaig was born in Jefferson County, Kentucky in May 1824. During the Mexian-American war McKeaig served in the First Kentucky Volunteers; he studied law was admitted to the Illinois bar. When he moved to Shawneetown he began his law practice. In October, 1862 he mustered in to the Union Army as Colonel of the 120th Illinois Infantry.

[1056] "Jimtown" was Jamestown and lay only a few miles north of Camp Butler. Today the town is known as Riverton. The bridge was a covered wooden bridge.

[1057] During November 1862 the men of the Regiment were stricken with small pox, measles, and pneumonia, losing 265 of 844 men. One hundred and forty-eight men were sent home with disabilities; thus 413 men out of 844 or nearly one half of the Regiment was either discharged or in the hospital.

forces of Nathan Bedford Forrest and Sturgis met Forrest's forces at Brice's Crossroads, Mississippi[1058] on June 10, 1864. Forrest's outnumbered forces easily defeated Sturgis' cavalry before the infantry could be brought up in support.[1059] Forrest attacked the retreating Union forces the following day at Ripley, Mississippi and forced the retreat of Sturgis' troops back to Memphis. The One Hundred and Twentieth aided in the repulse of Forrest's attack on Memphis on August 21 and then moved in September to Paducah, Kentucky. From Paducah the Regiment moved south through Clifton, Tennessee and then up the Tennessee River to Florence, Alabama. After skirmishing with rebel forces on October 8, the Regiment moved downriver to Pittsburg Landing, Tennessee. October 11 the Regiment moved on to Johnsonville Tennessee where the Regiment was ordered to Memphis. The One Hundred and Twentieth remained on Provost duty at Memphis until mustered out on September 7, 1865.[1060,1061]

The collection of the State of Illinois has one National and one Regimental flag carried by the One Hundred and Twentieth Illinois Volunteer Infantry Regiment.

This 35-star national flag is devoid of inscriptions. The star pattern is typical of flags produces by Evans & Hassall and by Horstmann Brothers & Company for the Philadelphia depot.

[1058] Also known as the Battle of Tishomingo Creek.
[1059] Colonel George McKeaig was shot in the breast and arm and captured by Colonel Jesse Forrest, a younger brother of Nathan Bedford Forrest.
[1060] Blackman, William S., *The Boy of Battle Ford and the Man,* The Egyptian Press Printing Co., Marion, Ill., (1906). The author was a Sergeant in Company F.
[1061] *Report of the Adjutant General of the State of Illinois,* H. W. Rokker, State Printer and Binder, Springfield, Ill., (1886), volume VI, pp 367-9.

The 34-star regimental flag is typical of flags manufactured for the New York depot.

No other flags carried by the 120th Illinois are known to survive.

121st Illinois Infantry

The One Hundred and First Illinois Infantry was never organized.

122nd Illinois Infantry

The One Hundred and Twenty Second Regiment Illinois Infantry was organized by Colonel John I. Rinaker[1062] at Camp Palmer, in Carlinville, Macoupin County, Illinois in the summer of 1862. The Regiment was mustered in for three years' service on September 4, 1862. The men of companies A, B, D, E, G, H, I and K volunteered from Macoupin County, the men for Company C volunteered from Greene County and the men for Company F volunteered from Macoupin and Montgomery Counties. On October 8 the Regiment boarded the *Chicago, Alton & St. Louis Railroad* for the trip to Alton, Illinois where they boarded the steamship *John Warner* for the trip downriver to Columbus, Kentucky. At Columbus the Regiment moved via the *Mobile & Ohio Railroad* to Trenton, Tennessee. November 10 the Regiment was divided—the right wing, Companies A, F, D, I and C going 12 miles south to Humboldt, Tennessee to guard the railroad and bridges and the left wing, Companies B, E, G, H and K remaining at Trenton. December 18 the Regiment was ordered to Jackson, Tennessee in anticipation of possible attack by forces of Nathan Bedford Forrest. The Regiment arrived at Jackson only to learn that Forrest had captured Humboldt and Trenton along with 60 enlisted men and "our Company flag."[1063] The Union forces returned to Trenton and Humboldt but not before Forrest had destroyed miles of rails. December 27 the One Hundred and Twenty Second joined in the pursuit of Forrest and on December 31 met Forrest at Parker's Crossroads[1064] or Red Mound north of Lexington, Tennessee. Although Forrest was surrounded and nearly captured, he fought his way out and escaped. In January 1863 the Regiment returned to Trenton and assisted in repair of the railroad and bridges. February 17 the Regiment was ordered to Corinth, Mississippi. While on duty at Corinth the Regiment was part of Dodge's Expedition to Tuscumbia, Alabama and back; during the march the Union forces skirmished at Rock Creek April 22 and at Town Creek on April 28. The One Hundred and Twenty Second remained on guard duty at Corinth until June 4, when ordered to Grand Junction, Tennessee to guard the *Memphis and Charleston Railroad* from Grand Junction to Middletown, Tennessee.[1065] On October 31 the Regiment marched east to Iuka, Mississippi as part of Dodge's Division of Sherman's army (16th Corps) and on November 4 marched to Eastport, Mississippi for guard duty. December 6 the Regiment boarded the steamboats *Sunny South* and *Mattie Cook* for the trip down the Tennessee River to Paducah, Kentucky for guard and provost duty. Leaving Companies C, H and K at Paducah, on January 19, 1864, the remainder of the One Hundred and Twenty Second boarded the *Rob Roy* for the trip to Cairo, Illinois. While stationed at Cairo, on March 16 a detail of fifty men from the Regiment were sent to Centralia, Illinois to suppress a strike by engineers of the *Illinois Central Railroad.*[1066] Forrest attacked Paducah in late March and Companies E, H and K were sent to Paducah to aid in the defense of the town. Having been relieved by the One Hundred and Thirty Ninth Illinois Infantry, on June 27 the One Hundred and Twenty Second boarded the steamship *Magenta*, moved downriver to Memphis, and then marched to Lagrange,

---

[1062] Colonel John Rinaker was born in Baltimore, Maryland in November 1830. He came to Sangamon County in 1837. When he was nineteen, he attended Illinois College and in 1850 attended McKendree College. He studied law in Carlinville, Illinois and was admitted to the Illinois bar in 1854. In 1862 Rinaker raised a regiment known as the 122nd Illinois Infantry and he was made colonel of the regiment. *Journal of the Illinois State Historical Society,* Vol 7. No. 4 (Jan. 1915). Page 417

[1063] "Reminiscences by Aaron L. Lanning 1845-1934", From the diary of Aaron L. Lanning, 122nd Infantry Regiment, Company D, contributed by Bobbie Athon, as reproduced on the Macoupin County ILL Gen Web, page 7

[1064] The Regiment lost 57 wounded and 19 killed.

[1065] Regimental Headquarters and Companies A, F, E and I were at Salisbury, Companies K and G were at Curve Station, Companies C and D at Porter's Creek and Companies B and H at Middletown.

[1066] Ibid, Lanning, page 10.

Tennessee. On July 5 the regiment was ordered south as part of Smith's Expedition to Tupelo, Mississippi. Forrest suffered his worst defeat at the Battle of Tupelo[1067] on July 14, 1864; however, Union General Andrew Smith failed to follow up on the victory and returned to Memphis. Under Smith's command, the Regiment marched from Memphis again on August 4 for Holly Springs, Mississippi and then on to Oxford, Mississippi. When the Expedition reached Oxford, the enemy was no longer there, and Expedition marched back to Memphis. The Regiment was ordered to Jefferson Barracks, Missouri in early September and then marched to Ironton, Missouri in pursuit of rebel forces of General Sterling Price. Price eluded the Union forces over the next two months and the Regiment returned to St. Louis in November. The One Hundred and Twenty Second was ordered to Nashville and took part in battle of Nashville on December 15 and 16[1068]; the following day the Regiment joined in the pursuit of the retreating rebel army as far as Eastport, Mississippi where the Regiment remained until February 1865. In February the Regiment was ordered to New Orleans and then on to Dauphin Island at the mouth of Mobile Bay, Alabama. In the campaign against Mobile the Regiment fought at Spanish Fort and Fort Blakely,[1069] March 26-April 9. Following the fall of Mobile the Regiment marched to Montgomery, Alabama and remained on duty there until June 5 and then returned to Mobile where the Regiment was mustered out July 15, 1865.[1070,1071,1072]

There are two flags—one national and one regimental—carried by the One Hundred and Twenty Second Regiment Illinois Volunteer Infantry in the State of Illinois Military Museum.

The National color of the Regiment is not inscribed. The shape of the canton and the star pattern is typical of flags manufactured in Philadelphia by Evans & Hassall and by Horstmann Brothers & Company.

[1067] The One Hundred and Twenty Second lost 10 killed and 33 wounded.

[1068] England born George Stokes, Corporal Company C was awarded the Medal of Honor for "Capture of Flag, Nashville, Tennessee, December 16, 1864." *Civil War Medal of Honor Winners from Illinois,* Civil War Centennial Commission, July 1962.

[1069] John H. Calahan, Private Company B received the Medal of Honor for "Capture of flag, Fort Blakely, Alabama, April 9, 1865." *Civil War Medal of Honor Winners from Illinois,* Civil War Centennial Commission, July 1962.

[1070] *Report of the Adjutant General of the State of Illinois,* H. W. Rokker, State Printer and Binder, Springfield, IL, (1886), Volume VI, pp 391-4

[1071] Ibid, Lanning pp 5-16

[1072] Ibbetson, William H., *William H. H. Ibbetson, Co. D, 122d Reg.Ill.,* (Diary), Transactions of the Illinois State Historical Society, No. 37 (1939) pp 236-73.

The Regimental flag is typical of those manufactured by Horstmann Bros & Co. of Philadelphia.

A flag was presented to Company D prior to leaving Camp Palmer. "Some of the friends of the boys visited the camp at times, and once there was a flag presented to Company D."[1073] There is no description of the flag in Lanning's diary and the flag was captured and destroyed by Forrest in December 1862.

No other flags carried by the 122nd Illinois Infantry are known to survive.

[1073]Ibid, Lanning page 6. This apparently refers to the flag presented to Company D and not the Regimental colors.

123rd Illinois Infantry

"Brave Monroe gave his last command: 'Charge, boys, right ahead,'
When a traitor's ball hit him, and left him numbered with the dead.
Beloved by all was our dear old Col. for he was true and brave,
We sent him back to Mattoon where he was laid in the grave."[1074]

James Monroe[1075] was Major in the Seventh Illinois Infantry but became a Colonel when the One Hundred and Twenty Third Regiment Illinois Infantry was organized at Mattoon in Coles County, Illinois. The Regiment was mustered in for three years' service at Camp Terry, Mattoon on September 6, 1862. The men of Company A volunteered from Coles County; Company B from Cumberland County; Company C from Coles County; Company D from Coles and Shelby Counties; Company E from Putnam, Clark and Cumberland Counties; Company F and G from Clark County and Company H, I and K from Coles County. After two weeks of training, the Regiment boarded freight cars of the *Illinois Central Railroad* and moved to Louisville, Kentucky to help fortify the city against possible attack by Confederate General Bragg. On October 1, as part of McCook's Corps, the One Hundred and Twenty Third marched south in pursuit of Bragg and on October 8 fought Bragg's rebel forces at the Battle of Perryville.[1076] Bragg did not press the battle the following day but retreated through the Cumberland Gap while the Union forces marched to Munfordville, Kentucky where they remained until late November. The Regiment marched to Bledsoe's Creek east of Gallatin, Tennessee in December and then pursued General John Morgan's forces north as far as Bear Wallow,[1077] Kentucky. In early January 1863, the Regiment abandoned the pursuit of Morgan and instead marched south through Bowling Green, Kentucky and Nashville, Tennessee to Murfreesboro, Tennessee. At Murfreesboro, the One Hundred and Twenty Third skirmished at Woodbury, Tennessee on January 24, scouted around Auburntown, Liberty and Alexandria, Tennessee February 3-5 and then skirmished at Cainsville, Tennessee. While on reconnaissance in March the Regiment was attacked by rebel cavalry at Breed's Hill, but the Regiment repelled the attack and retreated to Murfreesboro. On March 20 the One Hundred and Twenty Third was part of a Brigade that was attacked and surrounded by Morgan's Cavalry near Milton, Tennessee. Morgan was wounded and the enemy driven from the field; the Regiment returned to Murfreesboro. May 6 the Regiment was transferred to Wilder's Lightning Brigade[1078] and was mounted and furnished Spencer repeating rifles. June 24 the Brigade as part of Sherman's Tullahoma (Tennessee) Campaign rode south out of Murfreesboro and seized Hoover's Gap; despite attacks by a superior number of enemy forces the Brigade held the Gap until re-enforcements arrived that evening. As the

---

[1074] Portion of a poem relating the death of Colonel Monroe, commander of the regiment. "My Experience while in the Army a Member of Co. E, 123rd Ill. Mt. Vol. Inft'y., Wilder's Brigade" by Jerome B. Quinn, Cumberland County.

[1075] Colonel Monroe was born in Green County Kentucky in 1832. He moved with his family to Charleston Illinois in 1835 and is listed as a merchant. Monroe was commissioed Captain of Commpany B 7th Illinois Infantry, then Major of the Regiment on March 21, 1862. He resigned on September 3, 1862 and was commissioned Colonel of the 123rd Illinois Infantry three days later.

[1076] The Regiment lost 36 men killed and 180 men wounded.

[1077] Bear Wallow is just east of present-day Cave City, Kentucky.

[1078] The Wilder Lightning Brigade consisted of the 17th Indiana Mounted Infantry, 72nd Indiana Mounted Infantry, 92nd Illinois Mounted Infantry (attached to Wilder's Brigade July 10, 1863), 98th Illinois Mounted Infantry, 123rd Illinois Mounted Infantry and the 18th Indiana Battery (Eli Lily's Battery).

Union troops advanced the Confederate forces abandoned Tullahoma and Wilder's Brigade destroyed the tracks, bridges, and depots of the *Nashville & Chattanooga Railroad.* Early July found the One Hundred and Twenty Third at Rowesville, Tennessee on the Duck River; during the next two months the Regiment foraged and captured horses and contraband in middle Tennessee from Manchester to Columbia to Pulaski. In mid-August the Regiment crossed the Cumberland and on August 17 a detachment from the Regiment rode into Chattanooga and secured the city until the remainder of the Brigade arrived. The Regiment crossed the Tennessee River on September 9 and led the advance to Tunnel Hill, Georgia. By September 18 the Lightning Brigade was in camp at Alexander's Bridge on Chickamauga Creek having skirmished continually en-route to the camp. In the next three days of heavy fighting, September 19-21, at the Battle of Chickamauga the One Hundred and Twenty Third was heavily involved as part of Wilder's Brigade.[1079] Despite valiant fighting the Union troops were forced to withdraw and the One Hundred and Twenty Third retreated to Friar's Island upstream from Chattanooga in the Tennessee River. During October the Regiment saw action at Hill's Gap, McMinnville, Farmington,[1080] and Sim's Farm near Shelbyville, Tennessee and then moved to camp at Maysville, Alabama. For two months the Regiment patrolled and foraged in the surrounding country and then moved to Pulaski, Tennessee in late December and to Mooresville, Alabama in January 1864. The Regiment remained in northern Alabama, on the Tennessee River until April when ordered to join Sherman's Atlanta campaign at Dalton, Georgia. During May the Regiment fought at Resaca and near Dallas, New Hope Church and Allatoona Hills. In June and July, they fought at Big Shanty, Noonday Creek, Kennesaw Mountain, Chattahoochee River, Stone Mountain Station and rode in a raid to Oxford and Covington. In the Siege of Atlanta, the Regiment primarily saw action at the Chattahoochee River Bridge and at Pace's and Turner's Ferries; following the fall of Atlanta, they rode in pursuit of Hood into north Georgia and Alabama. On November 1, 1864, the Regiment was dismounted and ordered to Louisville, Kentucky. While at Louisville, the Regiment was remounted and left Louisville December 28 for a raid through Nashville to Gravelly Springs, Alabama. In March 1865, the One Hundred and Twenty Third as part of the Cavalry Corps Military Division of the Mississippi, rode in General James Wilson's expedition through Alabama and Georgia. The raid resulted in the defeat of Nathan Bedford Forrest and the capture of Selma, Montgomery, Columbus, and Macon. On May 23, 1865, the Regiment was ordered to Chattanooga and from there to Nashville. On June 27 the Regiment was mustered out and the recruits with less than three years' service transferred to the Sixty First Regiment Illinois Infantry.[1081,1082]

There are two national flags carried by the One Hundred and Twenty Third Regiment in the collection of the State of Illinois. A photo was taken of the two flags when they were presented to the Adjutant General of Illinois in 1884. Accompanying the photo is a handwritten note by James A. Connolly, "Late Major of the 123rd Ills" in which he states that the first national flag shown below "is the original Flag of the Regiment, carried in the Battle of Perryville, Ky., and numerous other battles, until it was too badly shattered to be unfurled."

[1079] Baumgartner, Richard A., *Blue Lightning, Wilder's Mounted Infantry Brigade in the Battle of Chickamauga,* Blue Acorn Press, Huntington, WV, (1997) This book gives a detailed history of the involvement of the One Hundred and Twenty Third at Chickamauga.

[1080] At Farmington, Tennessee on October 7, in a battle with Confederate General Joseph Wheeler and Colonel James Monroe, the organizer and commander of the One Hundred and Twenty Third was killed.

[1081] *Report of the Adjutant General of the State of Illinois,* H.W.Rokker, State Printer and Binder (1886), Volume VI, p p416-421.

[1082] Connolly, Major James A., *Three Years in the Army of the Cumberland*, Indiana University Press, Bloomington, (1959) When Lincoln was assassinated, Connolly was one of the men assigned to escort Lincoln's body back to Springfield.

The flag is inscribed: PRESENTED BY THE LADIES OF CHARLESTON 123 INF REGIMENT. Charleston is the county seat of Coles County and Mattoon is a "sister" town. In his letters Connolly is presumably referring to this flag when he writes, "on the 27th the ladies of Mattoon, Ill., presented a fine silk regimental flag to my regiment and I had the honor of being selected to receive the flag;"[1083]

The second national flag of the Regiment is a flag of Philadelphia manufacture and does not have any inscriptions. Of this flag Connolly states, "The Flag...was procured and carried to the close of the war. Its staff is almost broken off by bullets."

There is a remnant of a flag of the 123rd Illinois Infantry at the Utah State Historical Society Museum in Salt Lake City Utah. John Morgan of Mattoon enlisted in Company I of the 123rd. At the Battle of Selma, Alabama (April 2,1865) "when an open attack was ordered against Selma's strong fortifications, Sergeant Morgan, like Sergeant Jasper, scaled the earth works and brought back the colors in triumph, seizing them from the hand of a dying color sergeant. The battle flag was shot to ribbons, and that evening at the conclusion of the battle, John Hamilton Morgan was presented by his commanding officer with the remnants of the tattered ensign as a memento of his heroism under fire. This treasured trophy is now in the possession of the Morgan family"[1084]

After the war John Morgan moved to Salt Lake City Utah. It is thought that the Morgan family donated the flag to the Museum in Utah after his death. [1085]

[1083] Connolly, Major James A., *Three Years in the Army of the Cumberland, the Letters and Diary of Major James A. Connolly,* Indiana University Press, Bloomington, Indiana (1959), p14.

[1084] "John Morgan" Utah, a centennial history,Lewis Historical pub. Co. January 1, 1949, page 478

124th Illinois Infantry[1086] -- "*Excelsior Regiment*"

The One Hundred and Twenty Fourth Regiment Illinois Infantry was organized at Camp Butler, near Springfield, Illinois in the summer of 1862 and mustered in for three years' service on September 10, 1862 under the command of Colonel Thomas J. Sloan.[1087] The men volunteered as follows: Company A from Henry County; Company B from Lodi[1088] and Batavia in Kane County; Company C[1089] from Sangamon and Jersey Counties; Company D (*Stanton Rifles)* from McDonough County; Company E and H from Aurora in Kane County; Company F volunteered from the area of Kewanee in Henry County,; Company G from Mercer County; Company I from McDonough, Adams and Shelby Counties and Company K from Sangamon and Wayne Counties. On October 6 the Regiment departed Camp Butler on the *Great Western Railroad* passed through Cairo, Illinois on the *Illinois Central Railroad* and arrived at Jackson, Tennessee on October 9. On November 2 the Regiment moved on the *Memphis & Ohio Railroad* to Bolivar, Tennessee as part of Grant's Central Mississippi Campaign. The men marched south into Mississippi as far as the Yacona River south of Oxford when on December 22, Grant learned that rebel forces had captured Holly Springs, Mississippi and destroyed the Union line of supply. The Union forces retreated to Memphis, Tennessee arriving there January 21, 1863. On February 20 the Regiment boarded the steamer *Platte Valley* for the trip downriver to Providence, Louisiana where a canal was being cut from the Mississippi River to Lake Providence to flank Vicksburg via the Mason and Tensas bayous. The Regiment boarded the steamer *Henry Von Phul* on March 20, sailed upriver and debarked at Berry's Landing; after a month in camp the Regiment sailed on the *Sioux City* downriver to a camp two miles above Milliken's Bend, Louisiana. The Regiment, as part of the 17th Army Corps, left Milliken's Bend on April 25 and moved south through Louisiana to cross the Mississippi river on the gunboat *Mound City* and land at Bruinsburg, Mississippi. The Regiment was engaged on May 1 at the Battle of Thompson's Hill[1090] where "we first smelt the powder in the presence of the foe."[1091] In the march toward Vicksburg the One Hundred and Twenty Fourth fought at Raymond on May 12, Jackson on May 14 and at Champion's Hill on May 16. The Regiment crossed the Big Black River on May 19 and joined the ill-fated assault on Vicksburg on May 22. During the siege of Vicksburg, Fort Hill was mined and blown up on June 25, in the subsequent "Slaughter Pen" Union troops rushed into the breach created by the explosion and the One Hundred and Twenty Fourth lost 49 men killed and wounded. Following the fall of Vicksburg on July 4, the Regiment occupied Vicksburg until November 1863 and during that time joined in expeditions to Monroe, Louisiana in August and to Canton, Mississippi in October. In early November the Regiment marched 14 miles east from Vicksburg to the Big Black River.[1092,1093] In early January 1864 the Regiment moved to Sunnyside, Arkansas and in February marched in Sherman's Meridian campaign designed to destroy the *Mobile & Ohio RR* and the *Southern Railroad of Mississippi* where the railroads met at Meridian, Mississippi. The Regiment

---

[1086] Also known as the *Excelsior Regiment, Sloan's Commercial Regiment* and *The Hundred and Two Dozen*

[1087] Colonel Sloan was a 26 year old phsician.He was relieved of command December 15,1863 forlack of leadership and respect to his officers.

[1088] Lodi or Lodi Station is now known as Maple Park

[1089] Company C was organized at Carpenter's Hall in Springfield and was known as the "*Springfield Company*."

[1090] Also known as the Battle of Port Gibson.

[1091] Howard, Richard L., *History of the 124th Regiment Illinois Infantry Volunteers, otherwise known as the "Hundred and Two Dozen,"* H.W. Rokker, Springfield, Ill, (1880), page 79.

[1092] On January 15, 1864, the commander of the Regiment, Colonel Thomas Sloan was dismissed from service following court martial. Judge Advocate General Holt reversed the order of dismissal, but Sloan was never reinstated.

[1093] On January 23, 1864, the Regiment won the prize banner "Excelsior" in a drill competition

reached Meridian on February 14, completed their work, returned to Vicksburg, and was placed on provost duty until February 1865. During the stay at Vicksburg the Regiment participated in expeditions to Benton and Yazoo City, Mississippi in May, to Pearl River, Mississippi in July and to Memphis in October 1864. On February 25, 1865, the Regiment boarded the *Grey Eagle* to sail downriver to New Orleans; March 12 the Regiment departed New Orleans on the ocean steamer *Guiding Star* and on March 16 debarked at Fort Gaines on Dauphine Island, Alabama. The Regiment boarded the *Peerless* and the *Thomas* on March 20 to sail across Mobile Bay and up the Fish River and debarked at Danby's Mills. The One Hundred and Twenty Fourth fought in the siege of Spanish Fort and Fort Blakely in March and April and at the capture of Mobile on April 12. In mid-April the Regiment was ordered to Montgomery for provost duty and in July to Vicksburg and then on to Chicago where the Regiment was mustered out at Camp Douglas on August 16, 1865.[1094,1095,1096]

There is one National and one Regimental flag from the One Hundred and Twenty Fourth Regiment Illinois Infantry in the military flag collection of the State of Illinois.

This is a "homemade" national flag. It is of unusual proportions and has the stars arranged in a star-shaped pattern. This is presumably the flag presented to Company C, "Saturday, September 6th (1862 ed.), this company was presented with a very fine silk flag, of which it was justly proud, and carried it as the flag of the regiment for many a weary mile. The money, over sixty dollars, for the purchase of this beautiful banner, was solicited by two friends of the company and cause, Misses Amelia Lloyd and Mary Devore."[1097]

Since this is the only National flag of the 124th Illinois in the Illinois collection, the following must also apply to this flag. "On the 10th of May (1865 ed.) our old, battle-scarred flag was inspected by Captain Scott, A.A.I. General, 3rd Division, 16th Army Corps, by order of Maj. Gen A.J. Smith, and ordered to be dropped from our returns, and deposited at the capital of Illinois. The following is its war history, deposited with it at Springfield:

"HEADQUARTEERS 124TH REG. ILL. INF'TY MONTGOMERY, ALA., May 10th, 1865,

---

1094 Ibid, Howard

1095 *Report of the Adjutant General of the State of Illinois,* H.W. Rokker, State Printer and Binder, Springfield, ILL., (1886), Vol. VI, pp446-8.

1096 Beck, Stephen C., A *True Sketch of His Army Life*, Edgar, Nebraska (1914)

1097 Ibid, Howard, page 5.

These colors were borne by the 124th regiment on the march of Gen. Grant's grand army from Lagrange, Tennessee, to the Yacona, Mississippi, and return to Lagrange, Tenn., in December, 1862.

"Also from Lagrange to Memphis, Tenn., in January, 1863; from Memphis to Lake Providence, La., in February, 1863, and from thence via Milliken's Bend and Bruinsburg to Port Gibson, Miss., in April 1863.

"They were carried in and through the following engagements: PORT GIBSON, May 1st, 1863; RAYMOND, May 12th, 1863; JACKSON, May 14, 1863; CHAMPION HILLS, May 16, 1863; SIEGE OF VICKSBURG, including the assault on the enemy's works May 22nd, and FORT HILL, June 26th, 1863.

"They were also borne on the Monroe, La., and Brownville, Miss., campaigns in August and October 1863, and in the great Meridian expedition under Gen. Sherman in February, 1864.

"They were also carried by the regiment on the occasion of the Champion Drill of the 3d (Logan's) Division, 17th Army Corps, when the 124th regiment won the Prize Banner, which was presented to it by Maj. Gen. McPherson, January 23rd, 1864, as the Excelsior Regiment of the 3d Division, 17th Army Corps. "During all this period the regiment constituted a portion of the 3d (Logan's) Division, 17th Army Corps.

"These colors were also borne by the regiment in the campaign against Yazoo City, and in the battle of BENTON, MISS. in May, 1864, under Gen McArthur.

"Also in the Jackson, Mississippi campaign, and in the battle of JACKSON CROSS ROADS, in July 1864, under Gen. Slocum, and in the White river and Memphis expedition in October, 1864.

'They were also borne by the regiment (though too much torn and shattered for actual use) in the great campaign against Mobile and Montgomery, Alabama, in March and April, 1865 under Gen.Canby, including the siege of SPANISH FORT, and the storming and capture of the enemy's works, April 8th, 1865, by the 3d Brigade, 3d Division, 16th Army Corps, of which the 124th regiment was a part.

"During the period including the foregoing operations, the regiment marched, by land and water, over 4,100 miles, and was engaged in 14 skirmishes, ten battles, and two sieges, of forty-seven days and nights, the thirteen days and nights, respectively, thus being under the fire of the enemy eighty-two days and sixty nights.

"J.J. HOWE

COL. By Brevet, Commanding Regiment"[1098]

[1098] Ibid, Howard, pp 347-9.

The Regimental flag is a 34-star flag and is in very good condition. It is not inscribed with battle honors.

In January 1864 the Regiment won the "Excelsior" banner as the best-drilled and finest Regiment in the Third Division, 17th Army Corps.[1099] However on April 5, 1864, when the Regiment, due to reorganization, was moved to the 19th Army Corps, the flag was returned to the Third Division of the 17th Corps. During the Atlanta Campaign, General Leggett's tent, uniform, and the "Excelsior flag" were captured and the flag was never recovered.[1100]

No other flags carried by the 124th are known to survive.

[1099] Ibid, Howard, p 167-77
[1100] Ibid, Howard, p 215

125th Illinois Infantry

"Our line of battle was soon formed and there we stood waiting for the signal to advance...Colonel McCook at the head of the brigade waved his sword and gave the command: 'Attention battalions, charge bayonets,' and with a rush and a cheer away we go. And now the battle commences. We have reached the wheat field, and at yon side are the rebel breast works...men are falling on all sides; there goes McCook down—quickly following him, Harmon, who was bravely urging his men on, falls, pierced through the heart. See! the colors have disappeared, but only for a moment when again they wave; the color sergeant had been shot down... It is too much, the works cannot be carried by assault, and our line, mangled torn and bleeding, falls back...Still the bullets, and the shells, and the solid shot fly, and still more brave boys are sealing with their life's blood their devotion to their country's flag. But why prolong the tale, the ground is soaked with blood; but with that love for the old flag which has floated so gallantly at our head over so many bloody fields, and under whose stars and stripes the weary and oppressed of every land have found a shelter, under whose protecting folds we have lived in peace and security, until driven by its enemies to war; with that love kindling in our breasts we stand ready to die, if need be, but never to dishonor its beauty and its grandeur."[1101]

A young Danville attorney, Oscar F. Harmon[1102] was the organizer and first Colonel of the One Hundred and Twenty Fifth Regiment Illinois Infantry. The men of Companies A, B, C, D, G, I and K volunteered from Vermilion County and the men of Companies E, F and H volunteered from Champaign County. The Regiment was organized and went into camp at the Danville fairgrounds, Vermilion County in the summer of 1862 and was mustered in for three years' service on September 3, 1862. Several days later the Regiment boarded open cattle cars of the *Indianapolis, Bloomington & Western RR* for the trip to Cincinnati and then marched south and crossed the Ohio River to camp in Covington, Kentucky. In late September the Regiment boarded steamers and were transported down the Ohio River to Louisville and then marched south to Perryville, Kentucky. At the Battle of Perryville, on October 8, the Regiment received its "baptism of fire." Following Perryville, the Regiment marched to Crab Orchard and Bowling Green finally arriving at Nashville, Tennessee on November 7. The Regiment remained in Nashville for training and for garrison duty until August 1863. As part of General Gordon Granger's Reserve Crops, the Regiment left Nashville in pursuit of the forces of Confederate General Bragg and marched toward Chattanooga via Franklin, Columbia, Athens, Huntsville and Lookout Mountain. The Corps went into camp at Rossville, Georgia on September 8, 1863. On September 19 and 20 the Regiment was held in a reserve position at the Battle of Chickamauga and following the battle was detached to guard Caldwell's Ford on the Tennessee River at the mouth of Chickamauga Creek. On the morning of November 16, 1863, a barrage from rebel batteries across the Tennessee River awakened the Regiment. The rebel battery was soon silenced but not before Chaplain Levi Sanders was killed. On November 24 the Regiment crossed the Tennessee River with Sherman's forces and on November 25 and 26 fought in the Battle of Missionary Ridge. The Regiment joined in the pursuit of the retreating enemy forces as far as Ringgold, Georgia but then was ordered north to relieve Burnside's forces at Knoxville. However, en-route word was received that the enemy had lifted the siege and the Regiment

---

[1101] Rogers, Robert M., *The 125th Regiment Illinois Volunteer Infantry. Attention Battalion!*, Gazette Steam Print, Champaign, IL, (1882), pp 91-2. A description of the Regiment's fighting at Kennesaw Mountain.

[1102] Colonel Harmon was born in New York in 1827. He was admitted to the New York state bar and began the practice of law. Several years later he moved to Danville Illinois and continued his law practice. Harmon enlisted in the Union army on September 4,1862 as the colonel of the 125th Illinois Infantry. Colonel Harmon was a close friend of Abraham Lincoln.

returned to Chattanooga arriving there on December 18. For the next several months the Regiment remained in camp on Chickamauga Creek at Lee and Gordon's Mills, but in May joined Sherman's Atlanta campaign. During May the Regiment saw action at Rocky Faced Ridge, battles at Resaca and Rome and at Dallas on Pumpkin Vine Creek. On the morning of June 27, the One Hundred and Twenty fifth led General Sherman's ordered assault on Kennesaw Mountain; in the battle the brigade commander, Colonel Daniel McCook and the Regimental commander, Colonel Oscar Harmon were killed.[1103] July 18 the Regiment crossed the Chattahoochee River and engaged the enemy the following day at Peachtree Creek; from July 22-August 25 the Regiment fought in the siege of Atlanta. In late August and early September, the One Hundred and Twenty Fifth fought at Jonesboro and Lovejoy Station and was then ordered to northern Georgia and Alabama in pursuit of rebel forces under General Hood. The Regiment returned to Atlanta in November to join Sherman's "March to the Sea". The One Hundred and Twenty Fifth moved from Atlanta to Savannah with little in the way of fighting by the Regiment; following the surrender of Savannah the Regiment remained in Savannah for a month. In January 1865 the Regiment marched north into South Carolina, bypassed Columbia in February and saw its first action in heavy fighting at Averysboro on March 15; its last battle of the war was at Bentonville, North Carolina on March 19. While camped on the south bank of the Neuse River in April, good news reached the Regiment that Lee had surrendered only to be followed a few days later by the appalling news that Lincoln had been assassinated. During May the Regiment marched via Richmond to Washington, D.C and marched in the Grand Review on May 24. The One Hundred and Twenty Fifth was mustered out on June 9, 1865. Those recruits whose term of service had not expired were transferred to the Sixtieth Illinois Infantry.[1104,1105,1106]

There is one National and one Regimental flag of the One Hundred and Twenty Fifth Volunteer Regiment Illinois Infantry in the flag holdings of the State of Illinois.

The arrangement of the stars in the canton of the National flag is typical of flags manufactured in Cincinnati. Although the flag is damaged it appears to have been a 34-star flag and thus was

---

[1103] Of the 240 men of the 125th Illinois Infantry who entered the battle, 4 officers were killed and 4 wounded, 43 enlisted men killed and 48 wounded and 5 missing—a total of 104 casualties or 43% of the Regiment. With the death of Colonel Harmon, Colonel James Langley of Champaign was named commander of the Regiment.

[1104] Behrens, Robert H., *From Salt Fork to Chickamauga,* The Urbana Free Library, Urbana, ILL, (1988), pp185-210.

[1105] *Report of the Adjutant General of the State of Illinois*. W. Rokker, State Printer and Binder, Springfield, ILL, (1886), Volume VI, pp 468-70.

[1106] Ibid, Rogers.

presumably manufactured in the early part of the War. A portion of the damage to the flag occurred at the Battle of Kennesaw Mountain; the flag is not imprinted with battle honors. In 1904 an inventory was taken of the South Dakota historic collections. The inventory reported "E. W. Eakin, late first sergeant Company D, 125th Illinois Volunteer Infantry, star torn from the battle flag of the 125th by rebel bullets at Kennesaw Mountain, Georgia." Edmund Eakin was a 27-year-old farmer from Georgetown, Illinois.

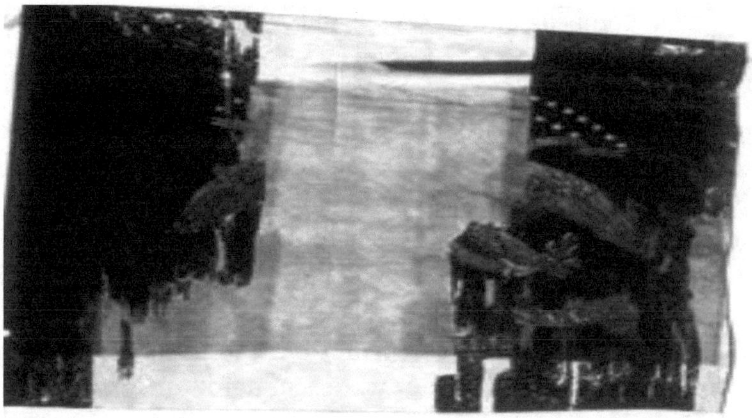

The Regimental flag is severely damaged, however, enough of the pattern remains to identify that the flag is of Philadelphia manufacture by Evans and Hassall. The Regimental flag is not inscribed with battle honors.

No other flags carried by the 125th Illinois Infantry are known to survive.

126th Illinois Infantry

During the summer of 1862, regiments for the Union Army were being organized throughout Illinois. In August there were seven companies at Camp Grant near Mattoon, seven companies at Camp Douglas in Chicago and six companies at Camp Butler near Springfield. Since no single Camp had enough companies to form a regiment, companies from each Camp were combined to form two regiments. The One Hundred and Twenty Sixth Illinois Infantry Regiment was formed at Alton, Illinois by combining five companies from Camp Grant at Mattoon (Companies A and C from Moultrie County, Companies D and F from Montgomery County and Company K from Shelby County) with five companies from Camp Douglas (Company B from Rock Island and Mercer Counties, and Companies E, G, H and I from Rock Island County).[1107] The Regiment, commanded by Colonel Jonathan Richmond[1108] of Mattoon, was mustered in for three years' service on September 4, 1862. The Regiment was initially in charge of guarding Confederate prisoners being held in the State Penitentiary at Alton; however, the prisoners were moved to Johnson's Island in Lake Erie and on November 20 the Regiment boarded the steamer *B. M. Runyan* and moved down the Mississippi River to Columbus, Kentucky. From Columbus the Regiment marched to LaGrange, Tennessee; December 19 six companies were sent north to Humboldt, Tennessee to guard the railroad bridge over Fork Deer River while four companies remained at LaGrange. On December 21, the detachment at Humboldt successfully resisted an attempt by Confederates to capture and destroy the bridge. From January 1863 until March of that year the Regiment was divided with 4 companies on duty at Jackson, Tennessee and 6 companies guarding the bridge at Humboldt, Tennessee. In late March the companies at Humboldt rejoined the Regiment at Jackson. On May 28 the Regiment moved by rail to Memphis, boarded the steamer *Luminary,* landed at Haynes Bluff above Vicksburg, Mississippi and moved to Snyder's Bluff. The One Hundred and Twenty Sixth Illinois remained at Snyder's Bluff assigned to picket the roads in the Oak Ridge area during the siege of Vicksburg[1109]. Following the fall of Vicksburg, the Regiment was ordered to Helena, Arkansas and joined General Frederick Steele's 12,000-man Little Rock Campaign. The Confederate forces abandoned Little Rock and the Union troops occupied the town on September 10 with little or no resistance. The One Hundred and Twenty Sixth was ordered back to Duval's Bluff, Mississippi in late October 1863 and remained there on post duty until August 1864. The Regiment was ordered to Pine Bluff, Arkansas on August 19, 1864, and performed post duty there until February 1865. In mid-February the Regiment moved to the mouth of the White River, Arkansas and remained there until June 4 when ordered back to Pine Bluff for muster out on July 12, 1865.[1110,1111,1112]

There is one National and one Regimental flag carried by the One Hundred and Twenty Sixth

---

[1107] Crowder, Rev. J. H. *Before and After Vicksburg,* The Otterbein Press, Dayton, Ohio, (1924), page 6.

[1108] Jonathan Richmond was born in Butler County, Ohio in 1823. He served in the Mexican-American War and at its close he returned home and opened a general store. He moved to Mattoon in 1860 and opened a dry goods store with his brother-in-law. In the summer of 1862 Richmond organized the 126th Illinois Infantry regiment and was named Colonel of the regiment.

[1109] The Regimental Surgeon, Charles Hunt, a prosperous physician and druggist from Urbana, Illinois developed "remittent fever" while caring for soldiers at the hospital at Haynes' Bluff, Mississippi. While traveling home for recuperation he was hospitalized at the General Hospital at Mound City where he expired on August 2, 1863. Cunningham, Judge Joseph O., *History of Champaign County,* reprint of 1905 edition by the Urbana Free Library, (1984), pp 954-6.

[1110] Ibid, Crowder

[1111] Eddy, T.M., DD, *The Patriotism of Illinois,* Clarke & Co., Publishers, Chicago, ILL, (1866), Volume II, pp 612-13.

[1112] Dyer, Frederick H., *A Compendium of the War of the Rebellion,* Thomas Yoseloff, Publisher, New York, NY, (1959), Volume III, page 1099.

Illinois Volunteer Infantry Regiment in the holdings of the State of Illinois.

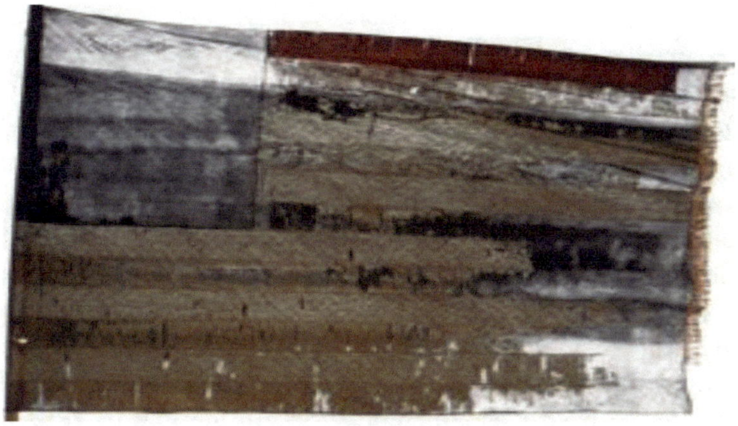

The National flag is in very poor condition and does not bear any inscriptions.

The Regimental flag contains 34 stars and is typical of flags manufactured in New York. The flag is in good condition and bears no inscriptions other than the Regimental designation in the lower scroll.

No other flags carried by the 126th Illinois Infantry are known to survive.

127th Illinois Infantry

The One Hundred and Twenty Seventh Illinois Infantry Regiment was raised during the summer of 1862 and mustered in at Camp Douglas on September 6 with John Van Arman[1113] of Chicago as Colonel. The men volunteered from the following counties, Company A—*(Wallace Guards)* from Kendall County; Company B—*(Woodworth Rifles)* one-third of the Company was from Cook County, but men also volunteered from Bureau, Champaign, De Witt, Iroquois, Kankakee, La Salle, Logan, Mercer, Stark, Stephenson and Whiteside Counties as well as from the states of Iowa, Indiana, Michigan and New York; Company C—*(Elgin Sharpshooters)*from Kane County; Company D—*(Grundy County Plough Boys*) from Grundy County; Company E—*St. Charles, Illinois Volunteers)* from Kane County; Company F—*(Little Rock Illinois Volunteers)* from Kendall County; Company G—*(Reserve Guards)*[1114] from Cook County; Company H—*(Lyons Scouts)* from Cook County; Company I—*(Eldridge Guards)* from Kane County and Company K—*(Kendall Guards)* from Kendall County.[1115] In the Fall of 1862 the regiment performed guard duty at Camp Douglas but on November 9 the Regiment departed on the *Illinois Central Railroad* for Cairo; at Cairo the men transferred to the steamer *Emerald* and arrived at Memphis, Tennessee on November 13. As part of the Army of the Tennessee the Regiment departed Memphis on November 20 on Grant's Central Mississippi campaign—the "Tallahatchie March"—south across the Tallahatchie River to Oxford, Mississippi.[1116] The Confederate forces withdrew and on December 8, at Oxford, Grant ordered two of his divisions to Memphis to join Sherman's Vicksburg campaign. The One Hundred and Twenty Seventh arrived at Memphis on December 13 and on December 20 boarded steamers for the trip down the Mississippi River and up the Yazoo River to Chickasaw Bayou, Mississippi at the rear of Vicksburg. The Regiment was engaged in fighting at Chickasaw Bayou and Chickasaw Bluff from December 25- until January 3, 1863. On January 4, General John McClernand assumed command and immediately ordered a blow at Fort Hindman—Arkansas Post—on the Arkansas River. The One Hundred and Twenty Seventh claimed to be "one of the first to plant its colors on the enemy's works"[1117] when the Fort fell on January 11, 1863. The Regiment was next ordered to Young's Point, Louisiana for picket duty and to aid in digging an ill-fated canal, which had been planned to by-pass Vicksburg. In March the One Hundred and Twenty Seventh joined in the expedition to relieve Porter's gunboat fleet trapped on Deer Creek and the Black and Steele's Bayous. The Regiment returned to Young's Point and on May 7 started on the march to Jackson, Mississippi via Grand Gulf. On May 14, at the Battle of Jackson, Union forces, including the One Hundred and Twenty Seventh Illinois, caused the Confederates to break and retreat towards Vicksburg. Two days later, May 16, the Federals attacked the retreating rebel army at Champion's Hill, defeated them and followed them into the fortifications at Vicksburg. The Regiment participated in the assaults of the enemy's works on May 19 and May 22; five members of the Regiment received the Medal of Honor for their bravery in the assault of May 22.[1118] Vicksburg surrendered on July 4, 1863; the Regiment joined in the pursuit of

---

[1113] Colonel John Van Arman was born in Plattsburgh, New York in 1820. He lived in Michigan before coming to Chicago in 1858 where he practiced law. Van Arman was Captain of the 1st Michigan Infantry in the Mexican-American War.

[1114] Also referred to as the "*Tourtelotte Guards*"

[1115] *Chicago Daily Tribune,* September 21, 1862

[1116] The purpose of Grant's campaign was to attack Confederate General Pemberton's forces and prevent them from returning to Vicksburg. This would in turn allow Sherman to enter Vicksburg with little resistance.

[1117] *Report of the Adjutant General of the State of Illinois,* H.W.Rokker, State Printer and Binder, Springfield, ILL., (1886) Volume VI, page 512.

[1118] During the assault of May 22, 150 volunteers were needed to attack the fort in front of the Union Fifteenth Army Corps. Since the possibility of returning alive seemed poor the attack was called "forlorn hope." Five men volunteered from the One Hundred and Twenty Seventh Illinois Infantry— Andrew Goldsbery (also spelled Goldsberry and Goldsburyß),

Confederate General Joe Johnston's forces to Jackson and then went into camp near the Big Black River. At camp the Regiment had less than 100 men fit for duty with 400 men in the hospital on Walnut Hills. In late September the Regiment was transported to Memphis and then marched the 300 miles overland to Chattanooga arriving there on November 15. En-route the Regiment was involved in operations on the *Memphis & Charleston Railroad* with fighting at Colliersville, Tennessee and Tuscumbia, Alabama. In the following weeks the One Hundred and Twenty Seventh fought in the battles of Lookout Mountain and Missionary Ridge, pursued the retreating rebel army as far as Graysville, Alabama and was then ordered to the relief of General Burnside's forces besieged at Knoxville, Tennessee. However, the siege of Knoxville was lifted before the Regiment's arrival and they returned to Bridgeport, Alabama in late December. In January 1864 the Fifteenth Army Corps, was stationed along the *Memphis & Charleston Railroad* with the One Hundred and Twenty Seventh Illinois garrisoned at Larkinsville, Alabama. In May the Regiment joined Sherman's Atlanta Campaign; the Regiment fought at Resaca on May 14-15 and at Dallas, New Hope Church, and Allatoona Hills May 25-June 5. On June 10 the Regiment, as part of the Fifteenth Army Corps, began the assault on Kennesaw Mountain culminating the Battle of Kennesaw Mountain on June 27. In July[1119] the Illinois regiment fought the retreating enemy forces at Nickajack Creek and the Chattahoochee River. During the siege of Atlanta, on July 28, Hood's forces assaulted the Regiment nearly breaking the Union line. Seeing the line was in danger of being breached, the Brigade commander, Brig. Gen. Joseph Lightburn, sent his 15-year-old orderly, Robinson Murphy to General John Logan for reinforcements. Murphy led two regiments into line for Lightburn's relief.[1120] By the end of July the Regiment had only 92 men fit for duty. In the flanking movement around Atlanta the One Hundred and Twenty Seventh fought at Jonesboro and Lovejoy Station in August and early September; following the fall of Atlanta, the Regiment joined in the pursuit of Hood into north Georgia and Alabama. In November the Regiment became a part of Sherman's "March to the Sea." Following the siege and capture of Savannah and Fort McAllister in December, the Regiment marched north into the Carolinas in January 1865. In the Carolina campaign the Regiment clashed with the enemy at Salkehatchie Swamps, South Edisto River and North Edisto River, South Carolina in February and fought at Bentonville, North Carolina in March. Following the surrender of Johnston's Confederate Army at Bennett's house on April 26, the One Hundred and Twenty Seventh marched to Washington D.C. Via Richmond and marched in the Grand Review on May 24. The Regiment mustered out on June 4, 1865.[1121,1122,1123]

The collection of flags in Springfield, Illinois contains a National and Regimental flag of the One hundred and twenty-seventh.

Company E; Andrew McCornack, Company I; Emmer Bowen, Company C; Theodore Hyatt, Company D; William Tomer, Company G. They received the Medal of Honor as cited— "Was a member of a "Volunteer Storming Party' which made a most gallant assault upon the enemy's works. Vicksburg, Mississippi, May 22, 1863." Beyer, W. F. and Keydel, O. F. *Deeds of Valor,* reprint by Longmeadow Press, Stamford, CT., (1993), and pages 190-7.

1119 On July 3, 1864, Sgt. Ira B. Whitney, Company B, captured the flag of the 6th Texas Infantry (*Austin Rifles*). The flag is now at the Arkansas Post Museum in Gillett, Arkansas.

1120 For his actions Robinson B. Murphy, Musician, Company A, received the Medal of Honor with this citation: "Being orderly to the brigade commander, he voluntarily led two regiments as reinforcements into line of battle, where he had his horse shot from under him. Atlanta, Georgia, July 28, 1864. *Civil War Medal of Honor Winners from Illinois,* Civil War Centennial Commission of Illinois July 1962, page 4. *Ledger--Sentinel,* (Oswego, Boulder Hill and Montgomery, Illinois) March 21, 2002.

1121 *Report of the Adjutant General of the State of Illinois,* H.W.Rokker, State Printer and Binder, Springfield, ILL., (1886) Volume VI, pp 512-5.

1122 Eddy, T. M. *The Patriotism of Illinois,* Clarke & Co., Publishers, Chicago (1866), Vol. II, 485-7.

1123 Dyer, Frederick H., *A Compendium of the War of the Rebellion,* Thomas Yoseloff, New York, (1959), Vol. III, pp1099-1100

The star pattern on this flag identifies its manufacturer as Longley & Brothers of Cincinnati. The flag measures 66" on the hoist by 74" on the fly.

The regimental flag is inscribed with battle honors. Based on the eagle, stars and scroll, it was probably manufactured by Longley & Brothers of Cincinnati. The following battles are inscribed on this flag: VICKSBURG; CHICKASAW BAYOU; ARKANSAS POST; JACKSON; MISSION RIDGE; RESACA; DALLAS; KENESAW MOUNTAIN; ATLANTA, July 22, 28 and Aug. 3; JONESBORO; SAVANNAH; BENTONVILLE

The Fort Douglas Military Museum in Salt Lake City has personal effects of 1st Sgt. George Theodore Hyatt. These include the remnants of a silk flag (red and white strips and a ragged portion of blue with portion of a gold star) a saber and the soldier's Medal of Honor. Accompanying the items is a hand written note: "Enclosed you will find shreds of one flag which we planted on the fort and defended for an entire day and brought off the remnants at night. Place these with my soldiers record. They will be sacred relics for our children. Keep them for I shall be glad to see them again to remind of the fearful scene of Friday the 22nd day of May 1863. It is about 3 PM on Sunday and the mail is just going out.

Yours Theodore"

I (ed) was contacted by a descendent of George Theodore Hyatt who has been researching the battle at Vicksburg. She states: "I'm researching history of my ancestor 1st Sgt. George Theodore Hyatt, who was in Sherman's Corps at Vicksburg (127th Illinois Volunteers). He later received the Medal of Honor for action on the 22nd of May 1863. He was part of the "Forlorn Hope" a volunteer storming party of 150 men who assaulted the Stockade Redan. It seems that the following day, General Ewing presented the battle flag that was carried in the assault to the survivors, who then cut it into pieces for souvenirs. The pieces that my family kept for generations consist of portions of both red and white stripes and a piece of blue field with a gold star."

One or possible two flags were captured by this Regiment at the Arkansas Post. In some respects, this guidon of "Hill's Cavalry" is similar to an unidentified guidon that reversed the pattern of the bars to white over red but retained a blue canton with 13 white stars-- 12 around a larger center star. The flag was captured by a member of the 127th Illinois Infantry but no identity for it has been established. One North Carolina flag, however, is similar but not cut in the form of a guidon.

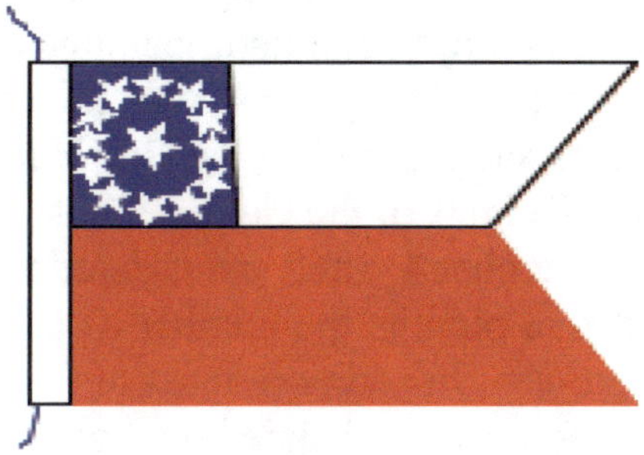

Unidentified Confederate cavalry guidon, by Wayne J. Lovett from a detailed sketch and notes by Howard M. Madaus

No other flags carried by the 127th Illinois Infantry are known to survive.

128th Illinois Infantry

The men for the One Hundred and Twenty Eighth Illinois Infantry volunteered from the southern Illinois counties of Williamson, Franklin, and Saline. Settlers who traveled to Illinois via the Mississippi River and down the Ohio River from southern states such as Virginia, Tennessee, Kentucky, and the Carolinas had settled the southern portion of Illinois. At the outbreak of the Civil War this part of the state had strong southern leanings and a large portion of Company G, Fifteenth Tennessee Regiment Volunteer Infantry CSA was raised in Williamson and Jackson Counties, Illinois.[1124] Many of the officers and privates of the One Hundred and Twenty Eighth Illinois Infantry were said to be members of the Knights of the Golden Circle.[1125] The Regiment was mustered in at Camp Butler near Springfield on November 4, 1862 and ordered to Columbus, Kentucky and attached to the 16th Army Corps. In the next five months the number of men in the Regiment were reduced from 861 to 161—mainly due to desertions. In addition, most of the officers were incompetent and had resigned. By Special Order, the Regiment was disbanded on April 1, 1863, and the remaining men consolidated into a detachment and transferred to the Ninth Illinois Infantry.[1126]

There is no record of any colors being issued to this Regiment.

[1124] Gleeson, Ed, Illinois *Rebels, A Civil War Unit History of G Company Fifteenth Tennessee Regiment Volunteer Infantry*, Guild Press of Indiana, Carmel, Indiana (1996).

[1125] Originally founded in 1854 to promote annexation of Mexico to extend the area of slavery. During the Civil War the organization provided the Confederacy with military intelligence and encouraged acts of sabotage by its members. The *Chicago Tribune* estimated that Illinois had 20,000 Knights. Faust, Patricia L. ed., *Historical Times Illustrated Encyclopedia of the Civil War,* Harper & Row, New York (1986) page 420.

[1126] *Report of the Adjutant General of the State of Illinois,* H. W. Rokker, State Printer and Binder, Springfield, ILL (1886) Volume VI, page 532

129th Illinois Infantry

"Sergeant Hess carried his flag in the charge upon the rebel works at Resaca, and with his precious burden was one of the first upon the enemy's breastworks, where he stood erect amid the rain of lead, and waved the flag, to cheer his comrades on. While holding it in his right hand a ball shattered that arm, but he immediately raised the fallen colors with his left, which arm soon shared the fate of his right. Still holding the flag in his broken and bleeding arms, he ran some rods to the first gun of the enemy, and leaping upon the smoking piece, was still displaying the flag when a bullet in the forehead ended his sufferings, and like Marco Bozzaris[1127], he died at the moment of victory. The gun upon which he was killed belonged to the 1st Georgia Artillery, a member of who fired the fatal shot."[1128]

The One Hundred and Twenty Ninth Illinois Infantry Regiment was organized during the summer of 1862 at Pontiac, Livingston County, Illinois with George P. Smith[1129] of Dwight as Colonel and mustered in for three years of service on September 8, 1862. The men of the Regiment volunteered from the following counties, Companies A, B, C, E, and G from Livingston County; Companies D, F, H and I from Scott County and Company K from Rock Island County. On September 22, 1862, the Regiment was ordered to Louisville, Kentucky; after two weeks in camp the Regiment, pursued rebel forces under General Bragg and then marched through Frankfort and Danville to Crab Orchard, Kentucky. Since the enemy had retreated, the Regiment was ordered to Bowling Green, Kentucky for drilling and guard duty. They remained there until November 20 when ordered to Mitchellville, Tennessee. A detachment[1130] was ordered to Buck's Lodge a station on the *Louisville & Nashville Railroad.* The remainder of Regiment was sent to South Tunnel, two miles north of Gallatin, Tennessee on February 2, 1863, and placed on guard duty and foraging details.[1131],[1132] In early June the Regiment was reassembled at Gallatin, Tennessee and moved to garrison nearby Fort Thomas until late August when ordered to Nashville, Tennessee. At Nashville, the Regiment as part of Major General Gordon Granger's Reserve Corps, encamped southeast of the city between Fort Negley and the Murfreesboro Pike. The One Hundred and Twenty Ninth remained on duty at Nashville from August 1863 until February 1864; on February 24 the men marched south out of Nashville en-route to Wauhatchie Valley (Lookout Valley), Tennessee, which lay southwest of Chattanooga. The Regiment remained in camp until May 2 when, as part of Sherman's Atlanta campaign, the men crossed Missionary Ridge, the Chickamauga battlefield and met the enemy at the battle of Resaca on May 13-15.[1133] During May and June the One Hundred and Twenty Ninth fought at Cassville, Burnt Hickory (Dallas), New Hope Church,

---

[1127] A Greek general and a hero of the Greek War of Independence.

[1128] Reed, Capt. Henry B., *The Back-Bone of Illinois in Front and Rear From 1861 to 1865 and Memorial Hall, Springfield, Illinois,* H.W. Rokker, Printer and Binder, Springfield, ILL, (1886) page 7. Captain Reed of Pontiac, Illinois was Captain of Company G of the 129th Illinois Infantry and undoubtedly witnessed the bravery of Color-Sergeant Hess of Company H.

[1129] Colonel George Smith was born in Wooster Ohio in 1826. He was educated in the local Wooster schools and then attended Pleasant college in Iowa.To prepare for medical practice he studied at the Eclecticc college of Cincinnati and then moved to Danville, Illinois to begin medical practice. With the outbreak of the civil war, Smith served as major of the Sixty-ninth Illinois Infantry. One year later he re-enlisted for three years and raised the One Hundred and Twenty ninth Illinois Infantry and was appointed colonel of the regiment.

[1130] In early December Companies A, D, F and K were detailed to guard the water tank and two railroad bridges over Drake's Creek, eight miles south of Mitchellville. The remainder of the Regiment was ordered to Buck's Lodge on December 22.

[1131] Companies C, D, F and K were detached for guard duty at Richland, Tennessee, fifteen miles north of Gallatin.

[1132] May 8, 1863, Colonel George Smith resigned and was replaced by Colonel Henry Case.

[1133] Colonel Case of the Regiment claimed that the colors of the One Hundred and Twenty Ninth were the first planted on the fort at Resaca. *O .R. Series 1—Volume XXXVIII/2 [s#73]*

Big Shanty station,[1134] Lost Mountain,[1135] Noyes Creek, Kolb's Farm, and Kennesaw Mountain. On July 3, the Regiment battled at Ruff's Station, crossed the Chattahoochee River on July 17 and crossed Peach Tree Creek on July 20 to fight in the bloody battle of Peach Tree Creek[1136]. During the subsequent siege of Atlanta, the One Hundred and Twenty Ninth was positioned near the Chattahoochee River Bridge. Following the fall of Atlanta, the Regiment occupied Atlanta until mid-November when they joined Sherman's "March to the Sea". On December 10 the Regiment entered Savannah and remained there until January 1865 when they marched into South Carolina. In January the men passed through Hardeeville, Robertsville, Lordensville, Branchville, Blackville, and Lexington before reaching Columbia on February 17; they met little resistance en-route. On March 4 the Regiment crossed the border into North Carolina and marched toward Fayetteville. North of Fayetteville, the Regiment met and defeated Confederate General Hardee's rebel forces at Averysboro on March 16 and Confederate General Johnston's troops at Bentonville March 19 and 20. Goldsboro was occupied on March 24 and Raleigh on April 14. Following the surrender of Johnston's army at Bennett's House on April 26, the One Hundred and Twenty Ninth marched to Washington DC via Richmond, Virginia and marched in the Grand Review in Washington on May 24, 1865. The Regiment was mustered out at Washington DC on June 8 and those enlisted men whose terms had not expired were transferred to the Sixteenth Illinois Infantry.[1137,1138]

There is one National and one Regimental flag carried by the One Hundred and Twenty Ninth Illinois Volunteer Infantry Regiment in the holdings of the State of Illinois.

This 34-star National flag is characteristic of those manufactured by Hugh Wilkens of Louisville, Kentucky. This is the flag referred to by Captain Henry Reed in the quote above and was the first flag to fly on the captured rebel works at Resaca. The center stripe is inscribed "129th REGIMENT ____"*. Unfortunately, the remainder of the inscription is lost. No battle honors are inscribed on this flag. The flag measures 68" on the hoist by 75" on the fly.*

[1134] Big Shanty station was located on the *Western & Atlanta Railroad.*

[1135] Also known as the battle of Gilgal Church.

[1136] In his history of the Regiment, Grunert states "Our division took seven flags from the enemy, (one of which was taken by our regiment;) the bearers of three had to be killed before the flags were taken." Grunert, William, *History of the One Hundred and Twenty-Ninth Regiment Illinois Volunteer Infantry,* R. B. Dedman, Winchester, ILL, (1866), page 86. Neither the Commander of the Regiment, Colonel Case in his report nor Grunert identify the Confederate regiment from which the flag was captured.

[1137] *Ibid,* Grunert

[1138] Dunlap, Leslie W. ed, *Your Affectionate Husband, J. F. Culver,* Friends of the University of Iowa Libraries, Iowa City, Iowa (1978).

The 34-star Regimental flag is typical of those manufactured by Evans & Hassall of Philadelphia. It measures 71" on the hoist by 77" on the fly.

No other flags carried by the 129th Illinois Infantry are known to survive,

130th Illinois Infantry

"On the 22d of the same month the Regiment was on the front line with its colors planted three times upon the rebel parapets, and as often was the flag staff shot off, while the color bearer was holding the colors up until ordered to fall back, done so in good order."[1139]

The One Hundred and Thirtieth Illinois Infantry was organized in the fall of 1862 at Camp Butler, near Springfield, Illinois and mustered into service on October 25 under the command of Colonel Nathaniel Niles.[1140]

The men volunteered from the following counties: Company A from Monroe County; Company B from Sangamon County; Company C from Alexander County; Company D from Christian County; Companies E and F from Bond County; Company G-one-third from Coles County, the remainder from Edgar, Bond, Clinton and Douglas Counties; Company H from Richland and Wabash Counties; Company I from Lawrence and Sangamon Counties and Company K from Clark County. The Regiment departed Camp Butler for St. Louis on November 11 on cars of the *Wabash Railroad*; at St. Louis the men boarded the steamboat *General Robert Allen* for the trip downriver to Memphis. The Regiment camped at Fort Pickering and served provost duty until March 27, 1863, when ordered to Milliken's Bend, Louisiana as part of the Vicksburg campaign. They remained at Milliken's Bend until April 15, then moved with the army south through Louisiana along the western side of the Mississippi River, crossed the River at Bruinsburg on May 1, and engaged the enemy at Port Gibson—the first battle for the Regiment. Following the battle, the Regiment moved north, fought at Champion's Hill on May 16 and in pursuit of the retreating enemy fought them again the following day at Big Black River Bridge. To quickly capture Vicksburg, two assaults were made on the Confederate works–on May 19 and 22--both failed and the Union forces began the siege of Vicksburg. Vicksburg surrendered on July 4, 1863, and the following morning the One Hundred and Thirtieth Illinois joined the line of march to Jackson, Mississippi where the enemy was entrenched; after a ten day siege the Confederate forces withdrew, burning the town as they left. The Regiment returned to camp at Vicksburg for rest and reorganization. At the end of August, the Regiment was ordered to New Orleans and on arrival went into camp at Carrollton, Louisiana. In early October the One Hundred and Thirtieth was ordered on a campaign into western Louisiana. The men landed at Algiers, across the River from New Orleans, and moved by cars of the *New Orleans, Opelousas & Western Railroad* to Brashear City. They then marched through Bayou Teche[1141] to New Iberia,[1142] Louisiana where they encamped until early December.[1143] On December 1 the Regiment was ordered back to New Orleans, boarded a Gulf

---

[1139] H. J. Conover, "A Personal Reminiscence" in *History and Roster, One Hundred and Thirtieth Illinois*, Advocate Book Print, Greenville, Illinois (1892) page 6-7. Henry J. Conover had been Sgt. Major of the Regiment, and the quote refers to the second assault on the lines at Vicksburg, May 22, 1863.

[1140] Nathaniel Niles was born in Plainfield Center, New York in 1817. He attended Albany Academy, served a clerkship in Albany and was admitted to the New York bar. In 1842 he moved to Belleville, Ilinois and served in the Second Illinois Volunteer Regiment in the Mexican-American War. At the onset of the civil war he first served as colonel of the Fifty-Fourth Illinois Infantry but resigned his commission when the regiment failed to fill the required number of men. In 1862 he was commissioned colonel of the 130th Illinis Infantry.*The United States Biographical Dictionary and Portrait Gallery of Eminent and Self Made Men: Illinois Volume (*Chicago: American Biographical, 1876), 486-87.

[1141] Bayou Teche is the area referred to In Longfellow's poem *Evangeline*

[1142] It was here that the Regiment produced a newspaper entitled *Unconditional Surrender—Grant*, printed on wallpaper. Johnson, Charles Beneulyn MD, *Muskets and Medicine or Army Life in the Sixties,* F. A. Davis Company, Philadelphia, (1917), page 150.

[1143] While at Iberia, Colonel Niles resigned and command devolved to Major John B. Reid.

steamer, and landed at Decros Point on Matagorda Peninsula. The Regiment remained in Texas until February 1864 when ordered back to Algiers, Louisiana, and the Red River Campaign. Under command of Major General William Franklin, the Regiment marched from Franklin to Alexandria, Louisiana through Bayou Teche. The Union forces advanced toward Shreveport but were met the afternoon of April 8, 1864, south of Mansfield[1144] by Brigadier General Alfred Mouton's Confederate Division. In "Mouton's Charge" the Union line crumbled and Brigadier General Camille Polignac's Texas brigade decimated the One Hundred and Thirtieth Illinois Infantry.[1145] The remainder of the Regiment retreated with the Union forces to the Mississippi River and was sent for garrison duty to New Orleans from June 1864 until January 1865. On January 14, 1865, the "One Hundred and Thirtieth Illinois Infantry was consolidated into and known as the Seventy-seventh Illinois Volunteer Infantry." Under Special Orders, No. 14. In August 1865, as part of the Seventy-seventh Illinois Infantry, the men of the One Hundred and Thirtieth participated in the Mobile campaign and fought in the attack and siege of Spanish Fort and Fort Blakely. Following the surrender of Mobile the Regiment went up the Tombigbee River to McIntosh Bluffs, 40 miles north of Mobile. On July 10, 1865, the Seventy-seventh Illinois Infantry was mustered out at Mobile and the remnants of the One Hundred and Thirtieth reorganized into a battalion of six companies under the command of Lt. Col. John B. Reid. The battalion was ordered to New Orleans in August and mustered out on August 15, 1865.[1146,1147]

The collection of the State of Illinois holds two national flags and one regimental flag carried by the One Hundred and Thirtieth Regiment Illinois Volunteer Infantry.

Charles Johnson in his history of the Regiment describes the presentation of two flags to Company E. The first presentation he describes as follows,[1148] "One moonless night in August, a little time before we left Greenville, our company was drawn up in front of the Court House to receive a beautiful flag, a present from the women whose husbands, brothers, sons and sweethearts were soon to see service at the front…a Miss Smith, a beautiful young woman, mounted the Court House steps, and in a few well-chosen words, spoken in a sweet voice, presented the flag…The flag was made of fine silk and most beautiful were its seven stripes of red, six of snowy white and delicate field of blue, studded with thirty-four immaculate stars…This flag we took with us when we went to the enemy's country, but unfortunately, during our various marches and transfers from one to another locality, it was misplaced, and never afterward found."

Johnson also relates the presentation of another national flag to Company E.[1149] "In the same month of August, 1862, another beautiful Bond County flag, the handiwork of the wives, sisters, mothers and sweethearts of the newly-enlisted men, was made at Pocahontas, my native village, and by one of its fair maidens, Miss Sarah Green, presented to an organization that later became Company E, 130th Illinois Infantry Volunteers. In due time this Pocahontas flag was carried to the enemy's country, and by his bullets its folds were more than once pierced during the Siege of Vicksburg. The war over, the flag was returned to the people from whence it came, and is today a highly cherished relic, in the care of

[1144] Known as the Battle of Sabine Crossroads
[1145] The Regiment lost 26 men killed or wounded and 9 officers and 223 men captured for an aggregate loss of 257. Among those wounded was Lieutenant Colonel John B. Reid. The captured soldiers were imprisoned at Tyler, Texas and not paroled until thirteen months later, just before the end of the War. The reports of April 12, 1864, list the total strength of the Regiment as 3 officers and 65 men. *O. R. Series I—Volume XXXIV/1[S#61] page 295.* The "flag of the Regiment" and the color-bearer were also captured. Wilkin, Jacob W.," Vicksburg" *Mollus, Illinois,* IV, page 232.
[1146] *Report of the Adjutant General of the State of Illinois,* H. Q. Rokker, State Printer and Binder, Springfield, ILL, (1886), Volume VI, pp 573-7.
[1147] Ibid, Johnson
[1148] Ibid, Johnson, page 36-7.
[1149] Ibid, Johnson page 37-8

J. W. Miles a Civil War veteran of Pocahontas."

Pocahontas Flag; Real "Old Glory."

(See page

This photo is from Johnson's book and depicts the flag "Old Glory" from Pocahontas, described above. The whereabouts of this flag today is unknown.

The national flag shown here is one of two national flags of the Regiment in the Springfield Illinois collection. It is "homemade" with the stars arranged in star-shaped pattern in the canton. The flag presumably was presented to one of the companies in the Regiment; however, no further information is known about the flag's origin or history.

The second National flag of the Regiment in the Illinois collection is typical of flags manufactured for the Philadelphia depot. Inscribed on this flag are the following battles: PORT GIBSON; CHAMPION HILLS; BLACK RIVER BRIDGE; VICKSBURG; JACKSON; MANSFIELD, La.; KANE RIVER; ATCHAFALAYA; SPANISH FORT; CAPTURE OF MOBILE.

Remnants of a third National flag were displayed at the Mississippi Valley Sanitary Fair in St. Louis in June 1864 as noted in the *War Democrat* a Wayne County Illinois newspaper.[1150] "Off in a miscellaneous department on the east side of the northern compartment and in a place not conspicuous for it is the flag of the 130th Illinois Infantry. It is reduced to a mere tattered remnant, having been carried through the battles of Jackson, Vicksburg, Champion's Hills, Port Gibson, Black River Bridge, and the late actions in Western Louisiana. Appended to it is the ambrotype of the color bearer Edward Dunn, the boy hero who was killed at Grand Ecore[1151] a few weeks ago."

The Regimental flag appears to be almost new. The eagle, the position and number of stars above the eagle and scrollwork below are all typical of flags manufactured for the New York depot. Other than the regimental designation in the lower scroll there are no inscriptions on this flag.

No other flags carried by the 130th Illinois Infantry are known to survive.

[1150] *War Democrat,* June 9, 1864.

[1151] The action referred to was Grand Ecore, Louisiana on April 3, 1864.

131st Illinois Infantry

The One Hundred and Thirty First Illinois Infantry was organized in the counties of southern Illinois adjacent to the Ohio River during July and August of 1862 and went into camp at Fort Massac in Massac County on the Ohio River. Unfortunately, between the time of organization and muster, measles infected the Regiment and nearly 100 men either died or were discharged because of the disease. The men volunteered from the following counties: Company A from Massac and Pope Counties; Company B from Hardin and Pope Counties; Companies C and D from Massac County; Company E from Gallatin County; Company F from Hardin County; Company G from Equality Township in Gallatin County; Company H from New Liberty in Pope County and Metropolis in Massac County; Company I from Hamilton County and Company K from Gallatin County. The Regiment was mustered in for three years' service on November 13, 1862, with George W. Neely[1152] of Metropolis as Colonel. On December 2, the Regiment moved by the steamboat *Iowa* to Cairo, Illinois and armed with poor quality Harper's Ferry flintlock rifles. On December 7, the Regiment was ordered downriver to Memphis where some of the Harper's Ferry weapons were exchanged for Springfield and Enfield rifles. The Regiment was assigned to the Thirteenth Army Corps of the Army of the Tennessee preparing for the Vicksburg campaign. December 20 the Regiment again boarded the *Iowa* and moved downriver to Milliken's Bend, Louisiana and then on December 27 moved up the Yazoo River with Sherman's forces. The One Hundred and Thirty First saw its first action of the war at Chickasaw Bayou December 26-28 and at Chickasaw Bluffs December 29. However, the Confederate lines held, and the Union troops moved back to Milliken's Bend in January 1863. General John McClernand[1153] took command of the Thirteenth Corps and ordered an expedition against Arkansas Post.[1154] The Corps moved up the Mississippi River to the White River, through a pass to the Arkansas River and disembarked three miles from Arkansas Post on January 10, 1863. The One Hundred and Thirty First was engaged in the fighting at the Post on January 11 and following the surrender of the Post returned to Young's Point near Vicksburg on January 23. During the next two months while the Regiment stood guard duty at Young's point many men were ill and many died due to measles, smallpox and pneumonia. On March 2 the Regiment was sent on the steamer *Westwind* to Memphis to recover and remained at Fort Pickering until May 1863. On May 10, the Regiment, now reduced in number to 400 men, departed Memphis on the steamboat *Golden Era*, arrived at Sherman's Landing[1155] on May 12 and then on May 17 moved to Milliken's Bend. The Regiment remained on guard duty there for one week and then returned to Sherman's Landing to serve on picket duty and to man mortar boats. On June 7 the Regiment along with the One Hundred and Twentieth Illinois Infantry was ordered back to Milliken's Bend to support Federal troops that had been attacked by a force of 1200 rebels. The enemy retreated at the sight of the Federal reinforcements and the One Hundred and Thirty First returned to Sherman's Landing. Following the fall of Vicksburg on July 4, the Regiment moved into the city for guard and picket duty and on July 20 was ordered to Paducah, Kentucky for picket duty and drilling. Due to the reduction in the number

---

[1152] Colonel George Washington Neely was born in Kentucky in 1830. He grew up on the Neely farm in Pope County Illinois, attended McKendree College in Lebanon, Illinois, was admitted to the Illinois bar and began practice in Metropolis. Masaac County, Illinois. Shelton, Galen, "The Enigmatic Col. George Washington Neely, *Metropolisplantet.com*; August 10,2023

[1153] John McClernand was a Douglas Democrat from Shawneetown, Gallatin County in southern Illinois.

[1154] Known as the Post of Arkansas by the Confederates and Fort Hindman by the Federals the fort was an earthen fort 190 feet on each side on a bluff 25 feet above the Arkansas River.

[1155] Sherman's Landing was near Sherman's Headquarters located on Tuscumbia Bend and downriver from Young's Point on the Mississippi River.

of men in the Regiment, on September 16, 1863, the Regiment was consolidated into a battalion of four companies and the Colonel, Major, Surgeon, Chaplain, ten officers of the line and all surplus non-commissioned officers mustered out; Colonel Richard A. Peter was given command of the Battalion.[1156] The Battalion remained on picket duty at Paducah until October 15 when ordered to Mayfield, Kentucky to join a company of cavalry and a battery of artillery, under command of Colonel Richard Peter; they moved to Murray, Kentucky to protect Union populace from raids by enemy guerrillas. Although the Battalion arrived at Mayfield, the cavalry and artillery did not. Nevertheless, the One Hundred and Thirty First moved on to Murray and on October 16, with the help of Negro labor, threw up defensive works. The enemy attack did not materialize and on October 20 the Battalion was ordered to Calloway Landing on the Tennessee River for transfer to Vicksburg on the steamboat *Masonic General*. On arrival at Vicksburg the Battalion was ordered into camp and on October 30, pursuant to Special Order No. 90, The One Hundred and Thirty First Battalion was consolidated with the Twenty Ninth Illinois Infantry and all camp equipage and quartermaster stores were turned over to the Quartermaster of the Twenty Ninth.[1157]

There are no flags of the One Hundred and Thirty First Illinois Infantry Regiment in the flag collection of the State of Illinois and no flags carried by the Regiment are known to exist elsewhere. Presumably the flags of the Regiment were turned over to the Twenty Ninth Illinois Infantry.

[1156] The Battalion strength was approximately 200 men.

[1157] *Report of the Adjutant General of the State of Illinois,* H. W. Rokker, State Printer and Binder, Springfield, ILL, (1886), Volume VI, 605-9.

132nd Illinois Infantry

The One Hundred and Thirty Second Illinois Infantry was mustered in for one hundred days service at Camp Fry in Chicago, Illinois on June 1, 1864, under the command of Colonel Thomas J. Pickett of Moline. The men volunteered from the following counties, Company A from Mendota Township in La Salle County; Company B from Cook and Fulton Counties; Company C from Kane County; Company D from Fulton and Peoria Counties; Company E from Canton Township in Fulton County; Company F from De Kalb County; Company G from Henry, Knox and Cook Counties; Company H from Rock Island County; Company I from Cook County and Company K from Cook, DeKalb and Winnebago Counties. The Regiment was ordered to Columbus, Kentucky on June 6 and on June 15 moved on to Paducah, Kentucky for garrison duty. While at Paducah the Regiment skirmished at Haddix Ferry[1158] Kentucky July 25-27 and then returned to Paducah and remained there until expiration of service. The Regiment returned to Chicago for muster out on October 17, 1864. There are no flags of the Regiment in the collection of the State of Illinois and no record of a flag presentation to this Regiment.

[1158] "Haddix Ferry" should be spelled "Haddox Ferry." On today's map the location of the skirmish occurred where Haddock (Haddox) Road crosses the Tennessee River in Livingston County, Kentucky.

133rd Illinois Infantry

The One Hundred and Thirty Third Illinois Volunteer Infantry was organized during the spring of 1864 and mustered into service for one hundred days at Camp Butler on May 31, 1864 under the command of Colonel Thaddeus Phillips of Carlinville. The men of the Regiment volunteered from the following counties, Company A (*Fuller Guard)* from Sangamon County; Company B from Morgan County; Company C from Greene and Macoupin Counties; Company D from Madison County; Company E from Sangamon and Macoupin Counties; Companies F and G from Macoupin County; Company H from Macoupin and Jersey Counties; Company I from Menard and Sangamon Counties and Company K from Champaign and Vermilion Counties. Soon after muster, the Regiment was ordered to the Rock Island Barracks to guard prisoners of war. Having completed its term of service the Regiment returned to Camp Butler and was mustered out on September 24, 1864.

There are two flags of the One Hundred and Thirty Third Illinois Infantry in the collection of the State of Illinois. The National flag appears to have been presented to Company A of the Regiment and is so inscribed: FULLER GUARD; Co. A 133 REGT ILL VOL . The flag measures 47"on the hoist by 68" on the fly.

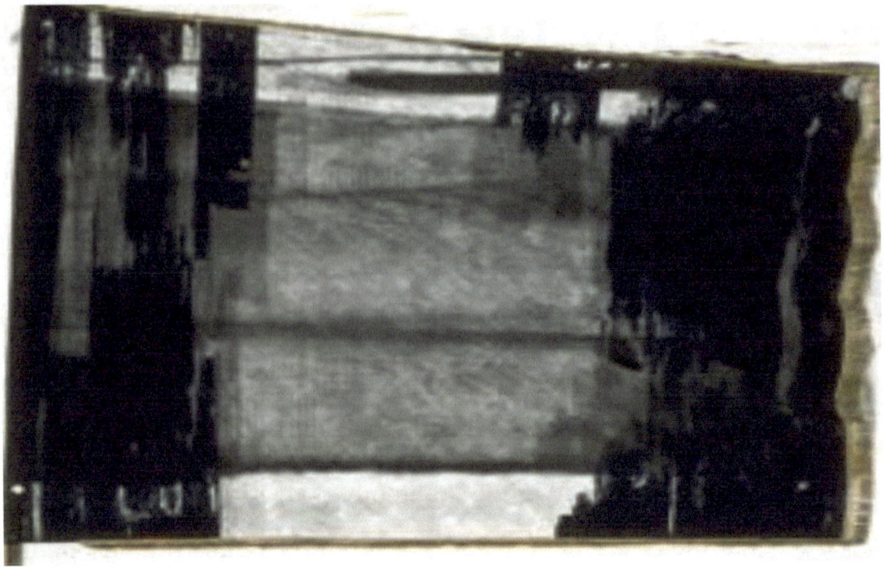

This is a remnant of what appears to have been a Regimental flag; however, the flag is in tatters

No other flags carried by the 133rd Illinois Infantry Regiment are known to survive.

134th Illinois Infantry

The One Hundred and Thirty Fourth Illinois Volunteer Infantry was organized in Chicago[1159] at Camp Fry and mustered in for one hundred days service on May 31, 1864, under command of Colonel Waters W McChesney. The men of Company A volunteered from Cook County; Company B from Iroquois County; Company C and D from Cook County; Company E primarily from Cook County with some volunteers from Winnebago, Livingston and Lake Counties; Company F from Cook County; Company G from Cook, Carroll and Iroquois Counties; Company H from Henry County and Company I and K from Cook County. On June 3, the Regiment was ordered to Columbus, Kentucky for garrison duty. Upon completion of one hundred days of service the Regiment returned to Chicago and was mustered out on October 25, 1864.[1160]

No flags carried by this Regiment known to exist.

[1159] Colonel W. J. Hancock, president of the Board of Trade, raised the Regiment but due to failing health he relinquished command to Colonel Waters W. Mc Chesney. Danforth, Willis, "How I Came to be in the Army, and General E. A. Paine's Plan of Federal Salvation, *MOLLUS, Wisconsin* Volume I, 324-39.

[1160] *Report of the Adjutant General of the State of Illinois,* H. W. Rokker, State Printer and Binder, (1886) Volume VII, page 54.

135th Illinois Infantry

The One Hundred and Thirty Fifth Illinois Infantry was organized in central Illinois and mustered into service for one hundred days at Mattoon in Coles County on June 6, 1864, under the command of John S. Wolfe a Champaign attorney. The men of the Regiment volunteered from the following counties, Companies A and B from Champaign County; Company C from Edgar and Effingham Counties; Company D from Effingham and Vermilion Counties; Company E from Vermilion County; Company F from Bond County; Company G from Douglas County; Company H from Crawford County; Company I from Champaign, Coles and Cumberland Counties and Company K from Vermilion County. The Regiment reported to Benton Barracks in St. Louis on June 10 where five companies remained under the command of Capt. Samuel G. McAdams of Company F. Five companies, commanded by Col. Wolfe, were sent to guard the *Missouri Pacific Railroad* at crossings over the Gasconade River and the Osage River. A short time later these five companies were ordered to guard duty at Jefferson City, Missouri where they remained until expiration of service. Company H, commanded by Lieutenant Peter Jones was ordered to Sulfur Springs, Missouri to guard the *Iron Mountain Railroad.* The Regiment served a total of 114 days and was mustered out at Camp Butler on September 28, 1864.

No flags carried by the One Hundred and Thirty Sixth Illinois Infantry are known to exist.

136th Illinois Infantry

The companies forming the One Hundred and Thirty Sixth Illinois Infantry went into camp at Centralia, in Marion County in early May 1864 and were mustered in for one hundred days service on June 1 under command of Colonel Frederick A. Johns. The men volunteered from the following counties, Company A from Franklin and Perry Counties; Company B from Washington County; Company C from Richland and Wayne Counties; Company D from Perry County; Company E from Marion County; Company F from Clay County; Company G from Richland County; Company H from Edwards County; Company I from Hardin, Massac and Pope Counties and Company K from Pope County. Immediately after muster the Regiment was sent via the *Illinois Central RR* to Cairo and was then sent by boat down the Mississippi River to Columbus, Kentucky to guard against attack by Confederate General Forrest. The Regiment remained at Columbus on garrison duty until mid-August when a report was received stating that Forrest was planning an attack on Paducah, Kentucky. The Regiment was ordered to intercept Forrest and marched from Columbus on August 12. However, on arriving at Mayfield, Kentucky word was received that Forrest had changed his course and was marching to Memphis. Part of the One Hundred and Thirty-Six was sent to Paducah, Kentucky and then moved by boat back to Columbus while the remainder of the Regiment marched directly back to Columbus. In early September, the Regiment's term of enlistment expired but the Regiment was asked to extend service for an additional fifteen days. The Regiment agreed to the extension and on September 26 departed Columbus by boat for Cairo. While enroute by rail from Cairo to Chicago, word was received that Confederate General Price was making a raid through Missouri and the destination of the Regiment was changed to Benton Barracks at St. Louis. From Benton Barracks the Regiment was divided and companies sent to various forts around St. Louis. On October 15 the Regiment was ordered to Camp Butler near Springfield, Illinois and mustered out of service on October 212, 1864, having served a total of 144 days.[1161]

There are no flags carried by the One Hundred and Thirty-Sixth Regiment Illinois Infantry in the collection of the State of Illinois Military Museum.

[1161] *Report of the Adjutant General of the State of Illinois*, H. W. Rokker, State Printer and Binder, Springfield, Ill, (1886) Volume VII, page 89.

137th Illinois Infantry

Washed in the blood of the brave and the blooming,
Snatched from the altars of insolent foes;
Burning with star-fires, but never consuming,
Flash its broad ribbons of lily and rose.[1162]

Colonel John Wood organized the One Hundred and Thirty-Seventh Illinois Infantry at Camp Wood in Quincy in Adams County, Illinois in May 1864[1163]. The Regiment was mustered in for one hundred days on June 5, 1864. The men volunteered from the following counties, Company A from Adams and Hancock Counties; Company B from Adams County; Company C from Knox and McDonough Counties; Company D from Adams, Fulton, and McDonough Counties; Company E from Scott County; Company F from Pike County; Company G from Adams, McDonough and Pike Counties; Company H from Brown and Pike Counties; Company I from McDonough County and Company K from Schuyler County. On June 9 the Regiment was ordered to Memphis, Tennessee for guard and Picket duty in Memphis. On August 21 the Regiment joined in the repulse of Forrest's attack on the city. The Regiment was mustered out of service a Springfield, Illinois on September 4, 1864.

The military flag collection of the State of Illinois includes a national flag carried by the One Hundred and Thirty-Seventh Illinois Infantry Regiment.

The 35-flag shows very little evidence of wear and the pattern of stars in the canton is typical of flags manufactured in Philadelphia. Inscribed in gold lettering on the center stripe— "137th Regiment Illinois Vols."

[1162] Castle, Henry A., *The Army Mule and Other War Sketches*, The Bowen-Merrill Co, Indianapolis, (1898), page 90.
[1163] John Wood established Adams County in 1825 and founded the town of Quincy. He served as mayor of Quincy, was elected to the State Senate and was chosen Lieutenant Governor of Illinois in 1856. Upon the death of Governor Wm Bissell, Wood served as Governor of Illinois from March 18, 1860, until newly elected Governor Yates was inaugurated ten months later.

138th Illinois Infantry

The One Hundred and Thirty-Eighth Illinois Infantry was organized at Camp Wood in Quincy, Adams County, Illinois during June 1864 by Colonel John W. Goodwin and mustered in for one hundred days' service on June 21, 1864. The men of the Regiment volunteered from the following counties, Company A from Monmouth Township in Warren County; Company B from Hancock and Knox Counties; Company C from Warren County; Company D from Oquawka Township in Henderson County; Company E from Knox and Warren Counties; Company F from Livingston County; Company G from Kankakee County; Company H from Grundy County; Company I from La Salle County and Company K from La Salle and Marshall Counties. On June 26 the Regiment was ordered to Fort Leavenworth, Kansas. Companies C and F were detached and sent to Weston, Missouri for garrison duty there from July 7 to August 3. The Regiment was mustered out at Springfield, Illinois October 14, 1864.

There are no flags carried by the One Hundred and Thirty-Eighth Illinois Infantry Regiment in the holdings of the State of Illinois.

139th Illinois Infantry

The One Hundred and Thirty-Ninth Illinois Infantry was organized at Peoria, Illinois under the command of Colonel Peter Davidson and mustered in for one hundred days on June 1, 1864. The men volunteered from the following counties, Company A from Bureau County; Company B from Putnam County; Company C from Woodford and Tazewell Counties; Company D from Galesburg in Knox County; Companies E and F from Peoria County; Company G from Bureau County; Company H from Bureau and Stark Counties; Company I from Mason County and Company K from Geneseo in Henry County. On June 8 the Regiment was sent by steamboat down the Illinois River to St. Louis and then on to Columbus, Kentucky. However, after one week the Regiment was ordered back upriver to garrison duty at Cairo, Illinois. In early August, guerillas captured steamboats near Paducah, Kentucky and took from them several hundred cattle and horses. In retribution, Colonel Davidson was ordered to take his Regiment and capture from rebel farmers living in northern Kentucky a like number of cattle and horses. The stock was recaptured along with several of the guerilla leaders and the expedition returned to Cairo. On September 25 the Regiment was moved to Peoria to be mustered out, the term of the Regiment having expired. However, on arriving in Peoria, it was learned that Confederate General Price was marching toward St. Louis, and there were not enough troops in St. Louis to defend the city. President Lincoln sent a dispatch to Colonel Davidson requesting him to move his Regiment to St. Louis even though the Regiment had already served more than one hundred days. The Regiment immediately moved by rail to St. Louis and then joined in the march toward Franklin, Missouri to meet Price's forces. Learning of the Union advance, Price retreated to southern Missouri. The One Hundred and Thirty-Ninth returned to Peoria and was mustered out on October 25, 1861.[1164]

There were no known colors issued to the One Hundred and Thirty-Ninth Illinois Infantry.

1164 *Report of the Adjutant General of the State of Illinois,* H.W. Rokker, State Printer and Binder, Springfield, IL (1886) Volume VII, page 40.

140th Illinois Infantry

The One Hundred and Fortieth Illinois Infantry was organized at Camp Butler, near Springfield in May and mustered in for one hundred days on June 18, 1864, under the command of Colonel Lorenzo H. Whitney of Dixon. The men volunteered from the following counties, Company A and B from Whiteside County; Company C from Galena in Jo Daviess County; Company D from Lee and Whiteside Counties; Company E from Lee County; Company F from Mercer County; Company G from Rock Island County; Company H from De Witt County; Company I from Ogle and Rock Island Counties and Company K from Madison County. On the day of muster the Regiment left by rail for Cairo and then moved by steamer downriver to Memphis. From Memphis the Regiment marched thirty miles east to Grand Junction near the Wolfe River. There the Regiment was formed into divisions to guard the *Memphis & Charleston RR* and the *Mississippi Southern RR* south as far as Holly Springs, Mississippi. In early October the Regiment returned to Memphis to perform guard duty until ordered to Camp Fry in Chicago for muster out. Although the One Hundred and Fortieth was mustered out on October 29, 1864, they were, at the request of Adjutant General Fuller of Illinois, reorganized and moved to Missouri in pursuit of the Confederate forces under General Price. This expedition occupied six weeks after which the Regiment returned to Camp Fry and dismissed, having served a total of five months.[1165]

There are no flags from this Regiment in the flag collection of the State of Illinois in Springfield.

[1165] *Report of the Adjutant General of the State of Illinois,* H.W. Rokker, State Printer and Binder, Springfield, IL (1886) Volume VII, page 157.

141st Illinois Infantry

The One Hundred and Forty-First Illinois Infantry Regiment was organized at Elgin, Illinois and mustered in for one Hundred days service on June 16, 1864, under the command of Colonel Stephen Bronson. The men volunteered from the following counties, Companies A, B, C and D from Kane County; Company E from Winnebago County; Company F from Kendall, La Salle and Mc Henry Counties; Company G from Kane County; Company H from Du Page County; Company I from Kane County and Company K from Kane and Mc Henry Counties. The Regiment was sent to Columbus, Kentucky for garrison duty. Upon completion of one hundred days of service the Regiment was mustered out on October 10, 1864.

There is no record of any flags being furnished to this Regiment and no flags from this Regiment in the State of Illinois flag holdings in Springfield.

142nd Illinois Infantry

The One Hundred and Forty-Second Illinois Infantry was organized at Freeport, Illinois and mustered in for one hundred days service on June 18, 1864, under the command of Colonel Rollin Ankeny. The men volunteered from the following counties, Company A from Cook and Stephenson Counties; Company B from Jo Daviess County; Company C from Boone County; Company D from Ogle County; Company E from Jo Daviess County; Company F from Stephenson County; Company G from Carroll, Cook and Stephenson Counties; Company H from Randolph and St. Clair Counties; Company I from Ogle County and Company K from Randolph County. On June 21 the Regiment was ordered to Memphis and from there to White's Station east of Memphis to guard the *Memphis & Charleston RR.* In October the Regiment moved to Chicago and was mustered out on October 27, 1864.

There are no flags of the One Hundred and Forty Second Illinois Infantry Regiment in the flag collection of the State of Illinois Military Museum.

143rd Illinois Infantry

The One Hundred and Forty-Third Illinois Infantry was organized at Mattoon, Illinois by Colonel Dudley C. Smith and mustered in for one hundred day's service on June 11, 1864. The men volunteered from the following counties, Company A from Charleston in Coles County; Company B from Cairo in Alexander County; Company C from Christian and Montgomery Counties; Company D from Clay and Effingham Counties; Company E from Fayette County; Company F from Effingham, Madison and Vermilion Counties; Company G from Shelby County; Company H from Coles and Montgomery Counties; Company I from Moultrie County and Company K from Coles, Effingham and Madison Counties. On June 16, the Regiment was ordered to Memphis for guard duty and remained there until July 27 when ordered to Helena, Arkansas for garrison duty. On September 10 the Regiment proceeded to Mattoon, Illinois where it was mustered out on September 26, 1864.

The military flag holdings of the State of Illinois Military Museum include a National and Regimental flag carried by the One Hundred and Forty-Third Illinois Infantry.

This National flag contains 34 stars in a square canton. A square canton is typical of flags manufactured in New York, but the pattern of the stars is not typical of New York manufacture, therefore the origin the flag is unclear-presumably homemade. The flag is inscribed: 143rd REGT ILL VOL INFY. *It measures 59" on the hoist by 73" on the fly.*

On the 34-star Regimental flag, the position of the stars above the scroll, the design of the eagle and the embellished scroll beneath the eagle are typical of flags manufactured in New York.

No other flags carried by the 143rd Illinois Infantry are known to survive.

144th Illinois Infantry

The One Hundred and Forty-Fourth Illinois Infantry Regiment was organized at Alton, Madison County, Illinois and mustered in for one year's service on October 21, 1864, under the command of Colonel Cyrus Hall. The men volunteered from the following counties, Companies A and B (*Alton Battalion*), C, D, E, F, G, H, I and K from Madison County, and in addition men from Greene County volunteered for Company E and men from Jersey County volunteered for Companies F, G and H. The Alton prison opened in 1833 as the first Illinois Penitentiary but was closed in 1860 and the prisoners were moved to Joliet. In 1862 the Alton prison was reopened to hold Confederate prisoners and other anti-Union "undesirables." The prison was first guarded by men from the Thirteenth U.S. Infantry and subsequently by men of the Seventy-seventh Ohio Infantry, the Thirty-seventh Iowa Infantry, the Tenth Kansas Infantry and finally by the One Hundred and Forty-Fourth Illinois Infantry. At the conclusion of the Civil War the prison closed on July 7, 1865, and the Regiment was mustered out on July 14, 1865. The prison was later torn down and the land became a city park.

The military flags collection of the State of Illinois holds only a Regimental flag carried by the One Hundred and Forty-Fourth Illinois Infantry Regiment. It contains 34 stars and is typical of flags manufactured for the New York depot.

No other flags carried by the 144th Illinois Infantry are known to survive.

145th Illinois Infantry

The One Hundred and Forty-Fifth Illinois Infantry was organized and mustered in for one hundred days' service at Camp Butler, near Springfield, Illinois on June 9, 1864, under the command of Colonel George W. Lackey. The men volunteered as follows, Company A from Vienna in Johnson County; Company B from Clinton and Mc Lean Counties; Company C from Morgan County; Company D from Logan County; Company E from Macon County; Company F from Jackson and Williamson Counties; Company G from Cass and De Witt Counties; Company H from Christian and Logan Counties; Company I from Mc Lean and De Witt Counties and Company K from Mc Lean County. The Regiment was ordered to St. Louis, Missouri and performed guard duty in that District until returned to Camp Butler for muster out on September 23, 1864.

The military flags collection of the state of Illinois contains a national flag carried by the One Hundred and Forty-Fifth Illinois Infantry Regiment. There are no inscriptions on the flag and the manufacturer is unknown.

No other flags carried by the 145th Illinois Infantry are known to survive.

146th Illinois Infantry

The One Hundred and Forty-Sixth Illinois Infantry Regiment was organized at Camp Butler in Springfield, Illinois under the command of Colonel Henry H. Dean and was mustered in for one years' service on September 18, 1864. The men volunteered from the following counties, Company A from Carroll County; Companies B and C from Winnebago County; Company D from Lake and Kendall Counties; Company E from Stephenson County; Company F from Bureau County; Company G from Mc Lean County; Company H from Bureau County; Company I from Peoria County and Company K from Mc Henry and Tazewell Counties. Companies A, E, H, I and K remained at Camp Butler to guard prisoners, volunteers, and substitutes while Companies C and B performed guard duties at Brighton, Illinois and Company F guard duties at Jacksonville. Illinois. When President Lincoln was assassinated in April 1865, the One Hundred and Forty Sixth Illinois Infantry was selected as the honor guard to accompany the body from the Illinois State House to Oak Ridge Cemetery. The Regiment was issued new uniforms, rifles and bayonets and drilled five hours a day for two weeks in readiness for the funeral procession.[1166] The Regiment was mustered out of service at Camp Butler on July 5, 1865.

The National flag and the Regimental flag of the One Hundred and Forty Sixth Regiment Illinois Infantry are in the collection of the State of Illinois. Both are in good condition and were presumably carried by the Regiment as they marched in Lincoln's funeral cortège.

There are no inscriptions on the flag and the manufacturer is unknown.

[1166] Kunhardt, Dorothy Merserve and Kunhardt, Philip B. Jr., *Twenty Days*, Harper and Row, Publishers, New York, (1965), pp281-2.

146TH ILLINOIS INFANTRY

The 34-star Regimental flag is typical of New York manufacture based on the position of the stars above the eagle and the embellished scroll beneath the eagle. The flag measures 71" on the hoist by 73" on the fly.

No other flags carried by the 146th Illinois Infantry are known to survive.

147th Illinois Infantry

The One Hundred and Forty Seventh Illinois Infantry was organized at Camp Fry, Chicago by Colonel Hiram F. Sickles and mustered in for one year's service on February 18 and 19, 1865. The men of the Regiment volunteered from the following counties, Company A from Winnebago County; Company B from Whiteside County; Company C from Kendall and La Salle Counties; Company D from Cook County; Company E from Stephenson County; Company F from De Kalb County; Company G from Whiteside, Lee and Ogle Counties; Company H from Cook and La Salle Counties; Company I from Lake and Mc Henry Counties and Company K from Cook and Kankakee Counties and from the state of Indiana. On February 25 the Regiment departed for Dalton, Georgia traveling through Louisville, Nashville and Chattanooga, Tennessee en-route. On March 12 a detachment of the Regiment joined in an assault on a nearby guerrilla camp and on March 20 seven companies of the Regiment marched to Spring Place, Georgia, east of Dalton to break up a guerilla camp. In late March the Regiment moved to Ringgold, Georgia for guard duty and to suppress guerrillas active in the area.[1167] In May the Regiment was ordered to Calhoun where they assisted in accepting the surrender and parole of the rebel forces in north Georgia. From August through November the One Hundred and Forty Seventh Illinois performed guard duty in southwest Georgia and then moved to Savannah where the Regiment was mustered out on January 20, 1866.

The military flags collection of the State of Illinois holds one National and one Regimental flag carried by the One Hundred and Forty Seventh Illinois Infantry.

Based on the pattern of the stars, and the shape of the canton, this 35-flag was probably manufactured by Wm. Scheible of Philadelphia.

[1167]Canadian born, Corporal Welsey J. Powers, Company F, from Blackberry in Kane County, was awarded the Medal of Honor for "Voluntarily swam the river under heavy fire and secured a ferryboat by means of which the command crossed. The action occurred on the Oostanaula, River in Georgia on April 3, 1865.

The 35-star Regimental color was manufactured by John S. Shilleto & Co. of Cincinnati.

No other flags carried by the 147th Illinois Infantry are known to survive.

148th Illinois Infantry

The One Hundred and Forty Eighth Illinois Infantry Regiment was organized at Camp Butler and mustered in for one years' service under the command of Colonel Horace H. Willsie on February 22, 1865. The men volunteered from the following counties, Company A from Bureau County; Company B from Adams and Fulton Counties; Company C from Peoria and Tazewell Counties; Company D from Adams County; Company E from Bureau County; Company F from Adams, Knox and Peoria Counties; Company G and H from Adams County; Company I from Henry and Pike Counties and Company K from Bureau County. On February 22 the Regiment was ordered to Nashville, Tennessee and from there to Tullahoma, Tennessee. The Regiment was on duty at Tullahoma from March 1 until June 18 and then divided into detachments—4 companies were sent to Decherd, Tennessee to guard the junction of the *Fayetteville Railroad* with the *Nashville & Chattanooga Railroad,* one company sent to McMinnville, Tennessee to guard the terminus of the railroad from Tullahoma and 4 companies detailed between Lombardy and Anderson's Station, Tennessee to guard the *Nashville & Chattanooga Railroad.* The Regiment was mustered out at Nashville, Tennessee on September 5, 1865, and returned to Springfield, Illinois.

The military flags collection of the State of Illinois holds one National and one Regimental flag carried by the One Hundred and Forty Eighth Illinois Infantry.

This National color has the star arrangement in the canton that is typical of those colors manufactured by Wm Scheible of Philadelphia

The unique pattern of the stars above the upper scroll, the position of the mid part of the lower scroll and the position of the eagle and shield are typical of flags manufactured by John S. Shilleto of Cincinnati.

No other flags carried by the 148th Illinois Infantry are known to survive.

149th Illinois Infantry

The One Hundred and Forty Ninth Illinois Infantry was organized at Camp Butler and mustered in for one years' service on February 11, 1865, under the command of Colonel William C. Kueffner. The men of the Regiment volunteered from the following counties, Company A from Sangamon County; Companies B, C and D from St. Clair County; Company E from Vermilion County; Company F from Douglas County; Company G from Brown, Cumberland, Jersey, and Morgan Counties; Company H from St. Clair County; Company I from Madison and St. Clair Counties and Company K from St. Clair County. On February 14, the Regiment was ordered to Chattanooga, Tennessee for railroad guard duty and then on May 2 was ordered to Dalton, Georgia and on July 6 to Atlanta. From July 25 until muster out the Regiment was assigned to guard duty in Allatoona, Georgia. The Regiment was mustered out at Dalton, Georgia on January 27, 1866.

The military flags collection of the State of Illinois has one Regimental flag carried by the One Hundred and Forty Ninth Illinois Infantry Regiment.

This 35-star Regimental flag was manufactured by John S. Shilleto & Company of Cincinnati.

No other flags carried by the 149th Illinois Infantry are known to survive.

150th Illinois Infantry

The One Hundred and Fiftieth Illinois Infantry Regiment was mustered in for one year's service under the command of Colonel George W. Keener at Camp Butler on February 14, 1865. The men of the Regiment volunteered from the following counties, Companies A and B from Mc Lean County; Company C from Edgar County; Company D from Iroquois County; Company E from Vermilion County; Company F from Clinton and Randolph Counties; Company G and H from Madison County; Company I from Macon County and Company K from Vermilion County. The Regiment was ordered to Bridgeport, Alabama on February 18 to garrison Forts 3 and 4 and the blockhouses on the *Nashville & Chattanooga Railroad* between Bridgeport, Alabama and Chattanooga, Tennessee. On March 24 the Regiment was ordered from Bridgeport to Cleveland, Tennessee and then on May 3 to Dalton, Georgia; the Right-Wing of the Regiment remained at Dalton while the Left-Wing garrisoned Spring Place, Georgia. The Regiment left Dalton on July 7 and arrived in Atlanta the following day. Five weeks later, August 14, the Regiment was ordered to Griffin, Georgia. The Companies of the Regiment were distributed in the District of Allatoona as follows—Companies A and E at Griffin; Company D at Jackson; Company C at West Point; Company F at Newman, Companies B and G at LaGrange, Company K at Greenville; Company H at Franklin; and Company I at Atlanta. The Regiment mustered out of service on January 16, 1866, at Atlanta, Georgia.

There are no flags of the One hundred and fiftieth Illinois Infantry in the holdings of the State of Illinois.

One Hundred Day Regiments

Early in 1864 the Governors of Illinois, Ohio, Indiana, and Iowa, believed that the War was nearly over based on the advantageous position of the Union forces along the Mississippi River and the impending campaign in north Georgia. The governments in these four states wished to hasten the end of the War and tendered to President Lincoln a volunteer force of 85,000 men to be mustered for one hundred days service. The purpose of the "One Hundred Day Regiments" was to relieve veteran soldiers who were guarding prisoners, forts, and arsenals, and allow veterans to join the forces in the field in a push to end the War quickly.

Illinois furnished thirteen regiments--the One Hundred and Thirty Second through the One Hundred and Forty Third Infantry Regiments plus the One Hundred and Forty Fifth Infantry Regiment as One Hundred Day Regiments. Illinois also furnished two companies including the *Alton Battalion* as a One Hundred Day Battalion. Five of the Regiments, upon completion of their one hundred days of service extended their term of service to aid Union General Rosecrans in his defense of Missouri. None of these Regiments were issued flags.

CAPTURED CONFEDERATE FLAGS

The State of Illinois has in its collection captured Confederate flags that were not returned to the former Confederate states. During the Civil War, there was no specific directive concerning the disposition of Confederate flags captured by the Union troops. In some cases, the flags were passed up through the ranks to army command while at other times the flags were retained by the men in the company or regiment who had captured them. By the end of the War there were at least 500 Confederate flags in the possession of the U. S. Government. In 1887 President Grover Cleveland ordered the return of the captured flags but the order was rescinded following political pressures from Union veteran organizations. However, in 1905 the government returned the captured Confederate flags to the Governor of each of the former Confederate states and to the appropriate border States. Those flags which were not associated with a specific State or where the State of origin was unknown were deposited with the antecedent organization of what would become the Museum of the Confederacy in Richmond. Many of the Confederate flags in possession of the Northern States were also returned over the years including after World War Two, but Illinois, Iowa, Minnesota, New York and Wisconsin along with the Chicago Historical Society retained possession of some of the captured flags. Ohio also still has a few.

The following Confederate flags are those that are still held by the State of Illinois.

First National Flag

Adopted on March 4, 1861 while the Confederate government was in Montgomery, Alabama, the flag's description from the Committee on Flag and Seal stated, "The flag of the Confederate States of America shall consist of a red field with a white space extending horizontally through the center and equal in width to one-third of the width of the flag, and red spaces above and below to the same width as the white, the union blue extending down through the white space and stopping at the lower red space, in the center of the union a circle of white stars corresponding in number with the States of the Confederacy."[1168] The Confederate flags in the Illinois collection are all First National Flags. The State does not have any Second or Third National flags.

This cotton flag measures 86 x 51" with a field of three horizontal bars, red (16"), white (16 1/2") and red (16 1/2"). The canton is blue, 34 1/2 on the staff by 28 1/2" on the fly. The eleven stars are white with ten in a circle of 22" in diameter (each star measures 5 1/2" in diameter); the center star is 8 1/2" in diameter. The leading edge of the flag has been hemmed 3/4" to serve as reinforcement to nail the flag to the staff. Mrs. M. E. Bane donated the flag to the Illinois National Guard 16 January 1901. No other history is available.

This is probably a garrison flag since it measures 167 x 109". The material is cotton (Madaus says it is wool bunting) bunting. The canton measures 72" on the staff by 63 1/4" on the fly. Each of the seven 5 pointed stars is 9 1/2" in diameter. The stars are set in a 10 1/2-diameter circle with the stars sewn on the reverse and the dark blue bunting cut away on the obverse to expose the star. There are three whipped eyelets that pierce the 1 3/4" white canvas heading. This flag is probably an early garrison flag of Fort Blakely, east of Mobile, Alabama taken 9 April 1865. This would have been replaced by the time of the battle with a Second National flag but this example was retained at the fort. According to the Illinois Adjutant General's records Captain Daniel Webster Sayer, from Charleston, Illinois, and

[1168] Congress of the Confederate States of America, March 3, 1861.

member of Co. C of the 8th Illinois Infantry captured this flag at the battle of Fort Blakely (Mobile Bay was the naval battle of August 1864). His son, Rockwell Trower Sayer of Chicago, Illinois donated the flag to the State of Illinois. German born Captain Henry Miller[1169], Company B, of Decatur received the Medal of Honor for "Capture of Flag" at Fort Blakely.

This is probably a garrison flag since it measures 168 x 107". The red material is cotton, and the white is flannel. The dark blue canton measures 75" on the staff by 68" on the fly. The 11 white stars are each 9" - 10" in diameter with ten of them set in a 54" diameter circle. The center star is 12 1/2 - 13" in diameter. The stars are sewn on the obverse and the dark blue bunting is cut away on the reverse to expose the white cotton star. There are brass grommets on each end of a 2 1/4" canvas header. Stenciled on the header in black is the inscription VAUGHN/MAKER and 18/COMMERCE St. [1170] This was the maker's mark of Irish immigrant, Henry Vaughan, a sail and flag maker from Mobile, Alabama.

Union forces commanded by Brig. Gen. John Palmer, on 7 October 1862 at LaVergne, Tennessee, took this flag. He presented the flag to the State of Illinois. In a Report of Brig. Gen James S. Negley; 9 October 1862: "at LaVergne, Tennessee on 7 October 1862:...The Confederates held their ground for thirty minutes and then fled in the wildest disorder, leaving 175 prisoners in our hands, among whom were 2 lieutenants, 2 lieutenant-colonels, and a number of line officers, three pieces of artillery, ordnance and quartermaster stores, a large amount of provisions, camp equipage, personal effects, stand of regimental colors, and three railroad cars which we destroyed. Their defeat was complete. Their loss in killed and wounded was about 60."[1171] This report details that the Confederate force that the Union troops faced was the Thirty-second Alabama Infantry who also lost their colors in this engagement.

This flannel flag measures 64 x 34" with JEFF DAVIS made of 6" high black

---

[1169] *Civil War Medal of Honor Winners,* published by the Civil War Centennial Commission of Illinois, Springfield, July 1962

[1170] There is one other Confederate flag identified in the heading as having been made by Vaughn. It is in the Museum of the Confederacy and is Accession No. 0985.2.72 formerly part of the Alabama Room Collection.

[1171] O.R, Series I, Vol. 16, pt 1, p. 1020-1

velvet outline letters sewn to obverse side only. The canton is dark blue, 23 x 23". The twelve stars are 4 1/2 " in diameter. The red bars are 11" wide and the white bar 12". The leading edge of the flag was doubled back over a white tape to form a reinforced hem, 3/4" wide, which was nailed to the staff.

According to the records of the Adjutant General of the State of Illinois this flag was captured at the Battle of Pea Ridge, Arkansas by the 37th Illinois Infantry Regiment and was sent to the State of Illinois.[1172] In a report of Col. Julius White, 37th Illinois Volunteers:[1173] "The force we encountered consisted of the Third Louisiana, Col. Hebert, a regiment formerly commanded by General McIntosh, Colonel Mitchell's and Colonel McRae's two regiments of Arkansans, and a large body of Indians, under command of Col. Albert Pike, with a reserve of several other regiments, all being under the chief command of General Benjamin McCulloch……. Their triumph was short-lived, however, for the Thirty-seventh Illinois immediately fired upon them and charged, routing their right wing at the same time that the First Brigade, under Colonel Pattison, came into action on our right, driving the left wing of the enemy in confusion from the field and retaking our guns." The Confederate forces involved in the Battle were the 3rd Louisiana Volunteers, the 2nd Arkansas Mounted Rifles, the 4th Arkansas Volunteers, 21st Arkansas Volunteers and the 14th Arkansas Volunteers. It is not known which Confederate Regiment carried this flag.

[1172] *Report of the Adjutant General of the State of Illinois*, H.W.Rokker, State Printer and Binder, Springfield (1886) p163
[1173] O.R., Series I, Vol 8, page 253 (dated March 6-6, 1862—Battle of Pea Ridge or Elkhorn Tavern, Ark.)

Battle Flags

The Confederates used several battle flags. The majority was patterned after those of Virginia - a square flag with blue cross and white stars. Four of the flags in the Illinois collection are of this type while the remaining two have other, very distinctive patterns.

According to an attached tag, this flag was "Captured together with two pieces of artillery, at Brookhaven, Miss. by the 2nd Cavalry Brigade, Col. J. G. Fonda com'd , August 18, 1864"[1174].

The obverse side of the Battle flag is inscribed "32 SHOTS. IN. GUNBOAT. GENL. BRAGG. JUNE 15. 1864. BARLOW'S BATTERY". The red cotton flag measures 51 x 37". The Southern Cross is blue with 13 stars. The inscription is sewn separately onto flag using white silk block letters. The flag has a sleeve 2 1/2" wide and red silk ties on the upper point of the sleeve.

The reverse side of the flag is inscribed TRUST IN GOD.

In the report of Brigadier-General Albert L. Lee, Commander Cavalry Division, Department of Gulf; 22 November 1864 "At 5 p.m., 17th Colonel Fonda, with 500 men of the Second Brigade, moved on Brookhaven, distance forty-six miles. At daylight on 18th he surprised the town, scattering a small infantry force stationed there and capturing a section of artillery, with caisson. The gunners were many

[1174] *Report of the Adjutant General of the State of Illinois,* H.W.Rokker, State Printer and Binder, Springfield (1886) Volume I, page 163

of them shot down at their pieces. He captured about 50 prisoner;..."[1175] Barlow's Battery was Barlow's Louisiana Battery commanded by Capt. William P. Barlow and was engaged at Brookhaven on November 18, 1864.

This red cotton flag measures 46 x 34" exclusive of 2" yellow fringe. The blue cross (faded) is 5" wide. The letters are painted, 4 1/4" high and read ROBERTSON CAVALIERS! GOD FOR THE RIGHT! The stars in clusters of three each measure 4" in diameter and the center star measures 5" in diameter. Although this flag is in the possession of the State of Illinois there is no record as to when and where this flag was captured. Unfortunately, a Confederate unit such as "Robertson Cavaliers" has never been found. This flag in design is similar to some that have come from Texas both in the secession era and unit colors so that offers a possible clue as to its identity.

This red banner, made by Charles W. Stiles under contract from the Charleston South Carolina Quartermaster Depot, is made of bunting, measures 44 1/2 x 46" and is the battle flag of the 46th Mississippi Infantry. The dark blue cross is 8" wide with a 1" white edge. Each star is 4 1/2" in diameter. The stars are sewn to both the obverse and reverse sides of the flag. There is a white 2 1/4" border around the flag. According to the records of the Illinois Adjutant General this flag was captured by Corporal Joseph Claucus, Company E, 2nd Illinois Cavalry, at Blakely, Alabama, 1 April 1865.[1176]

[1175] O.R. Series 1, Volume 45, part 1, pp. 3-5

[1176] *Report of the Adjutant General*—Volume 1, page 164

In a report of Lieutenant-Colonel A. B. Spurling, Commander of 2nd Maine Cavalry and 2nd Brigade, Lucas' Cavalry Division, near Fort Blakely, Alabama, April 7,1865:[1177]"...The enemy slowly retired before this advance, taking shelter behind fences and everything that could afford protection. They retired in this way for over a mile, contesting every inch of the ground. I had moved the Second Illinois up the road and kept its advance on a line with the Second Maine, and judging that the favorable opportunity had arrived I charged the enemy with that regiment. It was a complete success. His whole force was routed and a portion of it captured. Two companies of the 2nd Illinois Cavalry pursued the fugitives within half a mile of the enemy's works at Blakely, from which a sharp fire was opened with artillery. In all, 74 men were taken prisoners, including 3 commissioned officers. Nearly all of them were members of the Forty-sixth Mississippi Infantry. The colors of that regiment were also captured:"

Due to this flag's soiled and torn condition in 1986 it underwent conservation. After appropriate cleaning it was backed with dyed silk crepeline, tears were sewn in place with crepeline thread and flag stitched to the crepeline backing for support where necessary.

Made by the Richmond, Virginia Quartermaster Depot, this Third Bunting issue flag is made from red bunting and cotton cambric flag measuring 49 x 49". The cross is dark blue 4 3/4" wide with a 1/2" white edge. The thirteen stars are white and each measure 3 3/4" in diameter. The flag has a 2" wide white border. The flag was attached with three whipped eyelets, which pierce the white twill weave cotton heading measuring 2 1/4" wide.

According to the Adjutant General's Report: "This flag was captured in the last charge of July 22, 1864, before Atlanta, Ga., made by the 66th and 64th Illinois volunteers, sharpshooters. The flag belonged to the Texas Rangers, and was only surrendered by its bearer in his death, to Capt. George W. Reid and Sergeant (now 1st Lieut.) Henry Miller, Co. D, 64th Illinois".[1178]

In a letter from G.M.G. of the 9th Illinois Infantry, "The withering fire of the 66th Illinois with their 16 shooters did much toward saving the 16th corps and finally making a complete victory. That regiment alone captured several scores of prisoners. During the fight one man alone brought in four rebels. One of that regiment captured the colors of the 66th Georgia."[1179] Based on this report the captured flag is the flag of the 66th Georgia and not the flag of the Texas Rangers. This is one of the

---

[1177] O.R., Series I, Volume 49, Part 1, page 311.

[1178] *Report of the Adjutant General of the State of Illinois,* Vol 1., p164

[1179] *Aledo Weekly Record,* (Mercer County, Illinois) Wednesday August 17, 1864

few examples on a Richmond Depot battle flag that served in the Confederate Army of Tennessee which carried its own style of rectangular battle flags in 1864.

This flag is similar to a South Carolina State Flag. It measures 77 x 52" and is cotton with painted border and painted design. The design is a green palmetto tree with two red-edged, yellow shields. There is a yellow rattlesnake at the base of the tree and five black arrows protruding from the grass beneath the tree. The scroll is white and the black outline letters read: CONSTITUTIONAL RIGHTS. This interesting flag was captured by Captain Patrick's Company C, 16th Regiment Illinois Volunteer Infantry from a party of rebels at Utica, Missouri on 14 June 1861.Thus the capture of this flag predates the First Battle of Bull Run. It is possible that the flag dates from the era of the Missouri-Kansas borders wars of the 1850's.

Utica, Missouri was the home of the 4th Division of the Missouri State Guard. Most of their flags bore the Missouri State seal; however, it is possible that this was a company flag of the Guard.

The Kansas State Museum has two South Carolina flags from pro-slavery forces that were captured in Kansas in the border wars in the late 1850’s. Apparently, several militia units from Southern states marched to Kansas to support the pro-slavery forces—among these forces were several from South Carolina. It is possible that this flag might have belonged to one of those units.

As a reflection of the Southern sentiments that existed in Kansas, Missouri and Nebraska during the early 1860s is an article reported in the Indianapolis “Daily Sentinel”[1180] –“ Nebraska City, Feb. 10. Old Fort Kearney was taken possession of last night by a party of disunionists, and this morning a palmetto flag waves over the fortress bearing the motto “Southern Right.”’ Based on the above information, it appears that the palmetto flag was relatively common in the West at that time.

[1180] *Daily Sentinel,* February 21, 1861

This flag of dark blue bunting flag measuring 39 x 33". There is a red cotton shield, 11 1/2" high by 10" wide with a 1/16" white bound edge. The white cotton letters ("M. A.") are 2 3/4" high, and the white stars each measure 1 3/4" in diameter. The 13 stars are set in a circle that is 18" in diameter. An old tag attached to this flag reads: "Confederate Artillery Flag. Newton McCan, a private soldier, Co. M, 2nd Regiment U.S. Regulars, captured this flag on the evening of July 3rd 1863 at Gettysburg, Pennsylvania. Presented to the State of Illinois on 14 July 1885."

The 2nd Regiment of U.S. Cavalry was part of the Cavalry Corps of the Army of the Potomac. According to Confederate records, Moody's Battery or Moody's Artillery was in the battalion of Reserve Artillery on the Confederate right on 2 July 1863 and may have accidentally left their flag in a night movement. See O.R. Series I, Vol. 27, Pt. 1, pp387-388 and 429-431. Moody's Battery was the Madison Light Artillery, a Louisiana Artillery battery, also called Madison Tipperarys. Capt. George V. Moody was the commanding officer, and it is assumed that the inscription on this flag stood for Moody's Artillery.

Unfortunately, there are problems with the documentation of this flag. 1) There is no accession record for it therefore, when and by whom the flag was donated is unknown. 2) The name Newton McCan or Newton McCann does not appear as a soldier in either the ranks of US Cavalry or in any of the regiments of Illinois Infantry, Cavalry or Artillery.

Nevertheless, this is clearly a Confederate flag, almost certainly captured at Gettysburg and probably that of Moody's Artillery. Research in the future may shed further light on the story of this flag.

Flag Manufacturers

Prior to the Civil War, all the United States army's flags were supplied from the Philadelphia depot at the Schuylkill Arsenal. The Quartermaster of the army purchased cloth for the flags but transferred the cloth to a Philadelphia contractor, Samuel Brewer to sew and paint the flags. In addition to Brewer, the Philadelphia depot also let contracts for flag production to Horstmann Brothers, Evans & Hassall, William Scheible and Albert Ertle.

However, with the influx of thousands of new troops and hundreds of new regiments at the outset of the Civil War, the Philadelphia depot was overwhelmed, and additional depots were established by the Union Quartermaster at New York and Cincinnati. Although all the contractors for these depots produced flags in keeping with the Federal regulations, the regulations did not define such items as the arrangement of the stars and shape of the canton on the National flag or the shape and position of the eagle on the Regimental flag. Furthermore, some states such as Wisconsin, purchased the flags for their state volunteer regiments directly from flag manufacturers and not through the Quartermaster Department of the Army. The state of Illinois did not purchase flags for their volunteer regiments, rather the Quartermaster Sergeant of the regiment or the Quartermaster of a Brigade ordered the flags from the Quartermaster Department of the Union Army. However, early in the War most of the flags carried by the Illinois Regiments were either purchased by or sewn by patriotic supporters of companies or regiments. Therefore, the flags carried by the Illinois Volunteers came not only from the Quartermaster Depots at New York, Philadelphia, and Cincinnati but also from manufacturers in Chicago and Louisville and "homemade" from patriots at home.

Although the national and regimental flags carried by Union troops were made to Federal "standards," there was latitude in the interpretation of the standards. This resulted in differences in national and regimental flags produced by the various manufacturers. The following information defines these differences so that of manufacturer of a flag can be determined based on unique findings in flag design.

The flags carried by Illinois regiments, now in the holdings of the state of Illinois include flags produced in Chicago, Cincinnati, Louisville, Philadelphia, and New York. Although some of the flags carried by Illinois Volunteers were possibly made in St. Louis and Baltimore, since none of these flags are in the collection of the State of Illinois these manufactures will not be included.

CHICAGO DEPOT

Four Chicago based flag manufacturers advertised in the Chicago newspapers in the 1860s—

1. Gilbert Hubbard & Co., 205, 207 South Water St.,
2. Foster & Hardenbergh, 242 South Water St.
3. Purington & Scranton, 209 South Water St.
4. G. Down & Company, 150 Lake Street

However, only flags produced by Gilbert Hubbard & Company are in the State of Illinois holdings and in flag presentations reported in the Chicago newspapers, only flags produced by Gilbert Hubbard & Company are mentioned. A name associated with Gilbert Hubbard & Company is the name of the man responsible for painting the flags-presumably the eagle on the regimental flags and standards—Charles L. Dubois. The major business of the first three manufacturers was not flag production but rather they were sail makers, dealers in twine and cordage and makers of tents, tarpaulins and awnings while A. G. Down & Company sold "Flag Material and Ladies Cloths."

This is the flag of the Eighty-eighth Illinois Infantry produced by Gilbert Hubbard & Company. As seen in this flag, the National colors manufactured by Hubbard had the following characteristics:

1) *Gold paint used to inscribe the unit designation on the center stripe*
2) *A rectangular canton with gold painted stars arranged in six horizontal rows with five or six stars per row. Some flags made in 1865 may have seven rows of stars with five stars per row.*
3) *Flag secured to staff with silk ties that passed through four small brass grommets in the hemmed leading edge of the flag*[1181]

This is the regimental flag of the Twentieth Illinois Infantry and was produced by Gilbert Hubbard & Company. The regimental flags of Gilbert Hubbard & Company have the following findings:

1) *Thirteen stars surround the upper scroll*
2) *Rays extend upward from the stars*
3) *Eagle painted gold with flat upper border of wings and head to left*
4) *Shield covering center of the eagle's body*

[1181] Madaus, Howard Michael, "Part I, Into the Fray: The Flags of the Iron Brigade, 1861-1865," *Wisconsin Magazine of History,* Vol. 69, Number 1/Autumn, 1985, p 12.

5) *Lower scroll with gold lettering as shown here or lower scroll in three sections with center section lowered and state name in capital letters in the center section.*

CINCINNATI DEPOT

There were two flag manufacturers for the Cincinnati Depot:

1. Longley & Brother, 164 Vine Street (in 1862) and at 143 Walnut Street (1863-1865)
2. John S. Shilleto & Company, 101, 103 & 105 West 4th Street.

A third flag manufacturer in Cincinnati during the Civil War was Hannibal G. Hamlin, 92 West 4th Street. He produced high quality embroidered flags, but none are in the Illinois collection.

This is the national flag of the Seventy-seventh Illinois Infantry and is typical of national flags produced in Cincinnati by both Longley & Brother, and John Shilleto & Company.

The characteristics of Cincinnati manufactured national flags are:

1) Rectangular canton
2) The gold stars are arranged in seven horizontal rows of 5,5,5,5,5,5,4 stars for thirty-four star flags or 5,5,5,5,5,5,5 stars for thirty-five star flags

Horstman Brothers, Evans & Hassall and William Scheible of Philadelphia may have also produced flags for the Cincinnati Depot. The characteristics of their flags are detailed under "Philadelphia Depot."

This regimental flag of the Forty-fifth Illinois Infantry is representative of flags produced by Longley & Brother in Cincinnati. The characteristics of flags by this manufacturer are:

1) Outline of eagle is clear, but the feathering is ill-defined
2) The "black eye" in the white feathery head is poorly defined
3) A double arc of gold stars over the upper scroll that extend to the tails of the scroll but do not extend beyond the tips of the scroll
4) Twenty-one stars in the upper arc over 13 stars in the lower arc in the 34-star flags and 21 stars over 14 stars in the 35-star flags.
5) A three-piece red scroll beneath the eagle with a lowered center section.

Flags produced by Longley & Brothers in Cincinnati and Evans & Hassall in Philadelphia are similar except for the feather detail of the eagle and the eye of the eagle.

Above is the regimental flag of the Thirty-sixth Illinois Infantry manufactured by John S. Shilleto & Company of Cincinnati. Flags manufactured by Shilleto have the following characteristics:

1) The eagle is well defined with detailed feathering and a realistic head with a well-defined eye

2) A double arc of stars over the upper scroll with 19 stars in the upper arc over 15 stars in the lower arc in 34-star flags and 20 stars over 15 stars in the 35-star flags
3) The upper arc of stars extend beyond the tails of the upper scroll and the lower arc of stars is at the edge of the tails of the upper scroll

LOUISVILLE DEPOT

Several Illinois Civil War regiments carried flags manufactured by Hugh Wilkins of Louisville, Kentucky.

This national flag of the Seventy-ninth Illinois Infantry is typical of those flags manufactured by Wilkins.

The flags have the following characteristics:

1) Square canton either dark blue background as in flag above or with light blue background
2) Stars either silver or gold
3) Stars arranged in six horizontal rows of 5,6,6,6,6,5 for 34 star flags and 6,5,6,5,6,6, for 35 star flags
4) The flags were supplied with a staff topped by a spear shaped sheet metal finial painted red

This is the regimental flag of the Eighty-fifth Illinois Infantry and is typical of one variant of regimental flag produced by Hugh Wilkins.

The flag has the following characteristics:

1) Eagle with head toward staff on the obverse of the flag
2) Thirteen stars above eagle and across wings
3) No upper scroll in eagle's beak
4) Eagle's wings are outstretched and flat
5) Five-piece lower scroll for unit designation

This is the regimental flag of the Eighty-fourth Illinois Infantry and is typical of the second variant of regimental flag produced by Hugh Wilkens.

1) Eagle perched on US shield
2) Head of eagle down turned and holding red banner
3) Thirteen stars surround eagle
4) Five piece lower red scroll bearing unit designation

PHILADELPHIA DEPOT

Manufacturers of flags for the Philadelphia depot were:

1) Evans (George) & Hassall (William), 418 Arch Street, Philadelphia. These men left Horstmann Brothers & Company in 1858 to form their own company.
2) Samuel Brewer-no flags in Illinois holdings manufactured by Brewer
3) Horstmann Brothers & Company, salesroom at 723 Chestnut St. with factory at 5th & Cherry, Philadelphia.
4) William F. Scheible, 49 S 3rd St., Philadelphia
5) Albert Ertle of New York also delivered flags to the Philadelphia depot. (For description of Ertle's flags see New York depot).

The flag of the Sixty-sixth Illinois Infantry is typical of flags produced by Evans & Hassall and by Horstmann Brothers & Company.

Their flags have the following:

1) Rectangular canton
2) A double oval star pattern with 20 stars in the outer ring, 9 stars in the inner ring, 1 star in each corner and 1 star in the middle for 34-star flags; 20 stars in the outer ring, 10 stars in the inner ring, 1 star in each corner and 1 star in the middle for 35-star flags

This national flag was carried by the Thirty-sixth Illinois Infantry and was manufactured by William F. Scheible.

The following are characteristics of the Scheible flags:

1) Rectangular canton
2) Double oval star pattern with 21 stars in the outer ring, 10 stars in the inner ring, a star in each corner and no center star for 35-star flags. There are no flags in the Illinois collection manufactured by Scheible with 34 stars.

Evans & Hassall produced this regimental flag of the Seventy-eighth Illinois Infantry. Its design is like the regimental flags produced in Cincinnati by Longley & Brother.

The characteristics of the Evans & Hassall regimental flags are:

1) Eagle head to right-away from header or staff on obverse of flag
2) The ribbon in the eagle's beak is "V-shaped"
3) There is a double arc of stars over the upper scroll with 21 stars over 13 stars for 34-star flags and 21 over 14 stars for 35-star flags. The stars in the upper arc go to the tip of the upper scroll but do not go outside of the tip
4) There is a three piece scroll beneath the eagle for unit designation with the center section of the scroll lowered

W. H. Horstmann & Sons produced this regimental flag of the Seventeenth Illinois Infantry. The following can identify regimental flags produced by Horstmann:

1) Excellent detail of the eagle with the head of the eagle facing away from the staff on the obverse of the flag
2) Flat scroll held in the eagle's beak
3) Double arc of stars above the eagle with the ends of each arc inside of the ends of the upper scroll. For 34-star flags there are 21 stars in the upper arc and 13 stars in the lower arc, for 35-star flags there are 21 stars in the upper arc and 14 stars in the lower arc
4) Three-piece scroll beneath the eagle for unit designation with the center portion of the scroll raised.

The regimental flags produced by Scheible are like those produces by Evans & Hassall. However, no regimental flags produced by Scheible or Brewer have been identified in the State of Illinois holdings.

NEW YORK DEPOT

The following companies produced flags for the New York depot:

1) Brandon Brothers, 4 Tryon row, New York
2) S. A. Hull (location unknown)
3) F. Fox & Company, New York
4) Albert Ertle, 545 Broadway, New York
5) Paton (William) & Company, 341 Broadway and 88Leonard, New York
6) Tiffany & Company, 550 Broadway, New York
7) William Scheible, 49 S. 3rd Street, Philadelphia

This is the national color of the Eighty-third Illinois Infantry. The national flags produced by Paton, Brandon Brothers and Ertle all had similar patterns:

1) Square canton
2) Stars arranged in five horizontal rows. For 34-star flags the rows from the top were 7,7,6,7,7. For 35-star flags there were 5 rows of 7 stars.

National flags produced by Scheible for the New York depot had square cantons. The stars were small and arranged in five vertical rather than horizontal rows. National flags produced by Tiffany were embroidered. No national flags produced by Scheible or Tiffany for the New York depot have been identified in the Illinois collection.

The regimental flag of the Sixty-second Illinois Infantry is typical of regimental flags manufactured for the New York depot. The regimental flags produced by Tiffany were embroidered and none of the Tiffany flags are in the Illinois collection. The flags of Hull, Brandon, Fox, Ertle, Paton and Scheible were similar to the flag depicted above and had the following characteristics:

1) A double arc of gold stars over the eagle. The upper arc contained 18 stars and the lower arc either 16 or 17 stars depending on whether the flag a 34-star or 35-star flag

2) An embellished lower scroll outlined in gold with a gold flourish on each end of the scroll

3) The center section of the three-piece lower scroll is raised.

Apparently, the New York Depot defined the design of the flags for the Depot, thus the similarity in the flags among the various manufacturers. However, further research may allow identification of a specific manufacturer based on subtle differences in the flags.

The book published by Time-Life books in 1991, *Echoes of Glory, Arms and Equipment of the Union* on pages 242-295 has an excellent discussion and photos of the various Union military flags. A paper presented by Howard M. Madaus, "Conservation of Civil War Flags: The Military Historian's Perspective" and published in the Papers presented at the Pennsylvania Capitol Preservation Flag Symposium, Harrisburg, PA, October 29-30, 1987, details flag design unique to manufacturers and is accompanied by illustrative drawings of the flags. Finally, Richard A. Sauers, in *Advance the Colors! Pennsylvania Civil War Battle Flags,* published in 1987 in Volume II pages 555-562 discusses the flag designs by the various flag manufacturers.

Local Names of Military Organizations

The Volunteer soldier, not the professional soldier, fought the Civil War. Companies and regiments were raised in individual towns and counties throughout the North and South. Consequently, the companies and regiments often had local designations or "nicknames" in addition to their official designation. In researching the material on Illinois Civil War flags, a list of popular or local names with their official designations has been accumulated from regimental histories, diaries, newspapers, county and city histories and other historical works. It was necessary to do this to accurately identify the unit receiving a flag. For example, a newspaper account of a flag presentation might only identify the company as the "Cass County Guards" rather than 19th Illinois Regiment Infantry, Company F. Although the list seems comprehensive, it will probably never be complete. The Artillery Batteries are particularly difficult to identify since there were multiple names associated with each battery and the name of the battery usually changed with the change of the commanding officer. Unless otherwise noted all designations refer to Illinois units. Many of these unit designations came from articles in local newspapers and spelling is always suspect.

Abbott's Artillery--1st Regiment Light Artillery, Battery E
Abolition Regiment – 92nd Regiment Infantry
Adams County Cavalry--also called Adams County Dragoons-2nd Regiment Cavalry Co. L (formerly Co. C)
Alexander's Infantry--21st Regiment Infantry
Algonquin Guards (or Rifles) --15th Regiment Infantry, Co. F
Alton Battalion--144th Regiment Infantry (one hundred days Regiment), Cos. A and B
Alton City Guards--9th Regiment Infantry, Co. G (three months regiment)
Amboy Guards--89th Regiment Infantry, Co. I
Amboy Volunteers –13th Regiment Infantry, Co. Ca
Anderson Rifles--67th Regiment Infantry, Co. D
Anti-Beauregards--19th Regiment Infantry, Co. I
Apple River Company—96th Regiment Infantry, Co. E
Applington Avengers—92nd Regiment Infantry, Co. E
Arthur Guards--88th Regiment Infantry, Col. B
Asay Guards--82nd Regiment Infantry, Co. I
Audubon Rifles--37th Regiment Infantry, Co. I
Aurora Dragoons—Originally attached to 36th Regiment Infantry as Co. A Cavalry, subsequently assigned to 15th Regiment Cavalry as Co. I
Aurora Guards--7th Regiment Infantry, Co. C (three-month Regiment)
Aurora Guards—36th Regiment Infantry, Co. B
Aurora Rifles--13th Regiment Infantry, Co. H
Avery Guards--95th Regiment Infantry, Co. A

Babcock's Infantry--103rd Regiment Infantry

Bacon's Infantry--30th Regiment Infantry
Barber's (Thomas C.) Battery--2nd Regiment Light Artillery, Battery K
Barker's Dragoons--Charles W. Barker's Independent Cavalry Co., Chicago (two months Company)
Barker's Dragoons--12th Regiment Cavalry, Cos. H and I
Barnard's Company or Guards--88th Regiment Infantry, Co. K
Barnett's (Charles M) Battery--2nd Regiment Light Artillery, Battery I
Barrett's (Samuel E.) Battery--1st Regiment Light Artillery-Battery B (old)
Batavia Guards—52nd Regiment Infantry, Co. D
Bateman Rifles--33rd Regiment Infantry, Co. K
Birge's Sharpshooters--J. W. Birge's Regiment Sharpshooters-14th Regiment Missouri Infantry; afterward 66th Regiment Illinois Infantry.
Blair's Escort--11th Regiment Cavalry, Co. G
Blind Half Hundred--50th Regiment Infantry
Bloomington Cavalry-1st Regiment Cavalry, Co. A
Bloomington Infantry--8th Regiment Infantry, Co. K (3-month Regiment)
Board of Trade Battery, Chicago--James H. Stokes' Independent Battery
Board of Trade Regiment, Chicago 1st--72nd Regiment Infantry
Board of Trade Regiment, Chicago 2nd--88th Regiment Infantry
Board of Trade Regiment, 3rd--113th Regiment Infantry
Bolton's (William H.) Battery--2nd Regiment Light Artillery, Battery L
Boone Rifles—15th Regiment Infantry, Co. B
Borries' (Hermann) Battery--2nd Regiment Light Artillery, Battery A
Bouton's (Edward) Battery--1st Regiment Light Artillery, Battery I
Bowen's Independent Battalion, Missouri Cavalry, Companies A-D--1st Battalion Missouri Cavalry attached to 13th Illinois Regiment Infantry; afterwards 9th Regiment Missouri Cavalry, Companies A-D; afterwards 10th Regiment Missouri Infantry Cavalry, Companies A-D
Boyle's Battery—comprised of volunteers from 107th Regiment Illinois Infantry, 80th Indiana Infantry and 13th and 33rd Kentucky Infantry.
Brackett's (Albert G.) Cavalry--9th Regiment Cavalry (also known as First Western Cavalry)
Brainard Guards--72nd Regiment Infantry, Co. G
Bridges' Battery--19th Regiment Infantry, Co. G, afterward Lyman Bridges' Independent Battery, afterwards 1st Regiment Light Artillery, Battery B
Bristol Company or Bristol Light Infantry--36th Regiment Infantry, Co. E (from Bristol and Little Rock, Kendall County)
Bryan Blues-105th Regiment Infantry, Co. F
Bruce Guards—65th Regiment Infantry, Co. A
Bunker Hill Company--7th Regiment Infantry, Co. F (three-month Regiment)
Bureau County Rifles—93rd Regiment Infantry, Co. B
Bureaus County Tigers—93rd Regiment Infantry, Co. H
Burnett's Company—96th Regiment Infantry, Co. H
Burnt Prairie Company--40th Regiment Infantry, Co. C
Bushnell Light Guards--55th Regiment Infantry, Co. F
Burton's (Josiah H.) Battery--1st Regiment Light Artillery, Battery F
Busteed's (Richard J.) Battery--Chicago Light Artillery, also 1st New York Light Artillery Regiment, Batteries B and G and 4th New York Independent Battery
Butternut Brigade—6th Regiment Cavalry, Co. D

Cameron Guards—65th Regiment Infantry, Co. E
Cameron's (Daniel) Highlanders--65th Regiment Infantry, also called Second Scotch Regiment
Campbell's (Charles C.) Battery--2nd Regiment Light Artillery, Battery, Battery D
Canton Cavalry Company—7th Regiment Cavalry, Co. K
Canton Guards--67th Regiment Infantry, Co. F
Canton Rifles-- 55th Regiment Infantry, Co A (originally the Fulton Cadets)
Captain Whiting's Company--89th Regiment Infantry, Co. G
Carlinville Invincibles—7th Regiment Infantry, Co. K
Carmichael's (Eagleton) Cavalry Company--29th Regiment Infantry-Cavalry Co. B; afterwards 4th Regiment Cavalry, Co. B; afterwards Warren Stewart's Independent Battalion Cavalry, Co. B; afterwards 15th Regiment Cavalry Co. B
Carroll Rifles—15th Regiment Infantry, Co. J
Cass County Guards--19th Regiment Infantry, Co. F
Cass County Guards--from Cass County, Michigan-42nd Illinois Regiment Infantry, Co. E
Centralia Cavalry--1st Regiment Cavalry, Co. H (also called Marion Cavalry)
Champaign Home Guards--66th Regiment Infantry, Co. E
Champaign Tigers—10th Regiment Cavalry, Co. I
Champaign Tigers—20th Regiment Infantry, Co. A
Champaign County Volunteers—20th Regiment Infantry, Co Aß
Champion's Company-96th Regiment Infantry, Co. K
Channel's (Joseph R.) Battery--1st Regiment Light Artillery, Bat. C.
Chapman's (Fletcher H.) Battery, 1st--14th Regiment Infantry, Artillery Company L; afterward 2nd Regiment Light Artillery, Bat. B
Chapman's (Fletcher H.) Battery, 2nd Regiment Light Artillery, Bat. D
Cheney's (John T.) Battery--1st Regiment Light Artillery, Bat. F
Chicago Cavalry—8th Regiment Cavalry, Co. G
Chicago Cavalry--16th Regiment Cavalry, Cos. A and B
Chicago, Burlington and Quincy Railroad Company--89th Regiment Infantry, Co. E
Chicago Dragoons--Barker's Dragoons, Charles W. Barker's Independent Cavalry Co., Chicago (two months); later 12th Regiment Cavalry, Co. H and I
Chicago Highland Guards – 19th Regiment Infantry, Co. E
Chicago Infantry--19th Regiment Infantry, Co. D
Chicago Jackson Guards—23 rd Regiment Infantry, Co. C
Chicago Legion--51st Regiment Infantry, also called Ryan Life Guard
Chicago Light Artillery, Company A--1st Regiment Light Artillery, Battery A-also called Morgan's Battery, Smith's Battery, Wood's Battery, Willard's Battery, Wilcox's Battery
Chicago Light Artillery, Company A—James Smith's Independent Company A
Chicago Light Artillery (three months)
Chicago Light Artillery, Company B--1st Regiment Light Artillery, Battery B (old)
Chicago Light Artillery, Company G--1st Regiment Light Artillery, Battery B (new)
Chicago Light Infantry--19th Regiment Infantry, Co. D-a Chicago militia organized before the CW
Chicago Mercantile Battery--Charles G. Cooley's Independent Battery
Chicago Sappers and Miners--59th Regiment Infantry, Co. K
Chicago Zouaves, Companies A and B--became Companies A, C and K of 19th Regiment Infantry

Chickering's Company--88th Regiment Infantry, Co. F

Cicero Volunteers-37th Regiment Infantry, Co. D
Clinton Zouaves—41st Regiment Infantry, Co. F
Clay City Company--40th Regiment Infantry, Co. K
Clayton Guards--a company of State militia organized in 1857-became Company B, 50th Regiment Infantry
Clinton Cavalry Company--15th Regiment Cavalry, Co. D
Clinton Rangers—See Gilbert's Cavalry Company
Cogswell's Battery--53rd Regiment Infantry, Artillery Company A (afterwards 1st Regiment Light Artillery, Battery M; later William Cogswell's Independent Battery)
Coler Regiment--25th Regiment Infantry
Coles County Volunteers--8th Regiment Infantry, Co. C (three-month Regiment)
Col. G. A. Smith's Independent Regiment of Illinois Volunteers—became 35th Regiment Infantry
Colvin's (John H.) Battery—53rd Regiment Infantry, Artillery Company A; afterwards 1st Regiment Illinois Light Artillery, Battery K (new)
Concordia Guards-- 82nd Regiment Infantry, Co. C (named after the Concordia Club of Chicago)
Cooley's (Charles G.) Battery--Chicago Mercantile Battery-Charles G. Cooley's Independent Battery
Cooper's (Charles S.) Battery--2nd Regiment Light Artillery, Battery D
Cooper's (Edgar H.) Battery--1st Regiment Light Artillery, Battery D
Corcoran Guards--67th Regiment Infantry, Co. B
Cram's (Orrin W.) Battery- -1st Regiment Light Artillery, Battery E
Crocker's Escort--11th Regiment Cavalry, Co. G
Crosby Guards-88th Regiment Infantry, Co. H
Cudney's (Albert) Battery--1st Regiment Light Artillery, Battery I
Curtis' (John) Independent Company Infantry--144th Regiment Infantry, Co. B; afterwards Alton Battalion Infantry, Co. B
Cushman's Brigade--the original 53rd Regiment Infantry with ten companies of infantry, one of cavalry and one of artillery

Davidson's (Peter) Battery--Peter Davidson's Independent Company, Peoria Light Artillery; afterwards 2nd Regiment Light Artillery, Battery A
Davis' Escort--an independent cavalry company originally attached to the 36th Infantry as Company B; after December 1862 became Company K of 15th Cavalry Regiment
Decatur Grays--41st Regiment Infantry, Co. A
Decatur Light Infantry, Company A--8th Regiment Infantry, Co. B (3 month regiment)
Decatur Light Infantry, 8th Regiment Infantry – Cos A and B (3 month regiment)
DeGress'(Francis) Battery--1st Regiment Light Artillery, Battery H
DeKalb Rifles--42nd Regiment Infantry, Co. K
Dement Phalanx--46th Regiment Infantry, Cos. D, E, H, and I
Detroit (Michigan) Jackson Guards-23rd Regiment Infantry, Co. A (organized in Wayne County, Detroit, Michigan)
Dick Oglesby's Regiment – 8th Regiment Infantry
Dirty First--31st Regiment Infantry
Dixon Regiment--105th Infantry Regiment
Dixon Volunteers-13th Regiment Infantry, Co. A
Dodson's (Christian B.) Cavalry Company-Kane County Cavalry; afterwards Dodson's Squadron Cavalry, Co. A; afterwards 15th Regiment Cavalry, Co. H

Doggett Guards--Chicago Mercantile Association Independent Battery-Charles G. Cooley's Independent Battery
Dollins' (James J.) Cavalry Company--31st Regiment Infantry (Cavalry Co. C); afterwards 4th Regiment Cavalry Co. C; then Warren Stewart's Independent Battalion Cavalry, Co. C and then 15th Regiment Cavalry, Co. C
Douglas Brigade, 1st Regiment--42nd Regiment Infantry
Douglas Brigade, 2nd Regiment--55th Regiment Infantry
Douglas Guards or LaSalle Guards--23rd Regiment Infantry, Co. F
Dresser's (Jasper M.) Battery--31st Regiment Infantry, Artillery Co. A, afterwards 2nd Regiment Light Artillery, Battery D.
Drummond Guards--42nd Regiment Infantry, Co. F
Dundee Freemen—52nd Regiment Infantry, Co. I
Du Page County Cavalry—8th Regiment Cavalry, Co. D
DuPage County Rifles--13th Regiment Infantry, Co. K
Du Quoin Company--89th Regiment Infantry, Co. B
Dutcher's Roughs--74th Regiment Infantry, Co. G

Earlville Guards or Earl Rifles--23rd Regiment Infantry, Co. D
Edwardsville Volunteers--9th Regiment Infantry, Co. I (three-month regiment)
Egyptian Regiment--40th Regiment Infantry
Eldridge Guards—127th Regiment Infantry, Co. I
Elgin Battery--George W. Renwick's 5th Independent Battery
Elgin Continentals--7th Regiment Infantry, Co. A
Elgin Greys—36th Regiment Infantry, Col A
Elgin Sharp Shooters--127th Regiment Infantry, Co. C
Ellis Rifles—15th Regiment Infantry, Co. C
Elmira Rifles--19th Regiment Infantry, Co. B
Elmwood Guards-8th Missouri Regiment Infantry Co. G (From Elmwood Twsp. Peoria County, I Illinois.)
Ellsworth Zouaves--67th Regiment Infantry, Co. E
Emmet Guard--23rd Regiment Infantry, Co. A
Evanston Dragoons—8th Regiment Cavalry, Co. F
Ever Ready—125th Regiment Infantry
Excelsior Regiment--124th Regiment Infantry

Fairfield Company--40th Regiment Infantry, Co. D
Fairview Independent Rifles—17th Regiment, Co. C
Farnsworth's Cavalry—8th Cavalry Regiment
Fayette Guards--8th Regiment Infantry, Co. H (3-month Regiment)
Fayette Company--11th Regiment Infantry, Co. E (3-month Regiment)
First Chicago Board of Trade Regiment--72nd Regiment Infantry
First Douglas Regiment--42nd Regiment Infantry
First Hecker Regiment--24th Regiment Infantry (also called the Old Hecker Regiment)
First Irish Brigade-23rd Illinois Regiment Infantry
First Methodist Regiment – 73rd Regiment Infantry
First Scotch Regiment--12th Regiment Infantry

First Western Cavalry—9th Regiment Cavalry (also known as Brackett's Cavalry)
Fitch's (John A.) Battery-1st Regiment Light Artillery Battery E
Flood's (James P.) Battery-2nd Regiment Light Artillery, Battery C
Florence Rifles--39th Regiment Infantry, Co. E
Flying Horse Artillery Battery—Chicago Board of Trade Battery
Ford's (William) Cavalry Company-53rd Regiment Infantry, Independent Cavalry Company A; afterwards 15th Regiment Cavalry, Co. L
Forsyth Guards--89th Regiment Infantry, Co. D
Fox River Regiment--36th Regiment Infantry
Franklin's (Angereau) Battery-6th Regiment Cavalry (artillery company), afterwards 1st Regiment Light Artillery K (old).
Freeport German Rifles—46th Regiment Infantry, Co. C
Fremont Dragoons--8th Regiment Cavalry, Co. G
Fremont Fencibles--51st Regiment Infantry, Co. K
Fremont Rangers, Illinois Cavalry Cos. A, B and C-3rd Regiment Missouri Cavalry, Cos.G, H, and I.
Fremont Rifles--37th Regiment Infantry (also known as Illinois Greyhounds)
Fremont's Grey Hounds--13th Regiment Infantry
Freeport Company--11th Regiment Infantry, Co. A (3 month Regiment)
Freeport German Rifles--46th Regiment Infantry, Co. C
Frontier Regiment-10th Regiment Cavalry
Fuller Guard—133rd Regiment Infantry, Co. A
Fuller Guards—65th Regiment Infantry, Co. F
Fuller Guards--67th Regiment Infantry, Co. I
Fuller Guards--72nd Regiment Infantry, Co. D
Fulton Blues--17th Regiment Infantry, Co H
Fulton Cadets--changed to Canton Rifles- 55th Illinois Regiment Infantry, Company A
Fulton Guards--103rd Regiment Infantry, Co. C
Fulton Tigers-52nd Regiment Infantry, Co. F

Galena Greys-45th Regiment Infantry, Co. D-
Galena Railroad Guards--89th Regiment Infantry, Co. K
Galena Union Artillery—organized in Galena in April 1864 but never mustered
Galena Volunteers-12th Regiment Infantry, Co. F
Galesburg Invincibles--17th Regiment Infantry, Co E
Gallatin County Cavalry--1st Regiment Cavalry, Co. B (also known as Gallatin Dragoons
German Guides--13th Regiment Cavalry
German Rifles—65th Regiment Infantry, Co. G
Gibson's (Thomas) Battalion--16th Regiment Cavalry
Gilbert's (Franklin T.) Cavalry Company--52nd Regiment Infantry, Cavalry Co. A; afterwards 12th Regiment Cavalry, Co. H, afterward 15th Regiment Cavalry, Co .G
Gilmore's (Ephraim M.) Cavalry Company-32nd Regiment Infantry Cavalry Co. A; afterwards 12th Regiment Cavalry, Co. F
Glengarry Guards-- 65th Regiment Infantry, Co. H
Goodwin Rifles-45th Regiment Infantry, Co. E
Governor's Guards--113th Regiment Infantry, Co. G
Governor Yates' Legion--6th Regiment Cavalry (also known as Governor's Legion)

Graham's (Robert H.) Company Independent Rangers, Illinois Cavalry—14th Regiment Missouri Infantry (Home Guards-Cavalry Company)
Grant's Escort--4th Regiment Cavalry, Co. A
Grenadier Company--24th Regiment Infantry, Co .D
Grundy County Plough Boys--127th Regiment Infantry, Co. D
Grundy Tigers--11th Regiment Infantry, Co. F (three month regiment)
Gumbart's (George C.) Battery-2nd Regiment Light Artillery, Battery E

Halleck Guards--103rd Regiment Infantry, Co. D
Hamilton County Company--40th Regiment Infantry, Co. A
Hamilton's Escort--a cavalry company originally attached to 36th Infantry, Co. A; later consolidated into the 15th Regiment Cavalry, Co. I.
Hancock Guards--72nd Regiment Infantry, Co. A
Hardin Light Guards-10th Regiment Infantry, Co. B (3 months Regiment)
Hartmann's (Theobald) Dragoons--13th Regiment Cavalry, 1st Battalion
Hart's (Levi W.) Battery--1st Regiment Light Artillery, Battery H
Havelock Guards--72nd Regiment Infantry, Co. B
Hecker Regiment, 1st--24th Regiment Infantry (also Hecker Jager Regiment)
Hecker Regiment, 2nd—82nd Regiment Infantry
Henry County Regiment—112th Regiment Infantry
Henshaw's (Edward C.) Battery--Edward C. Henshaw's Independent Battery
Hickory Rifle Guards-34th Regiment Infantry, Co. I
Higgins Light Guard--51st Regiment Infantry, Co. G
Highland Guards (also called Chicago Highland Guards) --an original state militia that became company E of 19th Illinois Regiment Infantry
Highland Rifle Company--11th Regiment Infantry, Co. K
Highlanders--65th Regiment Infantry
Hillsboro Volunteers--9th Regiment Infantry Co. H (three month regiment)
Hoffmann's Dragoons--16th Regiment Cavalry, Co. C
Holden Guards--88th Regiment Infantry, Co. E
Hooker's Escort--15th Regiment Cavalry, Co. K
Hopkins' (Caleb) Battery—3rd Regiment Light Artillery, Battery C
Hopkins' (Caleb) Battery--10th Regiment Infantry (3month Regiment), Co. I; afterwards 2nd Regiment Light Artillery, Battery I
Houghtaling's (Charles) Battery-10th Regiment Infantry (3 month); afterward Ottawa Light Artillery, then 1st Regiment Light Artillery, Battery C
Hubbard Guards--88th Regiment Infantry, Co. G
Hulaniski's (Thaddeus C.) Battery--2nd Regiment Light Artillery, Battery L
Hundred and Two Dozen—124th Regiment Infantry
Hunter Guards—52nd Regiment Infantry, Co. K
Huntley's Cavalry Company—Oscar H. Huntley's Independent Company A Cavalry; afterwards 1st Regiment Cavalry, Co. K
Hutchens' (William D.) Cavalry Company-27th Regiment Infantry, Cavalry Co. E; afterwards 15th Regiment Cavalry, Co. E.

Illinois Cavalry—Fremont Rangers, Company A, B and C—3rd Regiment Missouri Cavalry, Companies

G, H and I

Illinois Cavalry, Independent Rangers, Robert H. Graham's Company—14th Regiment Missouri Infantry (Home Guards), Cavalry Co. L
Illinois Greyhounds--37th Regiment Infantry (also known as Fremont Rifles)
Illinois Rifles—24th Regiment Infantry, Co. C
Illinois Rangers--2nd Regiment Cavalry, Co. A
Indiana Company, Mechanics' Fusiliers—Co. C of disbanded 56th Regiment Infantry,
Irish Brigade--23rd Regiment Infantry (also called Mulligan's Brigade or 1st Irish Brigade)
Irish Dragoons, Patrick Naughton's--23rd Regiment Infantry, Co. I ;subsequently became 5th Iowa Cavalry Regiment, Co. L
Irish Legion--90th Regiment Infantry
Iroquois Light Guards--20th Regiment Infantry, Co. I

Jackson Guards--23rd Regiment Infantry, Co. C (originally organized as a State militia in 1853)
Jackson Guards, Detroit Michigan—23rd Regiment Illinois Infantry, Co. A
Jenks (Albert) Cavalry Company or Dragoons-- a cavalry company originally attached to 36th Regiment Infantry, Co. A; later consolidated into the 15th Regiment Cavalry, Co. I.
Jersey County Blues-61st Regiment Infantry, Co. C.
Jersey Guards--14th Regiment Infantry, Co. F
Jo Daviess Guards--12th Regiment Infantry, Co. F (3 month Regiment)
Jo Daviess Guards—15th Regiment Infantry, Co. E

Kane County Cavalry—8th Regiment Cavalry, Co. I
Kane County Cavalry--15th Regiment Cavalry, Co. H
Kane County Guards--52nd Regiment Infantry Co. I; also known as the Webb Guards
Kaneville Rifles-52nd Regiment Infantry, Co. A
Kankakee Blues—124th Regiment Infantry, Co. F
Kankakee Guards--124th Regiment Infantry, Co. A
Kankakee Valley Guards--42nd Regiment Infantry, Co. D
Keith's (Charles W.) Battery-- 2nd Regiment Light Artillery-Battery I (formerly Battery C)
Kendall Guards--127th Regiment Infantry, Co. K
Kendall Guards--89th Regiment Infantry, Co. H
Kentucky Brigade--54th, 60th, 62nd, and 63rd Regiments Infantry
Kewanee Volunteers--42nd Regiment Infantry, Co. A
Kimbark Guards--88th Regiment Infantry, Co. D
Kinmundy Company--40th Regiment Infantry, Co. B
Knoxville Dragoons-1st Regiment Cavalry, Co. D
Knoxville Rifles--Originally 9th Missouri Infantry Regiment Infantry, Co. A-which became 59th Illinois Regiment Infantry, Co. A
Koerner Regiment--43rd Regiment Infantry
Kowald's Independent Company Infantry-Gustav Kowald's Company, 3-month company attached to 1st Brigade Volunteers

Lacon Company--11th Regiment Infantry, Co. B (three months Regiment)
Ladies Life Guard--an unofficial company of ladies to work in "the hospital or in the field"
Lafayette Rifles--37th Regiment Infantry, Co. B

Lake County Guards--37th Regiment Infantry, Co. C
Lake County Rifles--37th Regiment Infantry, Co. F
Lake County Union Guards—15th Regiment Infantry, Co. I (also known as Union Rifles)
Lamoille Guards-52nd Regiment Infantry, Co. B
LaSalle County Regiment-104th Regiment Infantry
LaSalle Guards or Douglas Guards -23rd Regiment Infantry, Co, F
Lawrence County Volunteers-8th Regiment Infantry, Co. I (three month Regiment)
Lee County Guards—12th Regiment Infantry, Co. F
Lead Mine Regiment, Washurne--45th Regiment Infantry
Lena Rifles—15th Regiment Infantry, Co. G
Liberty Home Guards--34th Regiment Infantry, Co. G
Light Dragoon--Oscar H. Huntley's Independent Cavalry, Company A; afterwards 1st Regiment Cavalry, Co. K
Lincoln Guards--7th Regiment Infantry, Co. E (three-month Regiment)
Lincoln Rifles-24th Regiment Infantry, Co. F; Regiment had prior action as part of the Cairo Expedition
Lincoln Rifles—52nd Regiment Infantry, Co. H
Lincoln Rangers—also known as the Lincoln Regiment, 52nd Regiment Infantry
Lincoln Regiment--52nd Regiment Infantry
Lisbon Rifles—36th Regiment Infantry, Co. D
Litchfield Guards--7th Regiment Infantry Regiment, Co. D (three-month Regiment)
Little Rock Illinois Volunteers--127th Regiment Infantry, Co. F
Lockport Artillery—organized under State Militia law as an amateur battery three years before the civil war. Never mustered into service. In 1861 sent to Cairo and assisted in river blockade. Disbanded August 6, 1861.
Lockport Guards--90th Regiment Infantry, Co. C
Logan's Escort--2nd Regiment Cavalry, Co. A
Loomis Dragoons-1st Regiment Cavalry, Co. A
Lowell's (John W.) Battery-- 2nd Regiment Light Artillery, Battery G
Lyon Color Guard--58th Regiment Infantry
Lyons Guards-55th Regiment Infantry, Co. G
Lyons Scouts--127th Regiment Infantry, Co. H

Macon County's Own or Macon Regiment--116th Regiment Infantry
Macoupin Invincibles--7th Regiment Infantry, Co. K (three-month Regiment)
Madison (Reily) Battery—1st Regiment Light Artillery, Battery D; afterwards 2nd Regiment Light Artillery Battery B
Madison Volunteers--9th Regiment Infantry, Co. K (three-month Regiment)
Manierre Rifles--37th Regiment Infantry, Co. D (these men were from Manistee, Michigan)
Marion Cavalry—1st Regiment Cavalry, Co. H (also called Centralia Cavalry)
Mason Regulators—17th Regiment, Co. K
Mattoon Guards--7th Regiment Infantry, Co. B (three-month Regiment)
Marshall County Guards—11th Regiment Infantry, Co. B
Mattoon Regiment-21st Regiment Infantry
McAllister's (Edward) Battery--1st Regiment Light Artillery, Battery D (originally 10th Regiment Infantry, Co. K-three months Regiment) also called Plainfield Artillery.
McArthur's Escort--11th Regiment Cavalry, Co. G

McClellan Body Guard-Sturgis Rifles, James Steele's Independent Rifle Company
McClellan Brigade--58th Regiment Infantry
McClellan Dragoons, Company A and B--12th Regiment Cavalry Cos. H and I
McClernand Guard Battalion or Body Guard--16th Regiment Cavalry, Cos. I, K and L
McClernand's Escort--3rd Regiment Cavalry, Co. L
Mc Henry County Cavalry—8th Regiment Cavalry, Co. H
McKean's Escort--2nd Regiment Cavalry, Co. M
McKendree Regiment-117th Regiment Infantry
McKinstry Guards—11th Regiment Cavalry
McLean County Regiment--94th Regiment Infantry
Mechanics' Fusileers--56th Regiment Infantry (old); on orders of the War Department the "original" 56th Infantry Regiment was mustered out days after muster in without any active service
Mendota Rifles--37th Regiment Infantry, Co. E
Mercantile Battery, Chicago—Charles G. Cooley's Independent Battery
Mercer Blues—17th Regiment, Co. I
Methodist Preacher's Regiment--73rd Regiment Infantry, also call the Preacher Regiment
Michigan Company Sharpshooters--John Piper's Company, 66th Regiment Illinois Infan- try Co. D
Mihalotzy Rifles--24th Regiment Infantry, Co. F (originally the Lincoln Rifles)
Miller's (John B.) Battery—1st Regiment Light Artillery, Battery M
Milton Cavalry—8th Regiment Cavalry, Co. E (Also known as Wheaton Dragoons)
Milwaukee Railroad Company--89th Regiment Infantry, Co. C
Mississippi Marine Brigade Ram Fleet--Rams *Switzerland* and *Monarch*
Missouri Cavalry—See Bowen's Independent Battalion
Missouri Infantry, 9th Regiment—became the 59th Regiment Illinois Infantry
Missouri Infantry, 14th Regiment--Birge's Sharpshooters, Western Sharpshooters,
Moline Rifles--19th Regiment Infantry Regiment, Co. H
Moline Guards--37th Regiment Infantry, Co. H
Monmouth Dragoons-1st Regiment Cavalry, Co. G
Monmouth Union Guards—17th Regiment, Co. F
Montgomery Guards--23rd Regiment Infantry, Co. B (originally organized as a militia in 1842)
Montgomery Guards—65th Regiment Infantry, Co. B
Morgan County Regiment-101st Regiment Infantry
Morgan's (Francis) Battery-1st Regiment Light Artillery, Battery A-also called, Smith's Battery, Wood's Battery, Willard's Battery, Chicago Light Artillery, Wilcox's Battery
Morgan's (George W.) Escort—3rd Regiment Cavalry, Co. A
Morgan's (James D.) Escort—7th Regiment Cavalry, Co. C
Morgan's Veteran Brigade—10th, 16th, 60th Regiments Illinois Infantry and 10th Regiment Michigan Infantry
Morris Guards—36th Regiment Infantry, Co. G
Mount Carmel and Xenia Company--40th Regiment Infantry, Co. I
Mount Carroll Independent Rifles—45th Regiment Infantry, Co. A
Mount Erie and Jeffersonville Company--40th Regiment Infantry, Co. E
Mulligan's Battery—1st Regiment Light Artillery, Battery L
Mulligan's (James A.) Brigade--23rd Regiment Infantry (also called Irish Brigade or 1st Irish Brigade)
Mulligan Guards--90th Regiment Infantry, Co. A

Mystic Rifles-42nd Regiment Infantry, Co. G

National Blues--local Peoria military organization that furnished fifty-seven field, line and staff officers including those to the Illinois 8th, 17th, 47th, 77th, 108th, 139th Infantry Regiments, 11th Cavalry Regiment, Colonel of an USCT Regiment and an ensign on *Miss Florida*
National Guards--Consolidated with the Princeton Regiment to become the 57th Regiment Infantry
Naughton (Patrick) Irish Dragoons—attached to the 23rd Regiment Illinois Infantry; afterward to the 5th Regiment Iowa Cavalry, Co. L
Neely's (John C.) Battery—1st Regiment Light Artillery Battery I
Nelson Guards--88th Regiment Infantry, Co. C
New Baltimore Company--40th Regiment Infantry, Co. G
Newark Rifles—36th Regiment Infantry, Co. F
Nichols' (Erastus A.) Battery—2nd Regiment Light Artillery, Battery L
Ninth Missouri-redesignated as 59th Regiment Illinois Infantry
Nispel's (George L.) Battery—2nd Regiment Light Artillery, Battery E
Normal Regiment--33rd Regiment Infantry -also called the Teachers' Regiment
Normal Rifles--33rd Regiment Infantry, Co. A
Northern Rifles--39th Regiment Infantry Regiment, Co. G
Northwestern Rifles--44th Regiment Infantry

Oconto (Wisconsin) Irish Guards-23rd Regiment Illinois Infantry (attached artillery company); afterwards 1st Regiment Light Artillery, Battery L
Ogden Guards (or Rifles) --23rd Regiment Infantry Regiment, Co. E
Ogle Guards-15th Regiment Infantry, Co. H
O'Harnett's Cavalry-30th Regiment Infantry, Cavalry Co. D (from Crawford County); afterwards Warren Stewart's Independent Battalion Cavalry Co. D; afterwards 15th Regiment Cavalry, Co. D
Ohio Companies, Mechanics' Fusileers—56th Regiment Illinois Infantry—old Companies B and H
Ohio Mechanics & Fusileers-Co. B and H of disbanded 56th Regiment Illinois Infantry
Ohio Sharpshooters, 1st, 2nd, and 3rd Independent Companies—66th Regiment Illinois Infantry (Western Sharpshooters or Birge's Sharpshooters), Cos. G, H and K
O'Leary's (Arthur) Battery—Arthur O'Leary's Siege Artillery; afterwards 1st Regiment Light Artillery, Battery G
Ogden Rifles--23rd Regiment Infantry, Co. E
Ogle Guards—15th Regiment Infantry, Co. H
Old Hecker Regiment--24th Regiment Infantry (also called the First Hecker Regiment)
O'Mahony Rifle-23rd Regiment Infantry, Co. G (also called Mahony Guards)
Oneida Guards-42nd Regiment Infantry, Co. C (also known as Oneida Rifles)
Ord's Escort—2nd Cavalry Regiment, Co. A
Osceola Rifles, Stark County--19th Regiment Infantry, Co. B
Oswego Rifles--36th Regiment Infantry, Co. I
Ottawa Artillery—10th Regiment Infantry, Co. F (three months Regiment)
Ottawa Battery—Charles C. Campbell's Battery Light Artillery; afterwards 2nd Regiment Light Artillery, Battery D
Ottawa City Guards--23rd Regiment Infantry, Co. H
Ottawa Light Artillery—Edward C. Henshaw's Independent Battery

Ottawa Light Artillery, Company A—1st Regiment Light Artillery, Battery C
Ottawa Light Infantry--11th Regiment Infantry, Co. H (3 month regiment)

Palatine Volunteers-became part of 19th Regiment Infantry, Co. C (also known as the Palatine Light Infantry)
Palmer's (John M.) Escort—7th Regiment Cavalry, Co. C
Payson Guards--organized in early 1861 at Payson Illinois-became part of Company C, 50th Regiment Infantry
Pekin Guards or Pekin Volunteers--8th Regiment Infantry, Co.F (3 month Regiment)
Peoria Blues--17th Regiment Infantry, Co A
Peoria Blues--8th Regiment Infantry, Co. E (three months regiment)
Peoria Guards--67th Regiment Infantry, Co.G
Peoria Light Artillery--Davidson's Battery, afterwards 2nd Regiment Light Artillery, Battery A
Peoria Zouaves--Illinois Company that fought with the 6th Missouri Zouave Regiment
Perce's (William) Infantry Company—William Perce's Independent Company Militia Infantry (thirty days)
Persimmon Regiment--73rd Regiment Infantry, also called Preacher's Regiment Or Methodist Preacher's Regiment
Phillips Invincibles--109th Regiment Infantry
Pike County Union Guards--8th Regiment Infantry, Co. G (three months regiment)
Pioneer Company-24th Regiment Infantry, Co. H
Piper's (John) Michigan Company, Sharpshooters-66th Regiment Illinois Infantry, Co. D
Plainfield Artillery--1st Regiment Artillery, Battery D (previously was 10th Regiment Infantry, Co. K) also called McAllister's (Edward) Battery
Plato Dragoons-Originally attached to 36th Regiment Infantry as Co. B Cavalry, Subsequently assigned to 15th Regiment Cavalry as Co. K
Pope's Bodyguard--17th Regiment Infantry, Co. A
Powell's (John W.) Battery—2nd Regiment Light Artillery, Battery F
Preacher's Regiment-73rd Regiment Infantry, also called the Methodist Preacher Regiment or Persimmon Regiment
Prentiss Guards-Joseph D. Walker's Independent Company of Infantry (three months Regiment)
Prescott's (Mark H.) Battery—1st Regiment Light Artillery, Battery C
Princeton Regiment-consolidated with the National Guards to become the 57th Regiment Infantry

Princeton Guards—93rd Regiment Infantry, Co. K
Princeton Light Infantry—93rd Regiment Infantry, Co. I

Quincy Cadets-Local militia organization from Quincy Illinois that became part of Company C, 50th Regiment Infantry
Quincy City Guards-10th Regiment Infantry, Co. E (3months)
Quinby's Escort—2nd Regiment Cavalry, Co. B
Quinby's Escort—2nd Regiment Illinois Cavalry, Co. E and 5th Regiment Missouri Cavalry, Co. C; afterwards 4th Regiment Missouri Cavalry Company F (new)
Quinby's Escort—4th Regiment Cavalry, Co. D

Railroad Regiment--89th Regiment Infantry

Ramsey's (Israel P.) Battery—1st Regiment Light Artillery, Battery B (old)
Reaper Company--75th Regiment Infantry, Co. I
Renwick's Battery--George W. Renwick's Elgin Independent Battery Light Artillery
Reserve Guards—127th Regiment Infantry, Co. G
Richardson's (George R.) Battery-- 2nd Regiment Light Artillery, Battery F
Rich's (Judson) Battery-- 2nd Regiment Light Artillery, Battery I
Richland Volunteers--8th Regiment Infantry, Co. D (three months regiment)
Robbins Rifles--42nd Regiment Infantry, Co. B
Roberts Guards-42nd Regiment Infantry, Co. G
Robinson's (George I.) Battery--became James H. Stokes' Independent Battery
Rockford Blues--11th Regiment Infantry, Co. D (three months regiment)
Rockford City Guards--67th Regiment Infantry, Co. A
Rockford Rifles—45th Regiment Infantry, Co. G
Rockford Zouaves--11th Regiment Infantry, Col D (three months regiment)
Rock Island Greys-- Illinois Company that fought with the 6th Missouri Zouave Regiment
Rock Island Rifles--37th Regiment Infantry, Co. A –also called the Moline Rifles
Rock River Regiment--34th Regiment Infantry
Rock Run Rangers—74th Regiment Infantry, Co. I
Rodgers' (Benjamin F.) Battery- initially attached to 32nd Regiment Infantry; afterwards to 2nd Regiment Light Artillery, Battery K.
Rodgers' (Henry A.) Battery-- 1st Regiment Light Artillery, Battery D
Rombauer's (Raphael G.) Battery-- 1st Regiment Light Artillery, Battery G
Rosecran's Escort—7th Regiment Cavalry, Co. B
Rosecrans' Escort- a cavalry company originally attached to 36th Regiment Infantry, Cavalry Co. A; afterward consolidated into the 15th Regiment Cavalry, Co. I
Rourke's (John) Battery—11th Wisconsin Independent Battery—attached to 23rd Regiment Illinois Infantry; afterwards 1st Illinois Regiment Light Artillery, Battery L
Rumsey Rifles--12th Regiment Infantry, Co. A (three months Regiment)
Ryan Guard—90th Regiment Infantry, Co. B
Ryan Life Guard--51st Regiment Infantry, also called the Chicago Legion

Salem and Vandalia County Company--40th Regiment Infantry, Co. H
Sandwich Guards--10th Regiment Infantry, Co. H
Sappers and Miners, Chicago--59th Regiment Infantry, Co. K
Scammon Light Infantry--89th Regiment Infantry, Co. A
Schambeck's (Frederick) Cavalry—Frederick Schambeck's Independent Cavalry Company; afterward Hoffman Dragoons, Co. B; afterward 16th Regiment Cavalry, Co. C
Schwartz's Battery, Missouri Artillery—Adolph Schwartz's Independent Battery Missouri Artillery; afterward 2nd Regiment Illinois Light Artillery, Battery E
Scotch Regiment, 1st--12th Regiment Infantry
Scotch Regiment, 2nd--65th Regiment Infantry, also called Cameron's Highlanders
Scott Dragoons--3rd Regiment Cavalry, Co. I
Scripps Guard--72nd Regiment Infantry, Co. C
Second Chicago Board of Trade Regiment--88th Regiment Infantry
Second Douglas Brigade or Regiment-55th Regiment Infantry
Second Hecker Regiment--82nd Regiment Infantry

Second Methodist Regiment—115th Regiment Infantry
Second Scotch Regiment--65th Regiment Infantry
Seventh Congressional District Regiment--21st Regiment Infantry
Shabbona Rifles--A Company raised in Shabonna, DeKalb County-became Company C of the 58th Regiment Infantry
Shelby Tigers--41st Regiment Infantry, Co. H
Shepard Guards--72nd Regiment Infantry, Co. F
Sherer's (Samuel B.) Cavalry Company or Dragoons-- an independent cavalry company originally attached to the 36th Regiment Infantry as Cavalry Company B; after December 1862 became Co. K of 15th Regiment Cavalry
Shields' Guards--23rd Regiment Infantry, Cos. I and K (originally organized as a militia unit in 1854)
Siege Artillery—Arthur O'Leary's Siege Artillery; afterward 1st Regiment Light Artillery, Battery G
Sigol Rifles--43rd Regiment Infantry, Co. D
Silversparre's (Axel) Battery—1st Regiment Light Artillery, Battery H
Sloan's Commercial Regiment-124th Regiment Infantry
Slocum's Escort--15th Regiment Cavalry, Co. K
Smith's (Gustavus) Independent Regiment of Illinois Volunteers-35th Regiment Infantry
Smith's (Henry A.) Cavalry Company- an independent cavalry company originally attached to the 36th Regiment Infantry as Cavalry Company B; after December 1862 became Company K of 15th Regiment Cavalry
Smith's (James) Battery—James Smith's Independent Company A, Chicago Light Artillery (three months); afterward 1st Regiment Light Artillery, Battery A
Smith's (Jason B.) Battery—1st Regiment Light Artillery Battery K (old)
Smith's (William S.) Escort—7th Regiment Cavalry, Co. B
Smith's Guerrillas—119th Regiment Infantry
Sparrestrom's (Frederick) Battery—2nd Regiment Light Artillery, Battery G, later Battery A
Spencer's (George W.) Battery—1st Regiment Light Artillery, Battery M
Springfield Artillery--2nd Regiment Light Artillery, Battery C
Springfield Artillery--attached to 10th Regiment Infantry-three months, Co. I, Light Artillery
Springfield Company—124th Regiment Infantry, Co. C
Springfield Light Artillery--Thomas F. Vaughn's Independent Battery
Springfield National Guards--7th Regiment Infantry, Co. G (three month Regiment)
Springfield Zouave Greys--7th Regiment Infantry, Co. I (three months Regiment)
St. Charles Illinois Volunteers--127th Regiment Infantry, Co. E
Stanton Rifles—65th Regiment Infantry, Co. D
St. Charles Cavalry—8th Regiment Cavalry, Co. A
St. Charles Guards—52nd Regiment Infantry, Co. G
St. Clair County Volunteers--9th Regiment Infantry, Cos. A, B, C, C, E, F (three month Regiment)
St. Joseph Volunteers (Michigan)--42nd Regiment Illinois Infantry, Co. G
Steele's Bodyguard--1st Regiment Cavalry, Co. K
Steele's Company Rifles--Sturges Rifles
Steele's Escort--3rd Regiment Cavalry, Co. D and 15th Regiment Cavalry, Co. H
Stenbeck's (Andrew) Battery-- 2nd Regiment Light Artillery, Battery H (later Battery B)
Stephenson County's Best, No. 7—46th Regiment Infantry, Co. G
Stephenson County Yeomen—46th Regiment Infantry, Co. B
Stephenson Farmers' Light Infantry--46th Regiment Infantry, Co. A

Stevenson's Escort--2nd Regiment Cavalry, Co. A (part)

Stewart's (Warren) Independent Battalion Cavalry, Cos. A-E and Recruits—15th Regiment Cavalry Cos. A-F

Stoke's (James H.) Battery-- James H. Stoke's Independent Battery, Chicago Board of Trade Battery Stolbrand's (Charles J.) Battery—2nd Regiment Light Artillery, Battery G; Stookey's (Simon J.) Independent Company—144th Regiment Infantry, Co. A (100 days); afterward Alton Battalion Infantry, Co. A (100 days)

Stuart Guards--42nd Regiment Infantry, Co. H

Sturgis Light Guard--51st Regiment Infantry, Co. E

Sturges Rifles-Independent Infantry Company-served as special bodyguard for George B. McClellan --raised and equipped by Solomon Sturges also known as Steele's Company Rifles, B

Sycamore Brass Band-became the Regimental band of the 13th Regiment Infantry (three months Regiment)

Sycamore Cavalry—8th Regiment Cavalry, Co. B

Sycamore Rifles-52nd Regiment Infantry, Co. C

Taylor's (Ezra) Battery-- 1st Regiment Light Artillery, Battery B (old)

Teachers' Regiment--33rd Regiment Infantry -also called the Normal Regiment

Thielmann's (Christian) Independent Battalion Cavalry, Cos. A and B-16th Regiment Cavalry, Cos. A and B

Third Board of Trade Regiment-113th Regiment Infantry

Timony's (James P.) Battery-- 2nd Regiment Light Artillery, Battery D

Tiskilwa Tigers—93rd Regiment Infantry, Co. E

Tornado Company—93rd Regiment Infantry, Co. F

Tucker Light Guard--51st Regiment Infantry, Co. B

Turner Union Cadets--24th Regiment Infantry, Co. G (also served in the Cairo Expedition—originally a German athletic club called the German Turnverein)

Turner Rifles--37th Regiment Infantry Regiment, Co. G

Union Brigade--formed from remnants of the 58th Regiment Illinois Infantry and 8th, 12th and 44th Regiments Iowa Infantry after the Battle of Shiloh

Union Cadets--Gustav Kowald's Company Infantry; attached to 1st Brigade Volunteers (three months Regiment) later became 24th Regiment Infantry, Co. A

Union Guards-10th Regiment Infantry, Co. G (three months Regiment)

Union Guards--22nd Regiment Infantry, Co. I

Union Railway Guard--51st Regiment Infantry, Co. A

Union Rifles--24th Regiment Infantry, Co. C

Union Rifles—15th Regiment Infantry, Co. I (also known as Lake County Union Guards)

Union Riflemen—45th Regiment Infantry, Co. B

Vaughn's (Thomas F.) Battery-Thomas F. Vaughn's Independent Battery also known as Springfield Light Artillery

Vermilion County Zouaves--37th Infantry Regiment, Co. K

Walker's Infantry Company--Joseph D. Walker's Independent Company Infantry (three months Regiment)

Wallace Guards--127th Regiment Infantry, Co. A
Wallace Guards—65th Regiment Infantry, Co. C
Washburne Lead Mine Regiment--45th Regiment Infantry
Washington Guards--55th Regiment Infantry, Co. E
Washington Independent Regiment--19th Regiment Infantry
Washington Infantry--19th Regiment Infantry, Co. D
Washington Light Cavalry--Frederick Schambeck's Independent Cavalry Company; afterwards Hoffman Dragoons; afterward 16th Regiment Cavalry, Co. C
Washington Light Infantry--67th Infantry Regiment, Co. H
Washington Rifle Company—47th Co. B
Waterhouse's (Allen C.) Battery--1st Regiment Light Artillery, Battery E
Wayne Guards-18th Regiment Infantry, Co. G
Wayne Rifles—36th Regiment Infantry, Co. K
Wayne Rifles—15th Regiment Infantry, Co. D
Webb Guards--42nd Regiment Infantry, Co. I also known as the Kane County Guards
Weide's (Alexander) Company, Illinois Cavalry—1st Regiment Western Cavalry, Fremont Hussars, Missouri Volunteers, Co. F; afterward 4th Regiment Missouri Cavalry Co. F, afterward Co. L
Western Cavalry, 1st Regiment--9th Regiment Cavalry
Western Sharpshooters--also known as Birge's Sharpshooters-J. W. Birge's Regiment Sharpshooters 14th Regiment Missouri Infantry; afterward-66th Regiment Illinois Infantry.
Wheaton Dragoons—8th Regiment Cavalry, Co. E (Also known as Milton Cavalry)
White County Dragoons--1st Regiment Cavalry, Co. I
Whiteside County Cavalry—8th Regiment Cavalry, Co. C
White's (Lyman A.) Battery-- 1st Regiment Light Artillery, Battery B (new)
White's (Patrick H.) Battery--Patrick H. White's Battery became Charles G. Cooley's Independent Battery
Whittemore's (Henry C.) Battery--2nd Regiment Light Artillery, Battery H
Wicker Guards—72nd Regiment Infantry Co. K and 88th Regiment Infantry, Co. A
Wilcox's (Edward P.) Battery--1st Regiment Light Artillery, Battery A-also called Morgan's Battery, Smith's Battery, Wood's Battery, Willard's Battery, Chicago Light Artillery
Wilder Lightning Brigade--92nd, 98th, 123rd Illinois Mounted Infantry Regiments and 17th, 72nd Indiana Mounted Infantry Regiments and 18th Indiana Battery
Wilder's (William C.) Cavalry Company—Independent Company, Kane County, Cavalry; afterwards 15th Regiment Cavalry, Co. H
Will County Regiment-100th Regiment Infantry
Willard's (Charles M.) Battery--1st Regiment Light Artillery, Battery A-also called Morgan's Battery, Smith's Battery, Wood's Battery, Chicago Light Artillery, Wilcox's Battery
Willis' (George A.) Cavalry Company--originally attached to 36th Regiment Infantry, Cavalry Co. A; afterwards consolidated into the 15th Regiment Cavalry, Co. I.
Wilson's Dragoons--Thielmann's (Christian) Cavalry Company-afterwards 16th Regiment Illinois Cavalry, Co. A
Winnebago County Guards--67th Regiment Infantry, Co. C
Winnebago, Rockford Cavalry—8th Regiment Cavalry, Co. M
Winnebago Sharpshooters--52nd Regiment Infantry, Co. E
Wisconsin Company, Mechanics' Fusileers—56th Regiment Illinois Infantry, Co. A (old)
Wisconsin Independent Battery, 11th—John Rourke's Battery, attached to 23rd Regiment Illinois

Infantry; afterwards 1st Regiment Illinois Light Artillery, Battery L
Wisconsin Mechanics & Fusileers-Co. A of the disbanded 56th Regiment Infantry
Wolverine Guards (from Albion, Michigan)-42nd Regiment Illinois Infantry -part of Co. H
Wood's (Andrew M.) Battery—George Renwick's Independent Battery
Wood's (Peter) Battery--1st Regiment Light Artillery, Battery A-also called Morgan's Battery, Smith's Battery, Willard's Battery, Chicago Light Artillery, Wilcox's Battery
Woodford Rifles--17th Regiment Infantry, Co. G
Woodson Guards--61st Regiment Infantry, Co. A
Woodstock Light Guards-15th Regiment Infantry, Co. A
Woodstock Rifles—36th Regiment Infantry, Co. H
Woodworth Rifles--127th Regiment Infantry, Co. B
Worden's (Asa) Battalion-14th Wisconsin, 81st and 95th Illinois Infantry Regiments
Wyanet Union Guards—93rd Regiment Infantry, Co. C

Yates Light Dragoons-1st Regiment Cavalry, Co. F
Yates Phalanx--39th Regiment Infantry
Yates Rangers--7th Regiment Infantry, Col. H (three months Regiment)
Yates Sharpshooters, 1st Battalion-64th Regiment Infantry, Cos. A-F
Y.M.C.A. Company--74th Regiment Infantry, Co. K
Young America Guards--36th Regiment Infantry, Co. C (from Warren County)
Young's Lions--46th Regiment Infantry, Co. G

Zouaves--9th Regiment Missouri Infantry; afterward-59th Regiment Illinois Infantry

GLOSSARY

Standard flag nomenclature is used throughout this book. The following is a glossary of the terms.

battle honors- General Order No. 19, issued 22 February 1862, stated "It has been ordered that there shall be inscribed upon the colors or guidons of all regiments and batteries in the service of the United States the names of the battle in which they have borne a meritorious part." As a general rule, battles were only to be inscribed on the flag by order of the commanding general. The honors were sometimes displayed on streamers attached to the flag.

border- Fabric or fringe sewn around the edges of the flag

bunting- Woolen or cotton material from which flags are made.

camp colors- According to Army Regulations of 1861, Camp Colors were defined as"...bunting, eighteen inches square; white for infantry, and red for artillery, with the number of the regiment on them. The pole eight feet long." However, during the Civil War the term often referred to various small flags including the above defined camp colors as well as general guide flags and flank markers. An excellent review of this subject an these flags has been written by Howard Madaus.(2)

canton- a square or rectangular area on the top hoist portion of the flag. On the United States national flag this area is called a Union.

color or colors- a badge, pennant or flag. The term is used broadly to include national, regimental or state colors. In general usage a "color" is used by the infantry while a "standard" is used by the cavalry. In this book the term color and colors is used interchangeably to mean one or more flags.

cord- Silk threads of multiple plies with tassels on each end and the midpoint attached to the finial of the flag staff. The colors of the cord were blue and white for infantry, red and yellow for artillery and red and white for cavalry.

ferrule- a metal tip or cap at the bottom of the flag staff used to plant the staff in the ground; a metal sleeve used to hold a multi part pole together.

field- the main part of the flag.

finial- the metal decoration on the top of the flag staff; in Civil War flags this was usually shaped as a spearhead, eagle or an axe head.

flank markers- eighteen-inch flags on an eight-foot staff carried at each end of an infantry regiment's line of battle or line or march.

fly-- refers to the length of the flag as measured from the outer pole sleeve to the free edge of the flag.

garrison flag-The national flag flown at military installations. At the time of the Civil War the garrison flag for the Union Army measured 20 feet by 36 feet while the garrison flag for the Confederate Army measured 18 feet by 28 feet.

guidon- 1) a small flag carried by an artillery battery or a cavalry company to identify company (i.e. Company A). The flag was often used for alignment. Some infantry regiments carried guidons as flank markers. The shape of a guidon could be either rectangular or swallowtail.

hoist- the edge of the flag next to the staff or the vertical size of the flag as measured on the hoist end.

GLOSSARY

national flag-in this book this refers to the flag of the United States-the "Stars and Stripes."

obverse- The view of the flag when the hoist end or the staff end is to the viewer's left; the "front" of the flag.

pike-a military term for staff.

regimental flag-the term may refer to any flag carried by a regiment or to the specific flag carried by an infantry regiment which was by regulation a regimental color "to be blue, with the arms of the United States embroidered in silk on the center."

reverse- The view of the flag when the hoist end or staff end is to the viewer's right; the "rear" of the flag.

sleeve-a tube of fabric sewn to the hoist of the flag through which the pike of staff is passed.

standard- the color used by the cavalry.

Index

Made in the USA
Monee, IL
05 November 2026

67806906R00328